AF616224

DEVELOPMENTS IN GEOTECHNICAL ENGINEERING 44

GROUND MOTION AND ENGINEERING SEISMOLOGY

COMPUTATIONAL
MECHANICS
PUBLICATIONS

Further titles in this series:
(Volumes 2, 3, 5, 6, 7, 9, 10, 13, 16 and 26 are out of print)

1. G. SANGLERAT – THE PENETROMETER AND SOIL EXPLORATION
4. R. SILVESTER – COASTAL ENGINEERING, 1 and 2
8. L.N. PERSEN – ROCK DYNAMICS AND GEOPHYSICAL EXPLORATION, Introduction to Stress Waves in Rocks
11. H.K. GUPTA AND B.K. RASTOGI – DAMS AND EARTHQUAKES
12. F.H. CHEN – FOUNDATIONS ON EXPANSIVE SOILS
14. B. VOIGHT (Editor) – ROCKSLIDES AND AVALANCHES, 1 and 2
15. C. LOMNITZ AND E. ROSENBLUETH (Editors) – SEISMIC RISK AND ENGINEERI DECISIONS
17. A.P.S. SELVADURAI – ELASTIC ANALYSIS OF SOIL-FOUNDATION INTERACTIC
18. J. FEDA – STRESS IN SUBSOIL AND METHODS OF FINAL SETTLEMENT CALCULATION
19. Á. KÉZDI – STABILIZED EARTH ROADS
20. E.W. BRAND AND R.P. BRENNER (Editors) – SOFT-CLAY ENGINEERING
21. A. MYSLIVEC AND Z. KYSELA – THE BEARING CAPACITY OF BUILDING FOUNDATIONS
22. R.N. CHOWDHURY – SLOPE ANALYSIS
23. P. BRUUN – STABILITY OF TIDAL INLETS. Theory and Engineering
24. Z. BAZANT – METHODS OF FOUNDATION ENGINEERING
25. Á. KÉZDI – SOIL PHYSICS. Selected Topics
27. D. STEPHENSON – ROCKFILL IN HYDRAULIC ENGINEERING
28. P.E. FRIVIK, N. JANBU, R. SAETERSDAL AND L.I. FINBORUD (Editors) – GROUN FREEZING 1980
29. P. PETER – CANAL AND RIVER LEVÉES
30. J. FEDA – MECHANICS OF PARTICULATE MATERIALS. The Principles
31. Q. ZÁRUBA AND V. MENCL – LANDSLIDES AND THEIR CONTROL. Second completely revised edition
32. I.W. FARMER (Editor) – STRATA MECHANICS
33. L. HOBST AND J. ZAJÍC – ANCHORING IN ROCK AND SOIL. Second completely revised edition
34. G. SANGLERAT, G. OLIVARI AND B. CAMBOU – PRACTICAL PROBLEMS IN SOI MECHANICS AND FOUNDATION ENGINEERING, 1 and 2
35. L. RÉTHÁTI – GROUNDWATER IN CIVIL ENGINEERING
36. S.S. VYALOV – RHEOLOGICAL FUNDAMENTALS OF SOIL MECHANICS
37. P. BRUUN (Editor) – DESIGN AND CONSTRUCTION OF MOUNDS FOR BREAKWATERS AND COASTAL PROTECTION
38. W.F. CHEN AND G.Y. BALADI – SOIL PLASTICITY. Theory and Implementation
39. E.T. HANRAHAN – THE GEOTECTONICS OF REAL MATERIALS. The E_g, E_k Meth
40. J. ALDORF AND K. EXNER – MINE OPENINGS. Stability and Support
41. J.E. GILLOTT – CLAY IN ENGINEERING GEOLOGY
42 A.S. CAKMAK (Editor) – SOIL DYNAMICS AND LIQUEFACTION
43. A.S. CAKMAK (Editor) – SOIL-STRUCTURE INTERACTION
45. A.S. CAKMAK (Editor) – STRUCTURES AND STOCHASTIC METHODS

DEVELOPMENTS IN GEOTECHNICAL ENGINEERING 44

GROUND MOTION AND ENGINEERING SEISMOLOGY

Edited by

A.S. CAKMAK

Department of Civil Engineering, Princeton University, Princeton, N.J. 08544, U.S.A.

ELSEVIER
Amsterdam – Oxford – New York – Tokyo 1987

Co-published with

COMPUTATIONAL MECHANICS PUBLICATIONS
Southampton – Boston – Los Angeles 1987

ELSEVIER SCIENCE PUBLISHERS B.V.
Sara Burgerhartstraat 25, P.O. Box 211
1000 AE Amsterdam, The Netherlands

Distributors for the United States and Canada:

ELSEVIER SCIENCE PUBLISHING COMPANY INC.
52 Vanderbilt Avenue
New York, N.Y. 10017

COMPUTATIONAL MECHANICS PUBLICATIONS
Ashurst Lodge, Ashurst
Southampton, SO4 2AA, U.K.

British Library Cataloguing in Publication Data

Ground Motion and engineering seismology.
1. Soil mechanics 2. Rock mechanics
I. Cakmak, A.S.
624.1′513 TA710

ISBN 0-905451-89-9

Library of Congress Catalog Card number 87-70780

ISBN 0-444-98956-0 (Vol.44) Elsevier Science Publishers B.V.
ISBN 0-444-41662-5 (Series)
ISBN 0-905451-89-9 Computational Mechanics Publications, UK
ISBN 0-931215-87-0 Computational Mechanics Publications, USA

Printed in Great Britain by Adlard and Son Limited, Dorking

PREFACE

The Earthquake Engineering Community has a long way to go, as despite advances in the field of Geotechnical Earthquake Engineering, year after year earthquakes continue to cause loss of life and property and leave continued human suffering in their wake in one part of the world or another.

We hope to provide the Earthquake Engineering Community with a forum to help develop further techniques and methods through the exchange of scientific ideas and innovative approaches in Soil Dynamics and Earthquake Engineering, by means of this volume and its companion volumes. This volume covers the following topics: Seismicity and Tectonics in the Eastern Mediterranean, Seismic Waves in Soils and Geophysical Methods, Engineering Seismology, Dynamic Methods in Soil and Rock Mechanics and Ground Motion and contains edited papers selected from those presented at the 3rd International Conference on Soil Dynamics and Earthquake Engineering, held at Princeton University, Princeton, New Jersey, USA, June 22-24, 1987.

The editor wishes to express sincere thanks to the authors who have shared their expertise to enhance the role of mechanics and other disciplines as they relate to earthquake engineering.

The editor also wishes to acknowledge the aid and support of Computational Mechanics Publications, Southampton, England, the National Center for Earthquake Engineering Research, SUNY, Buffalo, NY, and Princeton University, in making this conference a reality.

A.S. Cakmak
June 1987

CONTENTS

SECTION 1: SEISMICITY AND TECTONICS IN THE EASTERN MEDITERRANEAN

SECTION 2: SEISMIC WAVES IN SOILS AND GEOPHYSICAL METHODS

SECTION 3: ENGINEERING SEISMOLOGY

SECTION 4: DYNAMIC METHODS IN SOIL AND ROCK MECHANICS

SECTION 1: SEISMICITY AND TECTONICS IN THE EASTERN MEDITERRANEAN

Geotectonics and Earthquake Risks in Jordan

Y.M. Masannat
Department of Civil Engineering, University of Jordan, Amman, Jordan

INTRODUCTION

Interest in earthquake engineering has just started recently in Jordan due mainly to the scarcity of large size structures and lack of specialists in this field. Special emphasis on the seismicity of Jordan began with the commencement of works in Khalid Bin El-Walid Dam on Yarmouk River in 1966 and in King Talal Dam on Zerqa River in 1972. With the increasing number of highway projects, dams, and multi-storey structures being built at locations close to the Jordan Valley there has been a growing awareness of the significance of seismic parameters in the design of such structures. In 1983 the Natural Resources Authority established the first Jordan Seismological Observatory in cooperation with the U.S. Geological Survey and the U.S. Agency for International Development. A network of 8 seismic stations is presently operational in Jordan extending from Makawir in the south to Jerash in the north. The network is presently expanded to thirty stations which are expected to be operational by the end of 1987 (El-Kaysi[1]). Most of Jordan is placed in Zone A of the Building Research Establishment Overseas Division[2] seismic risk maps (Figure 1). This zone includes areas in which MM intensity VII or greater earthquakes could occur within the lifetime of permanent buildings.

It is the objective of this paper to discuss the effects of regional geology and tectonism on the seismicity of Jordan, and to emphasize the importance of local geology in seismic zoning as well as in establishing the seismic design criteria for major structures. Various earthquake-related instability features are also discussed where

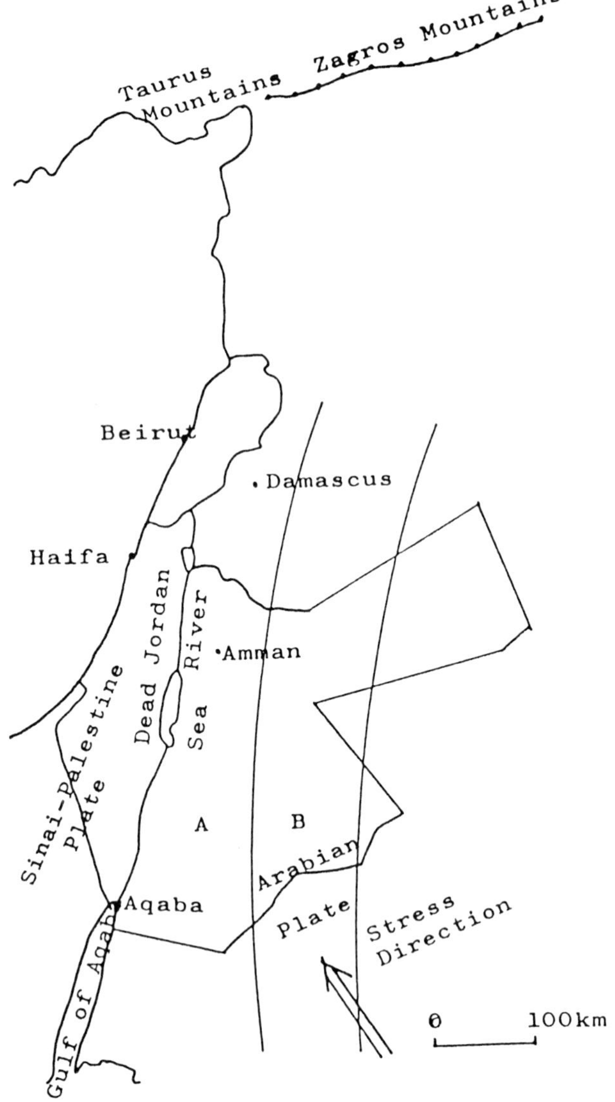

Figure 1. Regional stress direction and seismic zoning.

further research work is suggested.

GEOLOGY AND TECTONISM

Wetzel and Morton (Bender[3]) referred the formation of the Wadi Araba-Jordan Graben to repeated subsidence since the Cretaceous whereby marginal fault zones emphasized the character of a "graben de subsidence". Bender[3] adopted this hypothesis and considered it more adequate than the assumption of a post -Cretaceous left hand shear of more than 100 km in explaining the anomalies in geological features across the Jordan Rift. However, most scientists now accept the hypothesis of the multi-stage northward lateral shear displacement of the Arabian plate by 70 to 110 km with respect to the Palestine-Sinai plate(Figure 1). Some scientists believe in the combination of vertical and horizontal tectonics to explain the development of the Jordan Rift and the off-settings in the geological and geomorphological features on its both sides. The left-lateral shear along the Dead Sea Rift is associated with displacements on enchelon strike-slip faults. These resulted in the development of basins or depressions (Quennell[4]) delimited by steep normal faults which bound the Dead Sea and Jordan River Rifts on both sides. The Dead Sea is the greatest of these depressions and is bordered from the west and east by the Jericho and Araba strike-slip faults (Ben-Menahem[5]).

The geological data and the seismic record of the earthquakes strongly suggest that the Dead Sea Jordan Rift which extends from the Red Sea in the south to the Taurus-Zagros mountainous range in the north (Figure 1) is the major source of earthquakes in the region. The Jordan Rift represents a major tectonic zone between the Palestine-Sinai plate in the north-west and the Arabian plate in the south-east. The development of this depression and its tectonic activity are related to the rotational movement of the Arabian plate towards the north-east, the widening up of the Red Sea, and the spreading of the floor of the Arab Sea. These movements are associated with the development of high in situ SE-NW stresses. These stresses keep the crustal rocks along the Wadi Araba-Dead Sea shear zone in a critical state, and thus a potential source of future earthquakes. These stresses could thus be considered responsible for the development of many of the geological and geomorphological features along both sides of the Jordan Valley viz. strike-slip and normal faults, local and regional uplift and subsidence, and the compressional structural belts in the East Bank of Jordan (Figure 2). Mikbel[6] indicated that South

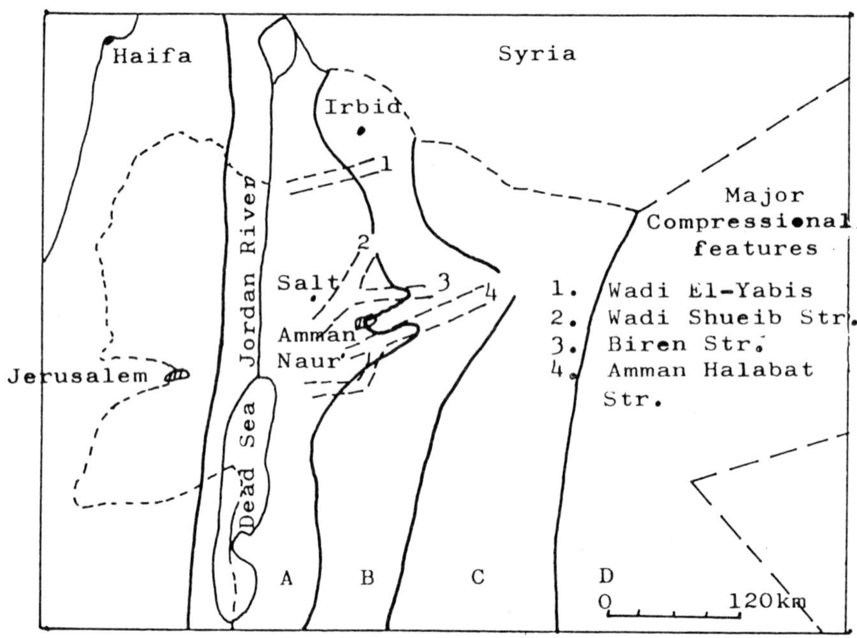

Figure 2. Modified proposal of seismic intensity zoning map of Central and North Jordan (Mikbel, 1985)

Amman, for example, is located on the Amman-Hallabat compressional structural belt that extends from Siyagha on the north-east corner of the Dead Sea to Qasr El-Hallabat in the east of Jordan. He referred the higher geomorphological and topographic levels of South Amman to the culmination of rock's successions along the ENE-WSW trending axes of these deformational belts. The author[7] has pointed to the characteristic unstable geomorphological and structural features of South Amman which change rapidly into more stable features within just few hundred meters to the south -east. The author[8] has also recommended, in his study of Jofeh landslide, that the seismicity of the area be given a proper consideration when establishing the design criteria of the contemplaled stabilization works.

SEISMICITY OF JORDAN

Earthquake prediction has received the attention of the scientists since a long time and still represents their major concern. The observation of changes in the electrical resistivity of rocks, fluctuation of water level in wells, variations in the rates of vertical and horizontal ground displacements in given locations, as well as abnormal changes in the behavior of animals are some of the tools being used by some scientists for earthquake prediction. However, experience shows that there is still no single method that can be relied upon to predict the occurrence of an earthquake with an acceptable degree of certainty that could help in taking emergency measures in advance to mitigate the catastrophic effects of strong earthquakes.

Thus, the long-term observation of the seismic activity in a given region as well as better understanding of the interrelationship between the geological setting and the tectonic activity represent currently the best available method for the assessment of earthquakes' risks.

El-Kaysi[1] indicated that during the period from September 1983 till September 1986, 2704 possible seismic events were recorded where approximately 88 percent of them were of magnitude 3.0 or less. Five felt events were recorded in 1984 and 1985 as shown in Table 1.

Table 1. Major earthquakes recently felt in Jordan

Date	Magnitude	Location
August 24,1984	5.2	Haifa area
October18,1984	4.7	Tiberius area
January25,1985	4.7	Near Jericho
December31,1985	4.6	North of Gulf of Aqaba
December31,1985	4.9	North of Gulf of Aqaba

Most of the earthquakes (78 percent) had a focal depth less than 10 km, and only 7 percent had a focal depth greater than 20km. Three years of earthquake monitoring in Jordan (El-Kaysi[1]) indicate that the East Bank of Jordan is significantly more active tectonically than the West Bank and that the epicenters of the recorded local events are concentrated along fault systems that are predominatly normal to the Jordan Valley.

Figure 3 shows the faulting pattern in central and North Jordan, the locations of the newly installed seismic stations, the approximate epicenters of earthquakes, and the major landslides along the highways.

According to Woodward-Clyde Consultants[2] at least five earthquakes within the 4000 year historical record have maximum estimated intensities of MM IX. These earthquakes occured in the years 1033, 1068, 1546, 1837 and 1927. According to Ben-Menahem[10] the catalogue of historical earthquakes for the Dead Sea Fault System between 2350 B.C. and 1980 A.D. contains 17 events with estimated magnitudes between 6.4 and 7.3. The July 11, 1927 earthquake was of an estimated magnitude $6\frac{1}{4}$ on Richter Scale, and was located about 5 km east of Jericho according to Nur and Reches[11]. This earthquake caused the collapse of many old buildings especially in the town of Salt, about 14 km from the Jordan Valley. Geologic evidence suggests that the Jordan Valley and associated active faults on both sides form the potential sources for such earthquakes.

Woodward-Clyde Consultants[9] conducted an extensive study for the evaluation of the seismicity of the area of the domestic water projects in Central and North Jordan which include the Jordan Valley and the northern highlands east of the Jordan Valley. They concluded that 1 earthquake with magnitude $6\frac{1}{2}$ every 100 years and 1 earthquake with magnitude $7\frac{1}{2}$ every 500 years are reasonable estimates for the operating and contingency-level earthquakes, respectively. The operating and contingency-level earthquakes were defined as those having 10 percent and 50 percent probability of occurrence within the lifetime of the facility (50 years), respectively. The magnitude-recurrence relationship used by them was:

$$\log N = 4.6 - 0.7\ M$$

where N is the number of earthquakes of magnitude M or greater occurring per 100 years. By plotting the mapped and inferred fault lengths given by Ben-Menahem et.al[12] for the Dead Sea fault on Tocher and Patwardhan's[13] graph of the relationship between fault rupture area and earthquake magnitude Woodward-Clyde Consultants[9] concluded that a magnitude 7 earthquake is probably the maximum

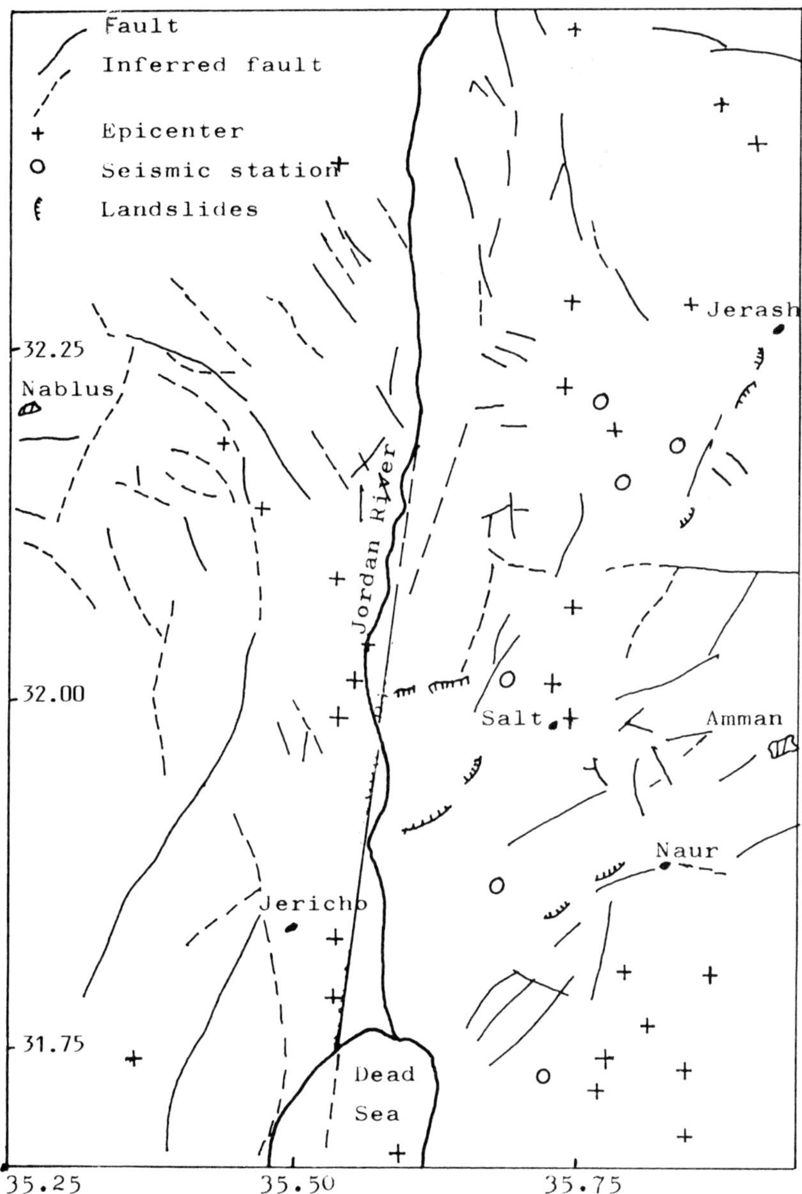

Figure 3. Faulting, landslides, and seismicity of Central and North Jordan.

earthquake likely to occur in the Jordan River-Dead Sea region.

The records of the historical and instrumented earthquakes are sometimes too short to provide reliable evaluation of the future anticipated earthquakes. Therefore, the study of earthquake deformations in Quaternary sediments has gained greater attention in recent years for the evaluation of the paleosismotectonics of a given region. El-Isa and Mustafa[14], for example, studied the decollement structures preserved in the Lisan deposits of the Pleistocene in a cross section representing about 1733 years of continuous deposition near the Araba fault that extends about 200 km between the Gulf of Aqaba and the southern Dead Sea region. These structures are assumed to have been formed by the local micro-folding and sliding of the soft sediments due to earthquake shaking. From their study they concluded that the seismic activity of the Araba fault has fluctuated within the study period but it was generally lower than that of the Dead Sea and the regions further to the north. They concluded also that the deduced earthquake magnitudes (with $M > 5.6$) corform to the frequency-magnitude relationship:

$$\log N = 5.24 - 0.68\ M$$

They gave average recurrence periods of 56 ∓ 3, 113 ∓ 7 and 340 ∓ 20 years for earthquake magnitudes equal to or greater than 5.6, 6.1 and 6.5 respectively which are much lower than those estimated by Ben-Menahem[5]. The uncertainties associated with this approach stem basically from the assumed shear strength parameters (cohesion and angle of shear resistance) for the deformed unconsolidated sediments which would introduce some errors in the calculation of the magnitudes of the past earthquakes.

SEISMIC ZONING

The construction of seismic risk maps which would be used by the civil engineers in establishing the seismic design criteria and parameters should include in addition to the study of the local historic record of earthquakes proper consideration of the local geology as well as the mapping of the major causative faults in the region.

According to the Jordan National Building Code the country is divided into four geopraphic zones viz. A,B,C, and D as shown in Figure 4. Table 2

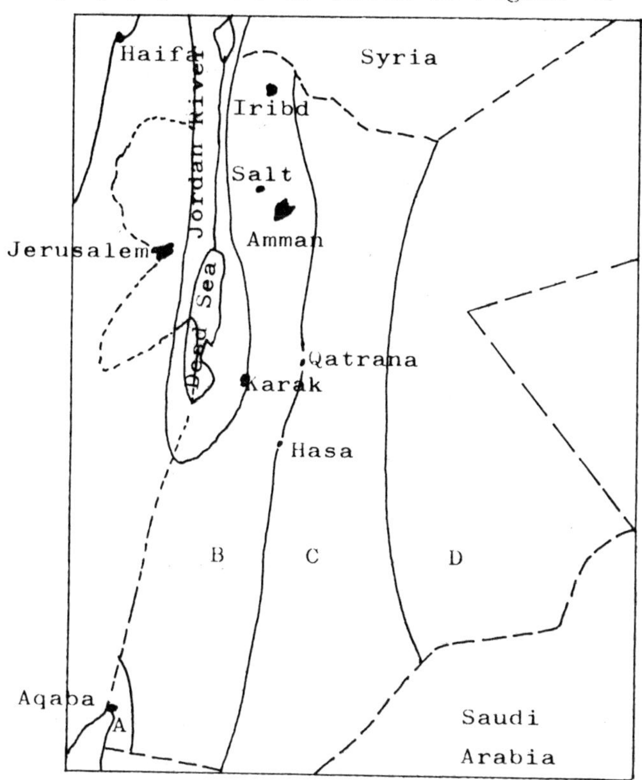

Figure 4. Seismic intensity zoning map of Jordan (National Building Code, 1985)

gives the values of the intensity factor (α) which is used in the calculation of the base shear force for each zone and the corresponding earthquake intensity according to Mercalli Scale.

Table 2. Intensity factor (α)

Zone	Earthquake intensity according to Mercalli Scale	Intensity factor (α)
A	>8	0.75
B	6-8	0.50
C	4-6	0.30
D	<4	0.10

It is worth mentioning that a high percentage of the population of Jordan live in cities and towns located close the Jordan Valley (<30 km) i.e. in Zones A and B of the seismic intensity

Zoning map (Figure 4) adopted by the Jordan National Building Code. It should be noted in this respect that this map is neither based on a thorough geotechnical evaluation of the different parts of Jordan nor on adequate data of instrumentally recorded events. It is rather based on the 4000 year historical record of earthquakes and on the proximity to the Jordan Valley and the South of Sinai. Thus, the boundaries between the different Zones should be considered as preliminary, and more refinements could be made in the future in view of the new gathered data about the geotectonic settings of the different areas in Jordan and about the recorded earthquakes especially after the installation of the new seismic stations. Mikbel[16] has suggested in this respect that the seismic intensity zoning map be modified as shown in Figure 2 in view of the explored weakness zones.

INSTABILITY FEATURES

Dynamic stresses resulting from strong earthquakes could result in different forms of soil instability such as landslides and liquefaction. The degree of distortion in the soil-rock masses depends primarily on the geological and geomorphological settings of the strata, their geotechnical characteristics, ground water conditions, and the dynamic loading conditions.

Based on the data and observations collected from 40 strong earthquakes Keefer (Talaganov[17]) presented the relationships between the magnitude of earthquake on one side and the area effected by landslides and the maximum epicentral distances within which landslide effects can be observed on the other side (Figures 5(a) and 5(b)). The relationship between the earthquake magnitude M and the maximum distance R (km) within which soil liquefaction was observed in the case of strong earthquakes ($M > 6$) was presented by Kuribayashi and Tatsuoka (Talaganov[17]) in the form:

$$\log R = 0.77\, M - 3.6$$

The above results indicate that for a magnitude $6\frac{1}{2}$ earthquake occurring in the Jordan Valley, liquefaction effects could be observed up to a distance of 25 km while landslides could occur within a distance of about 100 km from the epicenter.

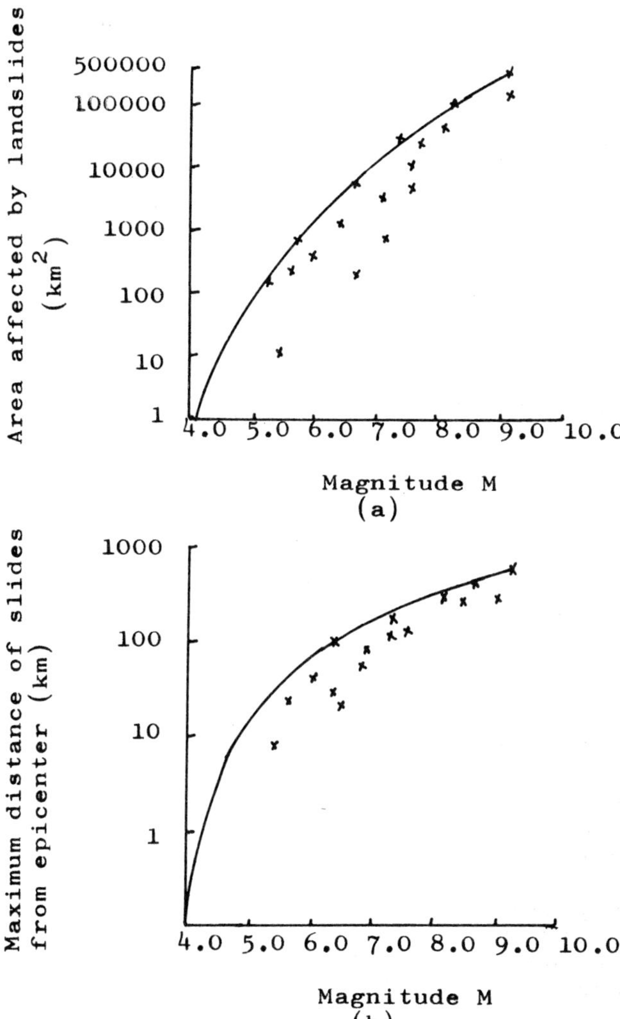

Figure 5. (a) Relationship between earthquake magnitude and area affected by landslides.
(b) Relationship between earthquake magnitude and maximum distance of slides.

The impoundment of water in the recently constructed earth dams reservoirs in Jordan such as King Talal Dam and Wadi El-Arab Dam or in the envisaged Maqarin Dam (storage capacity $>2.5\times10^8 m^3$) all of which are located at sites close to Jordan Valley, could have a serious impact on the seismicity of the region. Shallow to moderately deep artesian aquifers (100 to 300 m) with high sulfate warm waters were encountered at the sites of these dams. These hydrogeological phenomena combined with the volcanic activity manifested by the multistage basaltic flows suggest that the construction of the above dams would have an impact on the future seismic activity in the Jordan Valley and adjacent areas. Water storage may not itself cause earthquakes, as these are mainly related to the geotectonic processes, but it surely accelerates its occurrence through the build up of shear stresses or the reduction of shear resistance of rock mass with the build up of pore water pressure.

Future earthquakes, even those of moderate magnitude, could trigger more landslides in the critically stressed zones which are located close to the compressional belts described before. Major landslides in the East Bank of Jordan seem to have been affected by the seismotectonics of the region. It can be stated without reservation that many of the areas overlooking the Jordan Valley and the major tributaries of the Jordan River have experienced either recent or old slide movements. Landslides are frequently encountered along the main road cuts in Central and North Jordan (Figure 3) e.g. Swueileh-Jerash road, Amman-Naur-Adasyia highway, Salt-Shuneh road, Salt-Arda road, ...etc. Most of these slides occur in the lower formations of Ajlun Group (A1-2, and A3) of Upper Cretaceous which are dominated by plastic marls.

The seismic stability of partly saturated cohesive soils on natural slopes has not been studied adequately due mainly to the difficulty of assessing their strength characteristics under intricate geological and hydrological conditions. The cut in the spillway excavation on the right abutment of King Talal Dam (Figure 6), for example, is a typical example in this respect.

The more than 100 m high cut was carried out in a weakly to moderately cemented fine to medium grained sandstone intercaleted with a thick shale layer dipping gently (4° to 6°) outward. The stability of the sandstone overlying the shale during rapid drawdown or earthquake was the major concern of the designers. A detailed stability analysis involving the assumption of many failure surfaces passing through the sandstone and the underlying shale layer was carried out. Undisturbed samples were recovered from the shale layer where the peak and residual strength parameters were determined. Many cracks were detected in the sandstone which were considered as preferential paths for surface water seeping downward towards the shale layer. Softening and expansion of the shale after excavation was considered a major cause of the potential instability of the cut. A drainage gallery was constructed

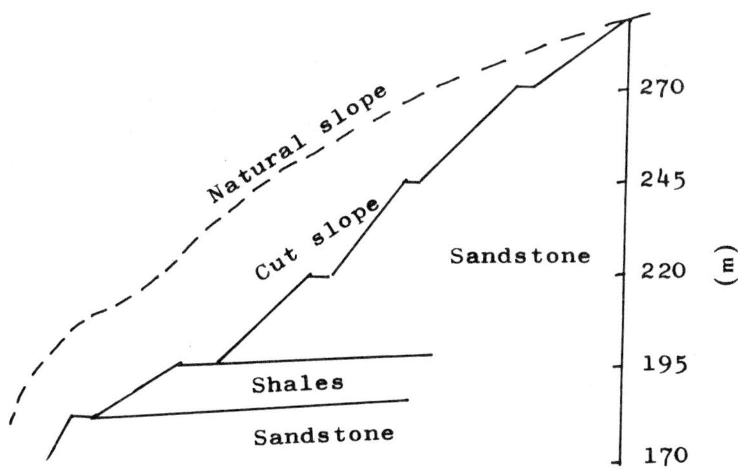

Figure 6. Typical cross-section in the spillway excavation of King Talal Dam.

to intercept water percolating through the sandstone towards the shale layer, and the slopes of the cut benches were considerably flattened to ensure the stability of the excavation.

The liquefaction potential of loose to moderately dense thick wadi deposits upon which large structures are being founded, as is the case in the downtown of Amman, has not been investigated thoroughly yet. The sides of the wadis are generally steep and thus the depth to bedrock could vary within a short distance. The author has frequently drawn the attention of local consulting firms to such a problem in some cases. It is fortunate that in most of the studied cases the wadi deposits were composed of a wide range of grain sizes and existed in a medium dense to dense state. They were not therefore considered vulnerable to liquefaction during moderate earthquakes. However, a new methodology to assess the liquefaction potential of such deposits during strong earthquakes is needed.

CONCLUSIONS

1- The Jordan Valley and its associated active faults on both sides could be considered the major source of most earthquakes in the region.

2- The earthquakes are characterized by their shallow focal depths.

3- The tectonic activity seems to be higher in Central and North Jordan than in South Jordan.

4- Many areas in Zone B of the seismic intensity zoning map of the Jordan National Building Code exist in a critically stressed condition.

5- Many of the major geological and geomorphological features of Jordan are related to the geotectonic setting of the region i.e. rotation of the Arabian plate, widening up of the Red Sea, and the spreading of the floor of the Arab Sea.

RECOMMENDATIONS

1- Most of the construction activities are currently carried out in areas close to the Jordan Valley (<30 km) where more intensive geological mapping and geotechnical evaluation of the projects sites are needed. Old buildings characterized by their poor design or low quality construction materials and located in areas close to the Jordan Valley should be either supported or demolished as the conditions may dictate.

2- Periodical revision of the adopted seismic intensity zoning map is needed in the light of the newly gathered information about the geology of Jordan and the instrumentally recorded earthquakes in the region.

3- More research is needed to assess the liquefaction potential of natural wadi deposits composed of sands and gravels with some fines, and also to evaluate the seismic stability of partly saturated cohesive soils on natural slopes.

REFERENCES

1. El-Kyasi, K. (1986), Introduction to Earthquakes, Workshop on Design of Earthquake Resistant Buildings, Amman, Jordan, November, 1986.

2. Building Research Establishment (1982), Earthquake Risk to Buildings in the Middle East, Overseas Building Note No 120, Garston, BRE.

3. Bender, F. (1974), Geology of Jordan, Gerbuder Borntraeger, Berlin.

4. Quennell, A.M. (1956). The Structural and Geomorphic Evolution of the Dead Sea Rift, Quarterly Journal of the Geological Society of London, V. CXIV, Part 1, pp. 1-24.

5. Ben-Menahem, A. (1981), Variation of slip and Creep Along the Levant Rift Over the Past 4000 Years, in The Dead Sea Rift, (Eds. Freund, R. and Garfunkel, Z.), Tectonophys. Vol 8o, pp 183-127.

6. Mikbel, S. (1986), Some Relevant Technoical and Geotectonical Considerations of South Amman Area, Jordan, Dirasat, Vol. XIII, No 7, Amman, Jordan.

7. Masannat, Y.M. (1984), Geotechnical Appraisal of the Stability of the Rock Cuts, Wadi Um Alrimam Vocational Center, (Unpublished Report), University of Jordan, Amman, Jordan.

8. Masannat, Y.M. (1985), Jofeh Landslide, Dirasat, Vol XII, No 9, Amman, Jordan.

9. Woodward-Clyde Consultants (1980), Domestic Water Project, North Jordan, Southern Region: Report Prepared for Stanley Consultants, Muscatine, Iowa.

10. Ben-Menahem, A. (1979), Earthquake Catalogue for the Middle East (92B.C.-1980A.D.), Bullettiondi Geofisica Teorica Ed Applicata, Vo. XXI, No 84, pp. 245-310.

11. Nur, A. and Reches, Z. (1979), The Dead Sea Rift-Geophysical, Historical, and Archaeological Evidence for Strike Slip Motion, EOS, Transactions AGU, 60, pp. 322, Abstract.

12. Ben-Menahem, A., Nur, A., and Vered, M. (1976), Tectonics, Seismicity, and Structure of the Afro-Eurasion Junction-The Breaking of an Incoherent Plate, Physics of the Earth and Planetary Interiors, Vol. 12, pp. 1-50.

13. Tocher, D.T. and Patwardhan, A. (1975), Relationship of Earthquake Magnitude and Length of Surface Rupture Based on Aftershock Zones, Geological Society of America (Abs.), Vol. 7, No 3, pp. 419.

14. El-Isa, Z.H. and Mustafa, H. (1986), Earthquake Deformation in the Lisan Deposits and Seismotectonic Implications, Geophys. J.R. Astr. Soc., Vol. 86, pp. 413-424.

15. Ministry of Public Works (1985), Loads and Forces Code, Earthquake Forces, Jordan National Building Code, Amman, Jordan.

16. Mikbel, S. (1985), Geological Applications of Remote Sensing in Jordan, Proceedings of the 1st National Symposium on Remote Sensing, 24-28 October, 1985, Baghdad, Iraq.

17. Talaganov, K. (1985), Seismic Soil Stability and Liquefaction Potential, International Seminar on Computer-Aided Design of Earthquake Resistant Engineering Structures, Skopje, Yugoslavia.

Seismic Hazard in the Eastern Mediterranean

M.Ö. Erdik, V. Doyuran
Earthquake Engineering Research Center, Middle East Technical University, Ankara, 06531, Turkey

INTRODUCTION

In probabilistic sense the seismic hazard can be defined as the probability of occurrence, at a given place and within a given time period, of ground motion due to an earthquake of a particular size capable of causing significant loss of value through damage or destruction. The seismic hazard is usually expressed in terms of the exceedance probability per year, called the annual hazard, or its reciprocal which is the mean return period in years, of a given ground motion intensity parameter for a specified site based on the available geological, seismological and statistical information. If the interest is in a region then seismic hazard maps can be developed through simultaneous hazard analyses for many sites in the region and constructing iso-acceleration contours for specified acceleration levels corresponding to given return periods.

The exist several efforts of probabilistic seismic hazard mapping throughout the world, such as: Algermissen et al.[2] for the Balkan region, Algermissen and Perkins [1] for the United States, Shah et al.[42] for Nicaragua, and Hattori[23] for all the seismic regions of the world.

The use of the probabilistic seismic hazard analysis provides greater logic and consistency over deterministic procedures by allowing for maximum utilization of geologic, geophysical, seismologic and historic data in a coherent manner. Furthermore, it yields the crucial information toward well-balanced engineering designs requiring a trade-off between costly designs for greater resistance and more economical designs with higher risks of economical loss.

The incorporation of the probabilistic seismic hazard assessment schemes into the earthquake resistant design, land use management and earthquake insurance activities has been under study for the last decade. For example, the probabilistic seismic hazard maps prepared by Algermissen and Perkins[1] have been adopted as the basis of the probabilistic maps provided in the "Tentative Provisions for the Development of Seismic Regulations for Buildings" (ATC-3[6]) for the determination of design spectra applicable to building structures. A similar approach is beign considered for use in the earthquake resistant design codes of the Balkan countries as well (UNESCO[44]).

This study aims at developing seismic hazard maps for the Eastern Mediterranean Region (Figure 1) by utilizing probabilistic procedures. These maps depict the probabilistic estimates of the maximum horizontal peak ground acceleration for given return periods. They are intended to form a basis for the assessment of the seismic hazard concerning civil engineering structures, in a manner similar to that in ATC-3[6].

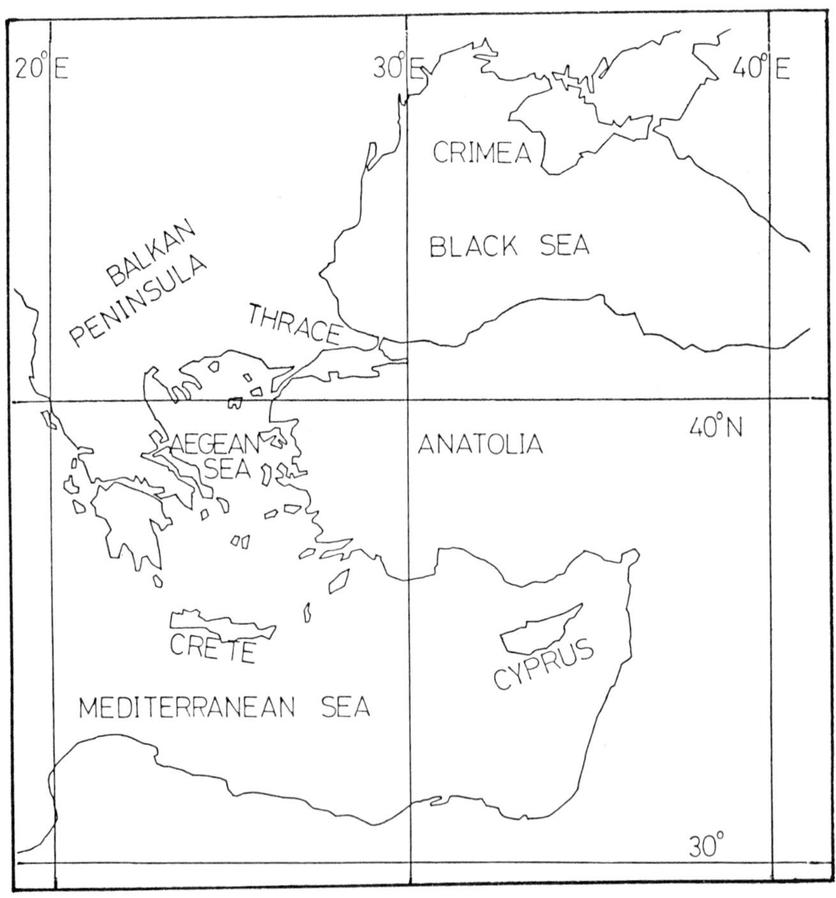

Figure 1. Region considered for this study

METHODOLOGY

Seismic hazard analyses aim at assessing the probability that the ground motion parameter at a site due to the earthquakes from potential seismic sources will exceed a certain value in a given time period. The probabilistic seismic hazard models developed by Cornell[11], Esteva[21] and Milne and Davenport[33] are called the "point-source models" since they are based on the assumption that the energy released during an earthquake is radiated from the focus of the earthquake and the intensity of the site ground motion is a function of the distance to the source. Although this assumptions may be acceptable for certain earthquakes and regions, it would not, however, be valid for large events where the total energy released is distributed along a long rupture zone. The "fault-rupture model" developed by Der Kiureghian and Ang[12] is based on the assumptions that the earthquake originates at the focus and propagates symmetrically on each side of the focus along a fault and the maximum intensity of the ground shaking at a site is determined by the rupture that is closest to the site.

Both of these models can be adequately represented by the so-called Total Probability Theorem (McGuire[27]):

$$Pr[A] = \iint Pr[A|s,r]\, f_S(s)\, f_R(r)\, ds\, dr \qquad (1)$$

where A represents the event that a specific measure of the ground motion is exceeded at the site during an earthquake of size s(magnitude of epicentral intensity) and distance r(fault, epicentral or focal distance) and S and R are continuous random variables representing the earthquake size and distance. The probability of occurrence of A, $Pr[A]$ is given by the integration of the conditional probability of A given s and r times the independent probabilities of s and r, $f_S(s)$ and $f_R(r)$, over all possible values of s and r. For the fault rupture model the s and $f_S(s)$ terms include the information on and distribution of the rupture length and the rupture location at the source.

The assessment of the probabilistic seismic hazard requires all the available information on seismicity and geotectonics of the region and on regional attenuation characteristics of the ground motion as well as the adoption of a stochastic model for the forecasting of future occurrences.

The key research steps and the parameters of the probabilistic seismic hazard assessment conducted in this study can be summarized as follows (Erdik et al[20]):

(1) Acquisition of seismic data. The studies in this step involved the compilation of data from the existing literature. The sources and mode of compilation of the seismic data utilized in this work will be elaborated under the heading "Earthquake Dase Base".

(2) Seismic source regionalization. Based upon the geologic evidence, geotectonic province, historic seismicity, geomorphic investigation and other relevant data seismic sources are identified. The inputs, procedure and the results of the seismic source zoning adopted in this work will be elaborated under the heading "Seismic Source Regionalization".

(3) Source seismicity information. This step involves the construction of the recurrence relationships and the assessment of the maximum magnitudes for each individual seismic source. The related studies are documented under the headings "Reccurrence Relationships" and "Maximum Magnitudes".

(4) Attenuation of the ground motion. For the attenuation of the peak ground acceleration the relationships developed for the western U.S.A. are adopted. The methodology and the findings of this step are covered under the heading "Attenuation Relationships".

(5) Recurrence forecasting. The essence of the probabilistic hazard analysis is that it incorporates a stochastic point process as the model of earthquake occurrences. The stochastic model used in this study assumes homogeneous Poisson process for the seismic occurrences. A brief explanation of the methodology is given under the heading "Seismic Hazard Model".

EARTHQUAKE DATA BASE

Instrumental recording of the earthquakes in the world began very late in the last century and continued for sometime with rather inadequate instruments resulting in very poor determination of earthquake foci. Although between 1920 and 1950 the accuracy of the location of the larger shocks, as reported by the International Seismological Summary (ISS) and the Bureau Central International Seismologie (BCIS), have improved, on the whole,the epicentral locations were still of low accuracy, especially for smaller shocks. With the inception of the World Wide Seismological Station Network (WWSSN) the location accuracy of the earthquakes has considerably improved after mid 60's. Thus, especially for the region considered, it can be stated that the reliable seismological data base for small magnitude events exists only for the last two to three decades.

The earthquake data utilized for the determination of the seismic sources and for the assessment of the source-specific frequency-magnitude characteristics are based on the NOAA Earthquake Data File (U.S.National Oceanic and Atmospheric Administration (NOAA) Data File and Bulletins of Preliminary Determination of Earthquakes, U.S.G.S., N.E.I.S.) covering the time period from 11 A.D. to 1986 A.D. In the NOAA Earthquake Data Files there exists an agglomerate of different magnitude estimates for pre-mid-1960 earthqukes. After mid-1960's the

NOAA Earthquake Data File starts to report the magnitudes with the identification of the scale used.

Figures 2, 3, and 4 show the seismicity of the region, respectively, for the periods of A.D.-1986, 1976 - 1986 and 1981-1986.

SEISMIC SOURCE REGIONALIZATION

The Eastern Mediterranean lies within the Alpine-Himalayan orogenic system. It is characterized by intense seismicity associated with complex tectonic activity. Figure 5 shows the neotectonic features in the region considered. This complex tectonic activity arises from the existence of a number of plates, large or small, each contributing to the present-day seismicity of the region. According to Morgan[36], Le Pichon[26], and McKenzie[31] the relative motion between the plates; namely, the African the Eurasian, and the Arabian accounts for most of the tectonic and/or seismic activity in the Eastern Mediterranean region. The existence of smaller but rapidly moving plates such as the Anatolian, the Aegean and the Black Sea (Figure 6) causes a local increase in seismicity, which further complicates the picture.

The Aegean plate comprises the southern part of mainled Greece, Crete, and the western part of Turkey. It is bounded in the north by the North Anatolian Trough (Allan and Morelli ; McKenzie[30]; Morelli et al[34]) and in the south by the Hellenic Trench and the Pliny-Strabo trench complex (McKenzie[30]; Dewey and Şengör[13]). The southern margin of the Aegean Plate is characterized by thrust faults indicating the overriding of this plate onto the African Plate (Morelli[34]; Woodside[46]). The relative motion rate between the African and the Aegean plates is in the order of 3.5 cm/year (McKenzie[30]), 2.5 cm/year (Papazachos and Comninakis[38]), or 2.6 cm/year (Le Pichon[26]).

The Anatolian plate contains most of Turkey and Cyprus. The northern boundary of the Anatolian plate is well-defined by the North Anatolian Fault (NAF), which is an active east-west trending right-lateral strike-slip fault. The southern boundary of the Anatolian plate forms an arcuate line, which starts from southwest of Turkey, follows south of Cyprus, and ends at the Gulf of İskenderun. From here on it joins with the East Anatolian Fault (EAF), a left lateral strike-slip fault, which forms the southern boundary of the Anatolian plate.

The Black Sea plate, seperated from the Anatolian plate by NAF, constitutes the southern extension of the Eurasian plate.

The neotectonic history of the Eastern Mediterranean starts with the collision of the Arabian plate with the Anatolian plate about 10 Ma ago along the Bitlis Suture Zone

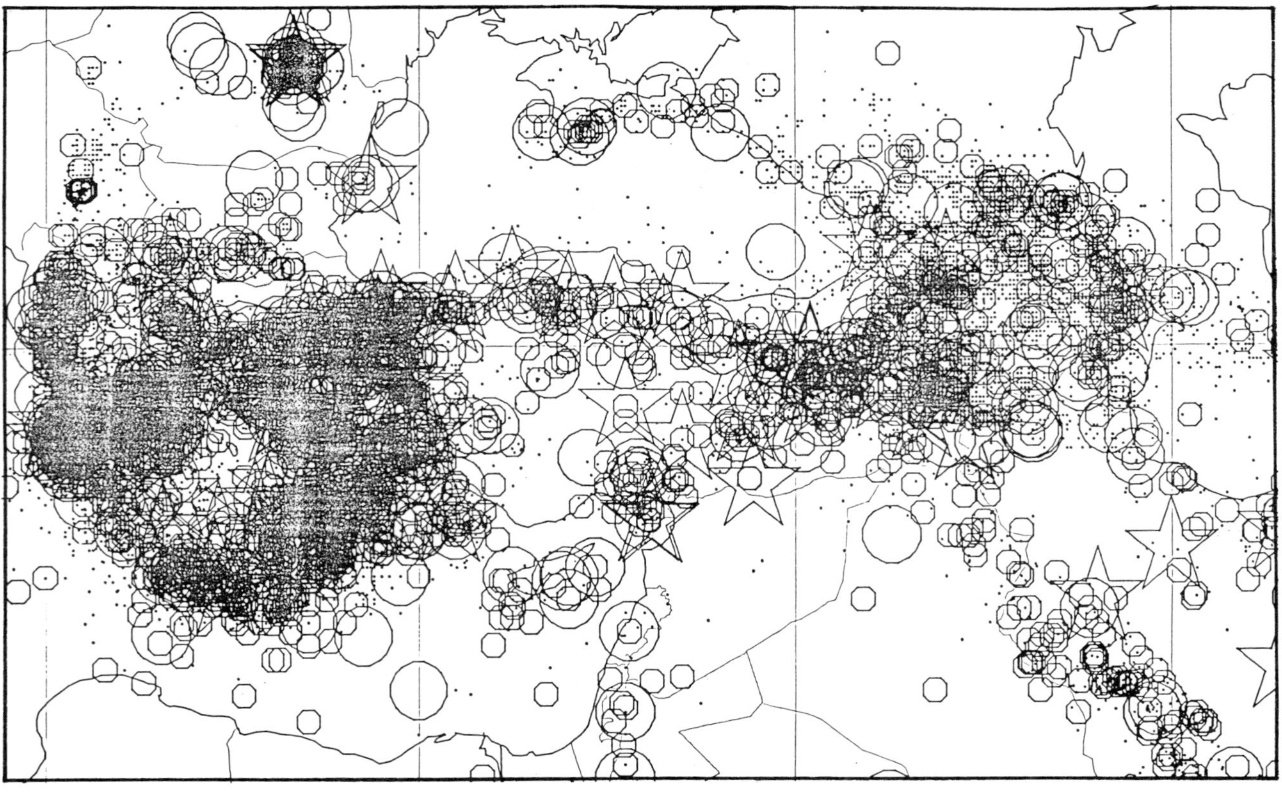

Figure 2. Seismicity of the region for the period 11 AD-1986. (Stars denote $M \geq 7$, circles denote $7 < M \leq 6$, octagons denote $6 < M \leq 5$ and points denote $M < 5$)

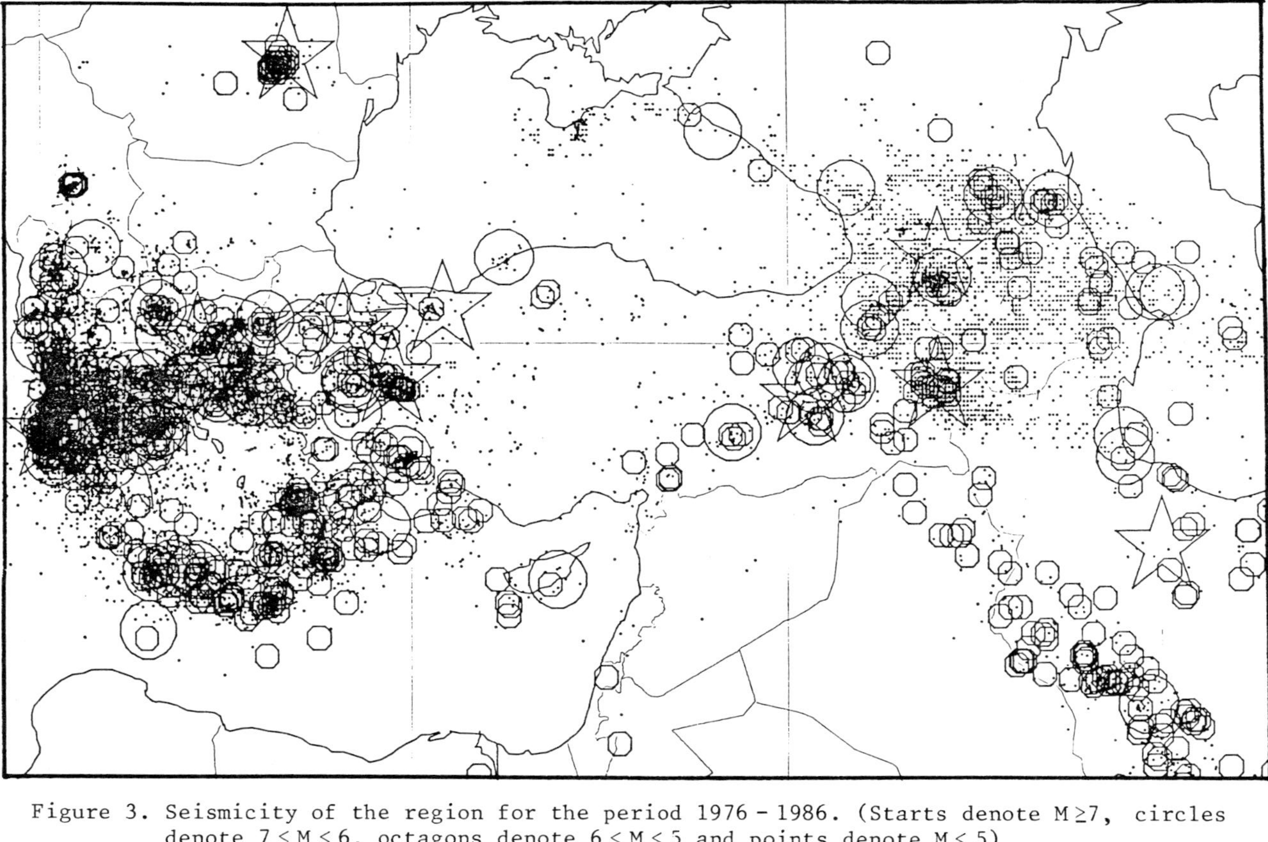

Figure 3. Seismicity of the region for the period 1976 - 1986. (Starts denote $M \geq 7$, circles denote $7 < M \leq 6$, octagons denote $6 < M \leq 5$ and points denote $M < 5$)

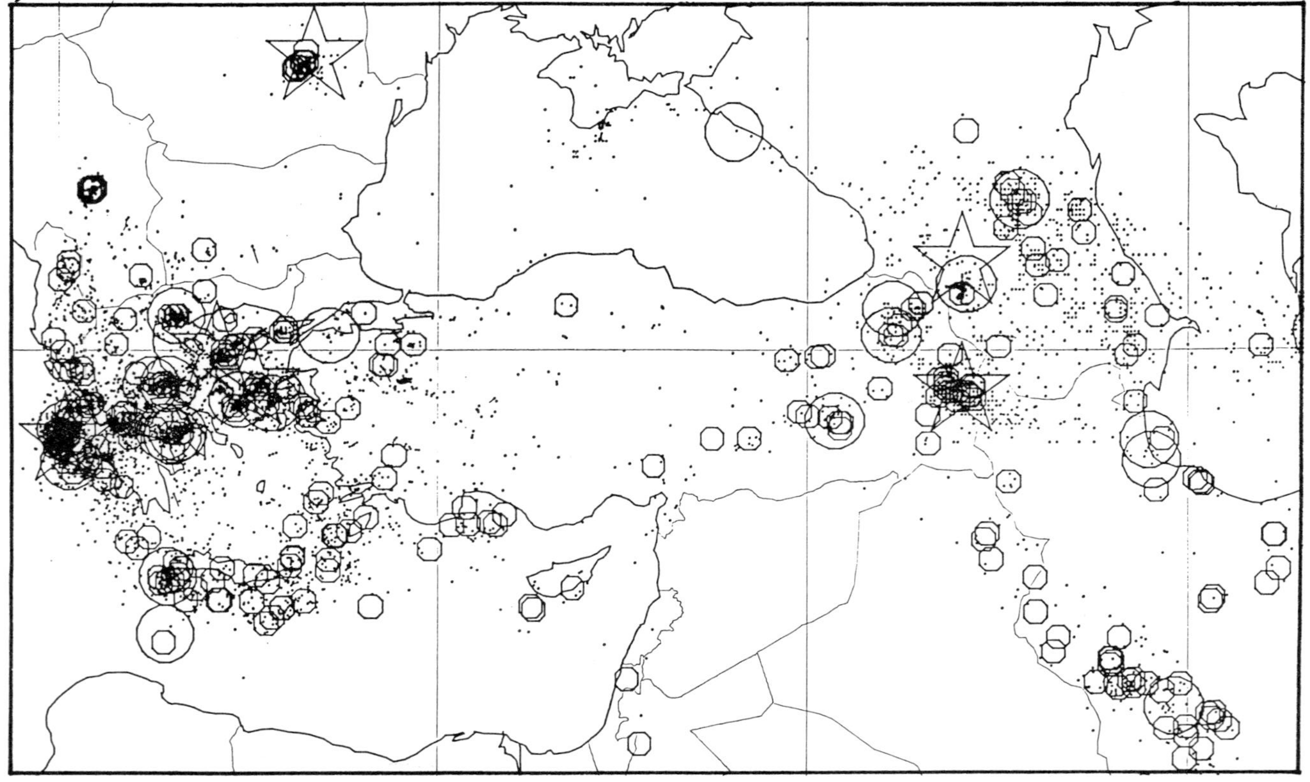

Figure 4. Seismicity of the region for the period 1981-1986. (Starts denote $M \geq 7$, circles denote $7 < M \leq 6$, octagons denote $6 < M \leq 5$ and points denote $M < 5$)

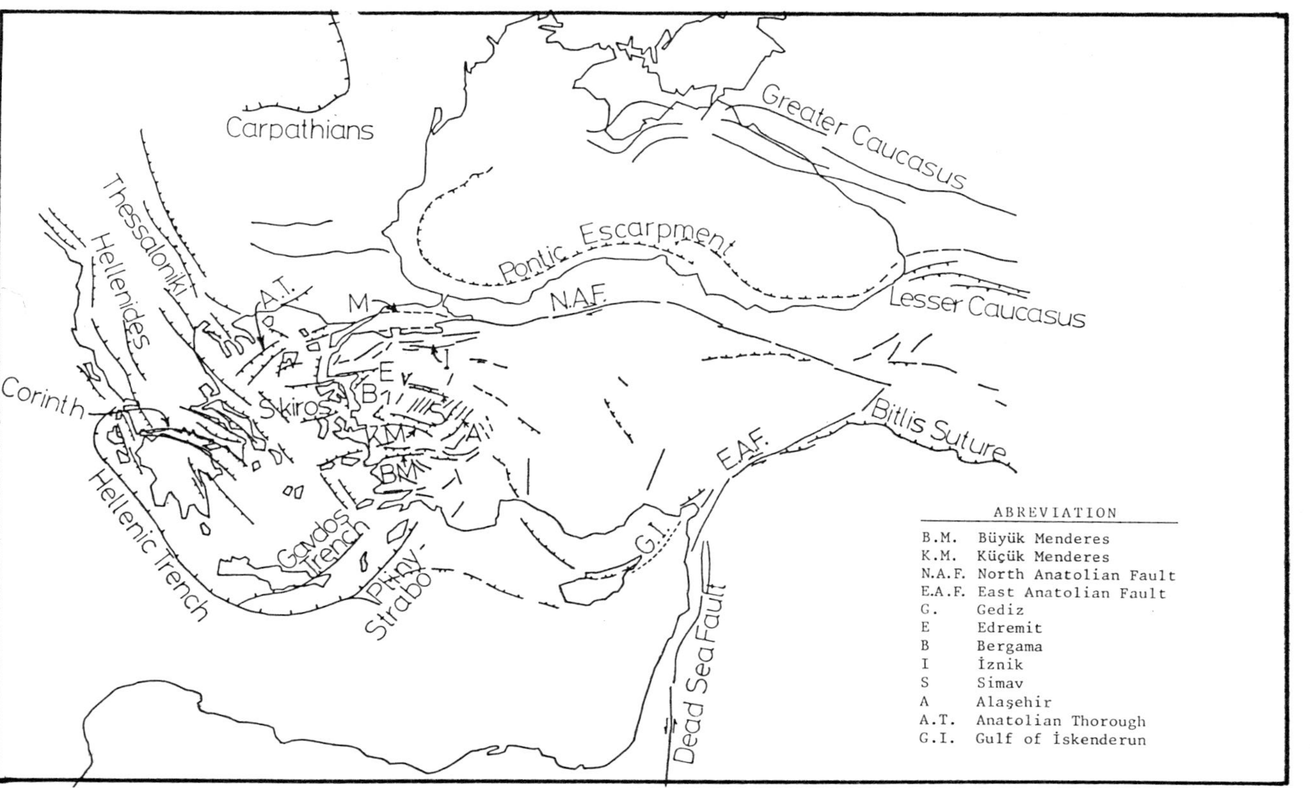

Figure 5. Neotectonic features of the region considered (After Erdik et al[20], UNESCO[45])

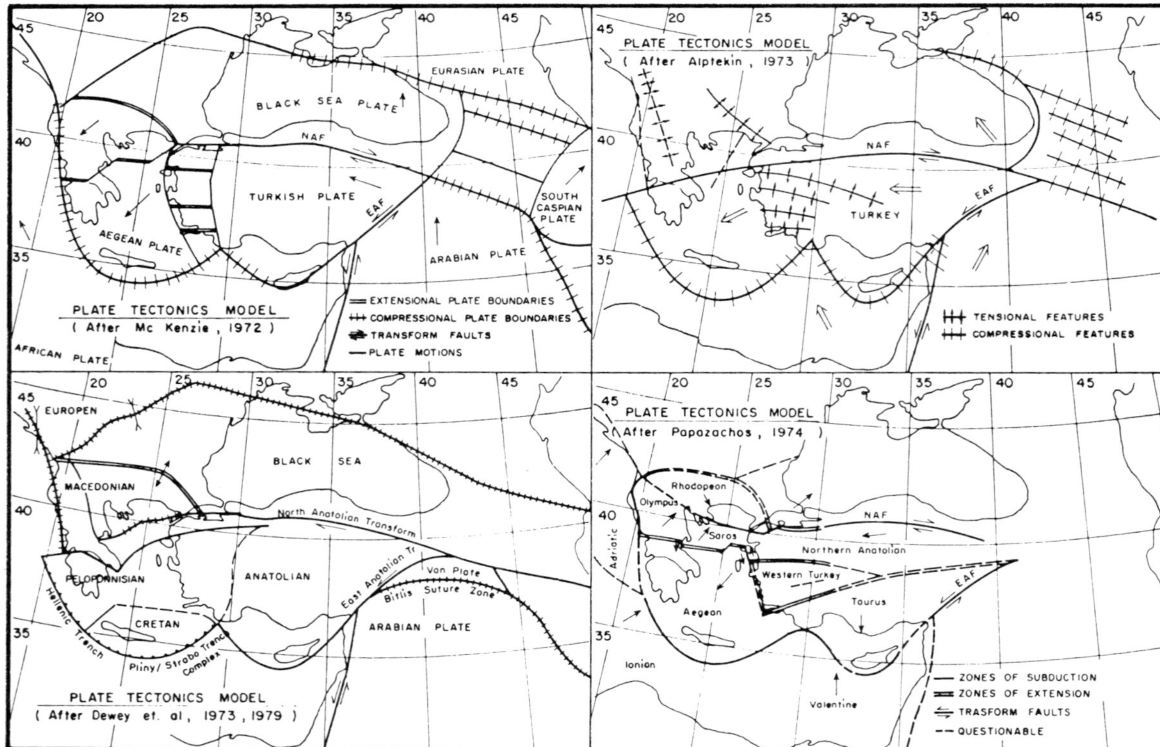

Figure 6. Various plate tectonics models proposed for the Eastern Mediterranean Region

(Figure 2). The Anatolian plate, because of buoyancy, began to move westward into the Aegean along the bounding NAF and EAF (Burke and Şengör[7]). This westward tectonic escape of the Anatolian plate relative to Africa is taken up by the subduction along the Hellenic trench. The complex pattern of rifting in Western Anatolia reflects and typifies the style of break up of continental and arc assembled Anatolia as it has moved into the larger general area of the Aegean Sea (Burke and Şengör[7]).

Widespread seismicity roughly encircling the Aegean region and the numerous active faults of Western Anatolia and Northern Aegean indicate that the Aegean is not a rigid integral part of the Anatolian plate, which itself is currently internally deforming along east-west oriented normal faults (Dewey and Şengör[13]). The fault plane solutions carried out by McKenzie indicated that the Western Anatolia and Northern Aegean are largely affected by N-S extension. The east-west grabens of the Western Anatolia, which close eastward and terminate westward, are associated with this extensional regions.

Based on the neotectonic considerations of the Eastern Mediterranean region, certain well-defined seismic sources may be delineated along the plate boundaries. Figure 7 depicts the proposed model for the seismic sources of the Eastern Mediterranean. Brief descriptions of the seismic sources are given below.

Anatolian Trough

The anatolian Trough is a well established plate boundary in the northern Aegean including Saros and Sporades Troughs and terminating at the Gulf of Corinth (Morelli et al[35]; McKenzie[30]). It corresponds to the western continuation of the northern strand of the NAF. It also includes the Sea of Marmara and the İzmit-Sapanca graben. In Central Greece, the Anatolian Trough exhibits shallow foci and a tensional stress field trending N-S (Drakopoulos[16]).

North Anatolian Fault

The NAF is a morphological distinct and seismically active right lateral strike-slip fault. The width of the shear zone ranges between a few hundreds of meters and a few kilometers. The maximum expected magnitude of earthquakes is 8.0 with shallow foci. For further information about NAF the reader is requested to refer to Erdik et al[20].

West Anatolian Graben Complex

The intense seismic activity of western Anatolia is closely related with roughly east-west trending grabens which are bounded by high angle normal faults. The İznik, Edremit, Bakırçay, Bergama, Simav, Gediz, Büyük Menderes, Küçük Menderes and Alaşehir are typical examples of this graben complex (Erdik

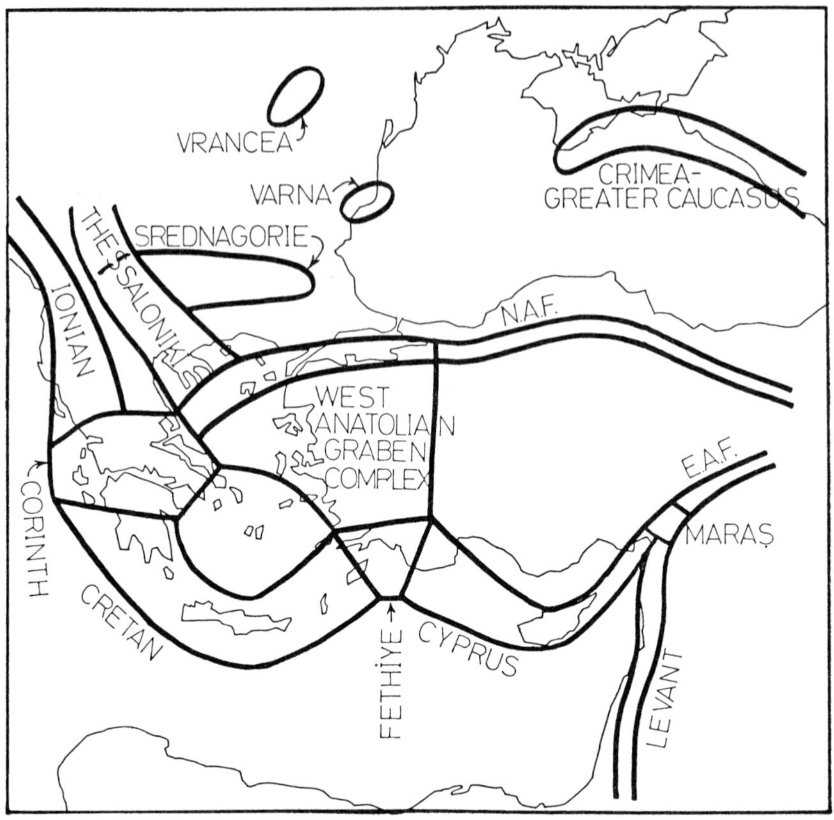

Figure 7. Seismic source regionalization considered in this study

et al.[20]). The grabens close eastward and terminate westward shortly after the western cost live of Anatolia. In the north of İzmir, however, the seismic zone extending into the Aegean Sea is related with the Skiros Trough. It is probably the western extension of the Gediz graben.

Thessoloniki

The seismic activity is associated with NW-SE normal faults with left-lateral strike slip component. They probably form the conjugate system of the NE-SW trending Anatolian Trough. The is characterized by shallow foci earthquakes.

Cretan

The seismic source referred to as Cretan corresponds to the Hellenic trench and the Pliny-Strabo trench complex, which also forms the southern boundary of the Aegean plate. Subduction of the Mediterranean lithosphere under the Aegean sea occurs along the Hellenic trench. The zone extends northwestward into Albania and southestward into Turkey forming a curvilinear tectonic belt.

Ionian

In this zone the compressional regime may be considered as the main cause of present seismic activity in western Albania. The zone is characterized by numerous longitudinal tectonic structures which are located in a relatively wide belt (Aliaj[31]).

Although it has different seismic potentials, the zone, in general, is characterized by medium seismicity ($5.0<M_{max}<6.0$).

Crimea-Greater Caucasus

The seismic source delinated along the northern margin of the Black Sea is defined on the basis of existing geological structures and of shallow focus earthquakes (Doyuran and Erdik[15]). According to McKenzie[30] this zone corresponds to the boundary between the Black Sea plate in the south and the Eurasian plate in the north. Between Crimea and the Caucasus the region is under the effect of compressional regime. The seismic activity along the northeastern part of the Black Sea is associated with the major structural features of the Greater Caucasus.

Vrancea

The seismic zone in the bend of the Eastern Carpathians (known as the Vrancea Massif) where the deep-seated foci responsible for the largest shocks in Moldavia is associated with deep-seated tectonic contact of the multistage-fault type (Drumya et al[17]). Most of the junction zones along which the displacements occurred are still active. The fault plane solutions for the Vrancea earthquake indicate a reverse fault (Enescu et al[18]). Major structural features of the Vrancea Massif suggest compressional regime in this area.

Cyprus

The southern boundary of the Anatolian plate runs in a large loop from the Gulf of İskenderun, south of Cyprus and to Rhodes (McKenzie[30]). The Eastern Mediterranean lithosphera underthrusts the Eurasian Lithosphere along the arcuate structure between the Ionian islands and Cyprus (Papazachos[37]). The Cyprus zone terminates on the east of the EAF and/or Bitlis Suture zone.

East Anatolian Fault (E.A.F.)

The EAF is an active left lateral strike-slip faults which extends all the way from Antakya to Karlıova, where it meets the NAF. The general trend of the fault is NE-SW. The fault zone is about 2-3 km wide and it consists of numerous parallel or subparallel, continuous and discontinuous or anastomosing fault traces. The EAF joins with the Dead Sea fault system (Levant) in the southwest, at the triple junction near Maraş.

Levant

The seismic source referred to as Levant accounts for the seismicity associated with the Dead Sea rift. The Dead Sea fault system, N-S trending left-lateral strike slip faults, is formed as a result of the relative motion between the African and the Arabian plates. It joins with the EAF at the triple junction near Maraş.

Corinth, Fethiye, Maraş

These seismic source correspond to the stress concentration areas at the junction of several sources as shown in Figure 6.

Sreanogorie

This block-fault mountain-valley region comprises fault-flexure zones of different orientations, which become active during Quaternary. The vertical neotectonic movements are characterized by various abrupt changes and increase from about 500m in the east to 1800 m in the west (Rizhikova[40]).

RECURRENCE RELATIONSHIPS

The empirical recurrence relationship for earthquake (Richter[38]):

$$\text{Log } N = a + bL \tag{2}$$

where N is the number of the earthquakes above the magnitude M in a given region and within a given period and a and b are regression constants, has been extensively used in many seismicity studies and has also been confirmed to hold for micro-earthquakes. The coefficient a is a constant that is dependent on the location and time of the sample used and b represents a constant thought to be characteristic of the region.

The earthquake catalogues are often biased due to incomplete reporting for smaller magnitude earthquakes in earlier periods. Thus, to fit the recurrence relationship to a region, one should choose among using:(1) a short sample that is complete in small events or (2) a longer sample that is complete in larger events or (3) a combination of the two data sets to complete the deficient data thereby obtaining a homogeneous data set. A direct attempt to fit these data to a regression relationship may result in quadratic or higher order expressions to accomodate the enherent bias and inhomogeneity of the data.

In the method used in this study, an artificially homogeneous data set is simulated through the determination of the period over which the data in a given magnitude group are completely reported (Stepp[43]). In this method, assuming that each magnitude class can be represented as a point process in time and the earthquake occurrence follows Poisson distribution, an unbiased estimate of the mean rate of occurrence of events per unit time interval is given by:

$$\bar{\lambda} = \frac{1}{N} \sum_{i=1}^{N} \lambda_i \tag{3}$$

where λ_i is the rate of occurrence of events per unit time interval for the i th subsample of the event set and N is the number of subsamples. The variance of the mean rate is given by the following expression:

$$\sigma^2_{\bar{\lambda}} = \bar{\lambda}/N \tag{4}$$

If the unit time interval is chosen as one year and T is the total time of the sample, the standard deviation of the sample mean is given by

$$\sigma_{\bar{\lambda}} = \sqrt{(\bar{\lambda}/T)} \tag{5}$$

A plot of standard deviation versus time will behave as $1/\sqrt{T}$ if the sample is unbiased and the earthquake process is stationary. Accordingly, the procedure consists of plotting the observed standard deviation versus time and determining the interval of time where the standard deviation behaves as $1/\sqrt{T}$. This time interval is then long enough to provide a reliable estimate of the mean rate of occurrence. The regression constants a and b of Eqn.(2) are obtained by applying a leastquare analysis on the adjusted mean annual rate of occurrence of events for each seismic source region.

Table 1 provides a list of the regression coefficients of the recurrence relationship for each seismic source zone based on the adjusted annual rates of occurrence.

MAXIMUM MAGNITUDES

The determination of the maximum magnitude expected in a source zone is a difficult and quite subjective problem. The following methods are used in determining the maximum magnitude expected.

Maximum historical earthquake procedure

For very active faults with long seismic history and short recurrence intervals between large magnitudes this method may provide reasonable assessments. In the standard rule-of-the-thumb practice the maximum historical magnitude (recorded or inferred) is increased by half a magnitude unit to yield the maximum magnitude.

Fault rupture length procedure

The method involves the use of regression analysis of the surface rupture characteristics with the magnitudes. The most common characteristic utilized has been the rupture length. The worldwide data (Cluff et al[10]) suggest that for faults of total length less than 200 km the maximum rupture amounts to about 1/6 of the total length and for faults of more than 1000 km in length 1/3 to 1/2 of the fault may rupture. A conservative fraction can be taken to be equal to one half of the total fault length.

TABLE 1

Log-linear recurrence relationship regression constants (adjusted using Stepp[43], procedure)

Seismic Source	a	b	Max. M_s
North Anatolian Fault	3.6	-0.80	8.0
Anatolian Trough	4.1	-0.78	7.8
West Anatolian Fault	4.7	-0.78	7.5
East Anatolian Fault	4.3	-0.78	8.0
Maraş	4.4	-0.81	7.5
Levant	2.5	-0.75	8.0
Cyprus	3.2	-0.72	7.5
Fethiye	4.4	-0.89	7.6
Cretan	5.1	-0.90	7.5
Corinth	4.6	-0.79	7.0
Ionian	4.1	-0.85	6.8
Thessaloniki	3.4	-0.73	7.2
Srednogorie	3.5	-0.76	7.0

The analysis of the rupture length versus magnitude data has yielded the following expressions for the North Anatolian Fault strike-slip earthquakes (Erdik et al[20]):

$$\log L = -6.4 + 1.1 M \tag{7}$$

where L is the rupture length in km's, M is the surface wave magnitude and regression of M on log L is conducted.

In the methodology used herein the fault rupture length procedure in connection with Eqn.(6) is utilized to determine the maximum magnitudes associated with the seismic sources encompassing strike-slip faults. The maximum historical earthquake procedure is applied to seismic sources where the tectonics is very complex and/or it is not possible to single out a prominent fault. Table 1 summarizes the maximum surface wave magnitudes assigned to each seismic source.

ATTENUATION RELATIONSHIPS

For a given earthquake the ground motion at the site can be obtained by either utilizing the so-called attenuation relationships or physical mathematical calculational models with varying degree of sophistication. The Peak Ground Acceleration (PGA) is the most commonly used single ground motion descriptor for the seismic hazard assessments. The authors of this work have based their seismic hazard map on the horizontal component of the PGA not because that it is the best parameter that describes the severity of the ground motion or governs the response of the structures, but because it has been sanctioned by the previous applications of similar nature and it has widespread usage.

Most of the existing attenuation relationships try to correlate the PGA with magnitude (M), distance (R) and with site soil conditions. The magnitude, by definition, is a band-limited measure of the ground displacement which necessitates an adherence to the definition used in the attenuation relationship in the compilation of the frequency-magnitude data and in assessment of the maximum magnitudes. It should be noted that, whereas most of the PGA attenuation relationships are based on the local magnitude (M_L), the magnitudes associated with the earthquakes in the Region considered are a conglomerate of body-wave, surface-wave and region-specific-local magnitudes.

The distance parameter (R) used in the attenuation relationships may be related to: (1) hypocenter, (2) epicenter, (3) center of energy release, (4) fault surface, and (5) fault trace. The definition of distance plays a great role especially in the near field and should be compatible with the regional characteristics. In regions like Turkey where most of the earthquakes are shallow and a associated with surface ruptures the use of the site to fault trace distance is physically and rationally justifiable.

In one type of attenuation relationships the PGA is correlated with M and R through one single equation. A critical compilation of this type of attenuation relationships is given by Idriss . The majority of the data base of these relationships pertain to the medium field (i.e.20-200 km) with limited near field and far field data and no physical constraints are introduced in the regression analysis. The relationships exhibit considerable dispersion. Typically, In PGA is proportional to 0.9 M with the standard deviation of about 0.7. Thus ±1 standard deviation zones almost overlap the mean estimates for earthquakes of two magnitude units apart, thereby decreasing the significance of such attenuation relationships tries to correlate the PGA with R for earthquakes of each given magnitude does not explicitly enter into the regression analysis and the distance dependence of the PGA does not have to conform to the same shape for each magnitude class. Schnabel and Seed[41] and Boore et al[8] have presented attenuation relationships of the second type.

The scarcity of the local strong-motion acceleration data in the region considered makes it indispensible to borrow the already developed acceleration attenuation relationships based on foreign data. After detailed considerations and comparisons the attenuation relationships of Schnabel and Seed with near field corrections from Campbell[9] are utilized for the seismic hazard maps developed in the present work.

Schnabel and Seed have studied the maximum horizontal ground accelerations on rock sites during earthquakes in the western part of the United States and presented the results of the variation of the PGA with distance from causative fault for different magnitudes in graphical format (Figure 8). In their study the available empirical data are treated with the theoretical understanding that the attenuation is governed by the combined effects of geometrical divergence and internal material damping. For events associated with finite faults the former factor is dominant at the near field and for events that can be modelled by point sources both factors would be comparable. The results presented seem to be in good accord both with the theory and observations and referenced for consideration in the U.S.Nuclear Regulatory Commission Standard Review Plans.

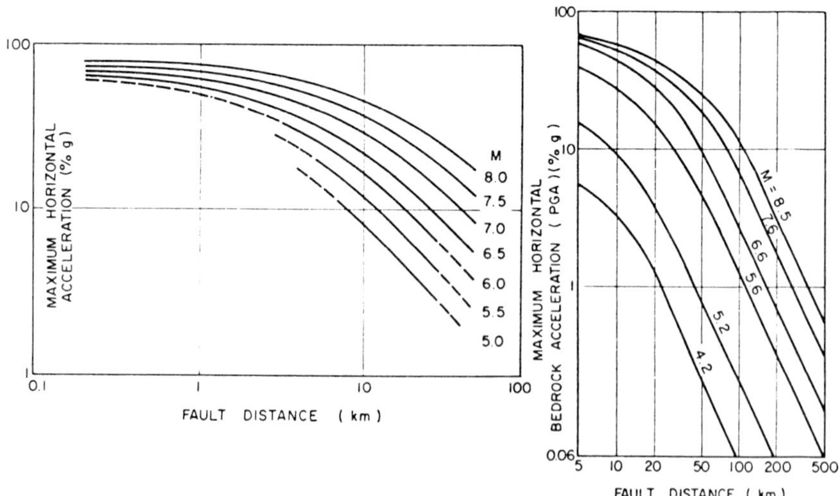

Figure 8. Attenuation relationships considered. (Respectively after Campbell[9] and Schnabel and Seed[42])

The main data base of the Schnabel and Seed attenuation relationship comes from the 1971 San Fernando earthquake which is rich in the medium field data but quite poor in the near field data. The three most recent significant events (1976 Gazli,

1978 Tabas and 1979 Imperial Valley) producing excellent near field data are elaborately treated by Campbell in which a regression equation is presented to analyse the near field PGA. The Campbell[9] near field attenuation relationship has been utilized, where necessary, to update the near field portion of the Schnabel and Seed relationships which are considered in the seismic hazard assessments developed in this study. Champbell's[9] near field PGA attenuation relationships are illustrated in Figure 8.

The PGA attenuation relationships utilized in this study are strongly controlled by the Western U.S.A. data. In order to provide a check on the applicability of these data the available Turkish strong motion data have been favorably compared with the Western U.S.A. data (Erdik et al[20]).

SEISMIC HAZARD MODEL

In this study the earthquake phenomenon is based on both (1) the "point source model" of Cornell[11] and (2) the "fault rupture model" of Der Kiureghian and Ang[12] depending on the source zone. In the point source model the earthquake energy is assumed to propagate isotropically from the source, whereas in the fault rupture model the source is assumed to be in-line with the fault rupture resulting in a non-isotropic propagation of seismic energy. The probabilistic earthquake hazard analysis can be expressed on the basis of both of these models by the "total probability theorem" (McGuire[27] given by Eqn.1).

Comparative studies of the point source model of Cornell and the fault rupture model of Der Kiureghian and Ang[12] indicate some differences between the results of the two methods (Erdik and Öner[19]). It has been observed that the point source model underestimates the seismic hazard in the vicinity of linear tectonic features such as the North Anatolian Fault compared to the fault rupture model.

The study conducted herein is based on the point source model, however, corrections representing the fault rupture model are implemented where applicable, although none of these corrections was found to be appreciable.

Numerous models for the forecasting of seismic hazard have been developed. The two major types of the stochastic models that are of common use are the Markov and Poisson Processes. The simplest stochastic model for earthquake occurrences is the Homogeneous Poisson Process (Cornell[11], Shah et al[42] and Der Kiureghian and Ang[12]).

The seismic hazard forecasting model employed in this study is based on the homogeneous Poisson Process. For earthquake events to follow this process, the following assumptions are in order:

(1) earthquakes are spatially independent;
(2) earthquakes are temporally independent; and
(3) probability that two seismic events will take place at the same location and at the same time approaches zero.

The first assumption implies that the occurrence of one event at a site does not affect the occurrences of other events. The second assumption implies that the seismic events do not have memory in time. Although there are theories and evidences suggesting that in certain source zones earthquakes occur in regular, predictable cycles, our general experience with limited time windows is the evidence of random unexpected occurrences of these events.

The non-Poissonian characteristics of the earthquake occurrences are mainly due to (1) clustering of events and (2) incompletetences of the earthquake catalogues. Specially for structures with failure criteria consitive to peak ground motions the secondary events in the cluster groups contribute only modertly to the overall seismic hazard, the main events providing the substantial contribution Merz and Cornell . The foreshock activity after the major main shocks in each seismic zone were detected through visual inspection and removed from the data at prior to the computation of the regression constants in the frequency-magnitude Recurrence Relationship of Eqn.2.

On the basis of rigorous stochastic testing methodologies, Gardener and Knopoff[22] and Yücemen[47] have shown, respectively for southern California and Turkey, that the sequence of earthquake events is indeed Poissonian when the clustering effects (aftershocks) are removed from the record. Furthermore, the Poisson Process indicates a constant hazard rate which is a quite descriptive phenomenon observed in seismic regions with a relatively uniform seismicity, such as most of the Aegean Region.

Thus it has been decided that it will be prudent to retain the Poisson Model on account of its simplicity and the fact that the seismic design decisions should be more sensitive to the mean number of events than to their temporal distribution. However, it should also be noted that, although the Poisson Model has been found to be adequate for assessing the earthquake hazard in regions where earthquakes occur frequently and their magnitudes are intermediate or small, for regions with large infrequent seismic events, the Poisson Model overestimates the probabilities of earthquake occurrences for short time of forecasts and underestimates the probability for long time of forecasts (Kiremidjian[25]).

For the actual computation of the probabilistic seismic hazard the computer programs developed by McGuire[27,28], Algermissen et al.[2] and by Algermissen and Perkins[1] have been utilized.

Both of these computer programs have been adopted on the basis of their applicability to different inputs and assumptions and also on the basis of their wide use for engineering and academic purposes.

SEISMIC HAZARD MAPS: RESULTS AND DISCUSSION

As the seismic hazard map an iso-acceleration contour map for the-return period of 475 years is provided in Figure 9. The hazard map provided is directly comparable, both in its method of compilation and its intended use, to the one provided in Algermissen and Perkins for the U.S.A.

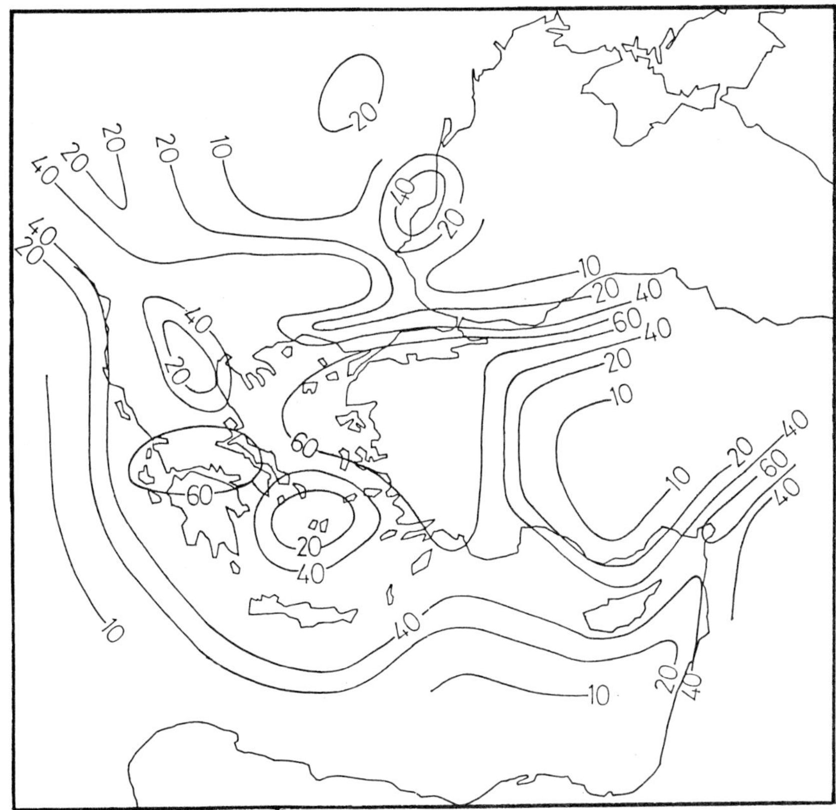

Figure 9. Iso-acceleration map for 475 year return period (in %g).

The following characteristics of this map should be noted:

(1) The contour levels of the acceleration hazard map range from 10% to 60%g.Below the 10%g level the ground shaking effects

are generally controlled by small magnitude events with unreliable recurrence rates and attenuation characteristics.

(2) The highest intensity contours encompass the North and the East Anatolian Fault Zone and the Aegean Grabens. The maximum mean peak ground acceleration in the immediate vicinity of these tectonic elements is predicted to be around 60%g. Note that these predictions in the vicinity of the fault zones are independent of the magnitude, at least for local magnitudes above 6.5.

(3) The isolated seismic sources in Bulgaria and Romania have experienced earthquakes with rather large recurrence intervals compared to the other regions. The peak acceleration values associated with these isolated regions are indicative of their low reccurrence interval rather than the size of the events that they have experienced.

(4) The peak ground acceleration attenuation relationships utilized are for mean maximum horizontal accelerations in bedrock. With soil deposits of soft-and medium-stiff sands and clays of appreciable depth the ground accelerations will be somewhat larger than those indicated on these hazard maps.

(5) In the seismic source regionalization methodology utilized in these maps, it has been tacitly assumed that in the future the locations of major seismic activity will be limited to the boundary and intraplate tectonics of the micro-plates as it has been the case over the course of the recorded history. In the past, certain parts of these boundaries remained locked for considerable periods of time and this indeed may be the case in future. However, there is no conceivable physical model that can accurately predict the locality and inception and termination times of these locking periods and the best one can do is to be on the conservative side and assume that these seismic sources will remain active in the future.

(6) In the delineation of the seismic sources the geological boundaries of the tectonic elements and their immediate vicinity have been considered. The seismic hazard near the source boundaries is directly and strongly affected by the changes in the delineation of these boundaries.

(7) The recurrence relationships used in the preparation of these maps are based on artificially complemented data set that can account for the small magnitude data deficiencies in earlier reporting periods. The net effect of such a procedure reflects itself in the increase of the slopes of the log-linear recurrence relationships and in the decrease of the relative number of earthquakes at the high magnitude end. The effect of such an approach on the hazard maps can be appreciable especially for high return periods.

(8) The maximum magnitudes assigned to each seismic zone have been based on realistic physical interpretations. It should be noted that, especially for lower maximum magnitude zones, adopting higher maximum magnitudes can have pronounced effects on the isoacceleration contours of these maps.

(9) The attenuation relationships utilized in these maps yield the mean peak horizontal ground acceleration with no statistical variations. The findings arrived at by Yarar et al. on the same source zone configuration and recurrence relationships indicate that the inclusion of a standard deviation of ln PGA in the order of 0.62 does not appreciably affect the high amplitude ground motion levels near the tectonic elements but may have pronounced effects (reaching up to 50%) on the low-level ground motion regions.

This study represents a rational attempt to estimate the probabilistic earthquake hazard in the Eastern Mediterranean Region and it is intended to serve as a reference for more advanced approaches and to stimulate discussion and suggestions on the data base, assumptions and the inputs, and to pave the path for the risk based assessment of the seismic hazard in the site selection and the design of nuclear power plants and in the design of common buildings and engineering facilities.

REFERENCES

1. Algermissen, S.T. and Perkins,D.M. (1976). A probabilistic estimate of maximum acceleration in rock in the contiguous United States, U.S.Geol. Surv., Open File Rep., No.76-416.
2. Algermissen, S.T., Perkins, D.M., Isherwood, W., Gordon, D., Reagor, G. and Howard,C. (1974). UNESCO Survey of the Seismicity-Seismic Risk Evaluation of the Balkan Region, U.S.Geological Survey, Golden, Colo., 32 pp.
3. Aliaj, Sh., (1982). General features of the neotectonic structure of Albania: Seismology, seismotectonics, Seismic Hazard and Earthquake Prediction; Earthquake Risk Reduction in the Balkan Region. Group A, Firal Report, UNESCO, A.15-A.21.
4. Allan, T.D. and Morelli, C.(1971). A geophysical study of the Mediterranean Sea. Boll. Geofis. Teor. Appl., 13(50): 99-142.
5. Alptekin,Ö. (1973). Focal mechanisms of earthquakes in Western Turkey and their tectonic implications. Ph.D. Diss. New Mexico Institute of Mining and Technology, 190 pp.
6. Applied Technology Council (1978). Tentative Provisions for the Development of Seismic Regulations for Buildings, NIS Spec. Publ., 510, Washington, D.C.
7. Burke, K., and Şengör, C. (1986). Tectonic escope in the evalution of the continental crust: Reflection Seismology: The continental crust, Geodynamics Series, 14, 41-53.
8. Boore, D., Oliver,A., Page,R. and Joyner,W. (1978). Estimation of ground motion parameters. U.S.Geol. Surv., Open File Rep., No.78.509.

9. Campbell, K.W. (1981). Near source attenuation of peak horizontal acceleration. Bull. Seismol. Soc. Am., 71: 2039-2070.
10. Cluff, L.S., Coppersmith, K.J. and Knuepfer, P.L.(1982). Assessing degrees of fault activity for seismic microzonation Conf., Seattle, Wash., II: 113-118.
11. Cornell, C.A. (1968). Engineering seismic risk analysis. Bull. Seismol. Soc. Am., 58: 1583-1606.
12. Der Kiureghian, A. and Ang,A.H. (1977). A fault rupture model for seismic risk analysis. Bull. Seismol. Soc. Am., 67: 1173-1194.
13. Dewey, J.F., and Şengör, A.M.C. (1979). Aegean and surrounding regions: complex multi-plate and continuum tectonic in a convergent zone. Bull. Geol. Soc. Am., 90: 84-92.
14. Dewey, J.F., Pitman, W.C., Ryan, W.B.F. and Bonnin,J. (1973). Plate tectonics and the evolution of the Alpine system. Bull. Geol. Soc. Am., 84: 3137-3180.
15. Doyuran,V., and Erdik,M., (1983). Geotectonics and seismicity of the Black Sea; Second Turkish Nuclear Power Plant Site Selection Investigations: Middle East Technical University, Earthquake Engineering Research Center, Ankara, 52p.
16. Drakopoulos, S. (1982). Seismic origin zones in Greece; Seismology, Seismotectonics, Seismic Hazard and Earthquake Prediction; Earthquake Risk Reduction in the Balkan Region, Working Group A, Final Report, UNESCO, A.83 - A.85.
17. Drumya, A.V., Ustinova,T.I., and Shchukin,Yu.K., (1976). Moldevia in "Seismic Zoning of the USSR"; S.V. Meduedeo (Ed.) Ch.2, 201-217, Isnel Program for Secientific Translation, Jerusalem.
18. Enescu, D., Cornea,I., and Mişicu,M., (1982). Mecanismal de Prodecere a Cutremurulur Din 4 Martie 1977 şı Efectele Associate de Directivitate, in "Cutremurul de Pamint din România de la 4 Martie 1977". Ş.Bâlan, V.Cristescu, and I.Curnea (Eds): Editura Academiec Republics; Socialiste România, Bucureşti, p.36-74.
19. Erdik, M. and Öner,S. (1982). A rational approach for the probabilistic assessment of the seismic risk associated with the North Anatolian Fault. In: A.Işıkara and Vogel (Editors), Multi-disciplinary Approach to Earthquake Prediction. Vieweg, Brauschweig-Wiesbaden, pp. 115-127.
20. Erdik, M., V.Doyuran, N.Akkaş and P.Gülkan (1985). A probabilistic assessment of the seismic hazard in Turkey, Tectonophysics, 117: 295-344.
21. Esteva,L.(1970). Seismic Risk and Seismic Design Decisions. In: R.Y.Hansen (Editor). Seismic Design for Nuclear Power Plants. Mass. Inst. Technol., Cambridge, Mass., pp.142-182.
22. Gardener, J.K. and Knopoff, L.(1974). Is the sequence of earthquakes in Southern California with aftershocks removed Poissonian? Bull.Seismol. Soc. Am., 64: 1363-1367.
23. Hattori,S. (1979). Seismic risk maps in the world, II.Bull. Int. Inst. Seismol. Earthquake Eng., 17: 33-96.
24. Idriss,I.M. (1978). Characteristics of erathquake ground motions. Proc. Specialty Conference on Earthquake Engineering and Soil Dynamics, ASCE, Pasadena, III: 19-21.

25. Kiremidjian,A. (1982). Stochastic models for seismic hazard analysis and their use in microzonation. Proc. 2rd Int. Earthquake Microzonation Conf., Seattle, Wash., pp.1205-1214.
26. Le Pichon, X. (1968). Sea floor spreading and continental drift.J.Geophys. Res., 73: 3661-3697.
27. McGuire, R.K. (1976). FORTRAN computer program for seismic risk analysis. U.S.Geol. Surv., Open File Rep., No.76-67.
28. McGuire, R.K. (1978). FRISK: computer program for seismic risk analysis using faults as earthquake sources. U.S.Dept. Interior, Geol. Surv., Open File Rep., No.78-1007.
29. McKenzie, D.P. (1970). Plate tectonics of the Mediterranean region. Nature, 226: 239-243.
30. McKenzie, D.P. (1972). Active tectonics of the Mediterranean region. Geophys. J.R. Astron. Soc., 30: 109-185.
31. McKenzie, D.P. (1978). Active tectonics of the Alpine-Himalayan belt, the Aegean sea and surrounding regions. Geophys. J.R. Astron. Soc., 55: 217-254.
32. Merz, H.A. and C.A. Cornell (1973), Aftershocks in Engineering Seismic Risk Analysis, Research Refort R73-25, Dept. of Civil Engineering, M.I.T., Cambridge, Mass.
33. Milne, W.G. and Davenport,A.G. (1969). Distribution of the Earthquake Risk in Canada. Bull. Seismol. Soc. Am., 59: 729-754.
34. Morelli, C. (1978). Eastern Mediterranean: geophysical results and implications. Tectonophysics, 46: 333-346.
35. Morelli, C., Pisoni,M and Gantar,C. (1975). Geophysical studies in the Agean Sea and in the Eastern Mediterranean. Boll.Geofiz., XVII (66): 127-168.
36. Morgan,W.J. (1968). Rises, trenches, great faults and crustal blocks. J.Geophys. Res., 73: 1959-1982.
37. Papazachos,B.C. (1974). Seismotectonics of the Eastern Mediterranean area. J.Solnes (Editor), Engineering Seismology and Earthquake Engineering, Noordhoff, Leyden, pp.1-32.
38. Papazachos, B.C. and Comninakis, D.E.(1971). Geophysical and tectonic features of the Aegean arc.J. Geophys. Res., 76(35): 8517-8533.
39. Richter, C.F. (1958). Elementary Seismology. Freeman, San Francisco, Calif.
40. Rizhikova, S. (1982). National report of Bulgaria: Seismology, Seismotectonics and Earthquake Prediction; Earthquake Risk Reduction in the Balkan Region, Group A, Final Report, UNESCO, A.47 - A.64.
41. Schnabel, P.B. and Seed, H.B. (1973). Accelerations in rock for earthquakes in the Western United States. Bull. Seismol. Soc. Am., 63, 501-516.
42. Shah, H., Mortgat,C., Kiremidjian,A.S. and Zguthy, T. (1975). A study of seismic risk for Nicaragua. The John A.Blume Earthquake Eng. Center Report No.1, Stanford University, Stanford, Calif.
43. Steep,J.C. (1973). Analysis of completeness of the earthquke sample in the Puget Sound area. In: S.T. Handing (Editor), Contributions to Seismic Zoning. NOAA Tech. Rep. ERL 267-ESL 30, U.S.Dep. of Commerce.

44. UNESCO, (1980). Earthquake risk reduction in the Balkan region, project No.RER/79/014.
45. UNESCO, (1982). Seismology, seismotectonics, seismic hazard and earthquake prediction, Earthquake risk reduction in the Balkan Region, W.G.A., Final Report, UNESCO.
46. Woodside, J.M. (1976). Regional vertical tectonics in the eastern Mediterranean. Geophys. J.R.Astron. Soc.47: 439-514.
47. Yücemen,M.S. (1980). Uncertainty analysis in the evaluation of seismic risk. Proc. 7th World Conf. Earthquake Eng., Istanbul, pp.301-308.

Space-Time Migration of Shallow Earthquakes in Eurasia and Implications for Earthquake Prediction

K. Kadinsky-Cade, M.N. Toksoz and A.A. Barka
Earth Resources Laboratory, Department of Earth, Atmospheric and Planetary Sciences, M.I.T., 42 Carleton St., Cambridge, MA 02142, U.S.A.

INTRODUCTION

The phenomenon of earthquake migration has been described primarily for shallow earthquakes in the Circum-Pacific region, in subduction zones (e.g., Mogi[1,2]; Kelleher et al.[3,4]; Sykes et al.[5]). Along these long continuous plate boundaries, the occurrence of large shallow earthquakes has often been observed to follow a systematic trend. The rupture zone of each large earthquake in a migratory sequence abuts that of the previous event with little or no overlap.

Migration is a very important characteristic of Eurasian earthquakes as well, although most of the events in question occur within continental areas. One of the earliest recognized instances of earthquake migration is the 1939-1967 sequence of 6 large or great earthquakes that occurred sequentially from east to west along the North Anatolian fault in Turkey, with total rupture length covering a distance of about 1200 km. By reviewing a number of studies that have been made during the last few years, we shall see that earthquake migration in Eurasia is not restricted to large or great events, but can occur at all scales.

The procedure followed in this paper will be to first review observations concerning Eurasian earthquake migration patterns in the cases of (1) plate boundary settings such as the Himalayan arc and the North Anatolian fault zone, (2) large or moderate-size earthquakes within intraplate settings, usually involving a sequence of three or four events, and (3) migration patterns observed in foreshock and aftershock sequences. Velocities and distances over which migration occurs in each case will be summarized in Table 1 for convenience. This review is not meant to be comprehensive, but instead to illustrate types of migration phenomena observed in Eurasia. Limitations to a comprehensive review result from lack of information about some of the earthquakes, many of which occurred before the advent of modern seismological instrumentation and analysis techniques.

Following the review section, we shall discuss the implications of earthquake migration for prediction studies. Of particular importance are: (1) some of the factors that influence earthquake migration patterns, such as rupture directivity or fault geometry, (2) an attempt to classify the different types of earthquake migration observed through our examples (viscoelastic processes, creep or stress corrosion processes), and (3) how the migration phenomenon affects earthquake recurrence times. Note that multiple rupturing of asperities within a single earthquake also follows a migration pattern, with seismic velocities.

REVIEW OF OBSERVATIONS

Earthquake migration along plate boundaries in Eurasia

Most of the 2500 km long Himalayan arc has ruptured in a series of large and great earthquakes since 1800. Seeber and Armbruster[6] have documented this rupture sequence in detail. It is not known whether the Himalayan seismicity of the last 180 years represents a cycle of enhanced activity to be followed by a period of quiescence, or if this seismicity is representative of the long-term average rate. In any case, a very clear migration pattern for large and great earthquakes can be observed along this arc for the period 1885 to 1950 (see Figure 1A). A linear migration velocity of about 40 to 50 km/year can be associated with these earthquakes (9 events, including the great earthquakes of 1905, M=8-8.4; 1934, M=8.3; and 1950, M=8.6). Some of the large earthquakes ($7 \leq M < 8$) shown in Figure 1A as solid circles lie within areas where much of the stress may have been released by great earthquakes earlier in the century (1803, 1833, 1897 earthquakes). According to Seeber and Armbruster[6], the large and great earthquakes take place along a nearly horizontal, northward-dipping detachment surface, which separates the Indian plate from an overlying sedimentary accretionary wedge. This detachment is about 200-300 km wide, and is assumed to extend along the full length of the arc. Seismicity patterns and focal mechanisms suggest that characteristics of slip along the Himalayan arc are similar to that in oceanic subduction zones (Seeber and Armbruster[6]).

A) Himalayan arc

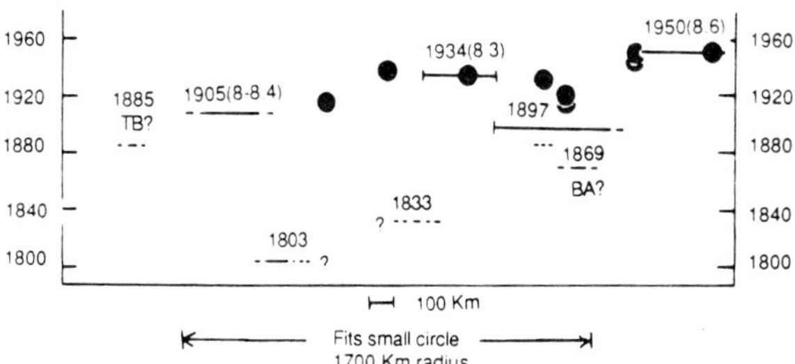

B) North Anatolian fault

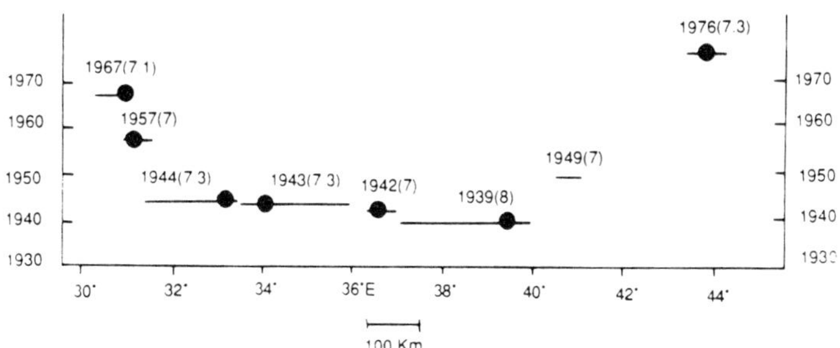

Figure 1. Earthquake migration along plate boundaries in Eurasia. (A) Himalayan arc, after Seeber and Armbruster[6], showing rupture zones of great earthquakes ($M \geq 8$; line segments) and locations of large earthquakes or epicenters for some of the great earthquakes (solid circles). TB=thrust belt, adjacent to the detachment on which most of the great earthquakes occur. Some of the large earthquakes may have occurred in the thrust belt. BA=Burmese arc. Data prior to 1897 are incomplete. (B) Migration of $M \geq 7$ earthquakes along the North Anatolian fault zone since 1939. Solid circles are epicenters and line segments are rupture zones. Data from Barka and Kadinsky-Cade[8].

The 1200 km long North Anatolian fault zone has also been subject to a clear earthquake migration during this century (see Figure 1B). The rate of migration along the central portion of the fault (approximately 32°E to 40°E) is a lot higher than at either end. This difference can be related most easily to the complexity of the fault zone. In the central part the fault zone consists of a single relatively continuous fault strand characterized by pure strike-slip motion, whereas at either end it is subject to interference by more complex stress regimes (extensional at the west end, compressional at the east end; for details see, e.g., Toksoz et al.[7]; Barka and Kadinsky-Cade[8]; Barka et al.[9]). The rupture zones are mostly separated from each other by geometric discontinuities in the fault trace. Note that along the North Anatolian fault, where epicenters are mostly well constrained by relocation studies (Dewey[10]), the epicenters are not necessarily located at the edge of the rupture zone for the previous earthquake.

Other examples of apparent plate boundary earthquake migration have been reported in the literature. These include the Hellenic arc (Wyss and Baer[11]; Purcaru and Berckhemer[12]), the Dinaric Coast, Yugoslavia (Purcaru and Berckhemer[12]), and even a possible migration along the Western European plate boundary (down the northern Mid-Atlantic ridge, and then eastward from the Azores to the Gibraltar area; Limond and Recq[13]). The Hellenic arc area is complicated by the presence of large intermediate-depth earthquakes which probably need to be included in a three-dimensional migration pattern. The Dinaric coast earthquake migration is based in part on events which are not necessarily well located, and requires further study. The Western European plate boundary migration is quite rapid (3-10 km/day) and is based on a statistical analysis of earthquakes covering a large magnitude range.

Intraplate earthquake migration

Three examples of moderate to large earthquake migration for intraplate events are shown in Figure 2. These include: a sequence of earthquakes with strong components of reverse and left-lateral strike-slip motion (Songpan, China; 1976), a series of earthquakes characterized by pure reverse faulting (Gazli, USSR; 1976-1984), and a sequence of normal faulting earthquakes (Corinth, Greece; 1981).

The Songpan, China earthquake sequence has been described by Jones et al.[14]).The three mainshocks of the sequence (Figure 2A) occurred on the Huya fault,

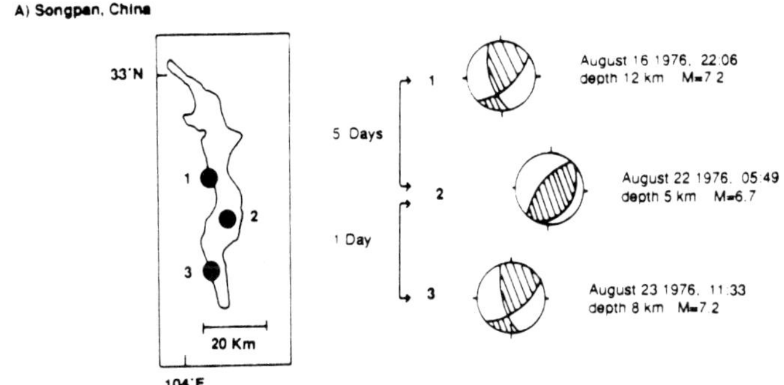

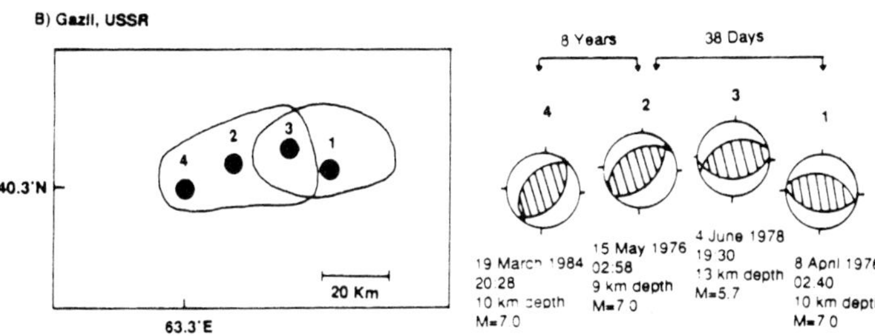

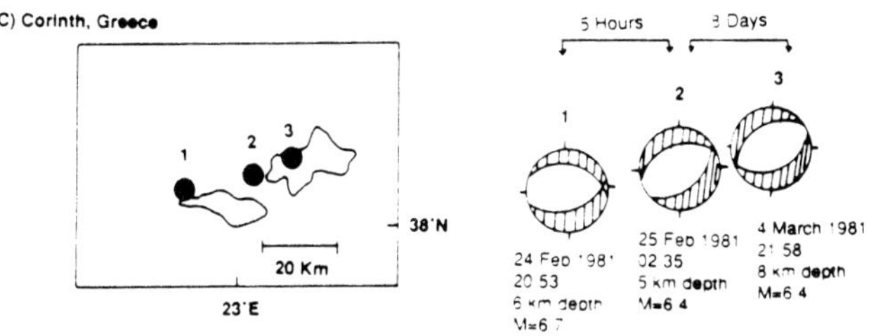

Figure 2. Intraplate earthquake migration. The numbering of the focal mechanisms in each case corresponds with the numbering in map view. The numbers reflect the time sequence of the earthquakes. The contour lines indicate zones of heaviest aftershock concentration. In cases A, B and C, this means total number of aftershocks. In case B, the two contours denote aftershocks of events 1 and 2 respectively. (A) From Jones et al.[14]. (B) From Eyidogan et al.[15]. (C) From Bezzeghoud et al.[16].

and were separated only by days. From the variations in focal mechanisms between the three shocks, Jones et al.[14] suggest that the first shock increased the normal and shear components of the stress on the fault segment corresponding to the second shock. This event in turn decreased the normal stresses on the fault segment corresponding to the third shock. They use this explanation to interpret the difference between the times needed to trigger the second and third shocks. They propose that variations in fault geometry may have been responsible for causing the fault to rupture as three separate events, and suggest that the second earthquake occurred in a right-stepping en-echelon offset area between the first and third earthquake fault segments (the first and third segments being characterized by a combination of left-lateral strike-slip and thrust motion).

It is not known whether the Gazli, USSR (Figure 2B) earthquakes (Eyidogan et al.[15]) occurred on a single arcuate fault or on separate faults in a preexisting zone of weakness. The progression from the first to the second shock was quite rapid. The third event (1978) is much smaller, may have just served to fill a small gap between the first and second shocks, and is not included in the interpretation of west-migrating earthquake activity. The fourth earthquake occurred several years later. An inversion of P waveforms for this event (Eyidogan et al.[15]) indicates a 13-second continuous rupture which broke two major asperities, and spread down-dip (to the northwest) and bilaterally. The rupture started at one of the asperities, as evidenced by the shape of the source time function used to model the earthquake.

The 1981 Corinth, Greece earthquake sequence has been studied by Bezzeghoud et al.[16]. The events shown in Figure 2C migrated rapidly from west to east. Waveform modeling for all three earthquakes (Bezzeghoud et al.[16]) yields source time functions that are remarkably similar, characterized by one or two precursory shocks of small amplitude preceding the main impulse by 2-3 seconds. According to Bezzeghoud et al.[16], these source time functions indicate triggering of the events by smaller asperities. The Corinth earthquake sequence may also be considered as a mainshock-aftershock sequence, with the second and third shocks triggered by the first event. The migration velocity shown in Table 1 that reflects the migration from event 1 to 3 is much slower than for event 1 to 2 (see Figure 2C). This suggests that the migration between 2 and 3 is slowed down by a barrier located between those two shocks.

Foreshock and aftershock migration sequences

The 1983-1984 Horasan-Narman, Turkey earthquakes are a clear example of aftershock migration. This sequence was studied by Eyidogan et al.[17]. During several months the aftershocks clustered in an area near the mainshock and its immediate moderate-size aftershock (events 1 and 2). Almost a year later, aftershock 3 occurred to the northeast. Aftershock 3 was followed by numerous aftershocks located in a cluster around it, and to the northeast of the main aftershock zone shown in Figure 3A. Earthquakes 3 and 4 may have occurred on a fault conjugate to the main fault, as suggested by surface geology. From the aftershock distribution alone (Eyidogan et al.[17]) we cannot distinguish between this possibility and that of a simple northeast extension of the main left-lateral strike-slip fault.

A) Horasan-Narman, Turkey aftershock sequence

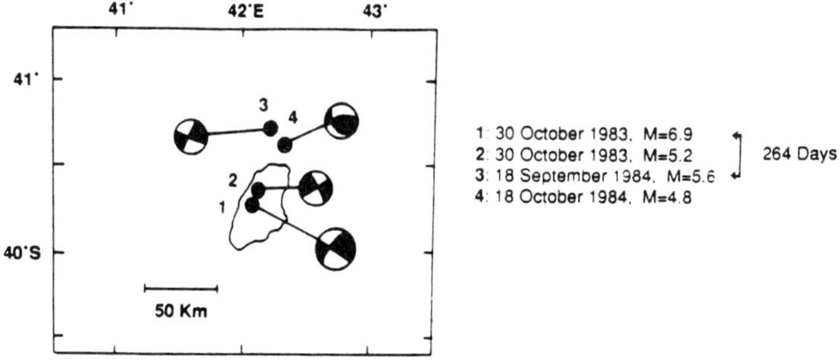

B) Southern Italy foreshock sequence

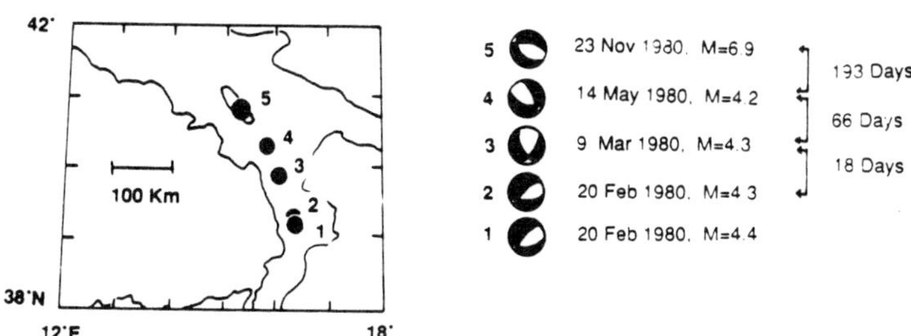

Figure 3. Aftershock and foreshock migration patterns. (A) From Eyidogan et al.[17]. (B) From Del Pezzo et al.[18]. Contours indicate main aftershock concentration.

The Southern Italy foreshock sequence has been discussed by Del Pezzo et al.[18] (see also Gasparini et al.[19]). According to these authors, normal faults in this area can be associated with an eastward migration of a rifting process which is active along the western (Tyrrhenian) margin of the Italian Peninsula. Historically, large earthquakes in Southern Italy have shown a tendency to cluster in time, with one event often triggering an adjacent one (e.g., 5 large earthquakes occurred in this area during an 18-year time span in the 17th century; Del Pezzo et al.[18]).

Migration observations for large shallow earthquakes outside Eurasia

For comparison with the Eurasian data, which mostly involve earthquakes in continental regions, migration rates for some characteristic shallow subduction zone earthquakes are shown at the end of Table 1. These include a series of great interplate earthquakes with intervening large events that migrated from east to west along the Aleutian arc (from Sykes et al.[5]), and a similar sequence of earthquakes that migrated from north to south along the southern part of the Chile plate boundary (from Kelleher et al.[3]). Note that other velocities of migration can be obtained for other portions of both the Aleutian and Chilean subduction zones. Velocities are generally less than 100 km/year in these areas. Earthquake migration is a clear and important phenomenon for many shallow interplate settings in shallow subduction zones. Several examples of large and moderate-size shallow earthquake migration (not all interplate thrust-type earthquakes) have been discussed for the Nankai trough and Japan trench areas by Mogi[2]. Migration velocities in these cases are all less than 100 km/year.

IMPLICATIONS OF EARTHQUAKE MIGRATION OBSERVATIONS IN EURASIA

Influential factors in earthquake migration patterns

The fact that earthquake migration occurs at all suggests that the rocks through which the migrating earthquakes propagate are already stressed almost to the point of failure. Lay and Kanamori[20] have described large fault zones as a collection of subfaults. These are subjected to a uniform regional loading stress to which incremental induced stress is added when adjacent subfaults fail. According to the model of Lay and Kanamori[20], the two factors that influence the degree of subfault

Table 1. Migration distances and rates

Where	quake ID*	magnitude	distance** (Km)	velocity (Km/year)	eq type
Himalayan arc	1885-1905	?-8.2	400	20	shallow thrust
	1905-1934	8.2-8.3	1,000	34	
	1934-1950	8.3-8.6	900	56	
North Anatolian fault	1939-1944	8-7.3	480	96	strike-slip
	1944-1967	7.3-7.1	140	6	
	1939-1976	8-7.3	530	14	
Songpan, China	1-2	7.2-6.7	14	1,022	reverse w/ srike-slip
	2-3	6.7-7.2	18	6,570	
Gazli, USSR	1-2	7.0-7.0	28	269	reverse
	2-4	7.0-7.0	16	2	
Corinth, Greece	1-2	6.7-6.4	18	31,536	reverse
	1-3	6.7-6.4	24	1,095	
Horasan-Narman, Turkey	1-3	6.9-5.6	50	69	strike-slip
Southern Italy	(1,2)-3	4.3-4.3	80	1,622	normal
	3-4	4.3-4.2	50	276	
	4-5	4.2-6.9	70	132	
Outside Eurasia:					
Aleutian arc	1938-1965	8.2-8.7	1,900	70	inter-plate boundary
Southern Chile	1880-1960	8.2-9.5	720	9	inter-plate boundary

* Earthquake Identification: see Figures 1-3. For areas outside Eurasia, references are given in text.
Note (1) Himalayan arc: only great earthquakes are included.
(2) North Anatolian fault: divided into western, central and eastern parts; distances measured along fault trace, not along the latitude projection shown in figure.
**Distance: for the Himalayan and North Anatolian cases, distance is measured between the mid-points of the rupture zones. Elsewhere (for smaller events) it is measured between the epicentral locations shown in the figures.

interaction are (1) the nature of the rupture process within individual subfaults and (2) the presence of transverse segmentation between subfaults. These factors have mostly been inferred from the study of subduction zone earthquakes.

In some of the cases described in the previous section we know that lateral segmentation plays an important role. Along the North Anatolian fault, for example, Barka and Kadinsky-Cade[8] have examined the fault geometry in relation to earthquake rupture, and have found that the rupture zones are separated from each other by geometric discontinuities. In this same area they have also found that fault geometry is a critical factor in determining where rupture initiates along a fault segment, and that this epicentral location does not necessarily coincide with the edge of the rupture zone for the previous earthquake. Geometrical effects also play an important role in delaying the rupture for the Songpan case, as mentioned earlier (Jones et al.[14]). Geometrical discontinuities can either occur within one fault segment or separate two neighboring segments.

It is not clear in the Eurasian examples how the rupture process influences the degree of subfault interaction or the direction of earthquake migration. Although work on subduction zone earthquakes sometimes suggests that fault rupture direction is closely linked with earthquake migration direction, there are a few counterexamples in the cases reviewed here (e.g., North Anatolian fault zone, Gazli earthquakes, or the Southern Italy foreshock sequence, in which the events are small enough to be considered as point sources). It is possible, however, that the previous rupture provides enough stress increment regardless of directionality to break a nearby asperity and start a new rupture (as suggested by the sharp onset of the source time functions in the Corinth and Gazli cases).

In general it would seem that several of the migrating earthquakes could have ruptured as one larger comprehensive earthquake if there had not been as much heterogeneity in the fault zone.

Earthquake migration processes

From a quick examination of Table 1, it can be seen that earthquake migration velocities in Eurasia vary by several orders of magnitude. In effect these velocities fall into three categories. First are velocities of 100 km/year or less, associated with delay times of years. These include the plate boundary-type earthquakes and several other cases as well. These velocities are similar to those observed in subduction zone earthquake migration patterns. They can be attributed to stress diffusion by postseismic viscoelastic rebound, reflecting dynamic coupling of the lithosphere with a viscoelastic asthenosphere (Anderson[21]; Lehner et al.[22]). According to this interpretation, postseismic increases in stress are transferred into the asthenosphere, then back into the lithosphere by relaxation.

The second category, with velocities of 100's of km/year, and delay times of weeks or months, cannot be associated with the same viscoelastic processes. Time-dependent rupture can, however, be associated in this case with stress corrosion phenomena (Das and Scholz[23]) or with fault creep (e.g., Bakun et al.[24]; Lindh and Boore[25]; Segall and Harris[26]; Wesson[27]). Stress corrosion can be explained as a stable, quasi-static crack propagation at velocities less than the sonic velocity of the medium (Das and Scholz[23]), and has been invoked to interpret various time-dependent rupture phenomena (e.g., aftershock occurrence). The interplay of creep and seismicity may be complicated. Accelerated creep (triggered, for example, by an earthquake) could load neighboring asperities within the fault zone that could fail in subsequent earthquakes.

The third category, with velocities of thousands of km/year, can also be explained by phenomena belonging to the second category. In this case, however, the events occur in small enough areas that the whole fault should probably have ruptured at once. The rupture zones corresponding to the migrating events were probably stressed to approximately the same level and the rocks had very similar strengths. The reason for multiple events rather than a single earthquake has to be related to small heterogeneities within the fault zone.

Note that it is possible for an earthquake on the main fault trace to trigger another earthquake on the same fault trace with relative ease, but also to trigger an event on a nearby fault with a longer time delay. If the main fault trace earthquake migration proceeds normally, and then at a later time the secondary fault segment ruptures without being recognized as separate from the main fault trace, an apparent reversal of direction in the migration pattern can occur.

Effects of migration on recurrence times

The earthquake migration process causes stress to be released along the whole fault zone all at once. As a result, the number of earthquakes and the total stress drop along a fault zone is higher during certain time periods. This has been observed in Turkey and in Israel(Ambraseys[28]), or in other parts of the world (e.g., Mogi[1]). In Eurasia periods of active seismicity (e.g., 1900-1918, 1935-1950, 1970-?) alternate with periods of quiescence (Purcaru and Berckhemer[12]). Any calculation of recurrence times needs to take this effect into account, particularly for intermediate-term prediction work.

REFERENCES

1. Mogi, K.(1968), Migration of Seismic Activity, Bulletin of the Earthquake Research Institute, Vol. 46, pp. 53-74.

2. Mogi, K. (1985). Earthquake Prediction, Academic Press. Tokyo, Orlando and London.

3. Kelleher, J., Sykes, L. and Oliver, J. (1973), Possible criteria for predicting earthquake locations and their application to major plate boundaries of the Pacific and the Caribbean, J. Geophys. Res., Vol. 78, pp. 2547-2585.

4. Kelleher, J., Savino, J., Rowlett, H. and McCann, W. (1974), Why and where great thrust earthquakes occur along island arcs, J. Geophys. Res., Vol. 79, pp. 4889-4899.

5. Sykes, L., Kisslinger, J, House, L. Davies, J. and Jacob, K.(1981), Rupture zones and repeat times of great earthquakes along the Alaska-Aleutian arc, 1784-1980. Earthquake Prediction: an International Review (Eds. Simpson, D. and Richards, P.), Maurice Ewing Series 4, American Geophysical Union, pp. 73-80.

6. Seeber, L. and Armbruster, J. (1981), Great detachment earthquakes along the Himalayan arc and long-term forecasting. Earthquake Prediction: an International Review (Eds. Simpson, D. and Richards, P.), Maurice Ewing Series 4, American Geophysical Union, pp. 259-277.

7. Toksöz, M.N., Shakal, A.F. and Michael, A.J. (1979), Space-time migration of earthquakes along the North Anatolian fault zone and seismic gaps, Pageoph., Vol. 117, pp.1258-1269.

8. Barka, A.A. and Kadinsky-Cade, K. (1987), Strike-slip fault geometry and earthquake activity in Turkey, submitted to Tectonics.

9. Barka, A., Toksöz, M.N., Kadinsky-Cade, K. and Gülen, L. (1987), The segmentation, seismicity and earthquake potential of the eastern part of the North Anatolian fault zone, in preparation.

10. Dewey, J. (1976), Seismicity of Northern Anatolia, Bull. Seism. Soc. Am., Vol. 66, pp. 843-868.

11.Wyss, M. and Baer, M. (1981), Earthquake hazard in the Hellenic arc. Earthquake Prediction: an International Review (eds. Simpson, D. and Richards, P.), Maurice Ewing Series 4, American Geophysical Union, pp. 153-172.

12.Purcaru, G. and Berckhemer, H. (1982), Regularity patterns and zones of seismic potential for future large earthquakes in the Mediterranean region, Tectonics, Vol. 85, pp. 1-30.

13.Limond, W. and Recq, M. (1981), Possible earthquake migration along the Western European plate boundary, Jour. Geophys. Res.,Vol. 86, pp. 11,623-11,630.

14.Jones, L., Han, W., Hauksson, E., Jin, A., Zhang, Y. and Luo, Z. (1984), Focal mechanisms and aftershock locations of the Songpan earthquakes of August 1976 in Sichuan, China, Jour. Geophys. Res., Vol. 89, pp. 7697-7708.

15.Eyidogan, H., Nabelek, J. and Toksoz, M.N. (1985), The Gazli, USSR, 19 March 1984 earthquake: the mechanism and tectonic implications, Bull. Seism. Soc. Am., Vol. 75, pp. 661-675.

16.Bezzeghoud,M., Deschamps, A. and Madariaga, R. (1986), Broad-band modelling of the Corinth, Greece earthquakes of February and March, 1981, Annales Geophysicae, Vol. 4 (B3), pp. 295-304.

17.Eyidogan, H., Toksöz, M.N., Gülen, L. and Nabelek, J. (1987), Aftershock migration following the 1983 Horasan-Narman earthquake, Earth Resources Laboratory unpublished report.

18.Del Pezzo, E., Iannaccone, G., Martini, M. and Scarpa, R. (1983), The 23 November 1980 Southern Italy earthquake, Bull. Seis. Soc. Am., Vol. 73, pp. 187-200.

19.Gasparini, C., Iannaccone, G., Scandone, P. and Scarpa, R. (1982), Seismotectonics of the Calabrian arc, Tectonophysics, Vol. 84, 267-286.

20.Lay, T and Kanamori, H. (1981), An asperity model of great earthquake sequences. Earthquake Prediction: an International Review (Eds. Simpson, D. and Richards, P.), Maurice Ewing Series 4, American Geophysical Union, pp. 259-277.

21.Anderson, D. (1975), Accelerated plate tectonics, Science, Vol. 187, pp. 1077-1079.

22.Lehner,F.K., Li, V.C. and Rice, J.R. (1981), Stress diffusion along rupturing plate boundaries, Jour. Geophys. Res., Vol. 86, pp. 6155-6170.

23.Das, S. and Scholz, C. (1981), Theory of time-dependent rupture in the earth, Jour. Geophys. Res., Vol. 86, pp. 6039-6051.

24.Bakun, W., Stewart, R., Bufe, C. and Marks, S. (1980), Implication of seismicity for failure of a section of the San Andreas fault, Bull. Seism. Soc. Am., Vol. 70, pp. 185-201.

25.Lindh, A. and Boore, D.(1981), Control of rupture by fault geometry during the 1966 Parkfield earthquake, Bull. Seism. Soc. Am., Vol. 71, pp. 95-116.

26.Segall, P. and R. Harris (1986), Slip deficit on the Parkfield, California, section of the San Andreas fault as revealed by inversion of geodetic data, Science, Vol. 233, pp. 1409-1413.

27.Wesson, R.(1987) Dynamics of fault creep, to be submitted to Jour. Geophys. Res.

28.Ambraseys, N.N. (1971), Value of historical records of earthquakes, Nature, Vol. 232, pp. 379-397.

Tectonics and Earthquake Risk of Iran

A.A. Nowroozi
Department of Geological Sciences, Old Dominion University, Norfolk, VA 23508, U.S.A.

INTRODUCTION

The tectonics and seismicity of Iran and surrounding countries are very complex. Many authors have made contributions (e.g., Stocklin[1]; McKenzie[2]; Nowroozi[3]; Berberian[4]), but still unresolved problems remain. The earth sciences have leaped forward as a result of studies related to magnetic field reversals, seafloor spreading, earthquake locations, focal mechanism determinations, remote imaging and the theory of plate tectonics. In recent years this theory has been applied to elucidate the geology of this complicated region. The theory of plate tectonics includes all the measurable plate motions on land and the ocean floor (Morgan[5]; LePichon[6]). When plates are in motion, their boundaries are deformed and characterized by earthquakes. At oceanic ridges where new oceanic crust is forming, and at fracture zones where plates pass each other, seismicity has a simple linear pattern. Focal mechanism solutions indicate normal faulting at ridges and strike slip faulting at fracture zones. The sense of motion on strike slip faults is in agreement with that of transform faults (Wilson[7]; Sykes[8]). At obduction or subduction zones where crustal material are crushed or consumed, earthquake mechanisms indicate thrust faulting.

In this paper, first I shall present a summary of the works related to the plate tectonics of the Middle-East. Then I shall concentrate on problems related to seismogenic faults, seismotectonic and seismic risks of Iran.

PLATE TECTONICS OF IRAN AND SURROUNDING COUNTRIES

Several plate tectonic models for this region are reported in the literature (Nowroozi[9]; Takin[10]; Jackson and McKenzie[11], and others). Basic data for construction of a model may include a. tectonic structures, such as active faults, folds, volcanoes; b. long duration seismicity; c. sense and amount of ground offsets following earthquakes; and d. type of earthquake mechanism, direction of various stress axes and slip vectors. Additional information may be obtained from gravity and magnetic anomalies (Snyder and Barazangi[12]). A revised plate tectonic model for Iran and surrounding countries is presented in Figure 1. This model is based on long term seismicity (Nowroozi[13]; Ambraseys and Melville[14]), structural styles, tectonic elements, faults, folds, volcanoes (Stocklin[15] and Gansser[16]), focal mechanism solutions (Nowroozi[9]; Jackson and McKenzie[11], Berberian[4]), and fault offsets (Nowroozi and Mohajer[17]). The model consists of several micro plates within a broad and seismically active zone between the Indian, African and Eurasian plates. Except the Arabian plate which is relatively large, the micro plates are: Lut, Sistan, Persian, Caspian Sea, and Turkish. The broad boundaries of the Caspian Sea, Persian and Lut micro plates are marked by several active or Quaternary volcanoes.

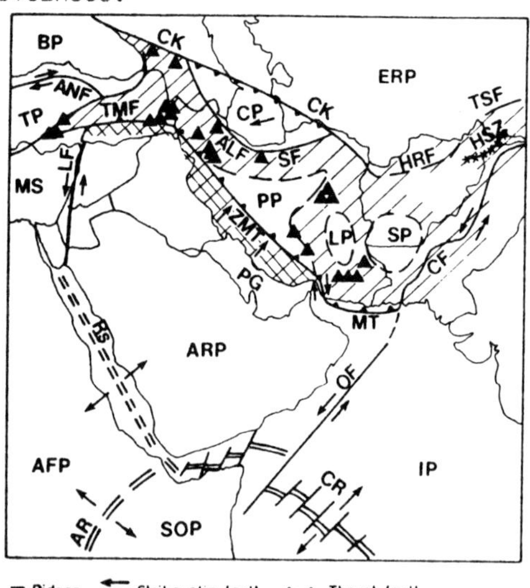

Figure 1. Plate Tectonic Model of the Middle East. Abbreviations are: AR: African rift; AFP: African plates; ARP: Arabian plate; ANF: Anatolian fault; ALF: Alborz fault; BP: Black Sea plate; CF: Chaman fault; CK: Caucasus-Kopet Dagh axis; CP: Caspian Sea micro plate; CR: Carlsberg ridge; ERP: Eurasian plate; HSZ: Kindu Kush seismic zone; HRF: Harir-rud fault; IP: Indian plate; LP: Lut micro plate; LF: Levant fracture; MS: Mediterranean Sea; MT: Makran thrust fault; OF: Owen fracture zone; PG: Persian

Gulf; PP: Persian micro plate; RS: Red Sea; SP: Sistan micro plate; SF: Shahrud fault; SOP: Somalian plate; TMF: Taurus Mountain fault; TP: Turkish plate; TSF: Tien Shan fault; ZMT: Zagros Main thrust. Hatched areas are broad deformational zones; cross-hatched areas are folded belt of Zagros-Taurus mountains.

Arabian plate The southern boundary of the Arabian plate is the seismically active Red Sea where earthquake mechanism solutions (Sykes[8]; McKenzie et al[18]; Ben-Menahem and Aboodi[19]) indicate that the Red Sea is opening from a series of en echelon ridges in a manner which is consistent with the northward motion of the Arabian plate. Its eastern boundary is the Owen fracture zone which is terminated near the coast of Makran where it is joined by the Makran thrust fault and the Chaman fault. The Owen fracture zone and the Chaman fault are seismically active. The earthquake mechanism solutions on them indicate a left-lateral strike slip motion in agreement with the northward motion of the Indian plate relative to the Arabian and Eurasian plates. The western boundary of this plate is the seismically active Levant fracture zone (Nowroozi[3]). Geological evidence indicates a left lateral strike slip motion on this fracture (Quennel[20]). The northern boundary of this plate is the Makran coast, and the folded series of the Zagros and Taurus mountains. The boundary limit is indicated by the thrust fault symbols in Figure 1. Folds axes directions are nearly parallel to the axis of the Red Sea and the direction of the Main Zagros thrust fault. This indicates that the folds are formed probably as a result of a northeast-southwest state of compression due to the northward motion of the Arabian plate. A large number of earthquake mechanism solutions presented by different authors (Nowroozi[9]; Jackson and McKenzie[11]; Berberian[4]; Chandra[21]; and others) indicate principally thrust faulting which is in agreement with the present continental collision between the Arabian plate and Iranian highland. The entire folded series are seismic, although seismicity is not uniform. The Zagros thrust has an important effect on the seismicity. Southwest of this thrust seismicity is sheet-like, relatively continuous with some localities where activities are high. Northeast of this thrust seismicity is sporadic.

Persian micro plate This is a small, roughly triangular micro-plate within the interior of the Iranian high plateau. The interior region has relatively less seismic activity. The southeast side is well marked by the Main Zagros thrust. Its other boundaries are less defined. Both Nowroozi[9], and Jackson and McKenzie[11] believe that a smaller microplate with less seismicity may exist within the Iranian high plateau. However, their proposed boundaries are not the same. Many writers have assumed a "median mass" in central

Persia. Stahl[22], Boeckh et al[23], Bogdanoff[24], and others have rejected it. Baier[25], Gansser[26], Stocklin[1]. The Persian, Lut and Sistan micro plates may be those elements within Iran and Afghanistan. The plates suffer from micro seismic activity due to the expected state of high stress concentration. However, no major earthquake within these plates has been reported so far.

Caspian Sea micro plate The northeast boundary of this small plate is the seismically active Caucasus-Kopet-Dagh axis. On the south it is bounded by the seismically active Shahrud-Alborz fault system which is tentatively extended into Caucasus. The plate covers the southern deeper Caspian Sea, the parts of Turkamania in the Soviet Union and the Gorgan Plain in Iran. The seismicity within the plate is very low, but all the boundaries are highly seismic. The crustal structure of the Caspian Sea is characterized by 25 km of disturbed sediments over a basaltic layer 15 to 20 km thick. There is no granitic layer. This composition suggests that the crust of the Caspian Sea micro plate is probably of oceanic type. The existing earthquake focal mechanism solutions indicate mostly a high component of thrust faulting on the entire southern boundary with some complications near the intersection points. Thus the Caspian Sea micro plate is underthrusting the Iranian highland. The direction of plate motion is in agreement with the crustal consumption at depth where molten crust has produced the Quaternary and presently active volcanoes.

Lut and Sistan micro plates The Lut micro plate has an oval shape; seismicity within it is very limited. Most major faulting and seismic activity of eastern Iran is confined to its broader boundaries which are indicated by cross hatching on Figure 1. Within the southern boundary zone of the plate lie the Taftan-Bazman volcanoes with andesitic flow, but mainly of olivine basalt composition. An active arc and convergent plate boundary appears to exist from the Makran coast of Iran to the southern boundary of the Lut micro plate where the oceanic crust of the Arabian Sea and the Sea of Oman underthrust eastern Iran and Pakistan. This type of oceanic-continental collision is, in general, characterized by a convergent plate boundary sequence (Burchfiel[27]) which is reported for the east Iran (Nowroozi and Mahajer[19]). The shape of the Sistan micro plate is not certain. It probably covers the Sistan basin and part of the central Afghan median mass (Shareq[28]). Again, its southern part is marked by Quaternary volcanoes and Alpine granites. The Lut and Sistan micro plates are separated by the north-south trending eastern Iranian mountains which are composed of sedimentary units with large blocks of melange (Stocklin[15]; Huber[29]).

Black Sea plate The northern boundary of this plate is the extension of the seismically active Caucasus-Kopet Dagh axis into Crimea and Rumania. The southern boundary is the right lateral strike slip seismogenic Anatolian fault. The eastern boundary is a complex series of volcanic rocks; the floor of the Black Sea has a sedimentary layer up to 16 km thick (Ross[30]).

Turkish plate This plate is marked by the Anatolian fault on the north and a chain of active and Quaternary volcanoes in the east. The shape, (Figure 1), is similar to that proposed by Jackson and McKenzie[11]. Focal mechanism solutions indicate a consistent right-lateral motion of the Anatolian fault which is in agreement with regional post earthquake observations.

Broad deformational zone The cross hatched area southwest of the Main Zagros thrust and south of the Taurus mountain is marked by the high seismicities of the folded series which were discussed before. This zone may be up to 200 km wide. Between the Taurus, Zagros, Makran and Chaman fault systems in the south and the Caucasus-Kopet Dagh, Harir rud, Tien Shan fault systems in the north lies a broad deformational zone marking the boundaries of the less deformed Caspian, Persian, Lut and Sistan micro plates. Within this zone are a number of active and Quaternary volcanoes. The most famous are the Alvand in Iran, the Elbrus in Caucasus, and the Ararat in eastern Turkey; and the Bazman Taftan series in southeastern Iran. The Hindu Kush intermediate focal depth seismic zone is also within this area. The geometry of this zone is discussed elsewhere by Nowroozi[31]. All major Iranian earthquakes have occurred within these broad boundaries. In Figure 2 all the known earthquakes with magnitudes larger than 5 are plotted. The sources of data are Ambraseys and Melville[14], Nowroozi[13], ISS, and PDE. Although the entire zone is seismically active, the rate of occurrences is not uniform. Nowroozi[13] has used geological data, regional trends, structural styles, distribution of salt domes, and active faults to deduce of seismotectonic boundaries or source zones. He divided Iran into 23 seismotectonic provinces whose boundaries are indicated in Figure 2.

SEISMIC RISK

The scope of this section is to estimate the peak horizontal ground acceleration across Iran for different levels of risk. The two basic approaches for assessments of seismic risk in a region are known as the deterministic and the probabilistic methods. Both techniques and their modifications are used for assessment of seismic risk in Iran. In the deterministic approach, the seismogenic fault zones and capable structures must be well known. Moreover, the earthquake epicenter must be well correlated with tectonic structures. In this case, the maximum potential magnitudes of the capable faults are often estimated from one of the empirical relations between fault length-magnitude, fault displacements-magnitude, fault width-magnitude, earthquake moment-magnitude, or a combination of these. For Iranian seismogenic faults these equations have been developed by Nowroozi[32]. Previous work by Mohajer and Nowroozi[33] needs revision in light of these recently developed local empirical relationships. When magnitude or intensity at a source is estimated, then ground acceleration at a site may be obtained if the attenuation of intensity or acceleration in the region is known. Attenuation of seismic intensities in Iran was reported by Chandra, McWhorter and Nowroozi[34]. This approach has been used by various authors for the siting of nuclear power plants.

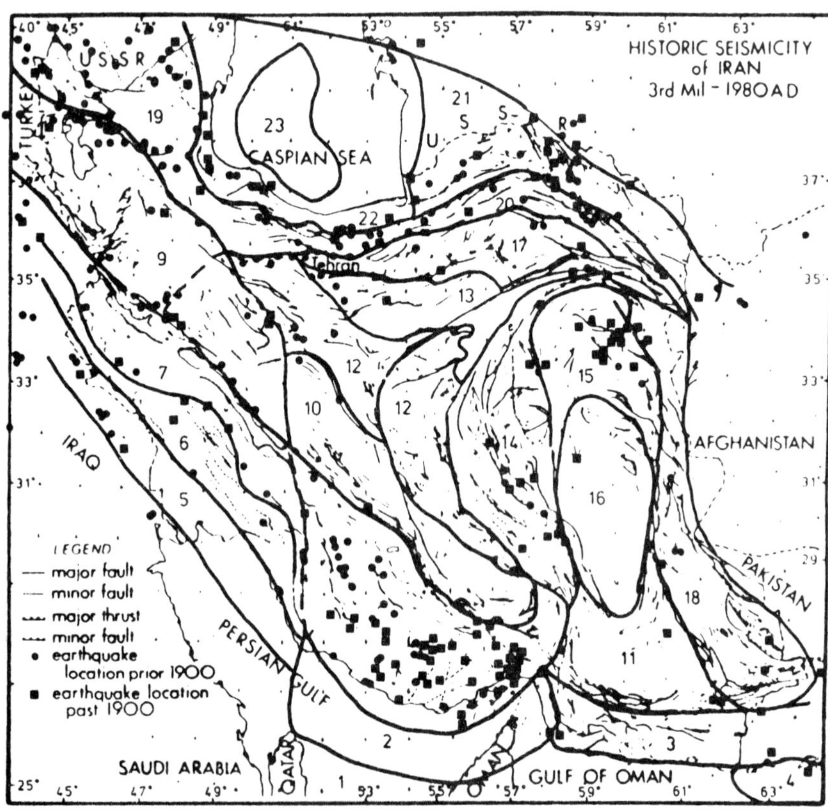

Figure 2. Seismicity and seismotectonic map of Iran, only earthquakes with a magnitude larger than 5 are included. Numbers refer to various seismotectonic provinces. The names of each province is given in Table 1.

In many parts of Iran, seismicity is diffused. Epicenters often cannot be correlated to a capable fault as can be seen from the seismicity map of Iran (Figure 2). A vivid example is seismicity associated with the folded belt of the Zagros, between the northern shore of the Persian Gulf and the main Zagros thrust. In this region, post earthquake field investigations have indicated that there is no apparent surface faulting even after a major earthquake. In eastern Iran, where well developed thrust and strike-slip faults are mapped, post-earthquake investigations have indicated substantial faulting. The fault map of Iran is given in Figure 3. When seismicity is not well correlated with tectonic structure as in many regions of Iran, a probabilistic approach can be advantageous. Requirements are a knowledge of all historic and instrumental seismicity together with seismotectonic provinces from which the frequency distribution of earthquake magnitudes or intensities is obtained. Then seismic risk may be calculated for a specific probability of occurrence during a given time exposure if regional attenuation laws are known (Nowroozi and Ahmadi[35]).

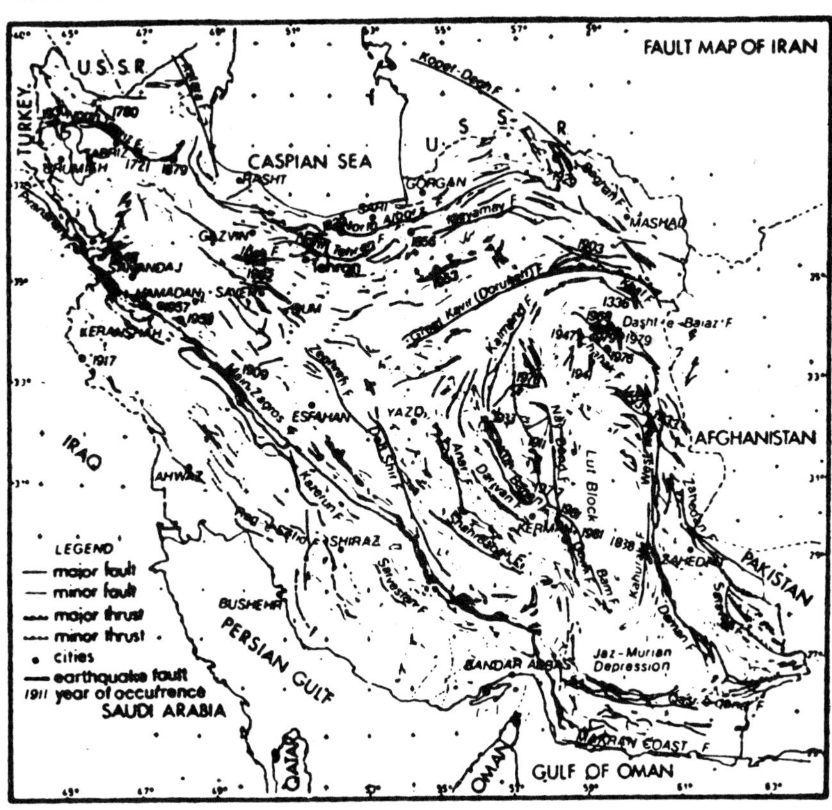

Figure 3. Fault map of Iran. Sources are discussed in the text.

Frequency distribution and return period A model of seismotectonic provinces of Iran was reported by Nowroozi[13]. The boundaries are marked on Figure 2 which contain all known earthquakes above magnitude 5. The provinces were used for risk analysis (Nowroozi and Rhmadi[35]). Earthquakes in each province were fitted to a log-linear and log-quadratic frequency magnitude relationships. To illustrate the method let us follow Richter and assume,

$$\text{Log } N = a - bM \tag{1}$$

where M is the magnitude of an earthquake, and N is the expected number of earthquakes with a magnitude greater than or equal to M per year. From data in each province, a and b are estimated by a least squared procedure. The results are given in Table 1.

The mean return period T is defined as the expected time interval for occurrence of one earthquake with magnitude greater than or equal to M, therefore:

$$T = \frac{1}{N}. \tag{2}$$

From equations (2) and (3),

$$T = 10^{-a+bM} \tag{3}$$

A more accurate definition of return period can be made by assuming a stocastic model for occurrence of earthquakes. Let us assume that the Poisson's process is valid, therefore:

$$P_n(M,t) = \frac{e^{-Nt}(Nt)^n}{n!} \tag{4}$$

$P_n(M,t)$ is the probability of having n events of magnitude M during a time exposure of t, N is the mean rate and Nt the expected number of earthquakes during time t. From equation (4), the probability of zero occurrence, n=0, is given by:

$$P_o(M,t) = e^{-Nt}. \tag{5}$$

From equations (1) and (5), we may write:

$$P_o(M,t) = \exp\left[-t(10)^{a-bM}\right]. \tag{6}$$

Risk is defined as the possibility of occurrence of an earthquake with magnitude greater than M during time t, therefore:

$$R(M,t) = 1 - P_o(M,t) \tag{7}$$

Table 1. The values of constant parameters a, b and estimated return periods in years for several magnitudes.

Provinces	a	b	M=5	M=6	M=7
1 Arabian Platform	3.8645	0.9677	9	87	811
2 Persian Gulf	4.9983	1.1035	3	42	532
3 Makran	4.5262	1.0194	4	39	407
4 Sea of Oman	2.7892	0.6486	3	13	56
5 Arvand-Shatt-al-Arab	4.5475	1.0550	5	61	688
6 Foothill Folded Series	6.1002	1.2026	1	13	208
7 Fars Folded Series	6.4654	1.1708	0.2	4	54
* 8 High Zagros Folded Series	4.8465	0.9623	1	7	69
	4.6752	0.9220			
9 Rezaiyeh	4.1007	0.8667	2	13	93
10 Esfahan-Sirjan	4.3337	1.1055	16	199	2540
11 Jaz-Murian	5.0667	1.0667	2	22	51
12 Central Iran	2.4947	0.6192	2	12	55
13 Kavir	3.1006	0.8214	10	67	446
14 Tabas	4.2626	0.8868	1	11	88
15 Ferdows	2.8295	0.6490	3	12	52
16 Lut	3.6952	0.8751	5	36	269
17 Shahrud-Doruneh	4.3742	0.9032	1	11	88
18 East Iran	2.5724	0.6321	3	16	78
	3.1637	0.7273			
19 Maku-Zandian	3.3338	0.8191	5	38	251
*20 Alborz	3.6935	0.7776	3	11	48
	1.8490	0.4944			
*21 Kopet-Dagh	3.5461	0.7502	3	11	45
	1.8506	0.4935			
22 Caspian Shore	3.5619	0.8059	3	19	120

*Estimations are based on two data bases.

The return period now is given by:

$$T = \frac{t}{R(M,t)} \qquad (8)$$

Results are given in Table 1. Accordingly, the Fars folded series is the most active region and the Esfahan-Sirjan the least active one; in the former province in every 0.2 years and in the latter in every 16 years a magnitude 5 earthquake is expected. For a magnitude 7 earthquake the most active regions are Kopet-Dagh, Alborz, Ferdows, and the Fars folded series, and again, the least active are the Esfahan-Sirian and the Arabian Platform.

Ground Accelerations Expressions for ground acceleration in terms of magnitude and focal depth have been developed by a number of authors (Esteva and Villaverde[36], Donovan[37], Bolt and Abrahamson[38]). The equations are in the form of:

$$A = a_1 e^{a_2 M}(R+a_3)^{a_4} \tag{9}$$

where a_1, a_2, a_3 and a_4 are empirical constants depending on site characteristic, and A, M and R are ground acceleration, magnitude, and focal distances respectively. The empirical values of a_1-a_4 are not established for Iranian conditions. In this work, the Donovan's values, $a_1 = 1.1$, $a_2 = 0.5$, $a_3 = 25$, and $a_4 = -1.32$ are used. These values are mostly derived from data of Californian earthquakes, although earthquake accelerations from other parts of the world were also used. The results for R = 20 km and return periods of 20, 50, 100 and 200 years are given in Table 2.

Table 2. Expected ground accelerations in units of 'g' for a focal distance of 20 km for several return periods.

Provinces	T=20Y	T=50Y	T=100Y	T=200Y
2 Persian Gulf	0.13	0.15	0.17	0.20
3 Makran	0.13	0.15	0.18	0.21
4 Sea of Oman	0.17	0.23	0.29	0.37
5 Arvand-Shatt-al-Arab	0.12	0.14	0.16	0.19
6 Foothill Folded Series	0.16	0.19	0.21	0.24
7 Fars Folded Series	0.20	0.24	0.27	0.31
* 8 High Zagros Folded Series	0.18	0.22	0.25	0.30
	0.19	0.23	0.27	0.32
9 Rezaiyeh	0.16	0.21	0.24	0.29
10 Esfahan-Sirjan	0.10	0.11	0.13	0.15
11 Jaz-Murian	0.14	0.16	0.17	0.20
12 Central Iran	0.13	0.15	0.17	0.20
13 Kavir	0.11	0.13	0.16	0.19
14 Tabas	0.17	0.21	0.25	0.29
15 Ferdows	0.17	0.24	0.30	0.38
16 Lut	0.13	0.16	0.19	0.22
17 Shahrud-Doruneh	0.17	0.21	0.25	0.29
* 18 East Iran	0.16	0.21	0.27	0.34
	0.16	0.21	0.25	0.31
19 Maku-Zandjan	0.12	0.16	0.19	0.23
* 20 Alborz	0.18	0.23	0.28	0.34
	0.18	0.26	0.36	0.48
* 21 Kopet-Dagh	0.18	0.24	0.29	0.36
	0.18	0.26	0.36	0.49
22 Caspian Shore	0.15	0.19	0.23	0.28

* Estimations are based on two data bases.

Esfahan-Sirian and Kavir provinces have the least accelerations and Kopet-Dagh, Alborz, Ferdows and Fars folded series have the most accelerations. Accordingly in a time of 20 years, it is expected that an acceleration of 0.18 to 0.20 g will occur in Kopet-Dagh, Alborz, Ferdows, Sea of Oman, High Zagros provinces while during the same 20 years, the provinces of Esfahan-Sirian, Kavir, and Arvand-Shatt-al-Arab will suffer an acceleration of about 0.10 g. The values of acceleration for other provinces are between 0.13 to 0.17 g.

Probabilistic Risks From the preceding equations, the probabilistic risks can be calculated. If M is eliminated between Equations (1) and (10), then the probability of occurrence of no earthquake with a peak ground acceleration greater than or equal to A in time t at focal distance R can be estimated. From equation 9 we obtain M:

$$M = \frac{1}{a_2} \log \left(A/a_1 \, (R+a_3)^{a_4}\right) \qquad (10)$$

Now, M, a and b and A are known. Then $N = 10^{a-bM}$, and the probability of no earthquake occurrence with ground acceleration greater or equal to A in time t at focal depth R is given by:

$$P_o(A,t,R) = \exp(-Nt). \qquad (11)$$

Similarly, the risk of at least one occurrence of an earthquake with acceleration greater than or equal to A in time t at distance R is:

$$R(A,t,R) = 1-P_o(A,t,R). \qquad (12)$$

Results are presented in Figure 4 for six provinces. During a 30 year exposure time, all provinces will experience at least an acceleration of about 0.05g with a very high probability. During the same period at the 10 percent probability level, the accelerations are about 0.12, 0.18, 0.26, 0.3, 0.5 and 0.4 the Esfahan-Sirjan province has the least risks, and Fars, Alborz and Ferdows provinces have the highest risks.

DISCUSSION

The peak ground acceleration is a very important parameter for construction design. When safety related components are constructed, studies must be site specific and often both deterministic and probabilistic approaches are used. For non-safety related components design codes may be sufficient. Seismic coefficients for Iranian anti-seismic codes are suggested (Adeli et al[39]), but not yet adopted by the authorities. Recently iso-acceleration maps of Iran were published (Rowshandel et al[40]; Mohajer and Bozorgnia[41]) which may be used for the design purposes of non-safety related structures as well. In this section comparisons are made between the results reported by our approach and those published by them. Mohajer and Bozorgnia[41] considered a large number of seismic sources comprising segments of capable faults and several area sources and then calculated the ground acceleration at a number of grid points for construction of their iso-acceleration map at 64 percent probability level and 50 years exposure time. The seismic risk at a site or a major city can be read by interpolation from this map. Roshandel et al[40] published a set of tentative probabilistic curves and iso-acceleration maps for several levels of probabilities and exposure times. For comparison purposes the seismic risk for several cities at 64 percent level from the iso-acceleration map of Mohajer and Bozorgnia[41] and the risk reported for those cities by Roshandel et al[40] are superimposed on our seismic risk curves for the appropriate provinces. We have shown our calculations for several focal distances. The cities of Esfahan and Sirjan are in the Esfahan-Sirjan province (Figure 5). The risk obtained by Rowshandel[43] is less than that reported by Mohajer and Bozorgnia[41] and the risk obtained by using the seismotectonic model.

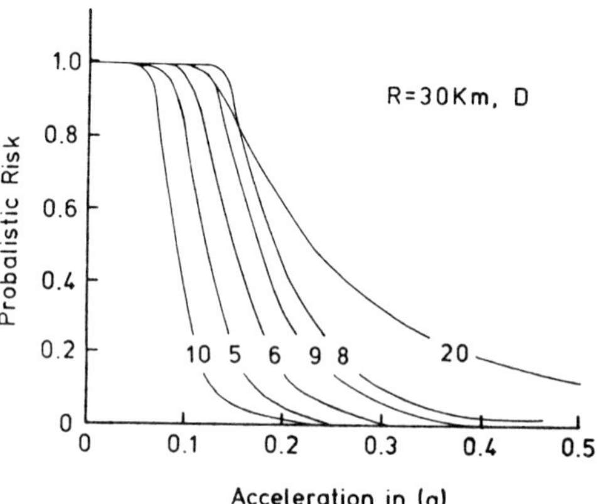

Figure 4. Probability of acceleration occurrences in 6 seismotectonic provinces. The provinces with increasing risks are: 10-Esfahan-Sirjan, 5-Arvand-Shatt-al-Arab, 6-Foothill folded series, 9-Rezaiyeh, 8-High Zagros folded series, and 20-Alborz.

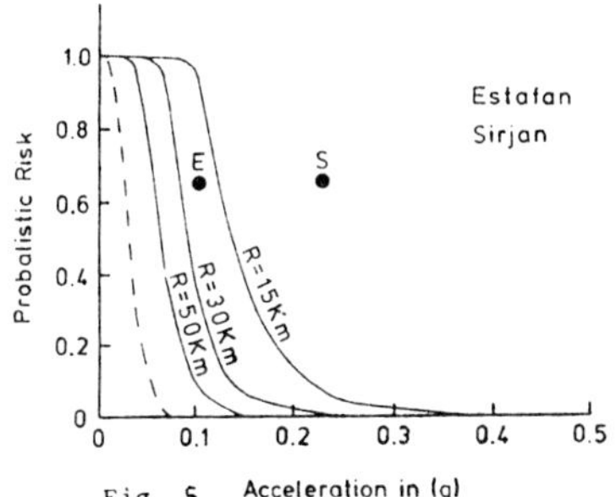

Fig. 5

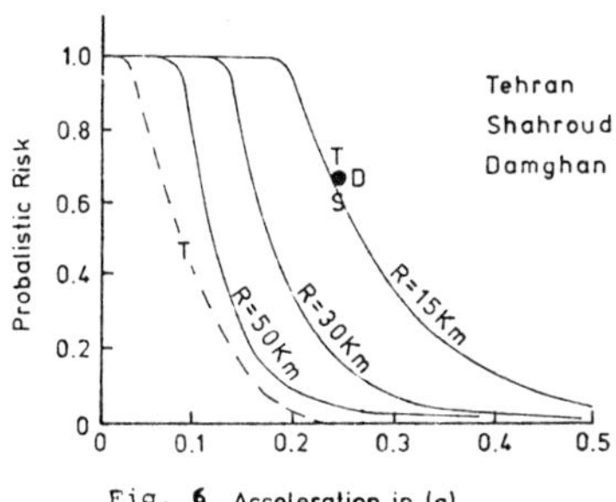

Fig. 6

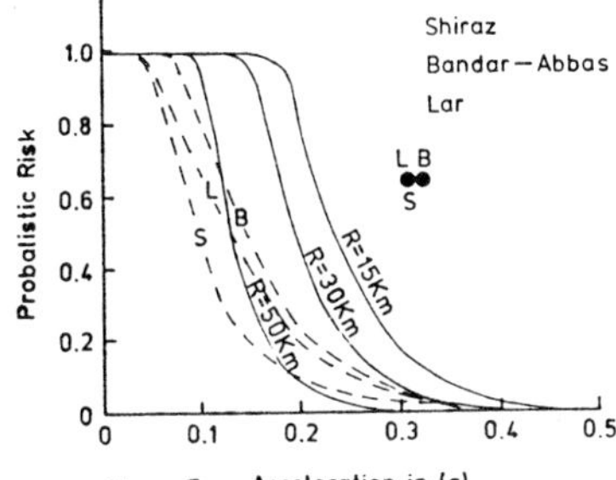

Fig. 7

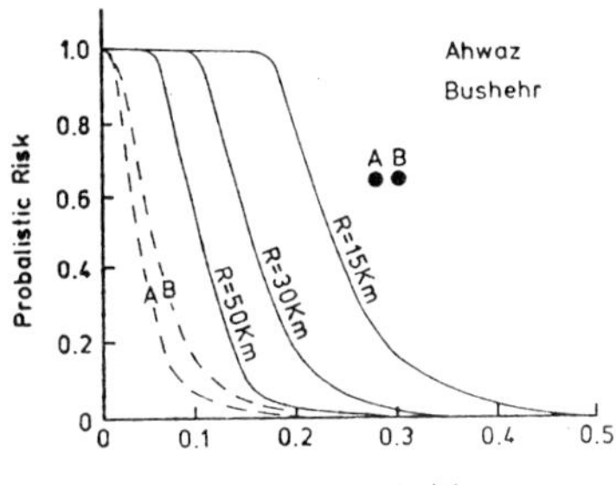

Fig. 8

Figures 5, 6, 7, and 8. Probability of acceleration occurrences for four Esfahan-Sirjan, Fars folded series, Shahrud-Doruneh, and Arvand-Shatt-al-Arab provinces. For comparison, acceleration occurrences determined by Rowshandel et al, dashed line, and Mohajer and Bozorgnia, dots are also given. The seismotectonic approach gives a more balanced acceleration.

However, their map indicates an acceleration of about 0.11g for Esfahan and 0.25g for Sirjan. The acceleration at Sirjan appears to be excessive. The cities of Tehran, Shahrud and Damghan are in the Shahrud seismotectonic province (Figure 6). The acceleration for these three cities as obtained from their map are very close to each other and fall approximately on our risk curve for a focal distance of 15 km. However, the risk curve of Roshandel et al[40] is considerably lower than the other values. The cities of Bandar Abbas, Lar and Shiraz are located within the Fars folded series of Zagros province (Figure 7). The acceleration values from their map for the cities of Lar and Shiraz are nearly the same, but that of Bander Abbas is slightly higher. In comparison with our results, their values appear too high. However, the risk curves published by Roshandel et al[40] are closer to our curve for a focal depth of 50 km, (Figure 7). The results for Ahwaz and Bushehr are given in Figure 8. Again risks assigned to these cities from their map are more than those calculated by our approach or Roshandel[43] et al. Mohajer and Bozorgnia used an average focal depth of 10 km for their calculation in most cases thus their maps appear to be too conservative for some areas. Roshandel et al[43], however, have used a limited number of seismic sources thus many cities may be too far from them. Therefore, their calculations have resulted in lower risks in many areas. The seismotectonic approach presented here appears to assign a more reasonable risk than the other two techniques, thus it may be appropriate for assessments of seismic risk in Iran, especially in the areas where seismic activities are diffused.

Conclusion Due to the complicated tectonic history of Iran, seismotectonics of this area are very complex. Present seismicity south of the main Zagros thrust zone is due to the northward motion of Arabian plate with respect to the Eurasian plate, and seismicity around the Caspian seashore is probably due to west-southwest motion of the Caspian plate with respect to the Persian plate. In east Iran and central Iran there are two smaller plates with relatively low background seismicity which probably are under a state of high stress. Most seismic activities of the country are attributed to the broad boundaries between the Lut, Persian, Caspian micro plates, and Arabian plate. The seismicity of the entire country is not uniformly distributed. It differs significantly from region to region. Results obtained by several investigators indicate that the Esfahan and Arvand-Shatt-al-Arab regions have the least and the Alborz, Kopet-Dagh, Ferdows, and Zagros provinces have the highest seismic risk.

Future work for construction of an isoacceleration map of Iran ought to be based on local magnitude-acceleration, intensity-acceleration, and fault length-magnitude relationships. Several useful equations for these purposes are developed by Nowroozi[32], and Chandra et al[34]. However, regional attenuation of acceleration with distance is one of the most important factor and least known parameters for this region and must be determined in the future.

1. Stocklin, J. (1968). Structural History and Tectonics of Iran: A review, Bull. Am. Assoc. Geol. 52, 1229-1258.
2. McKenzie, D.P. (1978). Active Tectonics of the Alpine-Himalayan Belt: The Aegean Sea and Surrounding Regions, Geophys. J.R. ast. Soc., 55, 217-254.
3. Nowroozi, A.A. (1971), Seismotectonics of the Persian Plateau, Eastern Turkey, Caucasus, and Hindu-Kush Region, Bull. Seis. Soc. Am., 61, 317-341.
4. Berberian, M. (1976). Contribution to seismotectonics of Iran. Publ. Geol. Surv. Iran, No. 39.
5. Morgan, J. (1968), Rises, trenches, great faults, and crustal blocks, J. Geophys. Res., 73-1959-1982.
6. LePichon, X. (1968), Sea floor spreading and continental drift, J. Geophys. Res., 73, 3661-3697.
7. Wilson, J.T. (1965), A new class of faults and their bearing on continental drift, Nature, 207, 343-347.
8. Sykes, L.R. (1968), Seismological evidence for transform faults, sea floor spreading and continental drift, History of the Earth's Crust (ed. Phinney, R.A.), Princeton University Press.
9. Nowroozi, A.A. (1972), Focal mechanism of earthquakes in Persia, Turkey, West Pakistan, and Afghanistana and Plate Tectonics of the Middle East, Bull. Seis. Soc. Am. 62, 822-850.
10. Takin, M. (1972), Iranian geology and continental drift in the Middle East, Nature, 235, 147-150.
11. Jackson, J. and McKenzie, D.P. (1984), Active tectonics of the Alpine-Himalayan Belt between western Turkey and Pakistan, Geophys. J. R. ast. Soc. 77, 185-264.
12. Snyder, D.B. and Barazangi, M. (1986), Deep crustal structure and flexure of the Arabian Plate beneath the Zagros collisional mountain belt as inferred from gravity observations, Tectonics, 5, 361-373.
13. Nowroozi, A.A. (1976), Seismotectonic provinces of Iran, Bull. Seis. Soc. Am. 66, 1249-1276.
14. Ambraseys, N.N. and Melville, C.P. (1982), A history of Persian earthquakes, Cambridge University Press.
15. Stocklin, J. (1977), Structural correlation of the Alpine ranges between Iran and central Asia, Mem. h. Ser. Soc. Geol. Fr., No. 8.
16. Gansser, A. (1964), Geology of the Himalayas, Interscience, London.
17. Nowroozi, A.A. and Mohajer, A. (1985), Fault movements and tectonics of eastern Iran: boundaries of the Lut Plate, Geophys. J. Roy. Ast. Soc. 83, 215-237.
18. McKenzie, D.P. Molnar, P. and Davies, D. (1970), Plate tectonics of the Red Sea and East Africa, Nature, 226, 243-248.
19. Ben-Menahem, A. and Aboodi, E. (1971), Tectonics patterns in the northern Red Sea region, J. Geophys. Res. 76, 2674-2689.

20. Quennel. A.M. (1958). The structural and geomorphic evolution of the Dead Sea rift. Quart. J. Geol. Soc. London. 114. 1-24.
21. Chandra, U. (1981). Focal mechanism solutions and their tectonic implications for the eastern Alpine-Himalayan Region, in Zagros, Hindu-Kush, Himalayas, Geodynamic evolution (ed. Gupta, H.K. and Delany, F.M.), Am. Geophysical Union, Washington, D.C.
22. Stahl, A.F. (1911), In Persian, Handbuch der Regionallen Geologie, Bd. 5, Abt. 6, Hft. 8, 1-46.
23. Boeck, H., de, Lees, G.M., and Richardson, F.D.S. (1929), Contribution to the stratigraphy and tectonics of the Iranian ranges, in structure of Asia (ed. Gregory, J.W.) Methuen, London, 58-176.
24. Bogdanoff, A.A. (1963), Sur le terme "etage structura", Rev. Geogr. Phys. et geol. Dynamique 5, 245-253.
25. Baier, E. (1938), Ein Beitrag Zum Thema Zwischenge-birge: Zentralbl. Mineral Geol. Abt. B11, 385-399.
26. Gansser, A. (1955), New aspects of the geology in Central Iran: World Pet. Congr. Proc. rth, Rome, 279-300.
27. Burchfield, B.C. (1983), The Continental Crust, Sci. Am. 249, 130-146.
28. Shareq, A. (1981), Geological observations and geophysical investigations carried out in Afghanistan over the period of 1972-1979, in Zagros, Hindu-Kush, Himalayas, Geodynamic evolution, (ed. Gupta, H.K., and Delaney, F.M.), Am. Geophys. Union, Washington, D.C.
29. Huber, H. (1977), Geological map of Iran, scale 1:1,000,000, National Iranian Oil Company.
30. Ross, D.A. (1971), The Red and The Black Seas, Am. Scientist 59, 420-424.
31. Nowroozi, A.A. (1976), Seismicity of Afghanistan and Hindu-Kush seismic zone, Proceedings, Acad. Nat. Dei Lincei, Rome, 21, 265-277.
32. Nowroozi, A.A. (1985), Empirical relations between magnitudes and fault parameters for earthquakes in Iran. Bull. Seis. Soc. Am., 75, 1327-1338.
33. Mohajer, A. and Nowroozi, A.A. (1978), Observed and probably intensity zoning of Iran, Tectonophysics, 49, 149-160.
34. Chandra, U., McWhorter, J.C. and Nowroozi, A.A. (1979), Attenuation of intensities in Iran. Bull. Seis. Soc. Am. 67, 237-250.
35. Nowroozi, A.A. and Ahmadi, G. (1986), Analysis of earth-quake risk in Iran based on seismotectonic provinces, Tectonophysics, 122, 89-114.
36. Esteva, L. and Villaverde, R. (1973), Seismic risk, design spectra and structural reliability. Proc. 5th World Conf. Earthquake Eng. Rome 2586-2597.
37. Donovan, N.C. (1973), A statistical evaluation of strong motion data including the February 9, 1971 San Fernando Earthquake Proc. 5th World Conf. Earthquake Eng. Rome, paper 155.

38. Bolt, B.A. and Abrahamson, N.A. (1982). New attenuation relations for peak and expected accelerations of strong ground motion. Bull. Seis.Soc. Am. 72, 2307-2321.
39. Adeli, H., Mohajer, A. and Nowroozi, A.A. (1982). Seismic zoning of Iran, in Proc. 3rd Inter. Microzonation Conf., Seattle, Washington.
40. Roshandel, B. Nemat-Nasser, S., and Adeli, H. (1979). A tentative study of seismic risk in Iran, Iran. J. Sci. Tech. 7, 211-241.
41. Mohajer, A. and Bozorgnia, Y. (1984). Ground acceleration distribution in Iran, A probabilistic approach. Proc. 8th World Conf. Earthquake Eng., San Francisco, USA.

The Seismicity of the Near East – Past and Present

A. Ben-Menahem
Department of Applied Mathematics, The Weizmann Institute of Science, Rehovot, 76100, Israel

ABSTRACT

A study of major earthquake occurrence along the Dead Sea transform (35.5°–36.5°E; 27.2°–37.5°N) during the past five millenia has been concluded. Geological, archaeological, biblical, historical and seismological evidence were integrated in an effort to quantify the seismological and engineering aspects of earthquakes in the said province. We have established: (1) tectonic patterns (2) variations of slip and creep (3) zoning and regionalization (4) long term seismicity cycles (5) seismic-risk evaluation (6) local scaling-laws for peak ground accelerations. The space-time distribution of seismicity (faulting and earthquake activity) appears to be stationary over time during the said time-window.

At the present time, a magnitude $6\frac{3}{4}$ earthquake is pending at the northern edge of the Levant rift, with its average recurrence interval (83 yrs) exceed by one standard deviation (32 yrs). If our observation that maximal magnitude earthquakes ($M_L = 7.3$) in the Holy Land are periodic (1500 yrs), is extrapolated into the future—the next event is not expected before the 23rd century.

INTRODUCTION

The Near-East is probably the only region on earth where evidence for earthquake activity has been documented in one way or another over the past five millenia. The interpretation of this valuable data, in the light of current tectonic models (Freund[1]; Ben-Menahem[2,3]), and with the aid of tools provided by modern seismology, calls for a comparative analyses of all available geological, historical, archaeological and seismological findings. The present study reports results from these four complementary disciplines. Our well-defined geological province extends along some 1140 km from offshore southern tip of the Sinai peninsula to the collision-zone of Levant rift and the East-Anatolian fault-system at the Kara-Su junction.

Major earthquakes with epicenters outside this region, such as the Zagros and north Anatolian fault systems, Aegean-Sea, Crete, Greece and the eastern Mediterranean basin, were also felt and even caused severe damage in the Near-East.

A schematic map of our region is shown in Figure 1. The numbers giving the displacements represent their integrated values over a period of about 20 million years. Figure 2 shows epicenters of earthquakes with $M_L \geq 7.4$ (black circles) that were felt in Jerusalem since 2100 B.C.

DATA SOURCES

I. Ancient seismicity (~3650 B.C. – 100 B.C.)
The main sources for this epoch are: Archaeological excavations, including submarine studies, stratigraphic and structural geological studies and evidence from the Bible.

II. Historical seismicity (100 B.C. – 1900 A.D.)
Talmudic and post-Talmudic literature. Roman, Arabic and medieval literature. Reports and books of pilgrims, historians, chronologists, travellers and priests. Eyewitness reports of the effects of earthquakes upon shrines, temples, palaces, monuments, ramparts, battlements, rivers, seas, mountains and cities. The data and its sources are given by Ben-Menahem[4–7].

III. Instrumental seismicity (1900 – present)
Seismic stations were established in the Near-East in the following order: Helwan (1899), Ksara (1910), Istanbul (1934), Jerusalem and Safed (1953), Elat (1968), Amman (1983). A network of portable stations was operating in Israel during 1976–1982 for the monitoring of microearthquakes in the magnitude range $0 \leq M_L \leq 3.6$.

DATA ANALYSIS (1976–1982) AND ITS RESULTS

The analysis of the data proceeded along seven steps in the following order:

I. Preparation of a comprehensive regional historical and contemporary catalogue of earthquakes.

II. Delineation of all epicenters along known active faults.

III. A detailed study of major 'calibrating pilot-earthquakes' (1927, 1969, 1979) for which instrumental and macroseismic data could be compared in order to derive empirical 'magnitude-intensity-acceleration-distance' and 'magnitude-slip-fault length' relations. These

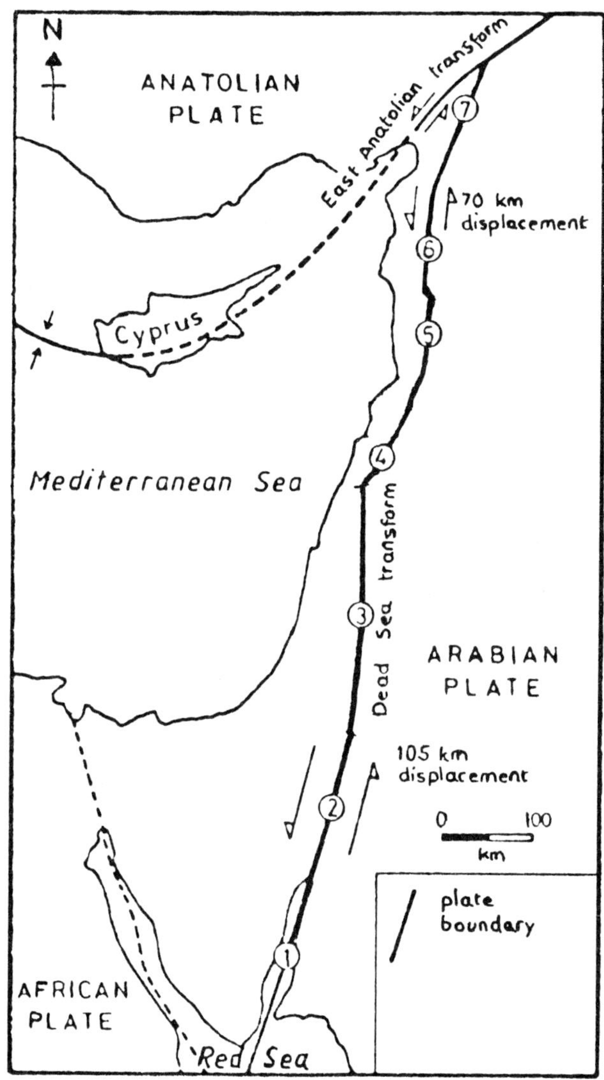

Figure 1:Segments of the Dead Sea transform: (1) Elat (2) Arava (3) Dead-Sea Jordan river (4) Beka'a (5) Southern Gharb (6) Northern Gharb (7) Kara-Su

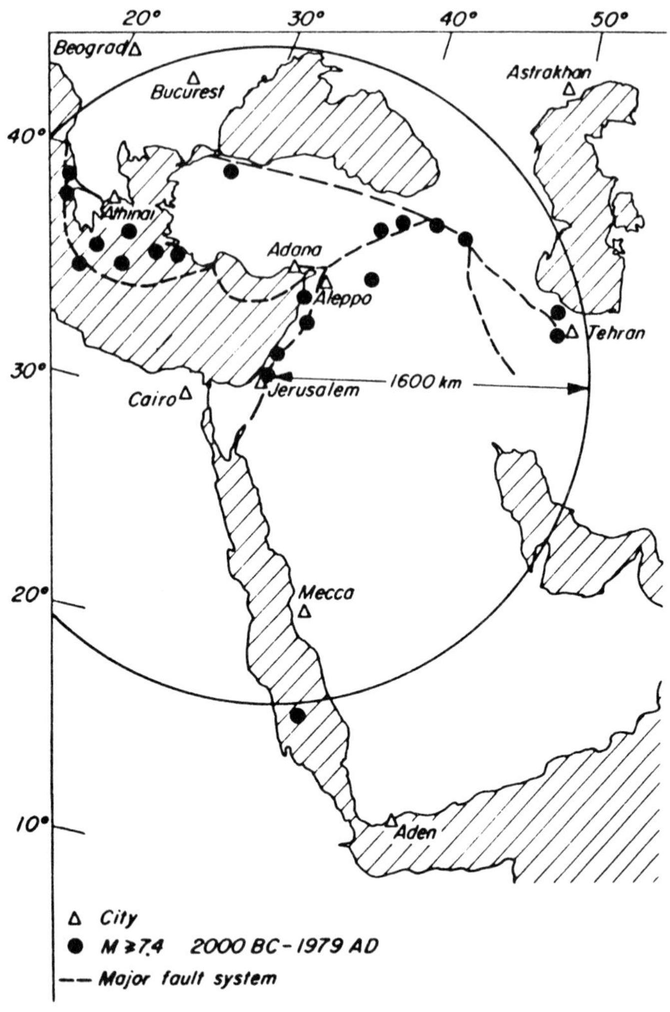

Figure 2:Epicenters of earthquakes with $M_L \geq 7.4$ (black circles) that were felt in Jerusalem since 2100 B.C.

relations were then used to assign local magnitudes to historical earthquakes.

IV. Source-mechanism studies of 63 earthquakes along the Dead-Sea transform during 1927–1979. These results were then used to effect the regionalization of the various fault systems along the transform, within the plate-tectonic scheme.

V. Microearthquakes study during 1976–1982 with the aid of 21 stations in Israel.

VI. Assignment of a frequency-magnitude relation to each fault-system (Table 1) and a maximal credible earthquake for each fault.

VII. Preparation of seismic-risk maps for the Dead-Sea transform (Ben-Menahem[8]).

The most salient features of our research are as follows:

(1) Studies of source-mechanism have confirmed the plate-tectonic concept of a megashear motion on the borderline of the Arabian and African plates. The evaluated rates of slip were in full accord with the observed displacements along the Dead-Sea transform.

(2) The Gutenberg-Richter frequency-magnitude law does accommodate a wide range of magnitudes from the smallest observed microearthquake ($M_L \sim 0$) to the extremal historical magnitude ($M_L = 8.0$). Indeed, when these frequency-magnitude relations are extended to 3600 B.C., one can account for the maximum magnitude earthquakes (MME) that echo in the Bible (time acts as a 'high-pass' magnitude-filter!) and are manifested in archaeological excavations. There is then no need to employ any corrections to the said law. The observed MME in each region agrees with their estimate on the basis of the integrated slip in each corresponding region.

(3) In the Dead-Sea region proper, most of the seismic activity since 2000 B.C., has been confined to the vicinity of its eastern shore with extremal seismicity at its southern tip near the prehistorical site of Bab-a-Dara'a (31°15'N, 35°32'E). This may constitute the first evidence that the Biblical 'cities of the plain' (Sodom, Gommorah etc.) were located there. Recent studies of earthquake deformations in the Lisan deposits (El-Isa and Mustafa[9]) near Bab-a-Dara'a, confirm our findings. Deep mud cores from the Dead-Sea bottom sediments may reveal (via its white layers caused by the whitening of the Dead-Sea during earthquakes) signatures of prehistoric earthquakes[10,11].

(4) The region extending south of the Dead-Sea up to the Red-Sea along Wadi Arava (31.0°–29.5°N) has been the least seismically active region during the past 3000 years. This does not exclude the possibility of enhanced seismic activity at earlier times. On the other hand,

Table 1: Regionalization and variation of seismicity along the Dead-Sea transform

	Segment	Total fault length L, km	Latitude spread °N	$\log_{10} N = a - bM_L$		Earthquake data		
				a	b	M_{max}	Mean return period, yrs	Last yr of occurrence
1	Gulf of Eilat and eastern Sinai	250	27.2–29.5	3.45	0.91	7.3	1500	1068
2	Arava Dead-Sea Jordan River	430	29.5–33.4	3.10	0.86	7.3	1500	746
3	Bekaa	140	33.4–34.6	2.80	0.86	7.3	3000	1201
4	Southern Gharb Northern Gharb Kara-su	320	34.6–37.5	3.75	0.86	8.0	1400	859
5	Secondary fault system	500	30.3–34.9	2.56	0.86	7.2	4300	1042
6	Off-shore and coastline fault system Gaza–Tripoli	330	31.5–34.5	3.35	0.94	7.2	2600	-524

Latest events:

Segment 1: $M_L \geq 5$ 1983, 1972, 1969, 1952
$M_L \geq 6.5$ 1969, 1588, 1312, 1068

Segment 2: $M_L \geq 6.4$ 1837, 1759, 1546, 1202, 1033, 746, 658, 362, -30, -435, -758, -853, -1250, -1560, -2027, -2150

Segment 3: $M_L \geq 6$ 1956, 1802, 1759, 1201, 1157, 991, 847, 565, 130

Segment 4: $M_L \geq 7$ 1872, 1656, 1408, 1157, 1170, 859, 528, 526, 500, 115, -63, -147

Segment 5: $M_L \geq 6$ 1182, 1151, 1105, 1042

Segment 6: $M_L \geq 6$ 1873, 1752, 1402, 1261, 1068, 881, 525, 502, 349, 306, 19, -141, -524.

the Northern Gharb was the most seismically active zone along the transform, especially during the past 2000 years. This fact is demonstrated in Table 2: the city of Antioch was destroyed and rebuilt 15 times during the past 23 centuries. Part of the earthquakes in the table originated on the SW tip of the East-Anatolian fault-system.

(5) Average return period for MME ($M_{max} = 7.3$) in the southern half of the transform (segments 1–3 in Table 1, 33.4°–27.2°N) is about 1500 years. It seems that MME tend to reoccur almost periodically every 1500 years (Table 1). Based on this premise, no MME is expected in this region prior to 2250 A.D.

(6) There is ample historical evidence for causative faults along the coastline and offshore Mediterranean, extending from Gaza to Tripoli (Table 1). These were responsible for many disasterous tsunamis that hit repeatedly the coastal cities of the eastern Mediterranean during the past 2500 years (excluding of course tsunamis which were generated from epicenters near Iskanderun, Cyprus, Rhodes, Crete and in the Aegean Sea). Indeed, recent geo-biological studies (Neev et al.[12]; Flemming[13]) of the coastal plane of Israel, revealed that it was downwarped and submerged, at least since the post-crusader period. After this the offshore-side remained submerged whereas the area east of it was uplifted to present day position. The associated tectonic events could explain the historical seismic activity. The earliest age of this process could not be more than 2500 B.C., the age of the oldest marine shells.

SPECIAL ASPECTS

Earthquake-engineering in antiquity

Excavations in Jericho, Hazor and other archaeological sites in the Near-East revealed clear evidence of collapses or tilt of walls, caused by earthquakes. These walls acted as seismoscopes through which one may infer the azimuth of the source. Kenyon[14] had noticed at Jericho a rather curious feature in the structure of the walls which was possibly intended as an anti-earthquake device: at intervals in the course of the wall were cavities about 3 feet wide running almost the whole thickness of the wall. The effect was certainly to localize the collapse. Also the bricks of the wall had been tied together by timber, possibly an anti-earthquake measure. Indeed, we read in Kings I: **6**, 36; **7**, 12 about King Solomon and his Temple: "... and he built the inner court with three rows of hewed stone, and a row of cedar beams."

Similarly, along the Arava fualt, we find three places where structures were fortified to withstand earthquakes: (1) The Nabatian temple at Aram (Gebel-E-Ram, 40 km east of Akaba) built ca 31–36 A.D., experienced a strong earthquake at 48 A.D. (2) Tel El Haleife, near Elat.

Table 2: The seismic history of Antakya (Antioch. 36.12°N, 36.10°E)

300 B.C.	Established by Seleucus I Nicator to commemorate the victory at the battle of Ipsus (301 B.C.).
148 B.C.	Destroyed.
64 B.C.	Destroyed. $M_L = 7.5$.
37 A.D.	Destruction. Emperor Caligula sent two senators to report the condition of the city.
115, Dec. 13	City destroyed during Emperor's Trajan's sojourn who was forced to take shelter in the circus for several days. Event mentioned in the Talmud. $M_L = 7.4$. Epicenter near Samandang.
334	Destruction. $M_L = 6.8$.
458, Sept. 14	City destroyed. 36.2°N, 36.1°E. $M_L = 6.5$. Allegedly 80,000 victims out of a population of 350,000.
500	City destroyed. 36.2°N, 36.1°E. $M_L = 7.5$.
526, May 20	City destroyed. 36.2°N, 36.1°E. $M_L = 7.0$. Allegedly 30,000 victims, including thousands of Christians who gathered to a great church assembly.
528, Nov. 29	City destroyed. 36.2°N, 36.1°E. $M_L = 7.0$.
588, Oct. 31	City destroyed. $M_L = 6.4$. Allegedly 60,000 victims.
713, Mar. 20	City destroyed. $M_L = 6.8$.
859, Apr. 8	Total destruction. 36.2°N, 36.1°E. $M_L = 8.0$. Earthquake felt in Mecca and caused damage in Jerusalem ($\Delta = 500$ km). $\Delta_f = 1500$ km!
972	Destruction. $M_L = 6.4$. Byzantine Emperor John Zimisces ordered 12,000 masons to rebuild the city.
1114, Aug. 10	Destruction. 36.5°N, 36.0°E. $M_L = 7.0$. Tsunami. Felt in Jerusalem $I_{MM} = 4$. Epicenter probably on East-Anatolian fault system.
1170, June 30	City demolished. 34.6°N, 36.2°E. $M_L = 7.9$. Many victims. St. Peter's cathedral collapsed over patriarch.
1408, Dec. 30	Heavy destruction. $M_L = 7.5$. No major destructive events for the next 414 years!
1822, Aug. 13	City destroyed again. 36.4°N, 36.2°E. $M_L = 6.9$. Tsunami at Iskanderun. Felt in Jerusalem. Population reduced to 5000 inhabitants.
1872, Apr. 02	Destruction. 36.2°N, 36.2°E. $M_L = 7.3$. 1800 victims.

(3) Petra, 80 km south of the Dead-Sea.

We thus see that the awareness and fear of earthquake hazard already troubled the minds of the inhabitants and 'engineers' of the Near-East more than 3000 years ago.

No timber and no cavities in walls could prevent, the eventual collapse of city walls by major earthquakes. Thus, the walls of Jericho collapsed not less then 17 times during the interval 2100 B.C. – 3100 B.C. If we give equal chance to destruction by erosion, floods, earthquakes and wars and assume that a least magnitude of 6.4 is needed to topple ancient wall fortifications within a distance of say 30 km from the city, we shall come up with about 4 such events in 1000 years. This is in general agreement with the frequency-magnitude relation for segment 2 in Table 1.

Risk analysis

The potential damage caused by earthquakes to human habitation and life-lines in the Near-East will certainly increase steeply as the region becomes more populated and more industrialized. Earthquakes such as happened in Antioch in 859, in Lebanon in 1202 and in Israel in 746 (Ben-Menahem[5]; Poirier and Taher[15]; Armijo et al.[16]) although not expected before the 23 century, can serve as an upper limit to possible destructiveness. Translating the effects of two of the above events into modern concepts, [by using Eqs. (5.1)–(5.9) of Table 3] we obtain for $M_L = 7.3$ ($L = 115$ km) a zone of heavy destruction (I_{MM} =8–9) within a radius of 90 km from the initial epicenter. The loss of life and property then depend on the size of largest city that happens to be "trapped" within this circle. A more adequate way to quantify the seismic risk is to prepare maps with equi-probability curves for the probability of occurrence of at least one earthquake that will induce a peak baserock acceleration exceeding some predetermined treshold. The equations of these curves, based on the assumption that earthquake-occurrence on a fault constitute a Poisson process in time, are given in Ben-Menahem et al.[8,17].

If the zone under consideration includes more than a single fault, the overall probability of exceedance or non-exceedance is the product of the respective individual probabilities for each fault. At the base of the risk calculation are empirical local scaling laws for ground accelerations.

Global empirical relations cannot in general be used for successful risk calculations in specific regions. However, *local* scaling laws, when carefully scrutinized, can still be used to make statistical prediction of the destructive effects of earthquakes in region where soil conditions, earthquake mechanism and seismic history are known. We have found that Eqs. (5.5)–(5.6) in Table 3 are compatible with measured ground

Table 3: Quantification of intensity attenuation and 'magnitude-intensity-acceleration-source parameters' relations on the Dead-Sea transform

Causative fault parameters

L	Fault-length (km)
W	Fault-width (km)
S	Fault-area (km^2)
U	Slip on fault (m)

Macroseismic data parameters

R	Hypocentral distance (km)
h	Hypocentral depth (km)
R_f	Limiting distance of human perceptibility
$I_{MM}(R)$	Intensity at hypocentral distance R on the modified Mercali scale
a_p	Peak horizontal acceleration (cm/sec^2)
$D(M_L)$	Duration of felt ground motion
M_L	Local magnitude
$T(M_L)$	Mean return period (average recurrence interval).

$$M_L = -0.5 + 0.36 \cdot 10^{-3} R_f \text{ (km) } + 2.5 \log_{10}(R_f + 25)$$
$$h \leq \text{ normal}; \quad R = (h^2 + \Delta^2)^{1/2} \tag{5.1}$$

$$M_L = -1.98 + 0.50 I_{MM}(R) + 2.5 \log_{10}(R + 25) + 0.36 \cdot 10^{-3} R(\text{km}) \tag{5.2}$$

$$M_L = 1.80 + 0.50 I_0 \quad (R = h = 7.5\text{km}) \tag{5.3}$$

$$M_L = 0.62 \log_{10}[U(m) S(\text{km}^2)] + 4.96 \tag{5.4}$$

$$\log_{10} a_p = 0.26 I_{MM} + 0.22 \tag{5.5}$$

$$a_p = 17.8 e^{1.21 M_L} (R + 25)^{-1.32} e^{-R/400}, \quad M_L \leq 7.5 \tag{5.6}$$

$$\log_{10} R_d(\text{km}) = 0.7 M_L - 3.20, \quad M_L \geq 5 \tag{5.7}$$

$$\log_{10} U(\text{m}) = 0.6 M_L - 2.18 \tag{5.8}$$

$$\log_{10} L(\text{km}) = 0.5 M_L - 1.60 \tag{5.9}$$

$$\log_{10} D(\text{sec}) = 0.4 M_L - 1.32 \tag{5.10}$$

$I_3 - I_4$	Isoseismal at $R = R_f$
I_8	Isoseismal at $R = L$ for $M_L \geq 6$
I_9	Isoseismal at $R = R_d = L - 25$ for $M_L \geq 6\frac{1}{4}$

$\log_{10} N = 3.94 - 0.86 M_L$ Frequency-magnitude relation for the entire Dead-Sea transform ($L = 1140$ km) (5.11)

Table 4: The fit of observed local peak acceleration (April 23, 1979†) with local and global empirical laws

Epicentral Distance Δ, km	Recorded Peak Acceleration cm/sec^2	LOCAL*	GLOBAL**
28	11.1 (DL)††	19.1	47.5
50	10.9 (D)	12.2	30.3
67	24.0 (A)	9.5	23.6
95	Less than 5 (K)	5.8	14.4

† USCGS, BISC, MB = 5.0–5.1;
Source parameters of this event: L = 5–7 km; W = 1–2 km;
U = 4–7 cm.

†† Rock type: DL – Dolomite and limestone; D – Dolomite;
A – Alluvium; K – Kurkar.

* $a_H = 17.8e^{1.2M_L}(R+25)^{-1.32}e^{-R/400}$ (Ben-Menahem et al.[8]).

** $a_H = 1080e^{0.5M_L}(R+25)^{-1.32}$ (Donovan[18]).

acceleration in Israel. In Table 4 we compare our local scaling laws with a global scaling law obtained by Donovan[18] from data in western north America and Japan. The comparison proves indeed that scaling laws derived from average world data cannot account for local lithological and geological conditions, nor do they take into account the correct frequency-magnitude relations appropriate for the geological province under study. Thus, we have found that the use of Donovan's scaling law lead to risk estimates which disagree with the seismic history of the Holy Land.

REFERENCES

1. Freund, R. (1965), A Model of the Structural Development of Israel and Adjacent Areas Since Upper Cretaceous Times, Geological Magazin, Vol. 102, pp. 189–205.

2. Ben-Menahem, A. and Vered, M. (1976), Tectonics, Seismicity and Structure of the Afro-Eurasian Junction, Physics of the Earth and Planetary Interiors, Vol. 12, pp. 1–50.

3. Ben-Menahem, A. and Aboodi, E. (1971), Tectonic Patterns in the Northern Red Sea Region, Journal of Geophysical Research, Vol. 76, pp. 2674–2689.

4. Ben-Menahem, A. (1977), Rate of Seismicity of the Dead-Sea Region Over the Past 4000 Years, Physics of the Earth and Planetary Interiors, Vol. 14, pp. P17–P27.

5. Ben-Menahem, A. (1979), Earthquake Catalogue for the Middle East (92 B.C. – 1980 A.D.), Bollettino Di Geofisica Teorica Ed Applicata, Vol. 21, pp. 245–310.

6. Ben-Menahem, A. (1981), Variation of Slip and Creep Along the Levant Rift Over the Past 4500 Years, Tectonophysics, Vol. 80, pp. 183–197.

7. Ben-Menahem, A. (1981), Micro- and Macroseismicity of the Dead-Sea Rift and Off-Coast Eastern Mediterranean, Tectonophysics, Vol. 80, pp. 199-233.

8. Ben-Menahem, A., Vered, M. and Brooke, D. (1982), Earthquake Risk in the Holy Land, Bollettino Di Geofisica Teorica Ed Applicata, Vol. 24, pp. 175–203.

9. El-Isa, Z.H. and Mustafa, H. (1986), Earthquake Deformation in the Lisan Deposits and Seismotectonic Implications, Geophysical Journal of the Royal Astronomical Society, Vol. 86, pp. 413–424.

10. Ben-Menahem, A. (1976), Dating of Historical Earthquakes by Mud Profiles of Lake-Bottom Sediments, Nature, Vol. 262, pp. 200–202.

11. Bloch, M.R. (1980), Dead Sea Whiteness and Its Origin, Proc. Israel Academy of Sciences and Humanities, No. 19, pp. 1–7.

12. Neev, D. et al. (1973), Recent Faulting Along the Mediterranean Coast of Israel, Nature, Vol. 245, pp. 254–256.

13. Flemming, N.C. (1979), Archaeological Evidence for Tectonic Activity in the Region of the Haifa-Qishon Graben, Israel, Tectonophysics, Vol. 52, pp. 177–178.

14. Kenyon, K. (1979), Archaeology in the Holy Land, Earnest Benn Ltd. London 4th Ed. 360 pp.

15. Poirier, J.P. and Taher, M.A. (1980), Historical Seismicity in the Near and Middle East, North Africa and Spain From Arabic Documents (7th–18th century), Bulletin of the Seismological Society of America, Vol. 70, pp. 2185–2201.

16. Armijo, R., Deschamps, A. and Poirier, J.P. (1985), Carte Seismotectonique De L'Europe et Du Bassin Mediterraneen, Institute De Physique Du Globe De Paris.

17. Ben-Menahem, A. (1986), Peak Accelerations From Subshear and Supershear Radiation of Kinematic Dislocation Models, Journal of the Physics of the Earth, Vol. 34, pp. 297–334.

18. Donovan, N.C. (1974), A Statistical Evaluation of Strong Motion Data, Including the February 9, 1971 San-Fernando Earthquake, Proc. 5th World Conf. on Earthquake Engineering, Rome.

SECTION 2: SEISMIC WAVES IN SOILS AND GEOPHYSICAL METHODS

Scattering of Plane Harmonic Waves by Multiple Dipping Layers of Arbitrary Shape

M. Dravinski, T.K. Mossessian
Department of Mechanical Engineering, University of Southern California, Los Angeles, CA 90089-1453, U.S.A.

INTRODUCTION

The extent of the damage in central part of Mexico City due to Michoacan Mexico earthquake of September 19, 1985, was remarkably severe considering the long epicentral distance of 350 km[1]. Nearly all the buildings that collapsed during the earthquake were located in the central portion of Mexico City, while the damage in the surrounding area was minimal[2]. The central part of Mexico City is founded on a lake bed formed by clay deposits which are considerably softer than the alluvium of the surrounding hilly region[2]. In the two second period, the acceleration response at the lake zone was about ten times greater than that of outlying districts[1,2]. Thus, it appears that subsoil structure played an important role in the resulting pattern and extent of damage in Mexico City during Michoacan earthquake.

Observations from previous earthquakes in the Valley of Mexico[3] show that in spite of differences in source mechanism and epicentral distances, accelerograms at the lake bed zone are similar with the low frequency oscillations known as resonant motion of the sedimentary basin[1,2]. Analysis of earthquake damage at other locations, e.g., San Fernando, California earthquake of 1971[4], Lima, Peru earthquake of 1974[5] indicate that the areas of intense damage atop alluvium can be highly localized. Since many other highly populated areas are located on sedimentary basins, the study of their response to seismic waves is of great interest in earthquake engineering and strong ground motion seismology.

Experimental studies by Ohta et al.[6], Kagami et al.[7], Kagami et al.[8], King and Tucker[9], and Tucker and King[10], as well as theoretical studies by Bouchon and Aki[11], Bard and Bouchon[12,13] and Dravinski,[14,15,16,17] have shown that presence of sediment filled valley may cause very large localized amplification of surface ground motion. Studies of microtremors and strong ground motion in deep alluvial basins[7,8,9,10,18] show the existance of a resonant-type motion of the valley. This can be explained only through interactions between horizontal and vertical waves within the valley. Thus, for thorough understanding of seismic response of deep sediment filled basins it is necessary to study amplification of the surface ground motion by using at least two dimensional models.

For scattering of seismic waves by irregularities with moderately shallow slopes Aki and Larner[19] developed a technique using the so called Rayleigh hypothesis. The method is applicable to wave lengths which are either larger than or of the order of the dimension of the irregularities. Bouchon[20] and Bard and Bouchon[12,13] used this method to study the seismic response of sediment filled valleys subjected to incident P, SV, and SH waves. Bouchon and Aki[11] extended the Aki-Larner method to study the near field of a seismic source in a layered medium with irregular interfaces.

Several other methods which are very effective for studying wave scattering in geotechnical problems are the boundary integral equation methods[14,15,16,17,21,22,23,24]. These methods require only discretization of the boundary of the scatterers[25,26] and the radiation conditions at infinity can be modeled exactly[27]. Still, at the present time the boundary integral equation methods are limited to problems involving linear , isotropic, and homogeneous materials[27].

Recently, Mossessian and Dravinski[28] examined a hybrid method for studying the scattering of elastic waves by dipping layers of arbitrary shape. This method combines the finite element technique and an indirect boundary integral equation method. The advantage of such approach is that it utilizes the versatility of the finite element method for detailed modeling of the near field and the effectiveness of the boundary integral equation methods in the far field. For wave propagation problems in geophysics and earthquake engineering this idea has been investigated by several authors. Chopra et al.[29] and Dasgupta and Chopra[30] proposed a technique for studying seismic response of foundations. The method combines the dynamic stiffness matrices of a viscoelastic half-space with the finite element modeling of the embedded foundation through a method known as the substructure method. However, they did not present any numerical results. Datta and Shah[31] studied scattering of SH waves by embedded cavities by combining finite elements with the eigenfunction expansion method. Shah et al.[32] considered diffraction of SH waves by surface and near surface defects using two different hybrid methods : First, they used the finite element method with an integral representation approach. Secondly, they employed the finite element method with a wave function expansion method. Mita and Takanashi[33] investigated the effects of back-filled sand on motion of an embedded foundation subjected to incident P or SV waves using a combination of finite element and an indirect boundary integral equation approach. Wong et al.[34] studied diffraction of P, SV and Rayleigh waves by embedded cavities and inclusions using a combination of finite element and eigenfunction expansion methods. However, their plain strain solution is not applicable to problems in which the scatterers are very close to the free surface. Recently, Alyagshi Eilouch and Sandhu[35] applied a finite element-boundary element coupling method for transient analysis of dynamic soil-structure interaction under SH motion only.

In all the papers cited above, the models considered are two-dimensional ones. For scattering of seismic waves by three dimensional subsurface irregularities there are very few solutions available at present time. In addition, most of the papers that have appeared in literature are limited to very simple geometries. Lee[36] presented a solution for problem of diffraction of elastic plane waves (P, SV and SH) by a semispherical canyon using the method of series expansion which he later[37] extended to study the case of a semispherical alluvial valley. His solution is applicable only for spherical-shape irregularities. Day[38] used a finite element technique to study scattering of seismic waves by cone-shape sedimentary basin only for axisymmetric conditions. Sanchez-Sesma[39] considered diffraction of vertical incident P wave by axisymmetric canyons using the c-complete functions[40]. Lee and Langston[41] investigated wave propagation in a three dimensional circular basin subjected to incident plane P and SH waves with various incident angles using a ray technique. Their method is limited to the high frequency range.

Recently, Avanessian et al.[42] used a hybrid method combining the finite element technique and the wave function expansion method to study seismic response of foundations for axisymmetric cases.

This paper illustrates an application of an indirect boundary integral equation approach in studying amplification of seismic waves due to dipping layers of arbitrary shape. Such subsurface irregularities can be found in various sedimentary basins such as the Los Angeles basin[43]. The paper is continuation of the work by Dravinski[17] and Dravinski and Mossessian[22]. The method originates in the work of Kupradze[44] with many extensions to the problems in strong ground motion seismology and earthquake engineering[14,15,16,17,22,23,24].

STATEMENT OF THE PROBLEM

Geometry of the problem is depicted by Fig. 1. Finite number of dipping layers of arbitrary shape are perfectly bonded together to form a layered half-space. The interfaces between the layers are assumed to be sufficiently smooth without any sharp corners. The spatial domain of the half-space is denoted by D_0 and that of the layers by D_j The interfaces between the layers are denoted by C_j, $j=1, 2, \ldots, R$. The layers are assumed to be weakly anelastic[45,46], homogeneous, and isotropic.

If the model is assumed to be of the plane strain type, then the motion of the medium can be described by a displacement vector $\underline{u}_j^T = (u_j, 0, w_j)$, $j = 0, 1, 2, \ldots, R$, where u and w represent the displacement components along the x and z axis, respectively and the superscript T denotes the transpose. The displacement field is related to the displacement potentials through

$$\underline{u}_j = \nabla\phi_j + \nabla \times (0, \psi_j, 0) \qquad (1)$$
$$j = 0, 1, \ldots, R,$$

where ϕ and ψ denote dilatational and equivoluminal displacement potentials, respectively. Throughout the paper, the subscripts 0, 1, 2, ..., R refer to D_0, D_1, D_2, ..., D_R, respectively. Summation over repeated indices is understood and underlined indices indicate that the summation is suppressed.

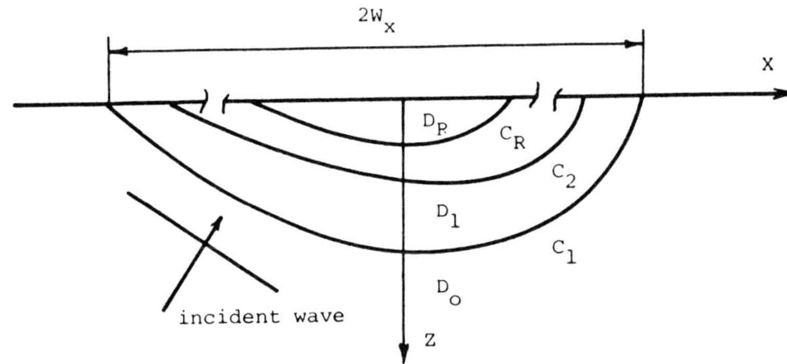

Figure 1. Problem model geometry.

Since the motion is of a steady state type the factor $\exp(i\omega t)$ is understood, where ω is the circular frequency. The displacement potentials satisfy the following equations of motion

$$M(h_j)\phi_j = 0 \tag{2a}$$

$$M(k_j)\psi_j = 0 \tag{2b}$$

$$M(*) \equiv \Delta + (*)^2. \tag{2c}$$

In Eqs. (2a-c) Δ is a two-dimensional Laplacian, h and k are the wavenumbers associated with P and SV waves, respectively and M() is the Helmholtz operator. Since the medium is assumed to be weakly anelastic one distinguishes two types of body wave velocities[45,46]: i) the intrinsic elastic velocities α^* and β^* in the absence of dissipation, and ii) the velocities of the waves α and β in the presence of dissipation. The attenuation is introduced through the factors Q_α and Q_β which are taken to be frequency independent[22]. Stress-free boundary conditions along the surface of the half-space are assumed. Usual radiation conditions should be satisfied by the scattered wave field at infinity.

A perfect bonding between the layers requires the continuity of the displacement and the traction fields along the interfaces C_j, j = 1, 2, ..., R according to

$$\underline{u}_{j-1}^{+} = \underline{u}_j^{-};\ \mathbf{x} \in C_j \tag{3a}$$

$$\underline{T}_{j-1}^{\nu +} = \underline{T}_j^{\nu -}; \ \mathbf{x} \in C_j \tag{3b}$$

$j = 1, 2, \ldots, R,$

where $\underline{\nu}$ is an outward unit normal on C_j, $\mathbf{x}$ is a position vector, and the superscripts + and - denote that the interface C_j is being approached from the outside and the inside, respectively. The incident waves are assumed to be plane harmonic P, SV, or Rayleigh waves[22].

SOLUTION OF THE PROBLEM

The total displacement field in the half-space and the dipping layers is specified by

$$\underline{u}_0 = \underline{u}^{ff} + \underline{u}_0^{s}; \ \mathbf{x} \in D_0 \tag{4a}$$

$$\underline{u}_j = \underline{u}_j^{s}; \ \mathbf{x} \in D_j, \ j=1, 2, \ldots, R \tag{4b}$$

where the superscripts ff and s denote the free and the scattered wave field, respectively. As the incident wave strikes interface C_1 it is partially reflected back into the half-space and partially transmitted into the first dipping layer. The reflection and the transmission of the scattered wave field in the first layer takes place along the interfaces C_1 and C_2. Similar process can be observed for the subsequent layers as well. Therefore, one can view each interface C_j as location of the sources which generate the scattered wave field throughout the layered medium. By expressing the displacement potentials in terms of single layer potentials[40,47] it follows then that

$$\phi_0^{s} = \int_{C_1^-} q_0(\mathbf{y})\phi_0(\mathbf{x},\mathbf{y})d\mathbf{y}; \ \mathbf{x} \in D_0 \tag{5a}$$

$$\phi_j^{s} = \int_{C_j^+} p_j(\mathbf{y})\phi_j(\mathbf{x},\mathbf{y})d\mathbf{y}$$

$$+ \int_{C_{j+1}^-} q_j(\mathbf{y})\phi_j(\mathbf{x},\mathbf{y})d\mathbf{y} \tag{5b}$$

$\mathbf{x} \in D_j; \ j = 1, 2, \ldots, R-1$

$$\phi_R^{s} = \int_{C_R^+} p_R(\mathbf{y})\phi_R(\mathbf{x},\mathbf{y})d\mathbf{y}; \ \mathbf{x} \in D_R \tag{5c}$$

$$\psi_0^{s} = \int_{C_1^-} q_0^{*}(\mathbf{y})\psi_0(\mathbf{x},\mathbf{y})d\mathbf{y}; \ \mathbf{x} \in D_0 \tag{5d}$$

$$\psi_j^{s} = \int_{C_j^+} p_j^*(\mathbf{y})\psi_j(\mathbf{x},\mathbf{y})d\mathbf{y}$$

$$+ \int_{C_{j+1}^-} q_j^*\psi_j(\mathbf{x},\mathbf{y})d\mathbf{y} \qquad (5e)$$

$\mathbf{x}\in D_j; j=1,2,\ldots,R-1$

$$\psi_R^{s} = \int_{C_R^+} p_R^*(\mathbf{y})\psi_R(\mathbf{x},\mathbf{y})d\mathbf{y};\ \mathbf{x}\in D_R \qquad (5f)$$

where p_j, p_j^*, q_j^*, and q_j^* are the unknown density functions. The functions $\phi_j(\mathbf{x},\mathbf{y})$ and $\psi_j(\mathbf{x},\mathbf{y})$ are the Green functions which correspond to a dilatational and equivoluminal line source in the half-space, respectively. The explicit solution for the Green functions can be found in the paper by Lamb[48].

Introduction of the auxiliary surfaces C_j^{+-} in the integral representation for the scattered wave field is the fundamental characteristic of the method used in the present paper. The surfaces C_j^+ and C_j^- are defined outside and inside of the corresponding interface C_j, respectively[40,44]. This eliminates the singularities in the kernels of the integrals in Eqs. (6a-f) as $\mathbf{x}$ approaches $\mathbf{y}$. Though introduction of the auxiliary surface simplifies the numerical procedure considerably, the auxiliary surfaces must be chosen carefully in order to obtain accurate results[23]. More on this is discussed in the papers by Dravinski[14] and Dravinski and Mossessian[22,23].

If the scattered waves are expressed in terms of discrete line sources, this leads to the scattered wave field of the form[22]

$$\phi_0^{s} = a_m^{0}\ \phi_0(\mathbf{x},\mathbf{x}_m) \qquad (6a)$$

$$\psi_0^{s} = a_m^{*\,0}\psi_0(\mathbf{x},\mathbf{x}_m) \qquad (6b)$$

$\mathbf{x}\in D_0;\ \mathbf{x}_m\in C_1^-;\ m = 1,\ 2,\ \ldots,\ M_1$

$$\phi_i^{s} = b_n^{i}\phi_i(\mathbf{x},\mathbf{x}_n) + a_m^{i}\phi_i(\mathbf{x},\mathbf{x}_m) \qquad (6c)$$

$$\psi_i^{s} = b_n^{*\,i}\psi_i(\mathbf{x},\mathbf{x}_n) + a_m^{*\,i}\psi_i(\mathbf{x},\mathbf{x}_m) \qquad (6d)$$

$\mathbf{x}\in D_i;\ i=1,2,\ldots,R-1;\ n=1,2,\ldots,N_i;\ m=1,2,\ldots,M_i;$

$\mathbf{x}_n\in C_i^+;\ \mathbf{x}_m\in C_{i+1}^-$

$$\phi_R^{s} = b_n{}^{R}\phi_R(\mathbf{x},\mathbf{x}_n) \tag{6e}$$

$$\psi_R^{s} = b_n^{*}{}^{R}\psi_R(\mathbf{x},\mathbf{x}_n) \tag{6f}$$

$$\mathbf{x}\in D_R;\ n=1,2,\ldots,N_R;\ \mathbf{x}_n\in C_R^{+}.$$

The unknown source intensities a_m^{j-1}, $a_m^{*\,j-1}$, b_n^{j}, and $b_n^{*\,j}$ are determined through the continuity conditions specified by Eqs. (3a-b). By choosing, say, L_j collocation points along each interface C_j, $j = 1, 2, \ldots, R$, to impose the continuity conditions, the source intensities are determined in the least-square-sense[22].

NUMERICAL RESULTS

For the sake of illustration of the method two numerical examples are considered both involving an incident plane harmonic Rayleigh wave. The amplitude of the incident wave is assumed to be the same as in the paper by Dravinski and Mossessian[22]. First problem involves three dipping layers of semi-circular shape. Second problem deals with a cross section of the Los Angeles sedimentary basin. For details on the testing the accuracy of the method reader is referred to the paper by the authors[22].

At this point it is convenient to introduce dimensionless variables for the parameters of the problems. The intrinsic shear wave velocity and the shear modulus of the half-space are chosen to be equal unity and the Poisson ratio for all the materials is assumed to be 1/3. Dimensionless frequency, Ω, is defined as a ratio of the maximum width ($2W_x$, see Fig. 1) of the subsurface irregularity and the wavelength of shear waves in the half-space.

Surface displacement amplitudes for three semicircular dipping layers and incident Rayleigh wave are depicted by Fig. 2. It is apparent from this result that presence of subsurface irregularity may cause locally very large amplification (reduction) of the surface ground motion. The amplification may change dramatically within a very short distance atop the surface of the half-space. Results corresponding to various incident P and SV waves lead to similar conclusions[22]. These results can be summarized as follows: The amplification of the surface ground motion depends strongly upon the nature and frequency of the incident wave, type of displacement component being observed, the impedance contrast between the layers, and the location of the observation station at the surface of the half space. It can be shown[22] that the geometry of the sediments and the maximum sediment depth plays very important role in amplification of the strong ground motion.

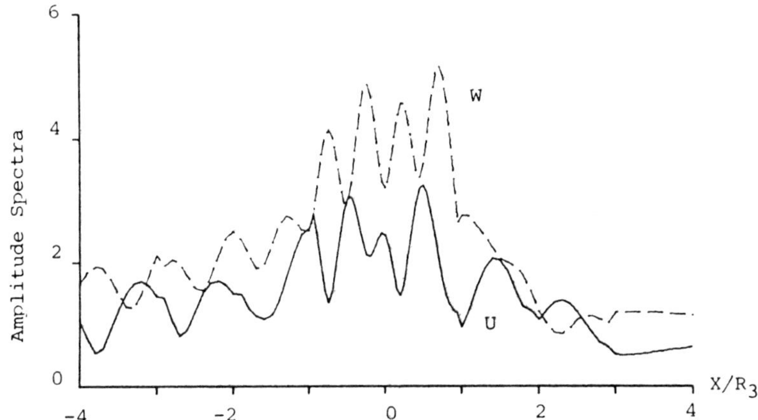

Figure 2. Amplitude for horizontal and vertical surface displacement spectra for three semi-circular dipping layers and incident Rayleigh wave. $\Omega=3$, $\mu_1^*=0.6$, $\beta_1^*=0.8$, $\alpha_1^*=1.6$, $\mu_2^*=0.4$, $\beta_2^*=0.7$, $\alpha_2^*=1.4$, $\mu_3^*=0.17$, $\beta_3^*=0.5$, $R_1=3$, $R_2=2$, $R_3 1$.

Strong ground motion of the Los Angeles basin is evaluated for a single section (EFG section according to Yerkes et al.[43]). For simplicity, all the distances are normalized with respect to a characteristic length, which is chosen to be 8 km (approximately the maximum depth of the basin). The total width of this section of the basin is 70 km (or 4.75 units) and the maximum thickness of the deposits is 8.5 km (or 1.06 units). The shape ratio of the cross-section is 0.24. The deposit is assumed to consists of a single layer with the impedance contrast between the bedrock and the deposits being 6.67. For incident Rayleigh wave the surface displacement amplitudes are depicted by Fig. 3. As before, presence of the sedimentary deposits produced locally very large amplification of the surface ground motion.

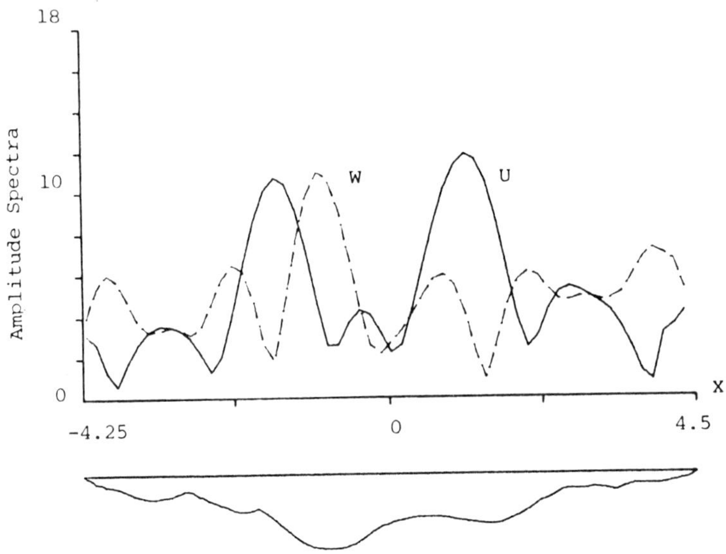

Figure 3. Surface displacement amplitude spectra for a cross-section of the Los Angeles basin and incident Rayleigh waves. $\Omega=0.91$, $\mu_1^*=0.045$, $\beta_1^*=0.3$, $\alpha_1^*=0.6$.

SUMMARY AND CONCLUSIONS

Plane strain model for amplification of strong ground motion by multiple dipping layers of arbitrary shape is considered by using an indirect boundary integral equation approach. Presented results indicate that the sedimentary basins may undergo severe amplification of the strong ground motion in comparison to the free field. The amplitude of the surface motion will in general depend strongly upon frequency and nature of the incident wave, location of the observation station at the surface, type of displacement component being observed, geometry of the layers, and the impedance contrast of the materials involved.

REFERENCES

1. Kobayashi H. Seo K. Midorikawa S. and Yamazaki Y. (1986). Measurement of Microtremors in and around Mexico D.F., Report on Seismic Microzoning studies of the Mexico City

Earthquake of September 19,1985, Parts 1 and 2, Tokyo Institute of Technology, Nagoya, Japan.

2. Anderson J. G. Bodin P. Brune J. N. Prince J. Singh S. K. Quaas R. Onate M. (1986), Strong Ground Motion from the Michoacan, Mexico, Earthquake, Science, Vol. 233, pp. 1043-1049.

3. Zeevaert L. (1964), Strong Ground Motion Recorded During Earthquakes of May the 11th and 19th, 1962 in Mexico City, Bulletin of the Seismological Society of America, Vol. 54, pp. 209-231.

4. Jennings P. C. (Ed.) (1971). San Fernando Earthquake of February 9, 1971, EERL-71-02, California Institute of Technology, Pasadena, California.

5. Repetto P. Arango I. and Seed H. B. (1980). Influence of Site Characteristics on Building Damage During the October 3, 1974, Lima earthquake, Report No. UCB/EERC 80/41, University of California, Berkeley, California.

6. Ohta Y. Kagami H. Goto N. and Kudo K. (1978). Observation of 1- to 5-second Microtremors and their Application to Earthquake Engineering. Part I: Comparison with Long-period Accelerations at the Tokachi-oki Earthquake of 1968, Bulletin of the Seismological Society of America, Vol. 68, pp. 767-779.

7. Kagami H. Duke C. M. Liang G. C. and Ohta Y. (1982). Observation of 1- to 5-second Microtremors and their Application to Earthquake Engineering. Part II. Evaluation of Site Effect upon Seismic Wave Amplification due to Extremely Deep Soil Deposits, Bulletin of the Seismological Society of America, Vol. 72, pp. 987-998.

8. Kagami H. Okada S. Shiono K. Oner M. Dravinski M. and Mal A. K. (1986). Observation of 1- to 5-second Microtremors and their Application to Earthquake Engineering. Part III. A Two Dimensional Study of Site Effect in the San Fernando Valley, Bulletin of the Seismological Society of America, Vol. 76, pp. 1801-1812.

9. King J. L. and Tucker B. E. (1984), Observed Variations of Earthquake Motion Across a Sediment-filled Valley, Bulletin of the Seismological Society of America, Vol. 74, pp. 137-151.

10. Tucker B. E. and King J. L. (1984). Dependence of Sediment-filled Valley Response on the Input Amplitude and the Valley Properties, Bulletin of the Seismological Society of America, Vol. 74, pp. 153-165.

11. Bouchon M. and Aki K. (1977), Near Field of a Seismic Source in a Layered Medium With Irregular Interfaces, Geophysical Journal of Royal Asreonomical Society, Vol. 50, pp. 669-684.

12. Bard P. -Y. and Bouchon M. (1980), The Seismic Response of Sediment-filled Valleys. Part I. The Case of Incident SH Waves, Bulletin of the Seismological Society of America, Vol. 70, pp. 1263-1286.

13. Bard P. -Y. and Bouchon M. (1980), The Seismic Response of Sediment-filled valleys. Part II. The Case of Incident P and SV Waves, Bulletin of the Seismological Society of America, Vol. 70, pp. 1921-1941.

14. Dravinski M. (1982), Influence of Interface Depth Upon Strong Ground Motion, Bulletin of the Seismological Society of America, Vol. 72, pp. 596-614.

15. Dravinski M. (1982), Scattering of SH Waves by Subsurface Topography, Journal of Engineering Mechanics Division, ASCE, Vol. 108, No. EM1, pp. 1-16.

16. Dravinski M. (1982), Scattering of Elastic Waves by an Alluvial Valley, Journal of Engineering Mechanics Division, ASCE, Vol. 108, No. EM1, pp. 19-31.

17. Dravinski M. (1983), Scattering of Plane Harmonic SH Waves by Dipping Layers of Arbitrary Shape, Bulletin of the Seismological Society of America, Vol. 73, pp. 1303-1319.

18. Bard P. -Y. and Bouchon M. (1985), The Two-dimensional Resonance of Sediment-filled Valleys, Bulletin of the Seismological Society of America, Vol. 75, pp. 519-541.

19. Aki K. and Larner K. L. (1970), Surface Motion of a Layered Medium Having an Irregular Interface due to Incident Plane SH Waves, Journal of Geophysical Research, Vol. 75, pp. 933-954.

20. Bouchon M. (1973), Effect of Topography on Surface Motion, Bulletin of the Seismological Society of America, Vol. 63, pp. 615-632.

21. Sanchez-Sesma F. J. and Rosenblueth E. (1979), Ground Motion at Canyons of Arbitrary Shapes Under Incident SH-Waves, Earthquake Engineering and Strustural Dynamics, Vol. 7, p 441.

22. Dravinski M. and Mossessian T. K. (1987), Scattering of Harmonic P, SV, and Rayleigh Waves by Dipping Layers of Arbitrary Shape, to appear in Bulletin of the Seismological Society.

23. Dravinski M. and Mossessian T. K. (1987), Amplification of Surface Ground Motion by an Inclusion of Arbitrary Shape, Numerical Methods for Partial Differential Equations, Vol. 3, pp. 9-25

24. Wong H. L. (1982), Diffraction of P, SV, and Rayleigh Waves by Surface Topographies, Bulletin of the Seismological Society of America, Vol. 72, pp. 1167-1184.

25. Cruse T. A. (1968), A Direct Formulation and Numerical Solution of the General Transient Elastodynamic Problem. II., Journal of Mathematical Analysis and Application, Vol. 22, pp. 341-355.

26. Cole D. M. Kosloff D. D. and Minster J. B. (1978). A Numerical Boundary Integral Equation Method for Elastodynamics, Bulletin of the Seismological Society of America, Vol. 68, pp. 1331-1357.

27. Kobayashi S. (1983) Some Problems of the Boundary Integral Equation Method in Elastodynamics, in Boundary Elements,(Ed. Brebbia C. A., Futagami T and Tanaka M.), pp. 775-784, Proceedings of the Fifth International Conference on Boundary Elements, Hiroshima, Japan, Springer-Verlag, New York.

28. Mossessian T. K. and Dravinski M. (1987), Application ofa Hybrid Method for Scattering of P, SV and Rayleigh Waves by Near Surface Irregularities, to appear in Bulletin of the Seismological Society of America.

29. Chopra A. K., Chakrabarti P. and Dasgupta G. (1976). Dynamic Stiffness Matrices for Viscoelastic Half-plane Foundations, Journal of Engineering Mechanics Division, ASCE, Vol. 102,

No. EM3, pp. 497-514.

30. Dasgupta G. and Chopra A. K. (1979), Dynamic Stiffness Matrices for Viscoelastic Half Planes, Journal of Engineering Mechanics Division, ASCE, Vol. 105, No. EM5, PP. 729-745.

31. Datta S. K. and Shah A. H. (1982), Scattering of SH Waves by Embedded Cavities, Wave Motion, Vol. 4, pp. 265-283.

32. Shah A. H., Wong K. C. and Datta S. K. (1982), Diffraction of Plane SH Waves in a Half-space, Earthequake Engineering and Structural Dynamics, Vol. 10, pp. 519-528.

33. Mita A. and Takanashi W. (1983) Dynamic Soil-structure Interaction Analysis by Hybrid Method, in Boundary Elements, (Ed. Brebbia C. A., Futagami T. and Tanaka M.), pp. 785-794, Proceedings of the Fifth International Conference on Boundary Elements, Hiroshima, Japan, Springer-Verlag, New York.

34. Wong K. C., Shah A. H. and Datta S. K. (1985), Diffraction of Elastic Waves in a Half Space. II. Analytical and Numerical Solutions, Bulletin of the Seismological Society of America. Vol. 75, pp. 69-91.

35. Alyagshi Eilouch M. N. and Sandhu R. S. (1986), A Mixed Method for Transient Analysis of Soil-structure Interaction Under SH Motion, Earthquake Engineering and Structural Dynamics, Vol. 14, pp. 499-516.

36. Lee V. W. (1978) Displacement Near a Three-dimensional Hemispherical Canyon Subjected to Incident Plane Waves, Iniversity of Southern California, Department of Civil Engineering, Report No. 78-16.

37. Lee V. W. (1984), Three-dimensional Diffraction of Plane P, SV and SH Waves by a Hemispherical Alluvial Valley, International Journal of Soil Dynamics and Earthquake Engineering, Vol. 3, No. 3, pp. 133-144.

38. Day S. (1977). Finite Element Analysis of Seismic Scattering Problem, Ph. D. Thesis, University of California at San Diego, California.

39. Sanchez-Sesma F. J. (1983), Diffraction of Elastic Waves by Three-dimensional Surface

Irregularities, Bulletin of the Seismological Society of America, Vol. 73, pp. 1621-1636.

40. Herrera I. (1985). Boundary Methods: An Algebraic Theory, Pitman Publishing, Inc., Boston.

41. Lee J. -J. and Langston C. A. (1983), Wave Propagation in a Three Dimensional Circular Basin, Bulletin of the Seismological Society of America, Vol. 73, pp. 1637-1653.

42. Avanessian V., Muki R. and Dong S. B. (1986), Axisymmetric Soil-structure Interaction by Global-local Finite Elements, Earthquake Engineering and Structural Dynamics, Vol. 14, pp. 355-367.

43. Yerkes, R. F., McCulloh, Schoellhamer J. E. and Veddes J. G. (1965), Geology of the Los Angeles Basin, California - An Introduction, Geological Survey Professional Paper, Vol. 420-A.

44. Kupradze V. D. (1963), Dynamical Problems in Elasticity, Progress in Solid Mechanics, Vol. 3, (Ed. Sneddon I. N. and R. Hill R.), North Holland, Amsterdam.

45. Aki K. and Richards P. G. (1980). Quantitative Seismology, Vol. 1, W. H. Freeman & Co., San Francisco.

46. Ben-Menahem A. and Singh S. J. (1981). Seismic Waves and Sources, Springer Verlag, New York.

47. Ursell F. (1973), On the Exterior Problems of Acoustics, Proceedings of Philosophical Society, Vol. 74, pp. 117-125.

48. Lamb H. (1904), On Propagation of Tremors Over the Surface of an Elastic Solid, Philosopical Transactions of the Royal Society of London, Ser. A359, Vol. 203, pp. 1-42.

Elastic Waves in Locally Inhomogeneous Layered Media

A.K. Mal, C.-C. Yin
Department of Mechanical Aerospace and Nuclear Engineering, University of California, Los Angeles, California 90024, U.S.A.

INTRODUCTION

The prediction of certain properties (e.g., amplitude, duration and frequency content) of the expected earthquake ground motion at a given site in a seismically active region is of great importance in developing earthquake resistant design codes and criteria for the region. These characteristics are known to be influenced by a variety of factors which include the rupture process at the source, the distance and geologic properties of the earth material between the source and the site and the mechanical properties of the soil in the vicinity of the site. A good understanding of the relative influence of these factors on the relevant characteristics of the earthquake ground motion is a prerequisite to a sound prediction strategy. Recent research on the physics of the earthquake phenomenon has provided a qualitative understanding of this influence.

In the immediate vicinity of the source, the ground motion is usually the strongest and rich in frequency content. The detailed rupture process associated with the source is most likely to exert the strongest influence on the characteristics of the motion in this so called "near field." However, the higher frequency components of the motion may still be significantly affected by the geometry and material properties of the local soil deposit provided the wavelengths of the associated waves are short compared to their propagation distance but comparable to the linear dimensions of the deposit.

In the intermediate distance range the ground motion is usually weaker than that in the near field, but is of longer duration. Furthermore, a significant amount of relatively high frequency energy still remains in the radiation and these can be strongly affected by local soil properties, resulting in amplification of the surface motion at certain frequencies. Thus the local soil conditions may have a strong influence on the relevant characteristics of ground motion in the entire range of distances where the shaking is relatively intense and where its frequency content covers the resonant frequency range of most engineered structures.

While the qualitative observations above underscore the importance of the influence of local soil conditions on the ground motion characteristics, the inclusion of this effect in engineering practice requires a more quantitative knowledge.

Analysis of historical earthquake records obtained on bedrock and deposit sites has been carried out by a number of researchers in an effort to discover the relationship between the dominant frequency and amplitude of the motion and the local properties of the site. The results appear to indicate that certain frequencies of ground motion are amplified considerably by thin low velocity surface sediments and that the overall spectral level of the motion increases with decreasing shear wave speed of the near surface material and with increasing sediment thickness at the site[1-4]. Furthermore these amplification effects appear to be independent of the properties of the source or the source site distance in the frequency range of engineering interest[5]. Empirical studies of distant and low level records of nuclear explosions[1] and microtremors[6] appear to be consistent with these observations. These are again somewhat qualitative results, due to the lack of a statistically significant data base for strong motion records and to the difficulty in quantitatively extrapolating results based on the study of motion generated by nuclear and microseismic sources which are clearly not representative of earthquake sources, especially in the distance and frequency ranges of engineering interest.

Theoretical modeling can be extremely helpful in removing some of the uncertainties associated with the empirical results. Theoretical models considered to date include one dimensional treatment of vertically propagating shear waves through soil layers with linear or nonlinear material properties[1] , finite difference calculations[9], exact wave theoretical calculations for simple geometries[10], approximate wave theoretical treatment of arbitrary two dimensional models[11] and calculations based on the boundary integral equation method[12-15].

In order to be applicable to real earthquakes, the theoretical model must simulate the wave phenomena that occur in an earthquake and this has proved to be extremely difficult to achieve. In addition, the lack of detailed subsurface data in most locations has limited the general applicability of the theoretical models. Finally, for reasonably realistic models, the computational tasks associated with the theoretical approach have proven to be enormous.

On the other hand, a great deal of progress has recently been made in developing wave propagation models of the earthquake phenomenon. The above mentioned studies and others have shown that through careful modeling it is possible to explain the major distinctive features of the strong ground motion recorded on deposit sites. It can be argued that the local soil conditions are likely to be the best known quantities of all the factors that are relevant in molding the characteristics of the ground motion at a chosen site, since these are usually either known or determined, at least to some depth and lateral extent in a major construction project. The availability of low cost, moderately powerful computers is also an encouraging factor and this situation is likely to improve in the near future. Finally, a quantitative evaluation technique, developed on the basis of sound theoretical principles may be of limited direct applications, but it can be of great value in extending results of the empirical studies to the quantitative characterization of ground motion on deposit sites in future earthquakes.

In what follows we briefly describe two recently developed theoretical techniques which appear to be applicable to very realistic models of this and other similar problems. Their computational aspects are discussed by means of a simple illustrative example.

SEMI-NUMERICAL METHODS

Let the sedimentary deposit occupy a volume V_S bounded by the surface S and be embedded in an earth model which consists of homogeneous plane parallel layers of rock overlying a uniform half space (Fig. 1). The soil may be inhomogeneous and the material properties everywhere are assumed to be linear isotropic and viscoelastic with frequency dependent dissipative properties. The seismic disturbances are assumed to be generated by sources external to the deposit.

Let $\underline{u}(\underline{x},\omega)$ denote the fourier time transform of the displacement vector at a point $\underline{x}$ within the solid, where the transform pair is defined through the equations

$$F(\omega) = \int_0^{\infty} f(t)\, e^{i\omega t}\, dt$$

$$f(t) = \frac{1}{2\pi} \int_{-\infty}^{\infty} F(\omega)\, e^{-i\omega t} d\omega \qquad (1)$$

Then $\underline{u}(\underline{x},\omega)$ is in general, a complex, frequency dependent function and is the solution of a time harmonic boundary value problem in which all field variables have the time dependence $\exp(-i\omega t)$. It should be noted that observed dissipative properties of the soil can be represented in the frequency domain

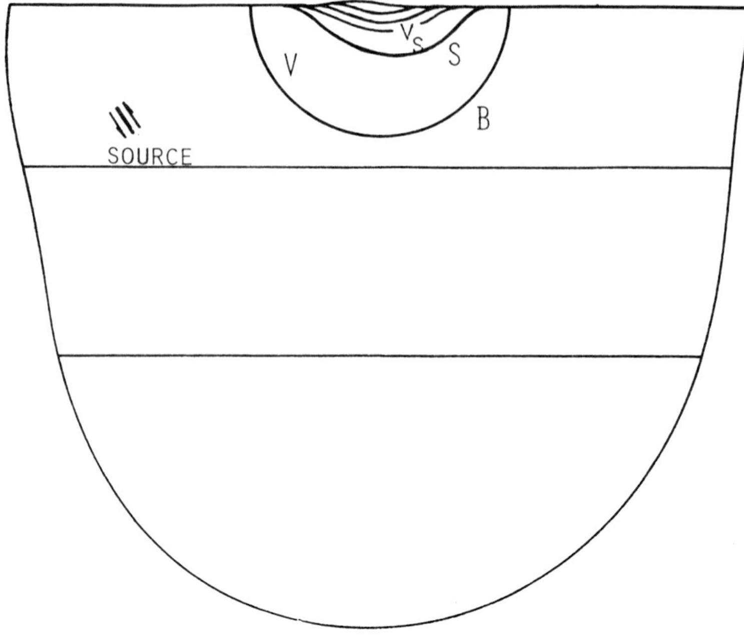

Figure 1. Geometry of the locally inhomogeneous layered solid.

through the use of complex, frequency dependent elastic constants. Time domain solution can be easily obtained by means of the highly efficient FFT algorithm. We shall consider time harmonic problems only, and omit the common time dependence as well as the frequency dependence in all field variables.

Let u_i denote the ith component of $\underline{u}$ and τ_{ij} the components of the stress tensor at $\underline{x}$ derived from $\underline{u}$ through the equation

$$\tau_{ij} = c_{ijkl}u_{k,l} \tag{2}$$

where c_{ijkl} is the elastic tensor which may in general, be a complex function of position and frequency. In Eq. (2) and all subsequent ones the comma notation and the summation convention are used. The displacement vector $\underline{u}$ satisfies the Navier's equation and is continuous everywhere except at the source, where it is discontinuous across the fault surface with specified form of the slip. The stress vector is continuous across all surface elements including those where a sharp transition in material properties may occur. In addition, the field must represent outgoing waves or satisfy a radiation condition at infinity.

The stated equations and conditions define a well posed boundary value problem of elastodynamics whose solution can in principle be obtained. In practice this is extremely difficult to accomplish due to the presence of the complex geometry and material properties of the deposit for realistic models of the earth. Clearly it is necessary to introduce some sort of discretization procedure and associated numerical schemes to deal with the inhomogeneous soil deposit.

Although the finite element method provides a convenient and powerful means in representing the deposit region, its conventional implementation presents difficulty in modeling the radiated field external to the discretized region. In order to examine the nature of this difficulty consider the finite element discretization of the volume V bounded by a surface B such that the deposit region V_s is contained within B (Fig.1). It can be shown that the displacement vector $\underline{u}$ minimizes the functional Π given by

$$\Pi = \int_V \{(\rho\omega^2 u_i u_i^* - \tau_{ij} u_{i,j}^*)/2 + f_i u_i^*\} dV + \int_B \tau_{ij} u_i^* n_j dS \qquad (3)$$

where ρ is the density of the material, a superstar implies complex conjugation, f_i is the ith component of the body force and n_j is the jth component of the unit normal vector to B.

In the conventional finite element technique, either the displacement u_i or the stress vector $\tau_{ij} n_j$ is prescribed on B. Application of the variational theorem results in a system of linear equations for the unknown nodal displacements within V and on B.

In the present problem neither the displacements, nor the stress vectors are known on the boundary B due to the presence of the unknown scattered field in the region external to the deposit. Thus the standard finite element formulation cannot be directly implemented in this case. An alternative would be to employ a time domain formulation in a finite region since in most practical applications the interest is in a limited time window. However, this could eliminate significant motions which may occur due to reflections from interfaces below the modeled region. Furthermore, the observed dissipative properties of most soils appear to have a very complex frequency dependence which are difficult, if not impossible to express in the time domain description of their constitutive equations, but are easily expressed in the frequency domain.

In recent years, alternative approaches have been proposed by several researchers[16-18] which have essentially removed the above mentioned difficulties and have made the finite element technique applicable to a large class of such problems.

The essential feature of this so called hybrid method is that the unknown terms in the boundary conditions on B are replaced by quantities which depend on the known "incident" field through the use of a representation of the scattered field in the form

$$u_i(\underline{x}) = u_i^o(\underline{x}) + \sum_j a_{ij} f_j(\underline{x}) \tag{4}$$

where $\underline{x}$ is a point outside V, $u_m^o(\underline{x})$ is the displacement that would be produced by the source in the layered solid in absence of the irregular deposit and a_{ij} are unknown coefficients. The functions f_j are members of a complete set of solutions of the Navier's equation satisfying the boundary and regularity conditions in the plane layered half space, in absence of the deposit.

Equation (4) is used to express the unknown nodal displacements and forces on B in terms of those due to the known incident field and the new unknowns a_{ij}, since they are continuous across B. Minimization of the discretized form of Π with respect to the unknowns yields a system of linear equations. Solution of these equations gives the unknown nodal displacements within V_s as well as the coefficients a_{ij}.

It should be noted that a deterministic solution of the equations requires that the number of these constants must be related to the number of elements on the boundary B. If a larger number of terms are retained in Eq. (4), then the solution must be obtained in a least square or other statistical sense. Furthermore, the success of the method depends critically on the ability to construct the set of functions $f_j(\underline{x})$. This has so far been accomplished for homogeneous, perfectly elastic, infinite[16] or semi-infinite[17,18] models of the external medium by means of eigenfunction and multipolar representation of the solution of Navier's equation. In order to apply the technique to more realistic problems, it will be necessary to develop methods for the construction of these functions for layered, visco-elastic media with empirically determined frequency dependent dissipative properties.

A second, more recently developed technique will be described next. This technique is based on an integral representation of the scattered field originally derived by Mal and Knopoff[19] in the form

$$u_m(\underline{x}) = u_m^o(\underline{x}) + \int_{V_s} \{\delta\rho\omega^2 G_{mi}(\underline{x},\underline{\xi})u_i(\underline{\xi}) - \delta c_{ijkl} G_{mi,j} u_{k,l}(\underline{\xi})\} d\underline{\xi} \quad (5)$$

where $\underline{x}$ can be either inside or outside of V_s, $\delta\rho$ and δc_{ijkl} are the perturbations (not necessarily small) in the density and the elastic tensor within the deposit region from the plane layered solid and the derivatives in the integrand are with respect to the integration variable $\underline{\xi}$. The displacement u_m^o has the same meaning as in Eq. (4) and $G_{mi}(\underline{x}, \underline{\xi})$ is the Green's function for layered solid, i.e., the mth component of the displacement at $\underline{x}$ due to a unit time harmonic point force at $\underline{\xi}$ in the ith direction.

For $\underline{x}$ in V_s, Eq. (5) is an integro-differential equation for the displacement field produced in V_s and can in principle be solved. For an inhomogeneous deposit with non-simple boundary, this can only be accomplished by numerical methods. Since the unknown functions in the equations are the displacements and strains within the deposit, it is convenient to discretize this region by means of finite elements. This results in a system of linear equations for the unknown nodal displacements within V_s. The elements of the coefficient matrix consist of integrals over the individual elements of certain products of the Green's function or its derivatives with the assumed shape functions. It should be noted that in two and three dimensional problems the diagonal and near diagonal elements of the matrix contain (integrable) singularities and special care must be taken in their evaluation, as in the case of the boundary integral equation method[14]. In the next section we discuss the solution of a class of simple one dimensional problems.

A SIMPLE EXAMPLE

We consider the reflection and transmission of elastic SH type waves across an inhomogeneous layer of thickness 2a occupying the region $-a < x < a$ bonded to an infinite homogeneous medium. The material properties in the external medium are described by its constant shear modulus μ_o, wave speed, β_o and density ρ_o while those in the layer by the variables $\mu(x)$, $\beta(x)$ and $\rho(x)$.

The Green's function for the infinite medium is given by

$$G(x,\xi) = -\exp(ik_o|x-\xi|)/(2ik_o\mu_o) \quad (6)$$

where $k_o = \omega/\beta_o$ is the wavenumber. The integral equation (5) becomes

$$w(x) = w_o(x) + \int_{-a}^{a} [\delta\rho(\xi)\omega^2 w(\xi)G(x,\xi) - \delta\mu(\xi)w,_\xi G,_\xi]d\xi \quad (7)$$

where $-a < x < a$ and

$$w_0 = \exp(ik_0x) \tag{8}$$

Before discussing the numerical solution of the above equation in the general case we consider a special case in which the material within the region $-a < x < a$ is also homogeneous, i.e., β, μ are constants. The solution of this problem is given by

$$\begin{aligned} w(x) &= w_0 + R \exp(-ik_0x), & x < -a \\ &= A \exp(ikx) + B \exp(-ikx), & -a < x < a \\ &= T \exp(ik_0x), & x > a \end{aligned} \tag{9}$$

where the constants can be easily calculated from the interface continuity conditions.

Note that the constants A and B can also be determined by substituting the expression for $w(x)$ in the second of Eq.(9) in the integro-differential equation (7). The constants R and T can then be calculated from (7) for x outside the layer. In other words the solution of the scattering problem can be obtained entirely from the integro-differential equation (7). For arbitrary properties of the layer, a numerical scheme must be employed. We used a standard finite element discretization of the region $-a < x < a$ by means of linear strain elements. This resulted in a system of linear equations for the nodal displacements which were solved for a range of input frequencies and material properties of the inhomogeneous layer. The numerical solution was compared with the exact analytical solution for several models of the layer material with piecewise constant properties. An example of the degree of agreement between the two is given in Table 1 in the case of a homogeneous layer. In the numerical solution using linear strain elements, sufficient accuracy was achieved with about 10 elements per wavelength, as can be expected.

The same general procedure can be applied to solve similar problems for the semi-infinite solid $x > 0$ containing an inhomogeneous layer (Fig 2). Let the region $0 < x < h$ be inhomogeneous with arbitrary variation in its properties while the remainder in $x > h$ be homogeneous. As before, the medium is assumed to be excited by a wave of unit amplitude incident from $x = -\infty$. The objective is to calculate the resulting wave field within and outside the inhomogeneous zone. The analytical solution of this problem can again be obtained for piecewise constant properties of the layer.

Frequency (Hz)	x	Re W	
		Exact	Numerical
0.5	-1.0	-1.0	-0.99998
	-0.5	-0.70711	-0.70740
	0.0	0.0	0.00006
	0.5	0.70711	0.70748
	1.0	1.0	0.99998
2.0	-1.0	1.0	0.99976
	-0.5	-1.0	-0.99976
	0.0	1.0	0.99977
	0.5	-1.0	-0.99976
	1.0	1.0	0.99976

Table 1: Comparison of exact and numerical solutions for the case of a homogeneous layer bonded to a homogeneous infinite medium. the material properties are $\beta_o = \mu_o = 1$, $\beta = 2$, $\mu = 8$, $a = 1$.

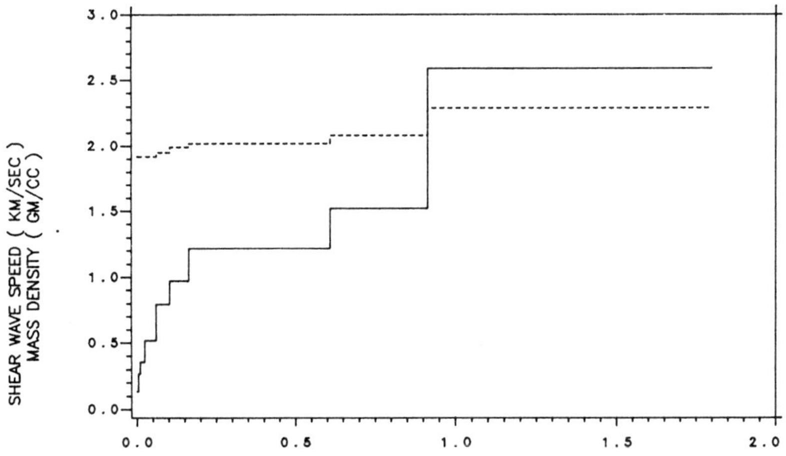

Figure 2: Material properties of the discretized medium. Solid line represents shear wave speed and the dashed line the density of the material. The properties are typical of deposit sites.

In order to apply the seminumerical methods it is necessary to construct a discretized model of the inhomogeneous region. This is shown in Fig. 2 for a specific case. It is of interest to note that the assumed velocity and density variations are representative of the near surface conditions in southern California and other regions, where a rapid increase in the shear wave velocity occurs within the top alluvial deposit. In the present calculations, the dissipative properties have been ignored. It should also be noted that the solution of this one dimensional discretized problem can be obtained by means of a Thomson Haskell type matrix procedure.

The Green's function for the half space is given by

$$G(x,\xi) = (i/2k_0\mu_0)[\exp(ik_0|x-\xi|) + \exp(ik_0(x+\xi))] \tag{10}$$

and the free term in the integral equation (5) is given by the sum of the incident and the free surface reflected waves, i.e.,

$$w_0(x) = \exp(-ik_0x) + \exp(ik_0x) \tag{11}$$

The numerical solution of Eq. (7) was obtained by means of the procedure indicated above. Representative results are presented in Figs. 3-6. It can be seen that the amplification of the surface motion is large and strongly dependent on the frequency. The time history shown in Fig. 6 was obtained by FFT inversion of the spectra for an incident pulse of unit amplitude and 1 sec. duration in the half space. The significant amplification and distortion of the pulse is caused by the inhomogeneous near surface material.

CONCLUDING REMARKS

The relative advantages of the methods should be clear from the above discussions. One advantage of the integral equation method is that the discretized region can be the deposit V_s itself, rather than the larger region V needed in the hybrid method. Also the number of unknowns in the integral equation method is always equal to the number of nodal displacements within the discretized region, as in the case of standard finite element method and the resulting system of linear equations is always determinate. In the hybrid method, the number of unknowns can be larger due the introduction of the coefficients a_{ij} and there is a certain amount of arbitrariness in selecting the functions f_i and their number. An advantage of the hybrid method is that once the coefficients a_{ij} have been determined, it is relatively easy to calculate the external scattered field from Eq. (4). In the integral equation method, it is necessary to re-evaluate the integral in Eq. (5) for x outside the region V_s, although asymptotic evaluation of

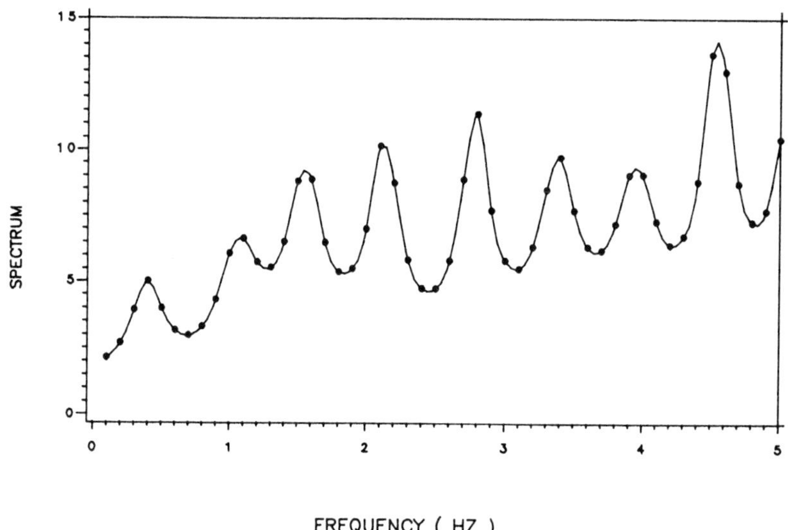

Figure 3. Numerical Solution for inhomogeneous half space shown in Fig. 2. Incident wave amplitude is unity.

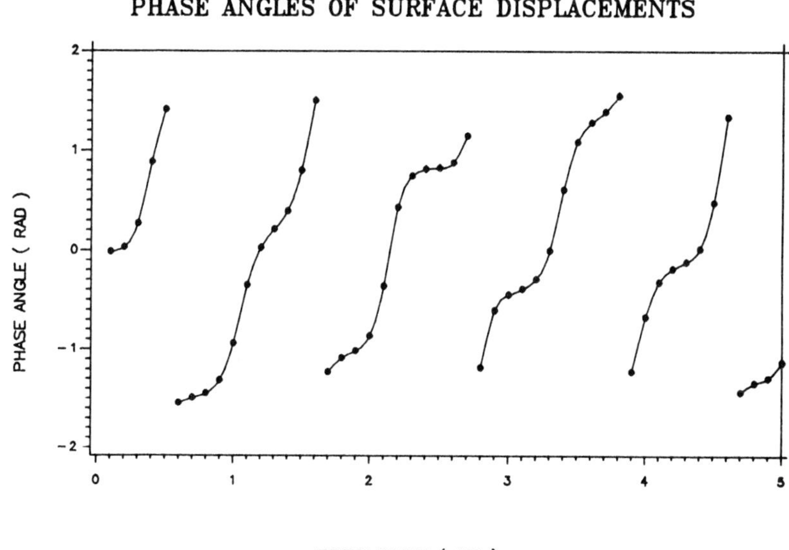

Figure 4. Same as Fig. 3.

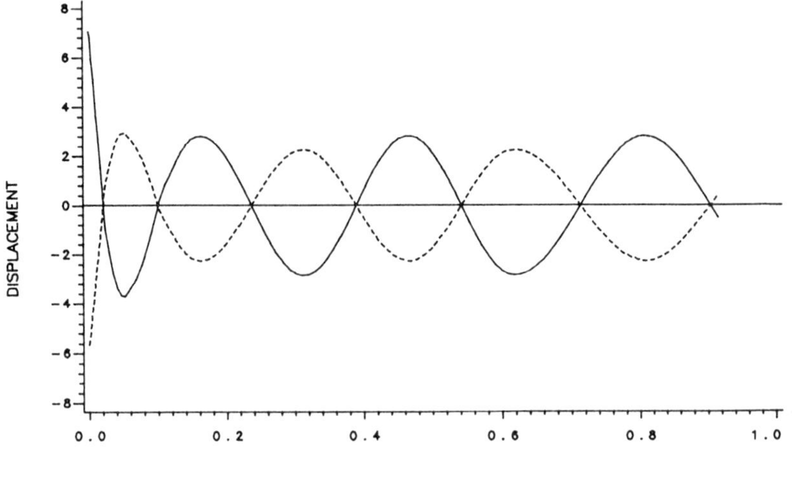

Figure 5. Same as in Fig. 3. Solid line is real part, dashed line the imaginary part of $\omega(o)$, at 4 Hz.

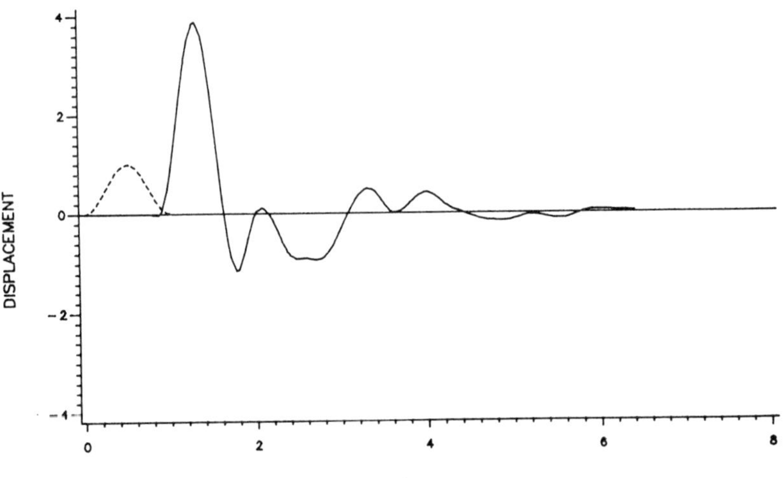

Figure 6. Solid curve represents the time history of surface displacement for the input shown in the dashed curve.

the integral can be obtained, if the interest is in the far field only. Both methods suffer from the fact that the coefficient matrix of the linear equations is non-Hermitian, at least for polynomial type shape functions.

Clearly, the success of these methods depends critically on the capability to construct the functions f_i in the hybrid formulation and the Green's functions in the integral equation formulation. They are given by simple, closed form expressions for homogeneous isotropic and unbounded (elastic or viscoelastic) solids. In the case of the semiinfinite or layered media, they can only be expressed in terms of certain integrals which must be evaluated numerically. In the integral equation approach, the discretization of Eq. (5) will require the evaluation of the Green's functions for a large number of such pairs and for a suite of frequencies. Thus it is necessary to have the means to perform these calculations in the most efficient manner. A computer code has recently been developed to accomplish this task; the details of this can be found in Xu and Mal[20]. It is hoped that the method can be used to provide solution for realistic two and three dimensional models in the near future.

REFERENCES

1. Rogers A.M., Tinsley J.C.and Borcherdt R.D. (1985), Predicting relative ground response, Evaluating Earthquake Hazards in the Los Angeles Region - an Earth-Science Perspective, U.S. Geological Survey Professional Paper 1360, pp. 221-248.

2. King J.L. and Tucker B.E. (1984), Observed variations of earthquake motion across a sediment filled valley, Bull. Seism. Soc. Am., Vol. 74, pp. 137-151.

3. Herrera I., Rosenbleuth E. and Rascon O. A. (1965) Earthquake Spectrum Prediction in the Valley of Mexico, Proc. 3WCEE, Vol.1, pp.161-164.

4. Munguia L. and Brune J.N. (1984), Local magnitude and sediment amplification observations from earthquakes in northern Baja-California-Southern California region, Bull. Seism. Soc. Am., Vol. 74, pp. 107-119.

5. Tucker B.E. and King J.L. (1984), Dependence of sediment-filled valley response on the input amplitude and the valley properties, Bull. Seism. Soc. Am., Vol. 74, pp. 153-165.

6. Kagami H., Okada S., Shiono S., Oner M., Dravinski M. and Mal A.K. (1986), Observation of 1 to 5 second microtremors and their application to earthquake engineering, part III. A two dimensional study of site effects in the San Fernando Valley, Bull. Seism. Soc. Am., Vol.76, pp.1801-1812.

7. Wong H.L. (1985), Effects of surface topography and site conditions, Strong Ground Motion Simulation and Earthquake Engineering Applications - A Technological Assessment, EERI Publication No. 85-02, Section 24, pp. 1-5.

8. Boore D.M. (1973), The effects of simple topography on seismic waves: implications for accelerations recorded at Pacoima Dam, San Fernando Valley, California, Bull. Seism. Soc. Am., Vol. 63, pp. 1608-1619.

9. Lavender A.R. and Hill N.R. (1985), P-SV resonances in irregular low velocity layers, Bull. Seism. Soc. Am., Vol.75 pp. 847-864.

10. Wong H.-L., Trifunac M.D. and B.D. Westermo (1977), Effects of surface or subsurface irregularities in the amplitude of monochromatic waves," Bull. Seism. Soc. Am., Vol. 67, pp. 353-368.

11. Bard P.Y. and Bouchon M. (1985), The two-dimensional resonance of sediment-filled valleys," Bull. Seism. Soc. Am. Vol. 75, pp. 519-541.

12. England R., Sabina F.J. and Herrera I. (1980), Scattering of SH waves by surface cavities of arbitrary shape using boundary methods, Phys. Earth. Planet. Int., Vol. 21, pp. 148-157.

13. Dravinski M. (1984), Evaluation of strong ground motion using boundary integral equation approach, Earthquake Source Modeling, Ground Motion and Structural Response, S.K. Datta (Ed.), ASME-AMD-Vol. 60, pp. 61-80.

14. Rizzo F.J., Shippy D.J. and Rezayat M. (1984), A Modern BIE Method for Elastic Wave Radiation and Scattering by Three Dimensional Objects, ibid, pp. 97-107.

15. Askar A., Cakmak, A.S. and Zheng X. (1984), Explicit Integration of the Boundary Integral Equation in the Frequency Domain for Scattering of Elastic Waves: Applications to Surface Topographies and Underground Structures, ibid, pp. 109-114.

16. Goetschel D.B., Dong S.B. and Muki R. (1982), A Global Local Finite Element Analysis of Axisymmetric Scattering of Elastic Waves, J. Appl. Mech., Vol. 49, pp. 816-820.

17. Wong K.C., Shah A.H., Datta S.K. and O'Leary P.M. (1984), Dynamic amplification of displacements and stresses around buried pipelines and tunnels, ibid, pp. 147-162.

18. Wong, K.C., Shah, A.H. and Datta S.K. (1985), Diffraction of elastic waves in a half space, Bull. Seism. Soc. Am., Vol. 75, pp. 69-92.

19. Mal A.K. and Knopoff L. (1967) Elastic wave velocities in two component systems, J. Inst. Math Applics., Vol. 3, pp. 376-387.

20. Xu P.-C. and Mal A.K. (1987), Calculation of the in-plane Green's Functions for a Multilayered Viscoelastic Solid, Bull. Seism. Soc. Am., (in press).

Seismic Wave Transmission Across Unbonded Frictional Interfaces

J.-M. Doong, D.A. Mendelsohn
Department of Engineering Mechanics, The Ohio State University, Columbus, Ohio 43210, U.S.A.

INTRODUCTION

It is recognized that the non-linear nature of dynamic soil-structure interaction requires further research. This includes the study of material non-linearities such as soil plasticity and coupled poro-elastic effects, and geometric non-linearities such as large deformations, and those due to changes in the contact geometry or stick-slip status of the soil-structure interface, Saxena[1]. The latter is the subject of the present research. Frictional stick-slip behavior between a structure and the rock or soil it rests in has been studied before, using models in which isolated support systems interact frictionally with the soil-foundation and specified ground motions act as input to the combined support-structure model, see e.g. Pan and Kelly[2], Kelly and Beucke[3] or Lee[4], or models in which a rigid mass-spring-dashpot system rests on a rigid foundation which drives the system, see e.g. Hundal[5], Crandall et al.[6], Westermo and Udwadia[7], or Mostaghel and Tanbakuchi[8]. These and similar studies do not treat the effect of the motion of the structure on the motion of the surrounding soil or foundation.

Representative examples of work which does treat this interaction, as well as interfacial separation and/or sliding include Wolf[9], Toki, et al.[10], Akiyoshi and Fuchida[11], Vaughan and Isenberg[12], Desai and Zaman[13], and Wolf[14]. All of these treat the soil as an elastic continuum and with the exception of Wolf[9], in which the structure is rigid, the structure is treated elastically too. Wolf[9], Toki, et al.[10], and Desai and Zaman[13] use the finite element method with joint elements representing the interface, Akiyoshi and Fuchida[11] present an analytical linearized solution, and Vaughan and Isenberg[12] use the finite difference method. The first use of the boundary element method appears in Wolf[14], in which an in-plane two-dimensional problem with separation, but no slip in the contact, is solved. The calculations in the present work and in all of these treatments of unbonded contact except Akiyoshi and Fuchida[11] are done in the time-domain requiring iteration to discern the contact and/or stick-slip boundaries. The present work extends the application of the boundary element method to treat the detailed stick-slip behavior in the contact for horizontally polarized (SH) motion. The time-domain formulation is based on Cole, et al.[15] and Mansur and Brebbia [16,17], while the treatment of the friction and stick and slip was developed by the authors.

While the method is formulated in general for an arbitrary structure interacting with the soil foundation, the technique is illustrated by studying in detail the response of a dissimilar layer or slab foundation with a free surface resting and pressed against a semi-infinite soil or rock mass. The contact is rough and Coulomb friction is assumed. The seismic disturbance is generated by a line-source treated as a body force. Several ideal time histories involving loading, unloading and then reloading of the interface are considered, and the regions of interfacial stick and slip are followed in time directly, and then used to calculate the surface motion of the layer. Studies are presented which illustrate the dependence of the layer motion on the ratio of foundation to layer wave speeds and on the coefficient of friction of the interface.

A TIME-DOMAIN BOUNDARY ELEMENT METHOD

We present here only the basic outline of the numerical method used. See Mendelsohn and Doong[18] for a precise description of the formulation. All direct boundary element techniques are based on a reciprocal theorem which relates, for a single body B with boundary ∂B, any two elastodynamic states (solutions of the equations of linear elastodynamics)[15]. Suppose one chooses as one state the unknown desired solution to, for example, a soil structure interaction problem, and as the other state the solution corresponding to a single concentrated source in an infinite medium, known as the Green's state. Then the reciprocal theorem may be simplified to yield a relation between the unknown displacement, a surface integral over ∂B involving the boundary tractions and displacements of both states, and a volume integral over B involving body forces and the Green's displacement. This relation is valid for any observation point in the body B. The choice of the infinite space Green's state is not restrictive and in fact is modified below to solve the layer-foundation problem.

Considering only SH motion, the surface and volume integrals become line and area integrals in 2D space, respectively. If the reciprocal relation is further specialized by placing the field or observation point on the boundary ∂B one obtains for a sufficiently smooth boundary, zero initial conditions, and a standard cartesian coordinate system (x_1, x_2)

$$\tfrac{1}{2}\, u(\underline{x},t) = \int_0^t \fint_{\partial B} [V(\underline{x},t;\underline{\xi},t_o)\tau(\underline{\xi},t_o)$$

$$- T(\underline{x},t;\underline{\xi},t_o)u(\underline{\xi},t_o)]ds(\underline{\xi})dt_o$$

$$+ \int_0^t \int_{\partial B} [V(\underline{x},t;\underline{\xi},t_o)f_B(\underline{\xi},t_o)]dA(\underline{\xi})dt_o, \qquad (1)$$

where V is the Green's displacement and T is the Green's boundary traction, and are both defined in the Appendix. Here, $\underline{\xi} = (\xi_1, \xi_2)$ is the source point in the Green's state, $\underline{x} = (x_1, x_2)$ is the observation or field point on ∂B, t and t_o are time, ds is a differential arc length, and dA is a differential area element. The unknown displacement and traction on the boundary are u and τ, respectively, and f_B is the body force distribution function. The notation on the line integral in Eq. (1) denotes that the integral is to be interpreted in the sense of a Cauchy principal value.

Next we introduce the set of nodes $\underline{r}_j$, $j = 1,2...,J$ distributed evenly on ∂B, and approximate the actual boundary with straight line elements, ΔB_j, centered at the nodes. We choose a constant time step Δt and define the incremental times $t_m = t_{m-1} + \Delta t$, $m = 1,2,...,M$. We also define the boundary nodal displacements and tractions at each time t_m, $u_j^m = u(\underline{r}_j, t_m)$ and $\tau_j^m = \tau(\underline{r}_j, t_m)$, respectively. For the purpose of integration, u and τ are assumed to be constant in space along each element and equal to the nodal value and u is assumed to be linear in time in each time step, while τ is taken constant in each time step. Then, taking $\underline{x} = \underline{r}_j$ and $t = t_m$, and approximating the boundary integral by the sum of generic integrals over the elements times the nodal values Eq. (1) may be written in the form

$$\tfrac{1}{2}\, u_j^m = \sum_{k=1}^{m} \sum_{\ell=1}^{J} [G_{j\ell}^k \tau_\ell^k + GN_{j\ell}^k u_\ell^k + GNP_{j\ell}^k u_\ell^{k-1}] + (\hat{f}_B)_j^m, \qquad (2)$$

where the integrals G, GN and GNP and the body force term f_B are given in the Appendix. Note that most of the terms of Eq. (2) involve known values of u_ℓ^k and τ_ℓ^k from previous time steps, $k < m$.

Next the boundary is divided into a region or regions on which displacements are prescribed as boundary conditions and a region or regions on which tractions are prescribed. Thus, along with the body force term, some of the current u_ℓ^m and τ_ℓ^m are known while others are unknown. Now separately at each time-step evaluate Eq. (2) at each j = 1,2...,J and separate all unknown terms from those that are known to obtain a matrix equation of the general form

$$[G]^D[\tau]^D + [GN]^T[u]^T = [f] \ , \tag{3}$$

where the vector [f] of length J contains all known quantities obtained from boundary conditions, the body force term, and the contributions from previous time steps. The matrix $[G]^D$ contains all J rows, but only those columns of the Green's displacement matrix which correspond to nodes at which displacements are prescribed and $[GN]^T$ contains all rows, but only those columns of the Green's traction matrix corresponding to nodes at which tractions are prescribed. The Green's displacement and traction matrices are found from Eqs. (A8,A9), respectively, by setting k = m. The vector $[u]^T$ contains the unknown displacements at the nodes at which tractions are prescribed and vice-versa for $[\tau]^D$. This linear system may be combined into a single square JxJ matrix system with $[u]^T$ and $[\tau]^D$ combined into a single vector of length J, which may be solved using standard Gaussian elimination. It should be noted that since V and T are non-zero only for field and source points sufficiently close to each other, by choosing the time step small enough, as suggested in Cole, et al.[15], the matrices $[G]^D$ and $[GN]^T$ may be made to be diagonal. This was done in the present analysis. Once all the boundary values are known the displacement anywhere in the body may be calculated from the original reciprocal relation before it was evaluated on the boundary to create the boundary integral equation, Eq. (1). This calculation involves the same integrals as in Eqs. (A8,A9,A10), but with R_j replaced with R.

Before going to the new applications involving frictional interaction between two bodies, the basic formulation just described was tested by comparing results to several analytical solutions of both time-harmonic and transient problems. Two of these problems were the scattering of a plane harmonic SH wave by a cylindrical cavity in an infinite medium, and scattering by a cylindrical canyon in a free-surface half-space. In both cases the agreement was excellent after several periods of motion, needed for the present transient solution to reach a steady-state. The method is easily extended to two bodies, done below for the frictional case and in Cole, et al.[15] for the bonded case. The present solution to the reflection and refraction of cylindrical SH waves from a point source by a bonded interface between two semi-infinite half-spaces also compares extremely well to an analytical solution. Once again, see Mendelsohn and Doong[18] for details on this or other matters in the formulation.

Let body B_1 represent the soil or rock mass and body B_2 represent the structure, and let the boundary displacement and tractions of each be denoted as,

$$\left.\begin{aligned}(u)_i &= \text{displacement}\\ (\tau)_i &= \text{traction}\end{aligned}\right\} \quad \text{of } B_i \text{ on } \partial B_i,\ i = 1 \text{ or } 2. \tag{4}$$

Over that portion of the boundaries of B_1 and B_2 that are not in contact we consider only standard displacement or traction boundary conditions, while on the interface or contact we introduce the boundary conditions and unilateral constraints appropriate to stick or slip given below. First, whether in a stick or slip zone, the tractions across the interface are continuous,

$$\hat{\tau} = (\tau)_1 = -(\tau)_2 \cdot \tag{5}$$

Define the relative interfacial motion or slip to be

$$g = (u)_1 - (u)_2 \cdot \tag{6}$$

and then in a slip zone we have according to Coulomb friction,

$$|\hat{\tau}| = \mu P = F \ , \tag{7a}$$

$$\mathrm{Sgn}\left(\frac{\partial g}{\partial t}\right) = -\mathrm{Sgn}(\hat{\tau}) \ , \tag{7b}$$

which implies the constraint that the slip velocity $\partial g/\partial t$ must be non-zero in a slip zone. Here μ is the coefficient of friction, P the normal stress on the interface, and F the maximum friction force the interface can withstand. In a stick zone we must have

$$|\hat{\tau}| < F \ , \tag{8a}$$

$$\frac{\partial g}{\partial t} = 0 \cdot \tag{8b}$$

These boundary conditions are used as follows. We take $(u)_1$ and $(u)_2$ to be unknown in a slip zone and prescribe $(\tau)_1 = \pm F$ and $(\tau)_2 = \mp F$, the sign depending on forward or backward slip. In a stick zone we take $(u)_1$ and $(\tau)_1$ to be unknown, and use the fact that although unknown $(\tau)_2 = -(\tau)_1$ and $(u)_2 = (u)_1 - g$, where g is the slip from the <u>previous</u> time step. Then by analogy from Eq. (3) applied once for each body B_1 and B_2, the following coupled system is arrived at

$$\begin{bmatrix} [GN]_1^{ST} & [G]_1^{ST} & [GN]_1^{SL} & [0] \\ [GN]_2^{ST} & [-G]_2^{ST} & [0] & [GN]_2^{SL} \end{bmatrix} \begin{bmatrix} [u]_1^{ST} \\ [\tau]_1^{ST} \\ [u]_1^{SL} \\ [u]_2^{SL} \end{bmatrix} = \begin{bmatrix} [f]_1 \\ [f]_2 \end{bmatrix} \tag{9}$$

The matrices $[G]_1^{ST}$ and $[G]_2^{ST}$ are those columns of the Green's displacement matrix corresponding to nodes in the current stick zone, while $[GN]_1^{SL}$ and $[GN]_2^{SL}$ are those columns of the Green's traction matrix corresponding to nodes in the current slip zone. The vectors $[u]_1^{SL}$ and $[u]_2^{SL}$ are the unknown displacements in the current slip zone, and $[u]_1^{ST}$ and $[\tau]_1^{ST}$ are the unknown displacement and tractions, respectively, in the current stick zone. Thus, combined the unknown vector is of length 2xJ, as is the right hand side vector made up of $[f]_1$ from Eq. (3) for B_1 combined with $[f]_2$ from Eq. 3 for B_2. Equation (9) is presented for the case that all other boundary conditions on those portions of ∂B_1 and ∂B_2 not on the interface are satisfied by the Green's state for B_1 and B_2, respectively. If they are not, then Eq. (9) must be modified to include the additional unknown boundary displacements or tractions. The additional sub-matrices and sub-right-hand-side vectors required are found from Eq. (3).

The stick-slip boundaries are unknown and change with each time-step, so that the remaining sets of Eqs. (7b,8a,9) must be solved iteratively at each time-step. This is done by taking the previous stick slip-configuration to be the current initial guess, checking constraint (8a) in the proposed stick zone and condition (7b) in the proposed slip zone, and updating the stick-slip boundary node by node until a solution is arrived at. While Eq. (9) must be resolved at each iteration, the elements of the matrices are obtained by rearrangement and need not be recalculated. As for the single body formulation above, the displacement at any point in either body may be found easily from the reciprocal relation evaluated at the point in question.

RESULTS

In order to illustrate the method, some initial numerical results are presented for an infinitely long slab foundation (B_2) of thickness H resting on a semi-infinite soil mass (B_1) shown in Fig. 1. The system is subjected to an anti-plane line source at $\underline{x} = \underline{x}_f = (0,-D)$ located in a depth D below the rough interface with coefficient of Coulomb friction μ. Recalling the form of the body force function, Eq. (A16) , we define

$$h_1(t) = h_{max} \left[\frac{4t(10\Delta t - t)}{(10\Delta t)^2}\right], \quad 0 < t < 10\Delta t, \tag{10}$$

which is shown as the positive portion of the non-linear function in Fig. 2. In all but Fig. 9 this is the time history used. For the study in Fig. 9, though, we also consider the additional functions $h_2(t)$, defined to be equal to $h_1(t)$ for $0 < t < 10\Delta t$ and equal to its mirror image about $t = 10\Delta t$ for $10\Delta t < t < 20\Delta t$, and h_3 and h_4 which are the linear counterparts of h_1 and h_2, respectively. All have the same maximum absolute value h_{max}. The body force integral in Eq. (A17) may be evaluated in closed form for the distributions h_1, h_2, h_3 and h_4. Note that any prescribed incident wave field on the interface may be included through the boundary conditions by superposition, rather than through the body force term as done here.

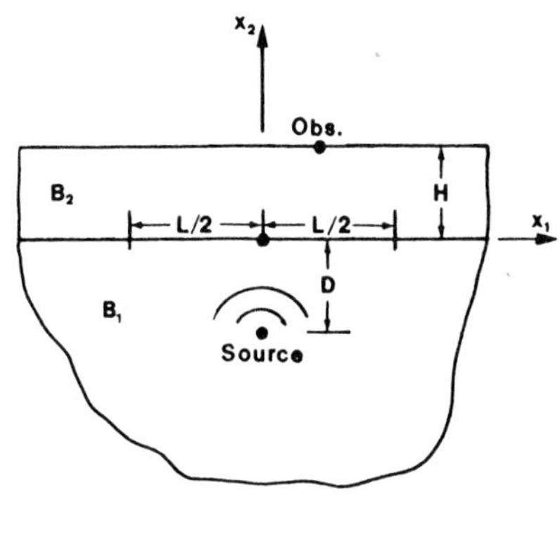

Figure 1. Geometry of slab foundation (B_2, C_2, ρ_2) of thickness H resting on a semi-infinite (B_1, C_1, P_1) soil or rock mass with a source located at a depth D.

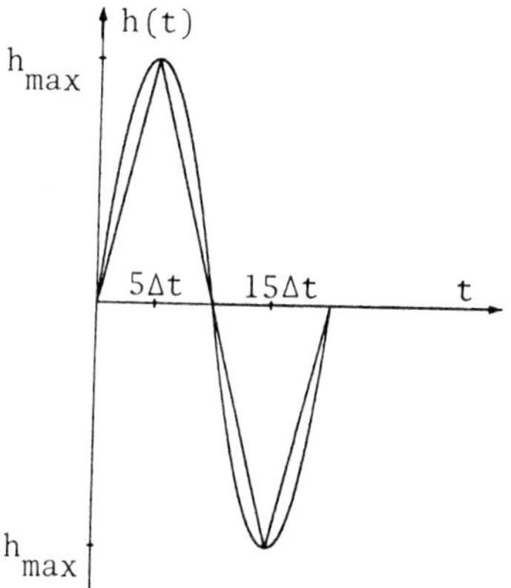

Figure 2. Source time histories used in the numerical calculations.

In order to satisfy automatically the condition that the surface of the layer be free of shear stress, the half-space Green's displacement and traction are used for the slab (B_2), and are also given in the Appendix, while the full-space functions are used for the soil mass (B_1). Thus, the only remaining boundary for both bodies is the interface which they share $(\partial B_1 = \partial B_2)$, and only the interface need be discretized. The length of the discretized portion is L and is evenly distributed about the x_1 axis.

The values of parameters common to all of the numerical results presented are given in Table 1. The choice of D, H and

h_{max} are somewhat arbitrary, however other choices showed the same qualitative behavior, but with different amplitudes.

$H = D = 30.48$ m (100 ft)
$\rho_1 = \rho_2 = 19.6$ kN/m^3 (125 lb/ft^3)
$c_1 = 762$ m/sec (2500 ft/s)
$\Delta t = .0025$ sec
$L = 366$ m (1200 ft)
$\lvert \Delta B_\ell \rvert = 7.6$ m (25 ft)
$J = 48$
$P = 598$ kN/m^2 (12500 lb/ft^2)
$h_{max} = 363 \times 10^3$ kN/m (25×10^6 lb/ft)

Table 1. Common Parameters used in all of the numerical calculations. See Fig. 1.

Figures 3-5 illustrate the frictional behavior on the interface. Figure 3 is a schematic of the entire progression of stick-slip configurations from initially undisturbed and sticking (1), with no relative slip in the interface, through the propagation of slip zones from $x_1 = 0$ outward and then returning to all sticking, but with residual relative slip in the interface (6). The denotations FS and BS refer to forward (positive) and backward (negative) slip, respectively. Initially a FS zone is formed about $x_1 = 0$ (2), and as it spreads and the incident wave subsides, sticking starts first at $x_1 = 0$, leaving two separate FS zones propagating outwards (3). Then as waves are reflected from the surface a BS zone is formed (4). Then, for all cases run, by the time this BS zone has spread and separated (5) the FS zones in (3,4) have disappeared due to attenuation, i.e. the driving shear falls below the critial value for slip. Note that the iteration algorithm does not depend on the knowledge of the expected configuration. Figure 4 shows examples of the shear traction, slip and slip velocity on the interface from Fig. 3, configurations (3) and (4). The interface has $J = 48$ nodes with $x_1 = -L/2 + |\Delta B_j|/2$ and $x_1 = |\Delta B_j|/2$ corresponding to nodes 1 and 25, respectively. Recall that slip velocities are zero in a stick zone and shear tractions are constant and equal to $\pm F$ in a slip zone. Also note that, as prescribed, the slip distribution does not change from the time of Fig. (4a) to that of (4b) in that region which remained sticking (nodes 22-23 and 26-27). While the occurrence of backslip has reduced the slip in-between (nodes 23-26).

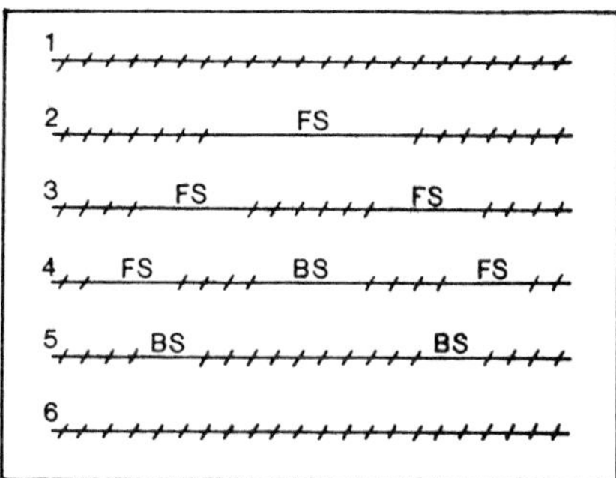

Figure 3. Schematic of the progression of the stick-slip configurations. BS and FS represent forward and backward slip respectively, and ////// denotes a stick zone.

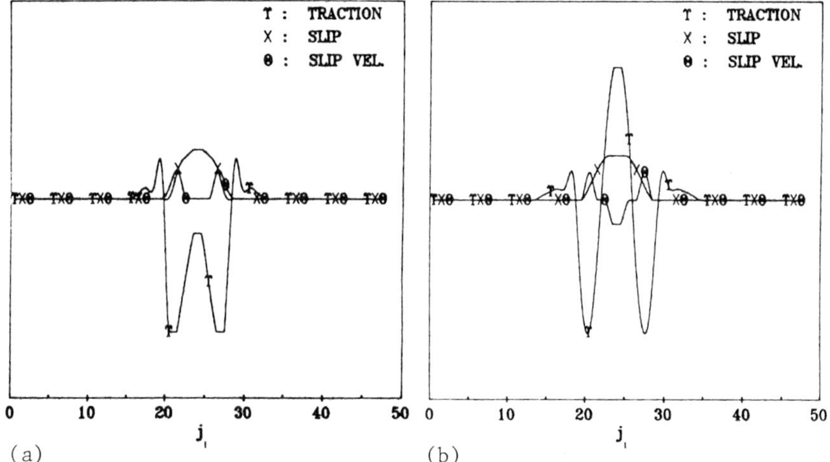

(a) (b)

Figure 4. Interface shear tractions, relative slip and slip velocity vs node j_I on interface: (a) at time step m = 24 and, (b) at time step m = 26, (c_1/c_2 = 1/4, μ = 8.0).

Figures 5-9 all show layer surface displacements $u_0 = u/u_{max}$, where u_{max} is the peak value in time and space of the surface displacement obtained for all the values of c_1/c_2 and μ that were run. This value occured at $c_1/c_2 = 1$ and $\mu = 8$ directly above the source and is u_{max} = .0017 m (.0056 ft). Figures 6 and 8 contain plots of u_0 at several times versus observation node number j_s on the surface, where the surface has been divided into 25 nodes between $x_1 = -L/2$ and $x_1 = L/2$ (node B is at $x_1 = 0$). Figures 5, 7 and 9 are plots of u_0 at j_s = 13 versus time step number m.

Figures 5 and 6 show the variation in surface displacement with coefficient of friction at a fixed soil to foundation speed ratio. The range of coefficient of friction was chosen to span interface behavior from very smooth to effectively bonded $\mu = \infty$; a value of $\mu = 80$ was sufficient for all of the calculations made. A marked reduction in displacement may be observed throughout as interface roughness is decreased.

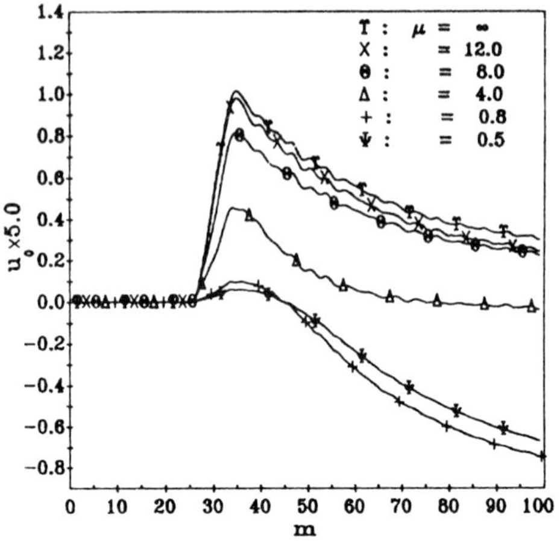

Figure 5. Normalized surface displacement u_0 at $j_s = 18$ versus time step m for various coefficients of friction μ, $(c_1/c_2 = 1/4)$.

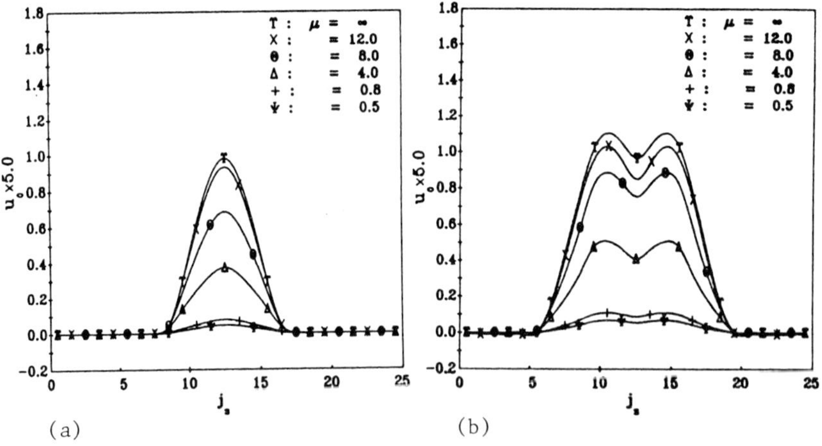

Figure 6. Normalized surface displacement u_0 vs node j_s on surface for various coefficients of friction: (a) at time step m = 25 and (b) time step m = 30, $(c_1/c_2 = 1/4)$.

While the limit case of $\mu = 0$ will yield identically zero surface displacements, the results indicate that for low enough coefficients of friction the layer will slide quite far back after the initial unloading and take a long time to come to rest since there is little resistance to its motion. It will take a wider mesh and more time steps to resolve the behavior for very small but non-zero μ. Figure 6 shows two snapshots of the distribution of displacement along the surface for the same parameters as in Fig. 5.

Figures 7 and 8 illustrate the effect of the mismatch in speed ratio across the interface for a fixed coefficient of friction.

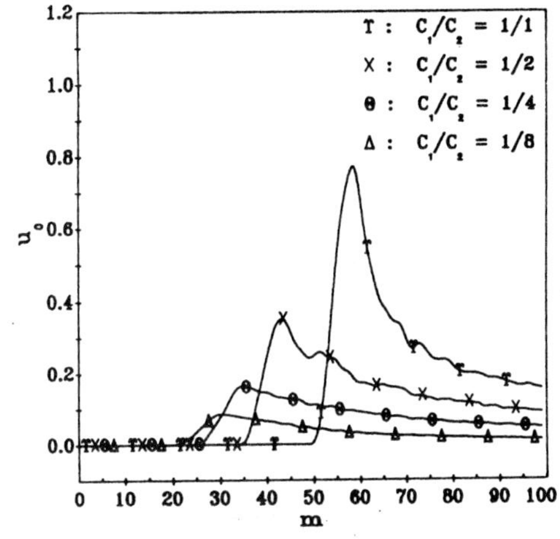

Figure 7. Normalized surface displacement u_0 at $j_s = 18$ versus time tep m for various half-space to layer speed ratios. ($\mu = 8.0$).

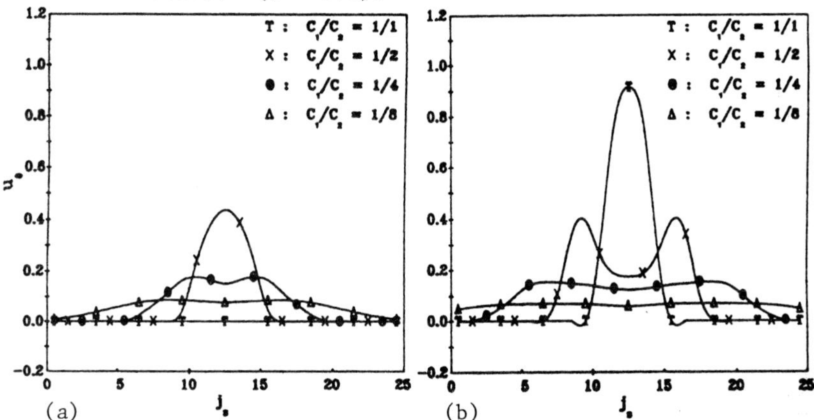

Figure 8. Normalized surface displacements u_0 versus node j_s on surface for various half-space to layer speeds ratios: (a) at time step m = 30 and (b) time step m = 38. ($\mu = 8.0$).

Figure 7 shows that the disturbance arrives at the surface sooner, and, due to increased reflection of energy from the interface into the half-space, the peak amplitudes are smaller the stiffer the layer becomes. Figure 8 contains snapshots of displacements at two times for the same parameters as in Fig. 7. Although it was mentioned above that the time-step was kept small enough to keep [G] and [GN] diagonal, in extending the calculations to $c_1/c_2 = 1/4$, $1/8$, keeping Δt fixed, this was not the case, and instead [G] and [GN] are diagonally banded.

Figure 9 illustrates the effects of some different source time histories. Here as in Figs. 5 and 7 the observation point is at node $j_s = 18$ on the surface. The time histories are defined above, Eq. (10), and are shown in Fig. 2. Note that h_1 and h_3 cause loading and unloading in one direction, and h_2 and h_4 add a reloading and unloading cycle in the opposite direction to h_1 and h_3 respectively. Figure 9 illustrates that this forces the layer back towards its initial position more quickly and, at least for the value of $\mu = 8$ used, appears to bring the layer to rest more quickly. There is no qualitative effect of a sharper transient in loading history (h_3 and h_4 compared to h_1 and h_2) and the reduced values for h_3 and h_4 are probably due to the smaller area under the load versus time curves for the linear histories, Fig. 2.

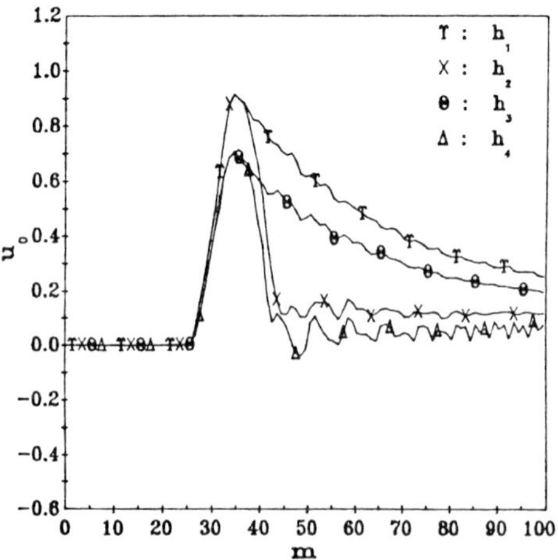

Figure 9. Normalized surface displacement u_0 versus time step m for various source time histories, ($c_1/c_2 = 1/4$, $\mu = 8.0$).

CONCLUSION

This paper presents initial results of a dynamic, time-stepping boundary element technique applicable to soil-structure interaction problems in which there is non-linear frictional stick-slip behavior at the interface. The method is quite general and contains an iteration algorithm capable of determining the a-priori unknown interface stick-slip configuration. Displacements and stresses elsewhere in the soil or structure may then be calculated from the state at the interface. Initial numerical results have been presented for the problem of a slab foundation resting on a semi-infinite soil or rock mass, disturbed by waves from an SH line source below the interface. The effect on the motion of the layer due to both friction on the interface and the material mismatch across it were studied. Results for several load time histories were also compared.

The present model will be extended to more realistic geometries, including multiple interactions, and load time histories. While only the first phase, due to the features of the present solution technique, natural extensions to in-plane motions with interface separation, 3D interactions, more realistic non-linear material behavior may also be added easily. Many of these extensions will require the use of the supercomputer, and the code is presently being adapted for use on one.

ACKNOWLEDGEMENT

The authors acknowledge the support of the National Science Foundation through grant CEE-8505274.

REFERENCES

1. Saxena, S.K. (Ed., Aug. 4-5, 1986). Siting and Geotechnical Program Focus and Directions, The Illinois Institute of Technology, PPS. 35-37, 44-46, 78-80, 85-89, 147, Chicago, Illinois.

2. Pan, T.C. and Kelly, J.M. (1983), Seismic Response of Torsionally Coupled Base Isolated Structures, Earthquake Engineering and Structural Dynamics, Vol. 11, pp. 749-770.

3. Kelly, J.M. and Beucke, K.E. (1983), A Friction Damped Base Isolation System with Fail-Safe Characteristics, Earthquake Engineering and Structural Dynamics, Vol. II, pp. 33-56.

4. Lee, D.M. (1980), Base Isolation for Torsion Reduction in Asymmetric Structures Under Earthquake Loading, Earthquake Engineering and Structural Dynamics, Vol. 8, pp. 349-359.

5. Hundal, M.S. (1979), Response of a Base Excited System with Coulomb Viscous Friction, Journal of Sound and Vibration, Vol. 64, pp. 371-378.

6. Crandall, S.H., Lee, S.S. and Williams, J.H. Jr. (1974), Accumulated Slip of a Friction Controlled Mass Excited by Earthquake Motions, J. Appl. Mech., Vol. 41, pp. 1084-1098.

7. Westermo, B. and Udwadia, F. (1983), Periodic Response of a Sliding Oscillator System to Harmonic Excitation, Earthquake Engineering and Structural Dynamics, Vol. II, pp. 135-146.

8. Mostaghel, N. and Tanbakuchi, J. (1983), Response of Sliding Structures to Earthquake Motions, Earthquake Engineering and Structural Dynamics, Vol. II, pp. 729-478.

9. Wolf, J.P. (1976), Soil-Structure Interaction with Separation of Base Mat from Soil, Nuclear Engineering and Design, Vol. 38, pp. 357-384.

10. Toki, K., Sato, T. and Miura, F. (1981), Separation and Sliding Between Soil and Structure During Strong Ground Motion, Earthquake Engineering and Structural Dynamics, Vol. 9, pp. 263-277.

11. Akiyoshi, T. and Fuchida, K. (1982), Soil-pipeline Interaction Through a Frictional Interface During Earthquakes (Ed. Cakmak, A.S., Abdel-Ghaffar, A.M. and Brebbia, C.A.), pp. 497-511, Proceedings of the Conference on Soil Dynamics and Earthquake Engineering, 1982, Southhampton.

12. Vaughan, D.K. and Isenberg, J. (1983), Nonlinear Rocking Response of Model Containment Structures, Earthquake Engineering and Structural Dynamics, Vol. 11, pp. 275-286.

13. Desai, C.S. and Zaman, M.M. (1984), Influence of Interface Behavior in Dynamic Soil-Structure Interaction, Proc. Eigth World Conference on Earthquake Engineering, 1984, San Francisco.

14. Wolf, J.P. and Darbre, G.R. (1986), Nonlinear Soil-Structure Interaction Analysis Based on the Boundary-Element Method in Time Domain with Application to Embedded Foundation, Earthquake Engineering and Structural Dynamics, Vol. 14, pp. 83-101.

15. Cole, D.M., Kosloff, D.D. and Minster, J.B. (1978), A Numerical Boundary Integral Equation Method for Elastodynamics, I, Bulleting of the Seismological Society of America, Vol. 68, pp. 1331-1357.

16. Mansur, W.J. and Brebbia, C.A. (1982), Formulation of the Boundary Element Method for Transient Problems Governed by the Scalar Wave Equation, Applied Mathematical Modelling, Vol. 6, pp. 307-311.

17. Mansur, W.J. and Brebbia, C.A. (1982), Numerical Implementation of the Boundary Element Method for Two Dimensional Transient Scalar Wave Propagation Problems, Applied Mathematical Modelling, Vol. 6, pp. 299-306.

18. Mendelsohn, D.A. and Doong, J.-M (1987), A Boundary Element Algorithm for Nonlinear Transient Elastic Wave Interaction with Frictional Interfaces, Proc. of IUTAM Symposium on Advanced Boundary Element Methods, 1987, San Antonio.

APPENDIX

The full or infinite space Green's state for SH motion is that due to an impulsive unit live-source at $\underline{x} = \underline{\xi}$ and $t = t_0$

$$f_B(\underline{x},t) = \delta(\underline{x}-\underline{\xi})\delta(t-t_0) \ . \tag{A1}$$

The solution for the displacement at any point $\underline{x}$ and time t is well known to be

$$V(\underline{x},t;\underline{\xi},t_0) = \frac{1}{2\pi\rho c^2} \frac{H(t-t_0-R/c)}{\sqrt{(t-t_0)^2 - R^2/c^2}} \tag{A2}$$

and the stress or traction is given by,

$$T(\underline{x},t;\underline{\xi},t_0) = \frac{n_\alpha y_\alpha}{2\pi c^2} \left[\frac{H(t-t_0 - R/c)}{\sqrt[3]{(t-t_0)^2 - R^2/c^2}} \right.$$

$$\left. - \frac{1}{(R/c)} \frac{\delta(t-t_0-R/c)}{\sqrt{(t-t_0)^2 - R^2/c^2}} \right] \ . \tag{A3}$$

where R is the distance between $\underline{x}$ and $\underline{\xi}$

$$R = |\underline{x} - \underline{\xi}| = \sqrt{y_1^2 + y_2^2} \ , \tag{A4}$$

where

$$y_\alpha = (x_\alpha - \xi_\alpha), \ \alpha = 1 \text{ or } 2 \ . \tag{A5}$$

Also ρ is the mass density, c is the speed of elastic shear waves and $\underline{n} = (n_1,n_2)$ is the outward normal of the surface on which T is being evaluated. The functions H( ) and δ( ) are the Heaviside and Dirac delta functions, respectively.

The free surface half-space Green's displacement and traction are obtained easily by the method of images and are

$$V_H(\underline{x},t;\underline{\xi},t_0) = V(\underline{x},t;\underline{\xi},t_0) + V(\underline{x},t;-\underline{\xi},t_0) , \tag{A6}$$

$$T_H(\underline{x},t;\underline{\xi},t_0) = T(\underline{x},t;\underline{\xi},t_0) + T(\underline{x},t;-\underline{\xi},t_0) \cdot \tag{A7}$$

The basic integrals over each time-step and element which appear in the discretized formulation in Eq. (2) are defined below

$$(2\pi\rho c^2)G^k_{j\ell} = \int_{t_{k-1}}^{t_k} \int_{\Delta B_\ell} \frac{H(\bar{t}_{mj})}{D_{mj}} ds(\underline{\xi})dt_0 , \tag{A8}$$

$$(2\pi c^2)GN^k_{j\ell} = \int_{t_{k-1}}^{t_k} \int_{\Delta B_\ell} \frac{(n_{\alpha\ell}y_{\alpha j})}{(R_j/c)} H(\bar{t}_{mj})x$$

$$[\frac{(1/\Delta t)}{D_{mj}} + (1 - \frac{t_k-t_0}{\Delta_t}) \frac{\bar{t}_{mj}}{D^3_{mj}}]ds(\underline{\xi})dt_0 , \tag{A9}$$

$$(2\pi c^2)GNP^k_{j\ell} = \int_{t_{k-1}}^{t_k} \int_{\Delta B_\ell} \frac{(-n_{\alpha\ell}y_{\alpha j})}{(R_j/c)} H(\bar{t}_{mj})x$$

$$[\frac{(1/\Delta t)}{D_{mj}} - (1 - \frac{t_0-t_{k-1}}{\Delta t}) \frac{\bar{t}_{mj}}{D^3_{mj}}]ds(\underline{\xi})dt_0 , \tag{A10}$$

where

$$\bar{t}_{mj} = t_m-t_0-R_j/c, \tag{A11}$$

$$D_{mj} = \sqrt{(t_m-t_0)^2 - R_j^2/c^2} , \tag{A12}$$

$$R_j = |\underline{\xi}-\underline{r}_j| = \sqrt{y_{1j}^2 + y_{2j}^2} , \tag{A13}$$

$$y_{\alpha j} = \xi_\alpha - r_{\alpha j} , \quad \alpha = 1 \text{ or } 2, \tag{A14}$$

$$n_{\alpha\ell} = \alpha \text{ component of normal to } \Delta B_\ell. \tag{A15}$$

The body force function and related integral are

$$f_B(\underline{x},t) = \delta(\underline{x}-\underline{x}_f)h(t) \ , \tag{A16}$$

$$(\hat{f}_B)_j^m = \int_0^{t_m} V(\underline{r}_j,t_m;\underline{x}_f,t_0)h(t_0)dt_0. \tag{A17}$$

Site Amplification in Mexico City (Determined from 19 September 1985 Strong-Motion Records and from Recordings of Weak Motions)

M. Çelebi, C. Dietel
U.S. Geological Survey, Menlo Park, CA 94025, U.S.A.
J. Prince, M. Onate, G. Chavez
Instituto de Ingenieria, UNAM, Mexico City, Mexico

INTRODUCTION

The Michoacan, Mexico earthquake of September 19, 1985 ($M_s = 8.1$) was one of the few earthquakes of this century that caused extensive loss of life and property (Rosenblueth and Meli[1], Bertero[2] and UNAM [Autonamous National University of Mexico][3]) While the epicenter of the earthquake was near the Pacific coast of Mexico (18.182°N, 102.573°W, origin time 13:17:47.8 UT) (NEIS[4], and Anderson and others[5]) and there was some damage on the coastal region, the main impact and destructiveness of the earthquake was experienced in the lakebed zone of Mexico City—approximately 400 km from the epicenter. The large distance to the principal area of destruction was one unique feature of the earthquake. A second unique feature is related to the subsurface conditions of Mexico City—in that there was substantial amplification of the low-frequency motions. In one respect, the second factor described is the cause of the first for if it were not for the subsurface conditions of the lakebed of Mexico City there would not have been the long-distance effect of the earthquake and the normal attenuation relationships would have applied.

The earthquake was recorded extensively at the epicentral coastal zone by the Guerrero array—a joint project of UNAM and UCSD (University of California, San Diego) (Anderson and others[5])—and at several sites in Mexico City.

Mexico has been prone to severe earthquakes in the past and Mexico City has repeatedly suffered from the long-distance effects of the earthquakes that originate at the subduction trenches near the Mexican Pacific Coast. The amplification of motions in Mexico City generated from strong motions originating at the Pacific coast subduction zone was recognized during and after past earthquakes—particularly after the 1957 earthquake during which Mexico City suffered extensive damage as it did in 1985. The earthquake of 28 July 1957 ($M_s = 7.5$) with its epicenter near Acapulco (~270 km from Mexico City), as widely reported (Duke and Leeds[6]; Merritt[7]; Herrera and others[8]; Rosenblueth[9]; Rosenblueth and Elorduy[10]), caused extensive damage in Mexico City. As a result, numerous studies of the structural and architectural aspects of the buildings in Mexico City and of the subsurface conditions were carried out. Herrera and others[5] studied amplification and resonant frequencies in the lakebed of Mexico City. Rosenblueth and Elorduy[10] published response spectra from records obtained at UNAM and other locations in the lakebed zone during the smaller 1962 (Zeevaert[11]) and 1964 earthquakes. These spectra clearly show spectral peaks between 1.5–2.5 seconds. Of particular interest are comparative displacement spectra from the 11 May 1962 earthquake obtained at the Latino–Americano building and Alameda Park only 600 meters away. The spectrum at the Latino Americano shows suppressed (by almost 50%) peaks because of preconsolidation realized through piles used in the periphery of the foundation.

These studies led to significant code changes then and later in 1976 to accommodate the unusual spectra of earthquake motions in Mexico City. For example, in the 1976 code, the design response spectrum peaks at a seismic coefficient of 0.24 between 0.8–3.3 seconds,—however, this coefficient then is reduced for ductility by a factor as much as 6—thus, in a sense nullifying the amplified seismic design coefficient. But the important point is that the amplification and associated resonant periods were recognized in 1957 and thereafter. The 1985 earthquake records provided concrete and detailed evidence of range of resonant periods and amplification of motion.

The purpose of this paper is to present quantified amplification ratios obtained from strong-motion records of the 19 September 1985 earthquake as well as weak motions recorded in January 1986. While structural design and construction problems existed, it should be repeated for emphasis that the main culprit in the destructiveness of this event was the unique subsurface conditions of Mexico City that gave rise to amplified seismic forces oscillating at resonant periods for long duration—some as long as 172 seconds or more (Anderson and others[5]).

GROUND MOTIONS—EPICENTRAL AND AT MEXICO CITY

The Guerrero array records (Anderson and others[5]) as well as the records obtained in the Federal District of Mexico City bring out the following facts:

- There were two events separated by 24 seconds.

- The peak accelerations at the epicentral area (Caleta de Campos, La Villita and La Union Stations), at Teacalco (~340 km from the epicenter and the only Guerrero array station close to Mexico City)(Figure 1), and at UNAM (~400 km from the epicenter) were on the order of 0.15 g, 0.05 g, and 0.035 g, respectively—indicating that the earthquake motions followed the normal attenuation relationships. This is clearly demonstrated in Figure 2 which presents the east-west components of the 19 September 1985 earthquake records starting from the epicentral coastal station of Caleta de Campos to UNAM in Mexico City. All of the coastal stations were on rock. UNAM station is on rock composed of lava overlying consolidated material.

- The peak accelerations (of five of the important stations in Mexico

City seen in Figure 3)—UNAM (rock), SCT (lakebed), VIV (transition zone), Tacubaya (rock), and CDA (lakebed)—were on the order of 0.035 g, 0.17 g, 0.042 g, 0.034 g, and 0.095 g, respectively—clearly indicating differences attributable to the unique subsurface conditions of Mexico City. The east-west SCT station acceleration component is shown in Figure 2 to demonstrate that this representative station in the lakebed zone recorded amplified motions as compared to UNAM station.

• The frequency contents of these motions indicate that at the lakebed zone as well as at the source there is significant 0.5 Hz energy (Singh and others[12]).

AMPLIFICATION OF MOTIONS IN MEXICO CITY

Singh and others[12], Kobayashi and others[13,14], and Ohta and others[15] present extensive results quantifying amplification of motions in the lakebed of Mexico City using both the strong-motion records of the 1985 event and microtremors. In summary, and in particular the work of Singh and others[12] report that in the lakebed as compared to the hill zones, the motions are amplified by 8 to 50 times and the motions at the hill zone sites in Mexico City when compared to the hard-rock coastal epicentral sites are amplified by a factor of 7.5 times at 0.5 Hz frequency—after correcting for the effect of distance.

In Figure 4, we present frequency dependent spectral ratios determined from strong motion records at some of the stations in Mexico City shown in Figure 3. These will be compared with spectral ratios determined from noise measurements made in January 1986 in different parts of Mexico City. The spectral ratios in Figure 4, all plotted with the same format and scale to provide comparative evaluation, are calculated from Fourier amplitude spectra of acceleration time histories of 19 September 1985 records obtained at the lakebed zone stations: SCT (Ministry of Telecommunications and Transportation) and CDAO (Central de Abastos Office Building), at the transition zone station VIV (Viveros) and at the hills zone station TAC (Tacubaya), all with reference to the UNAM (Autonamous National University of Mexico, Institute of Engineering Patio) station. The surficial geological formation of these stations are given by Anderson and others (1986a and b) as very soft soil (clay) for SCT and CDAO, soft soil for VIV, hard soil for TAC and rock (basalt) for UNAM. The amplification of motions in the lakebed zone (SCT and CDAO) as compared to the rock site (UNAM) was as much as 7–10 times in the horizontal direction at 0.4–0.5 Hz and 6 times in the vertical direction at 1.5 Hz. In the transition zone (VIV), amplification is about 4.5 times at 2 Hz in the horizontal direction. In the hills zone (TAC) compared to UNAM, no amplification can be claimed. Corresponding Fourier spectra for each of the three components of all stations for which spectral ratios are provide are shown in Figure 5. The figure substantiates again the dominant low frequencies of the motion in Mexico City. All components of all stations exhibit dominant frequencies between 0.3–0.8 Hz.

Now we present similar frequency dependent spectral ratios in Figure 6 obtained from weak motions (traffic noise) for comparable stations for which spectral ratios from strong motions are presented in Figure 4. While

UNAM and CDAO stations are the same as before, three new stations are identified: USA (American Embassy basement), SFO (garden of a house on San Fransisco street) and TLA (Tlatelolco—a government-sponsored social housing complex. The Nuevo Leon Building of which one block had overturned was within this complex). Relative locations of these stations are shown on the map in Figure 3. Stations USA and SFO are both within the boundaries of the transition zone. Station TLA is in the lakebed zone. The spectral ratios shown in Figure 5 exhibit several distinctive characteristics. They all have amplitudes significantly larger by an order of magnitude than those spectral ratios from strong motions. This is to be expected because the weak motions are not from the same source and their travel paths are not the same as those of the strong motions. Second, the SCT/UNAM spectral ratios peak at frequencies that correlate well with those from strong motions—at 0.5 Hz for horizontal and 1.5 Hz for vertical. On the other hand, the CDAO/UNAM spectral ratios from weak motions, while exhibiting peaks at frequencies of 1 Hz or less, does not show good correlation with those from strong-motion records. However, there is clear evidence that at CDAO the weak motions did not have sufficient energy to excite the lower frequency which is apparent in the plots. The USA/UNAM spectral ratios exhibit amplification at 0.8–0.9 Hz in the horizontal direction while SFO/UNAM spectral ratios tend to have amplification between 1.0–1.3 Hz in the horizontal direction. On the other hand, TLA/UNAM spectral ratios clearly show that amplification occurs between 0.5–0.7 Hz in the horizontal direction. The TLA/UNAM spectral ratio is representative of the spectral ratios obtained from the other stations at Tlatelolco.

CONCLUSIONS

The spectral ratios from strong motion records and weak motions exhibit the resonant frequencies for which amplification of motions were experienced in Mexico City during the 19 September 1985 Michoacan earthquake, the epicenter of which was approximately 400 km away near the Pacific coast of Mexico. While the spectral ratios from strong motions provide the resonant frequencies and amplitudes of amplification of motions at one location with respect to another, the spectral ratios from weak motions satisfactorily identify the resonant frequencies. Because the energy, path and source of the weak motions are not same as those of the strong motions; therefore, the weak motion spectral ratios should only be used to identify the resonant frequencies in Mexico City.

REFERENCES

[1.] Rosenblueth, E., and Meli, R., 1986, "The 1985 Earthquake: Causes and Effects in Mexico City," *Concrete International*, pp. 23–34.

[2.] Bertero, V. V., 1986, "Observations on Structural Pounding," talk given at the ASCE–International Conference on the 1985 Mexico Earthquake, Sept. 19–21, 1986, Mexico City, Mexico.

[3.] UNAM (1985), Efectes de los siesmios de September de 1985 en las consructtiones de la Ciudad de Mexico. Aspectos Estructurales, Segundo informel del Inst. de Ing. de la Univ. Nac. Auton. de Mexico, Nov. 1985.

[4.] "Preliminary determination of epicenters, No. 38–85" (National Earthquake Information Service, National Oceanic and Atmospheric Administration, Boulder, CO, 10 October 1985).

[5.] Anderson, J. G., *et al.*, 1986, "Strong Ground Motion from the Michoacan, Mexico, Earthquake," *Science*, v. 233, pp. 1043–1049, Sept. 1986.

[6.] Duke, M. C., and Leeds, D. J., 1959, "Soil Conditions and Damage in the Mexico Earthquake of July 28, 1957," *Bull. Seismol. Soc. Am.*, v. 49, no. 2, pp. 179–191.

[7.] Merritt, F. S., 1957, "Earthquake Revealed Defects in Design," *Engineering News*, pp. 38–44, Aug. 15, 1957. Also, "Mexico City Buildings Hit by Earthquake," Aug. 1, 1985 issue (p. 27) and "Learning from Disasters"–Editorial of August 22, 1957 issue (p. 128).

[8.] Herrera, I., Rosenblueth, E., and Rascon, O. A., 1965, "Earthquake Spectrum Prediction for the Valley of Mexico," Third World Conference on Earthquake Engineering, New Zealand, pp. 61–74.

[9.] Rosenblueth, E., 1960, "The Earthquake of 28 July 1957 in Mexico City," Second World Conference on Earthquake Engineering, Tokyo, Japan, pp. 359–379.

[10.] Rosenblueth, E., and Elorduy, J., 1969, "Characteristics of Earthquakes on Mexico City, Clay," in *Nabor Carillo: Elhundimiento de la Ciudad de Mexico, Proyecto Texcoco*, pp. 287–328.

[11.] Zeevaert, L., 1964, "Strong Ground Motions Recorded During Earthquakes of May the 11th and 19th, 1962 in Mexico City," *Bull. Seism. Soc. Am.*, 54, pp. 209–231.

[12.] Singh, S. K., Mena, E., and Castro, R., 1986, "Some Aspects of Source Characteristics of the 19 September, 1985, Michoacan Earthquake and Ground Motion Amplification in and near Mexico City from Strong Motion Data," submitted to BSSA Journal (courtesy S. K. Singh).

[13.] Kobayashi, H., Seo, K., Midorikawa, S., and Kataoka, S., 1986a, "Measurements of Microtremors in and Around Mexico D.F.," Part I—Report of Tokyo Institute of Technology, Yokohama, Japan, 97 pages.

[14.] Kobayashi, H., Seo, K., and Midorikawa, S., 1986b, "Estimated Strong Ground Motions in Mexico City due to the Michoacan, Mexico Earthquake of September 19, 1985 Based on Characteristics of Microtremor," Part II—Report of Tokyo Institute of Technology, Yokohama, Japan, Feb. 1986, 34 pages.

[15.] Ohta, T., *et al.*, 1986, "Research on the Strong Ground Motion in Mexico City During the Earthquake of September 19, 1985 Michoacan–Guerrero, Mexico" KITC Report No. 68, Kajima Institute of Construction Technology, Tokyo, Japan (46 pages).

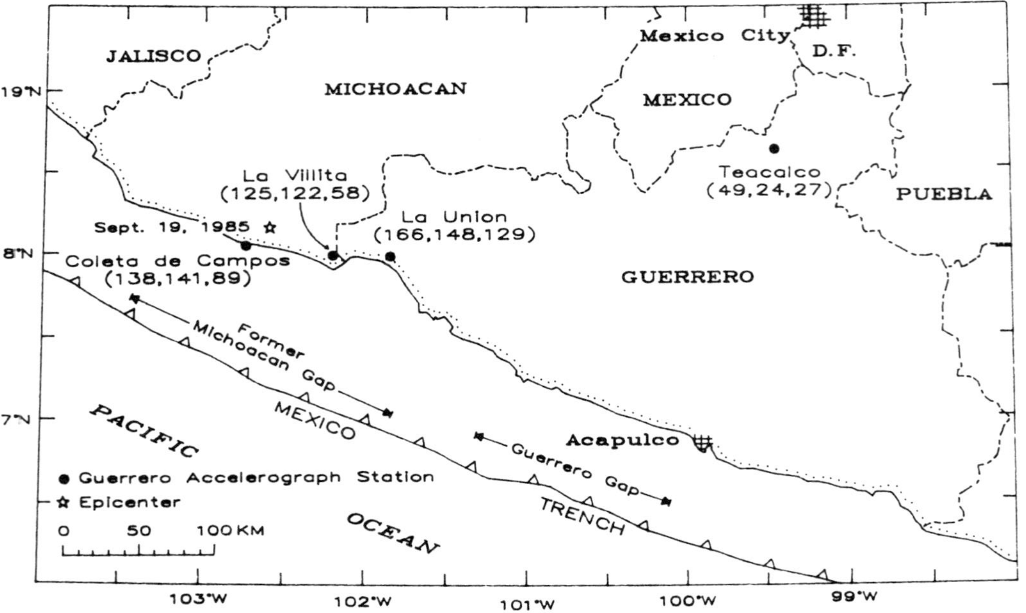

Figure 1. General map of part of the Pacific coast of Mexico (revised and adopted from Anderson and others,1986) showing the epicenter of the 19 September 1985 ($M_s = 8.1$) Michoacan Earthquake. Three of the several coastal stations and the Teacalco station (closest to Mexico City) of the Guerrero array are shown with peak accelerations in paranthesis for the NS , EW and vertical components, in that order.

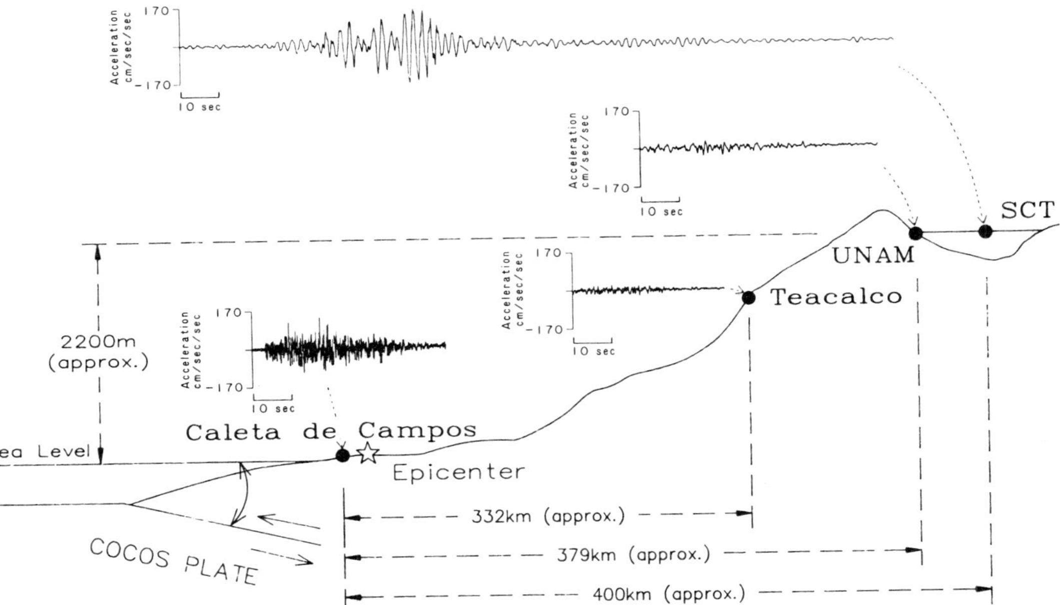

Figure 2. Schematic section showing relative locations of the epicentral station at Caleta de Campos, Teacalco station (closest to Mexico City), and Mexico City stations, UNAM (hills zone) and SCT (lake zone). The seismograms are east-west components of acceleration time-histories (all plotted to the same scale) recorded at respective stations and demonstrate the attenuation of motions with distance from the coast as well as amplification of motions at the lakebed of Mexico City.

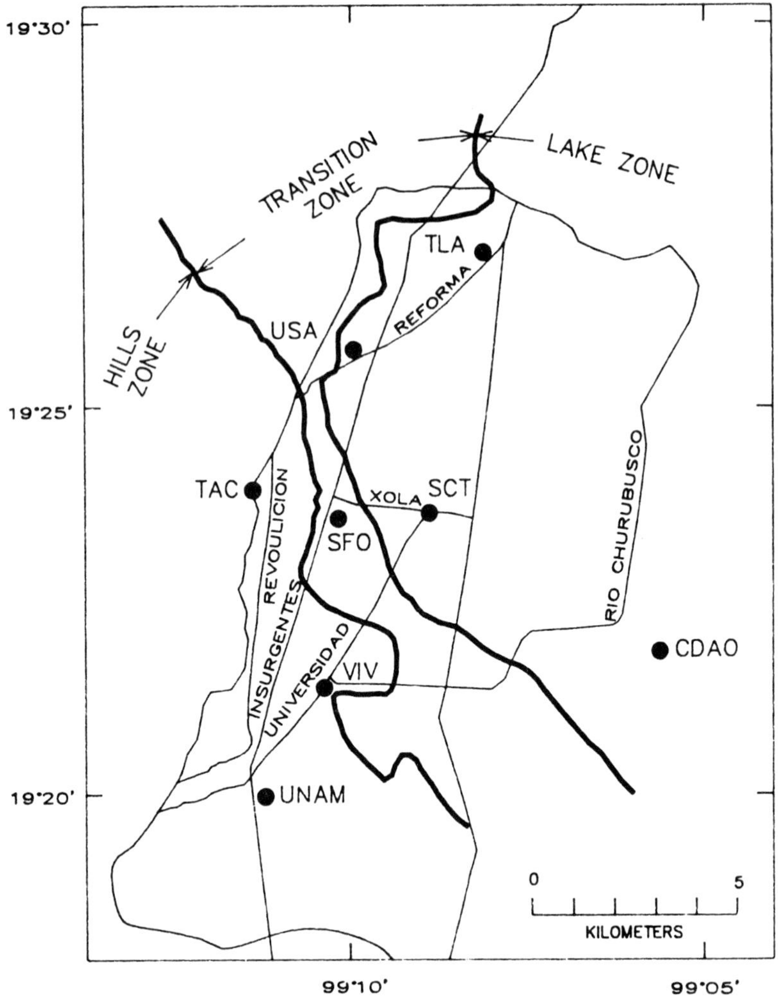

Figure 3. Map showing the three zones of Mexico City as well as the locations of stations discussed in the manuscript. UNAM, SCT, CDAO, VIV and TAC are strong motion stations. SFO, USA and TLA are the temporary stations established in January 1986 to facilitate recording of weak motions.

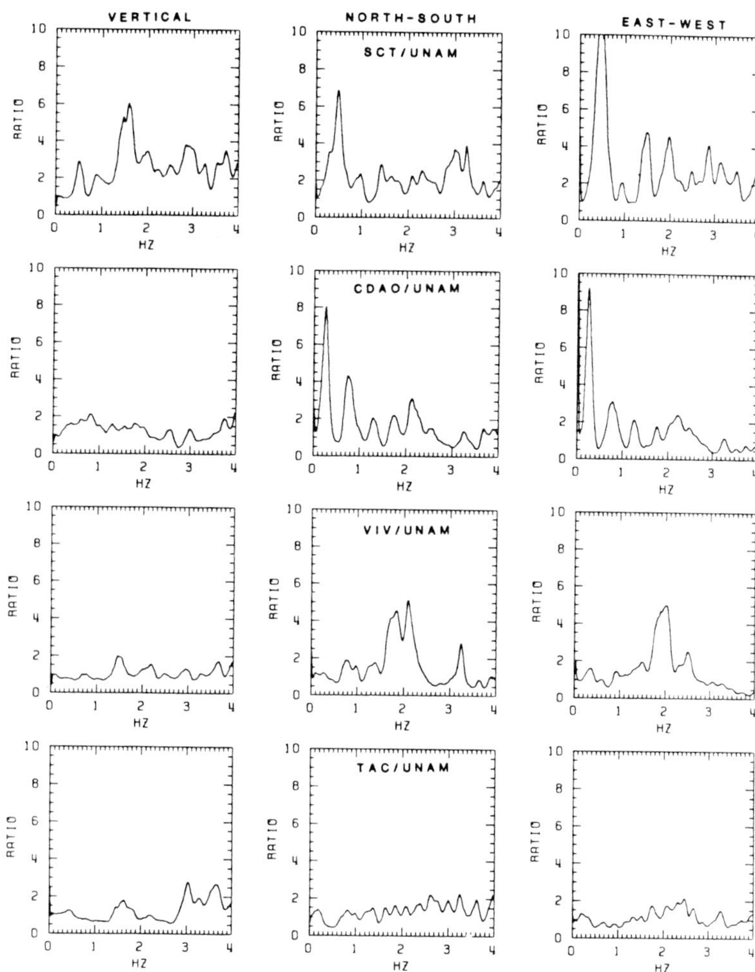

Figure 4. Spectral ratios for the vertical and horizontal components (NS and EW), respectively, derived from the strong motion records of the 19 September 1985 earthquake. Ratios shown are for stations SCT, CDAO, VIV and TAC with respect to UNAM. All plot have same format and scale to provide easy comparison. SCT and CDAO stations are in the lake zone, VIV is in the transition zone and TAC and UNAM are both in the hills zone. The plots clearly and quantitatively show the frequencies and amplitudes of amplification of motions experienced in Mexico City.

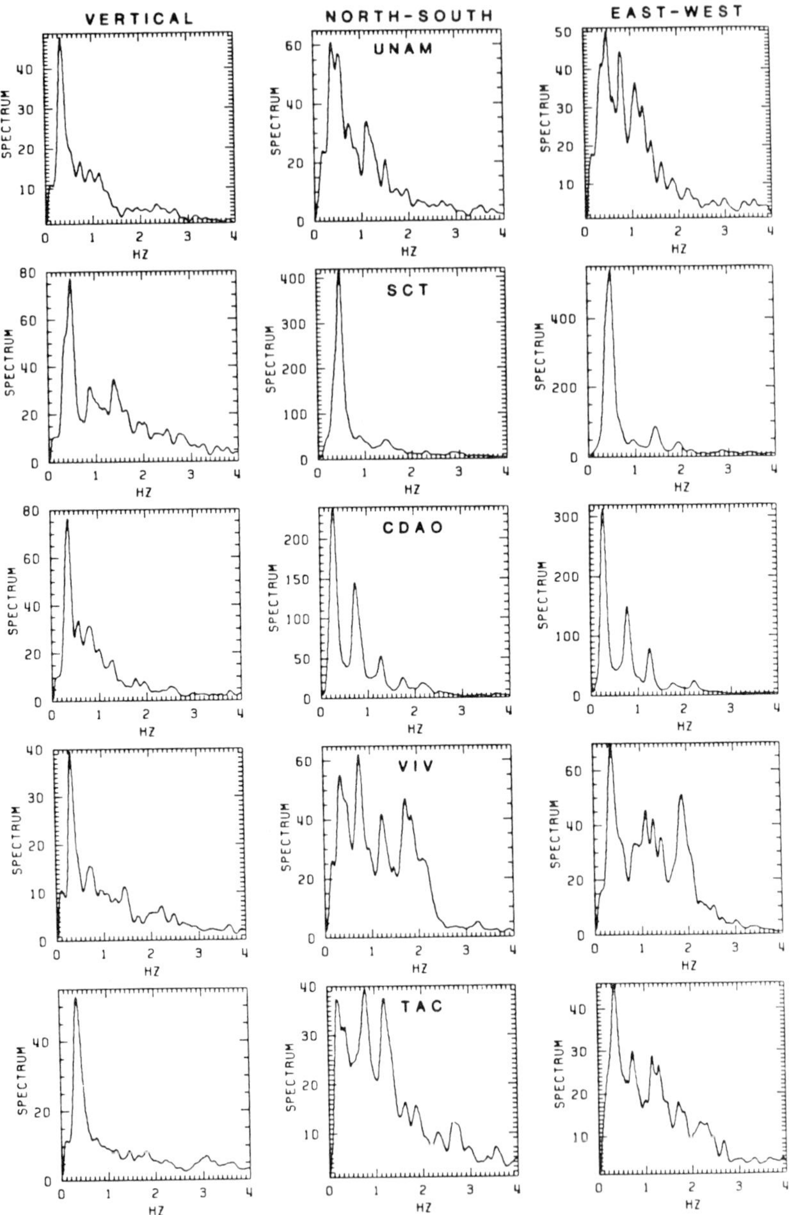

Figure 5. Fourier spectra for the vertical and horizontal components (NS and EW), respectively, derived from the strong-motion records of stations UNAM, SCT, CDAO, VIV and TAC in Mexico City. All plots demonstrate the significant low-frequency energy at all stations.

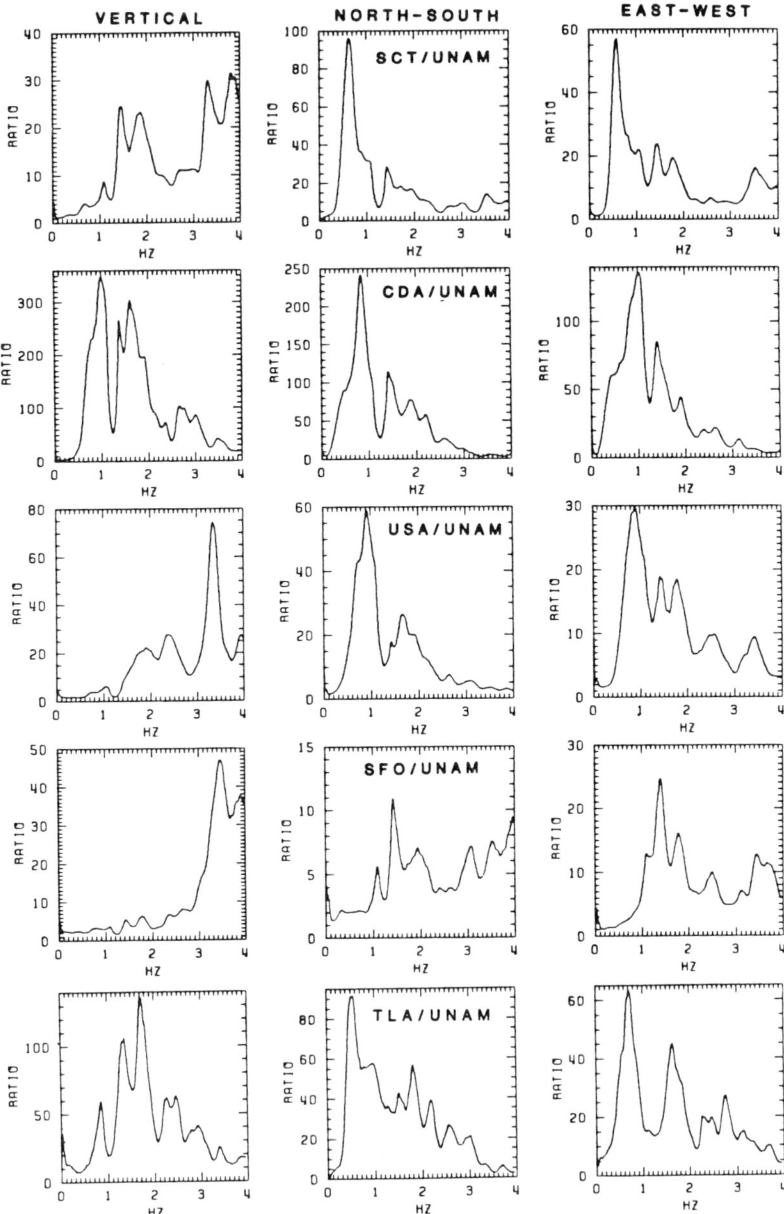

Figure 6. Spectral ratios for the vertical and horizontal components (NS and EW), respectively, derived from weak motions recorded at stations CDAO, SCT, USA, SFO, TLA and UNAM. All plots are made with respect to UNAM. CDAO, SCT and UNAM stations are same as the strong-motion stations.

Numerical Evaluation of Wave Propagation in Heterogeneous Viscoelastic Media

J.M. Crepel
Coyne et Bellier, Scientific Computing Department, 5, Rue d'Héliopolis, 75017 Paris, France
A. Pecker
Géodynamique et Structure, 6, Rue Eugène Oudiné, 75013 Paris, France

INTRODUCTION

Solutions to wave propagation problems are of primary importance to many situations among which the following are commonly encountered in engineering practice: earthquake type excitation of soil deposits, wave propagation from an explosive source, simulation of earthquake source mechanism, etc. Besides, some other situations occur for which refined analysis of wave propagation could be valuable. Among these, the need for a detailed evaluation of cross hole tests was recognized by Electricité de France - Service d'études et production thermique et nucléaire (EDF-SEPTEN) as a means of studying wave attenuation caused by material damping. Therefore, a comprehensive program based on field tests and theoretical works was undertaken to assess the possibility of measuring in situ material damping at low strains.

This paper describes the development of an efficient numerical tool capable of dealing with the three dimensional nature of the problem in horizontally layered soil deposits. After a description of the mathematics involved in the development of the computer program, various applications of the methods are presented: comparison of computed and measured particle velocities induced in the soil by harmonic excitation of a square plate, vibrations induced by sheet piles driving, propagation of a Ricker wavelet in a three layers system, generation of synthetic accelerograms. Applications to measurement of in situ material damping will be the next step as soon as field measurements are made available.

PRINCIPLES OF THE METHOD

Definition of the transformed domains

In the frequency domain, the elastodynamic equation can be written in vectorial form:

$$(\lambda + \mu) \text{ grad } [\text{div}(u)] + \mu \, \Delta u + \rho \, \omega^2 \, u + f = 0 \qquad (1)$$

where u is the displacement field in the soil and f is the volumic force field. It is known that results as function of time can be found using a reverse Fourier Transform (F^{-1}) relative to the frequency variable ω. λ and μ are the Lame moduli, whereas ρ stands for the volumic mass. Assuming that the soil is horizontally layered (fig. 1), these coefficients are only functions of the vertical space variable z (using a cylindrical set of coordinates (r, z, Θ)).

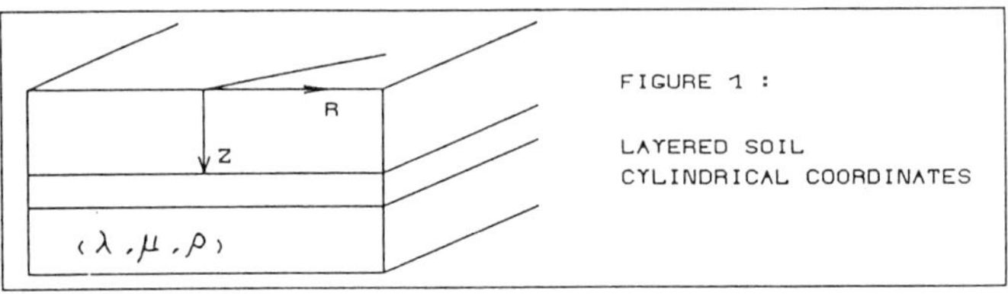

FIGURE 1 :

LAYERED SOIL
CYLINDRICAL COORDINATES

Let us now define the Fourier-Bessel Transform (H) in terms of the horizontal coordinates (r, Θ):

$$H[\psi(r,\Theta)] = \hat{\psi}(k, m) = \iint \psi(r, \Theta) \exp(-im\Theta) J_m(kr) \, r dr \, d\Theta \qquad (2)$$

where ψ is any field defined over the soil domain. Applying this transform to equation (1), all horizontal derivatives no longer appear. Then, we get a differential equation of the variable z, in the spectral domain. The results in the frequency domain will be calculated using the reverse Hankel Transform (H^{-1}):

$$H^{-1}[\hat{\psi}(k, m)] = \psi(r, \Theta) = (1/2\pi) \sum_{m \in \mathbb{Z}} \int \hat{\psi}(k, m) \exp(im\Theta) J_m(k, r) \, kdk \qquad (3)$$

In order to simplify, we will leave out the ^ notation, each time there is no ambiguity.

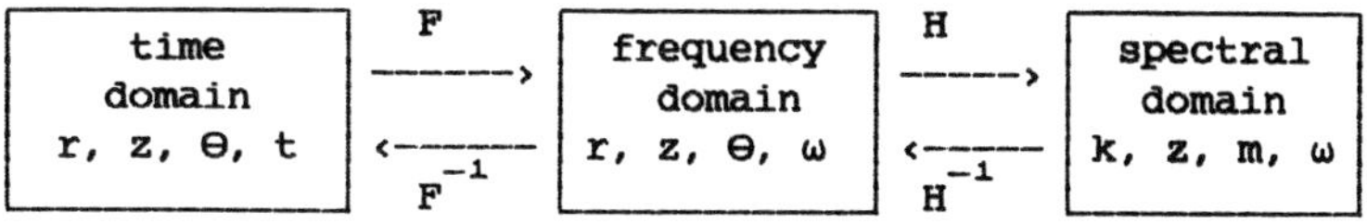

time domain r, z, Θ, t	$\xrightarrow{F}$ $\xleftarrow[F^{-1}]{}$	frequency domain r, z, Θ, ω	$\xrightarrow{H}$ $\xleftarrow[H^{-1}]{}$	spectral domain k, z, m, ω

FIGURE 2: THE 3 DOMAINS OF ANALYSIS

Form of the equations in the spectral domain

Making sure that the horizontal divergence and the rotational vertical component are expressed, we will display uncoupled equations in the spectral domain, which physically represent the propagation of the P-SV waves on the one hand, and of the SH waves, on the other hand.

$$\begin{bmatrix} v_1 = \operatorname{div}(u_H) \\ v_2 = \operatorname{rot}(u_H) \cdot e_z \\ v_3 = u \cdot e_z \end{bmatrix} \qquad \begin{bmatrix} t_1 = \operatorname{div}(\sigma_{Hz}) \\ t_2 = \operatorname{rot}(\sigma_{Hz}) \cdot e_z \\ t_3 = \sigma_{zz} \end{bmatrix} \tag{4}$$

with $u_H = u_r \; e_r + u_\Theta \, e_\Theta$

$\sigma_{HZ} = \sigma_{rz} \; e_r \; + \; \sigma_{\Theta z} \, e_\Theta$

Applying the H-transform on equation (1) and using above variables, we obtain the following uncoupled system:

P-SV waves:

$$\partial_z \begin{bmatrix} v_1/k \\ v_3 \\ t_1/k \\ t_3/k \end{bmatrix} = \begin{bmatrix} 0 & A & B & 0 \\ C & 0 & 0 & D \\ E & 0 & 0 & -C \\ 0 & F & -A & 0 \end{bmatrix} \begin{bmatrix} v_1/k \\ v_3 \\ t_1/k \\ t_3 \end{bmatrix} - \begin{bmatrix} 0 \\ 0 \\ g_1/k \\ g_3 \end{bmatrix}$$

SH waves:

$$\partial_z \begin{bmatrix} v_2/k \\ t_2/k \end{bmatrix} = \begin{bmatrix} 0 & G \\ H & 0 \end{bmatrix} \begin{bmatrix} v_2/k \\ t_2/k \end{bmatrix} - \begin{bmatrix} 0 \\ g_2/k \end{bmatrix}$$

with $A = k \qquad E = 4\,\mu\,(\lambda + \mu)\,(\lambda + 2\mu)^{-1} k^2 - \rho\,\omega^2$

$B = \mu^{-1} \qquad F = -\,\rho\,\omega^2$

$C = -\,\lambda(\lambda + 2\mu)^{-1} k \qquad G = \mu^{-1}$

$D = (\lambda + 2\mu)^{-1} \qquad H = \mu\,k^2 - \rho\,\omega^2$

For further details, the reader can refer to Kennett[4] or Tsakalidis[7]. We will write this system using the symbolic form:

$$\partial_z B = A B + G \tag{5}$$

Vector B components are, in the spectral domain, the transformed components of displacement and of stress vector on a horizontal plane. B is currently called "Thomson-Haskell vector".

Formal solution in source-free regions

Equation (5) is analytically solved diagonalising matrix A. Calculation of the eigenvalues brings out:

P-SV waves: $\gamma_s, \gamma_p, -\gamma_s, -\gamma_p$

SH waves: $\gamma_s, -\gamma_s$

with $\gamma_s = (k^2 - \omega^2 \rho \mu^{-1})^{1/2}$

$\gamma_p = [k^2 - \omega^2 \rho (\lambda + 2\mu)^{-1}]^{1/2}$

Let M be the matrix of the eigenvectors. We then define the wave vector V as:

$$B = M V \tag{6}$$

and system (5) gives the following scalar equations

$$\partial_z V = (M^{-1} A M) V - (M^{-1} G) \tag{7}$$

The general solution within a layer can be expressed as follows:

$$B(z) = M . L (z - z_o) . M^{-1} B(z_o) \tag{8}$$

where L is a diagonal matrix:

P-SV waves:
$L(h) = \mathrm{diag}\,[\exp(\gamma_s h), \exp(\gamma_p h), \exp(-\gamma_s h), \exp(-\gamma_p h)]$

SH waves:
$L(h) = \mathrm{diag}\,[\exp(\gamma_s h), \exp(-\gamma_s h)]$

$B(z_o)$ here stands for the six constants of integration. The B continuity on each interface enables generalizing the previous relation by defining a "Thomson-Haskell propagation operator" through the layered system:

$$B(z_2) = P(z_2, z_1)\ B(z_1) \qquad (9)$$

Adding the free-surface condition and the radiation condition in the halfspace, the problem can be analytically solved everywhere as function of z, k, m and ω.

Sources

When a source is located between z_1 and z_2, equation (9) becomes (see Kennett[4]):

$$B(z_2) = P(z_2, z_1)\ B(z_1) - \int_{z_1}^{z_2} P(z_2, \chi)\ G(\chi)\ d\chi \qquad (10)$$

Let us first consider the case of a point source at level $z = z_s$:

$$G(z) = \delta(z - z_s) \cdot G'$$

Using (10), we notice that the effect of this kind of source leads to a discontinuity of B at the source level:

$$B(z_s+) - B(z_s-) = -G' \qquad (11.1)$$

Let us now suppose the source is uniformly distributed between level $z_s - h$ and $z_s + h$. From (10) we get:

$$B(z_s + h) = P(z_s + h, z_s - h)\ B(z_s - h) - G' \int_{z_s - h}^{z_s + h} P(z_s + h, \chi)\ d\chi$$

and if we define $\bar{B}(z_s+)$ and $\bar{B}(z_s-)$ as:

$$\bar{B}(z_s+) = P(z_s, z_s + h)\ B(z_s + h)$$

$$\bar{B}(z_s-) = P(z_s, z_s - h)\ B(z_s - h)$$

we obtain:

$$\bar{B}(z_s+) - \bar{B}(z_s-) = -G' \int_{z_s - h}^{z_s + h} P(z_s, \chi)\ d\chi \qquad (11.2)$$

The right hand side term represents the equivalent point source due to a distributed source between z_s+h and z_s-h.

With such formula, there is actually six types of sources (fig. 3), more specifically dedicated to cross hole simulation.

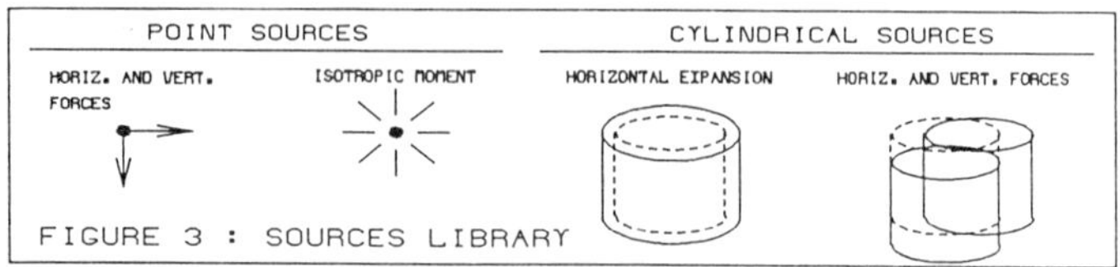

FIGURE 3 : SOURCES LIBRARY

Reflection - transmission (RT) coefficients

Direct implementation of these calculations gives rise to important problems of stability (Track[6], Kennett[4], Tsakalidis[7]). As a matter of fact, L matrix becomes rapidly ill conditioned as the thickness of the layers or the frequency increases. It is then preferable to partition the V wave vector into upgoing and downgoing waves:

$$V = \begin{bmatrix} V_u \\ V_d \end{bmatrix} \qquad (12)$$

The RT coefficients between two points z_1 and z_2 ($z_1 < z_2$) are defined as:

$$V = \begin{bmatrix} V_u(z_1) \\ V_d(z_2) \end{bmatrix} = \begin{bmatrix} T_u & R_d \\ R_u & T_d \end{bmatrix} \begin{bmatrix} V_u(z_2) \\ V_d(z_1) \end{bmatrix}$$

Continuity of B vector across an interface enables calculating the elementary (r_u, r_d, t_u, t_d) coefficients. We can bring out similarly, into a homogeneous layer of thickness h, the following relations:

$$r_u = r_d = 0; \quad t_u = t_d = \exp(\gamma.h)$$

By another way, recursive scheme can be set to calculate the global RT coefficients between any two points. In particular, $R_d^B(z)$ [resp. $R_u^F(z)$] will stand for the global reflection coefficient on the bedrock [resp. under the free surface] at level z.

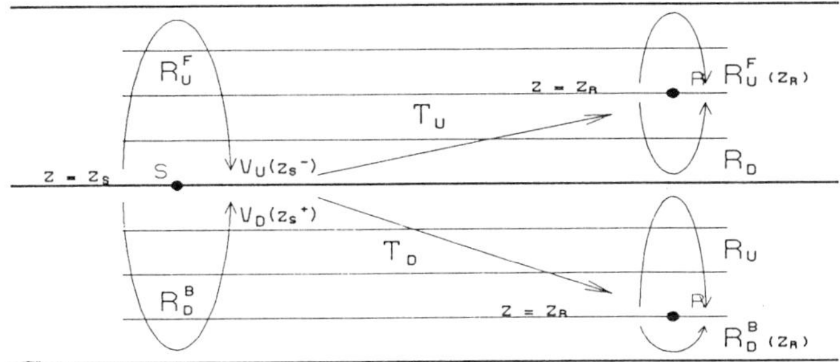

FIGURE 4: GLOBAL RT COEFFICIENTS USED IN THE SOLUTION

If the source is equally partitioned into upgoing and downgoing parts:

$$M^{-1} G = \Sigma = \begin{bmatrix} \Sigma_u \\ \Sigma_d \end{bmatrix} \qquad (14)$$

then it is possible to calculate the solution on both sides of the source, and then for any receiver (fig. 4).

$$V_u (z_s -) = [I - R_d^B R_u^F]^{-1} [R_d^B \Sigma_d - \Sigma_u]$$

$$V_d (z_s +) = [I - R_u^F R_d^B]^{-1} [\Sigma_d - R_u^F \Sigma_u]$$

if $z_r < z_s$

$$V_u (z_r) = [I - R_d R_u^F]^{-1} T_u V_u (z_s -)$$

$$V_d (z_r) = R_u^F (z_r) . V_u (z_r)$$

if $z_r > z_s$

$$V_d (z_r) = [I - R_u R_d^B]^{-1} T_d V_d (z_s +)$$

$$V_u (z_r) = R_d^B (z_r) . V_d (z_r)$$

HARMONIC EXCITATION OF A SQUARE PLATE

In order to evaluate the incident vibrations induced by a vibrating table and an impact machine on an electronic microscope, impact as well as harmonic tests of a square plate (1.20 m² area) were performed on site. The plate was installed at the level of foundation of the planned equipment and particle velocities induced in the soil and on the microscope were measured.

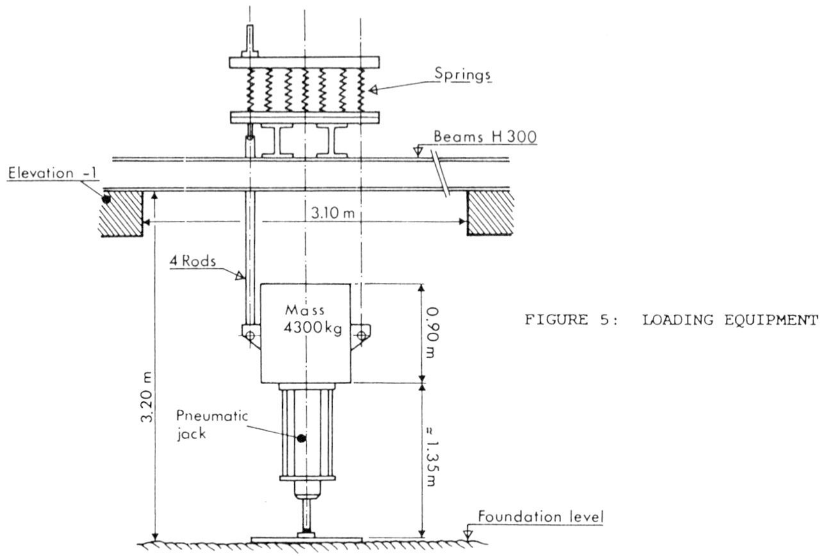

FIGURE 5: LOADING EQUIPMENT

In the following, we will just refer to the harmonic tests. A cross section of the plate with the loading equipment is given on figure 5; measurement of soil particle velocities were taken in a borehole, about 52 m away from the source at depths of 2.0 m and 7.00 m with three directional geophones clamped to the PVC casing with a pneumatic packer. Dynamic properties of the soil were obtained from downhole tests performed in three boreholes. Average values are gathered in figure 6, representing a cross section of the experimental arrangement. In the computer simulation, the soil profile was modelled by a two layers system overlying an half space. The first layer is seven meters thick with a shear wave velocity of 280 m/s, a unit mass of 1 800 kg/m³ and a damping ratio (hysteretic damping) of 2%. The second layer is 4 m thick, has a shear wave velocity of 400 m/s, a unit mass of 1 900 kg/m³ and a damping ratio of 2%. The halfspace has a shear wave velocity of 500 m/s, a unit mass of 2 000 kg/m³ and a damping ratio of 2%.

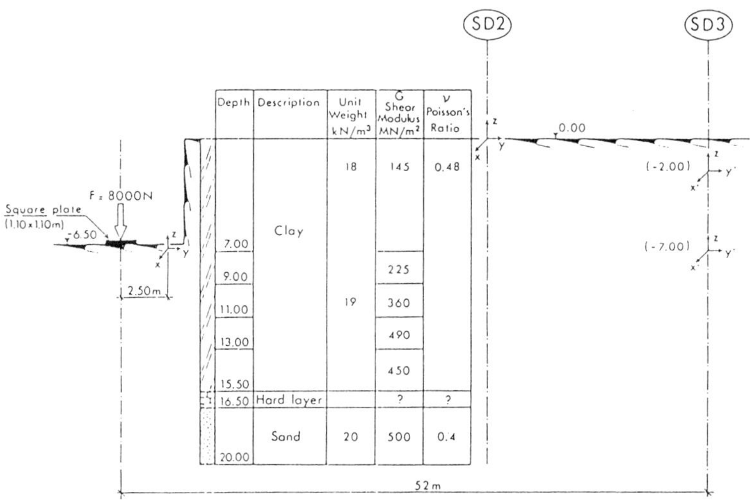

FIGURE 6: CROSS SECTION THROUGH EXPERIMENTAL ARRANGEMENT

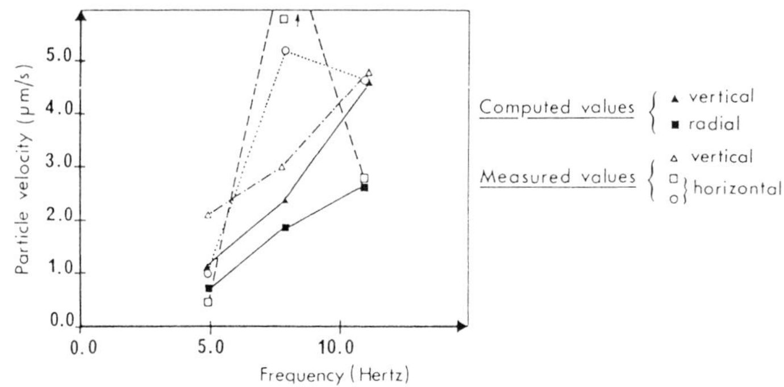

FIGURE 7: PARTICLE VELOCITIES IN BOREHOLE SD3
AT DEPTH 2.0 M

Harmonic tests at frequencies of 5, 8 and 11 Hz with a force amplitude of 8 000 N were performed. Measured particle velocities in the vertical and the two horizontal directions are given in figure 7 at a depth of 2 m. Note that orientation of horizontal components in the holes are unknown. Computed velocities induced by a point force at a depth of 6.5 m within the soil profile are given as crosses on the same figure. In that case, horizontal component corresponds to the radial component since the tangential one is zero. Very good agreement is obtained for the vertical components at all three frequencies.

For the horizontal directions, agreement is also fairly good, except for the 8 hertz frequency where the measurements give an unexplained peak.

For the deeper measuring point (7 m) very good agreement was also obtained for the vertical component; however, for the horizontal components, computed values were at least four times larger than the measured ones. The reason for this discrepancy is presently unknown and parametric studies would have to be performed to assess the importance of the soil damping ratios, the effect of the 1 m thick hard layer encountered at a depth of 15.5 m and ignored in the model.

VIBRATIONS CAUSED BY SHEET PILE DRIVING

Driving of sheet piles with a DELMAG D12 hammer, through a 15 m thick alluvium layer down to a rock formation can induce vibrations in the bedrock when the sheet pile hits the rock surface (fig. 8). The problem is to evaluate the vibration amplitudes 50 m away from the sheet piles where a sensitive old structure is founded on the rock. First of all, the force induced at the sheet pile tip was computed using the wave equation analysis developed by Smith[5]. It is shown that the force time history can be closely represented by a half sinusoid with an amplitude of 5 MN and a period of 3.10^{-3} to 4.10^{-3} second. This function is approximated through a Fourier Transform with 2 harmonics with frequencies 250 Hz and 500 Hz. Response to the two harmonic functions is computed in the frequency domain for the vertical and radial components (figure 9 gives for example, the modulus of the radial displacement versus the distance from the source for the 250 Hz harmonic); an inverse Fourier Transform gives the velocity time history from which the maximum particle velocity is found to be of the order of 1 μm/s. According to various standards for old structures, the value is small enough and no detrimental effects of sheet pile driving is expected on the structure.

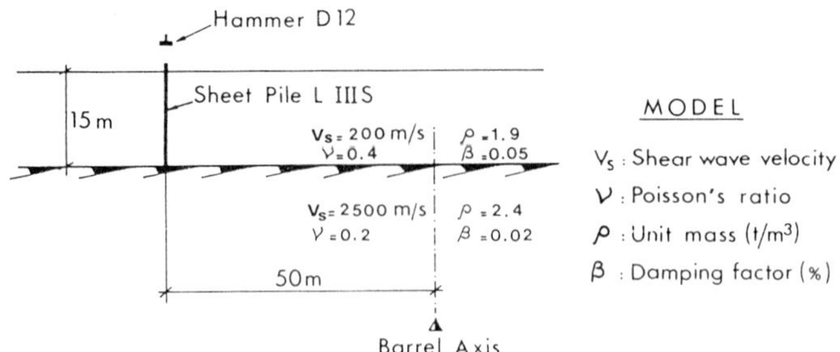

FIGURE 8: CROSS SECTION FOR THE PILE DRIVING TEST

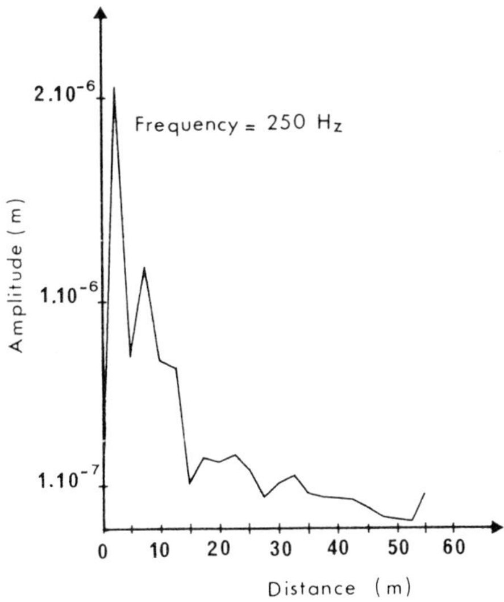

FIGURE 9: SOIL DISPLACEMENT AMPLITUDE IN RADIAL DIRECTION

PROPAGATION OF A RICKER SIGNAL

This model has been used by Dietrich[3], and is made up of three layers over an elastic halfspace. One difficulty is the fact that a soft layer is comprised between two hard layers (velocity inversion). The source is a vertical point force at the surface whose time-dependency varies as a Ricker signal. Vertical seismic sections are shown on figure 10 for three horizontal offsets. All seismograms have been amplified in proportion with time and, then normalized, so that the dominant wave has a unit amplitude (figure 10). For offset 0, we clearly see the propagation of the direct P-wave. We also notice reflected signals, with change of phase for the harder layers, and without change for the softer layer. With increasing offsets (1 000 m, 2 000 m), we can see a surface wave which travels slower than direct P or SV waves. Conversion between P and SV waves at each interface is equally shown.

SYNTHETIC SEISMOGRAMS - COMPARISON WITH EXPERIMENTS

This last example shows seismograms calculated with a real soil profile. Two layers overlie the elastic halfspace encountered at depth 2 200 m. P-wave velocity varies from 2 000 m/s to 5 100 m/s, while S-wave velocity varies from 1 000 m/s to 3 000 m/s. The source is a buried isotropic moment. Horizontal and vertical soil velocities have been measured at

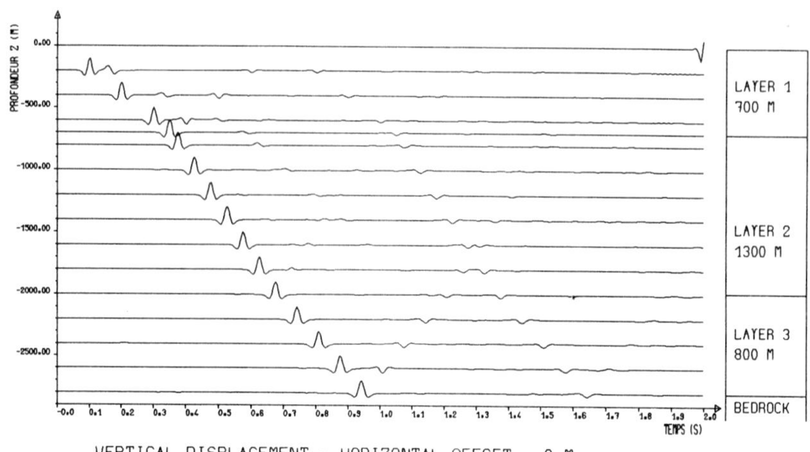

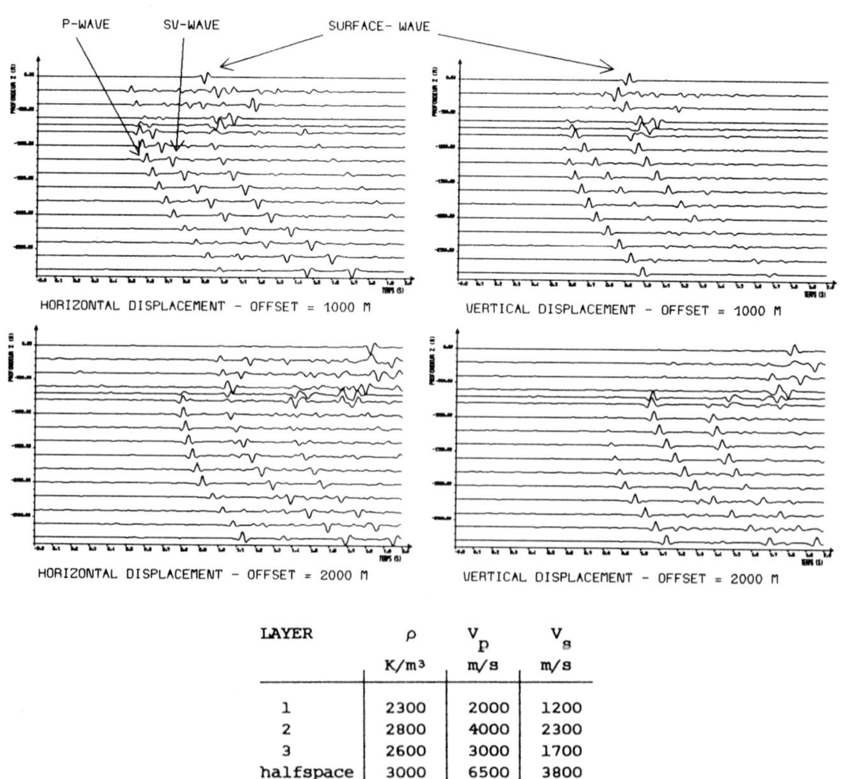

LAYER	ρ K/m³	V_p m/s	V_s m/s
1	2300	2000	1200
2	2800	4000	2300
3	2600	3000	1700
halfspace	3000	6500	3800

FIGURE 10: SIMULATION OF THE PROPAGATION OF A RICKER SIGNAL THROUGH A 3-LAYERED MEDIA

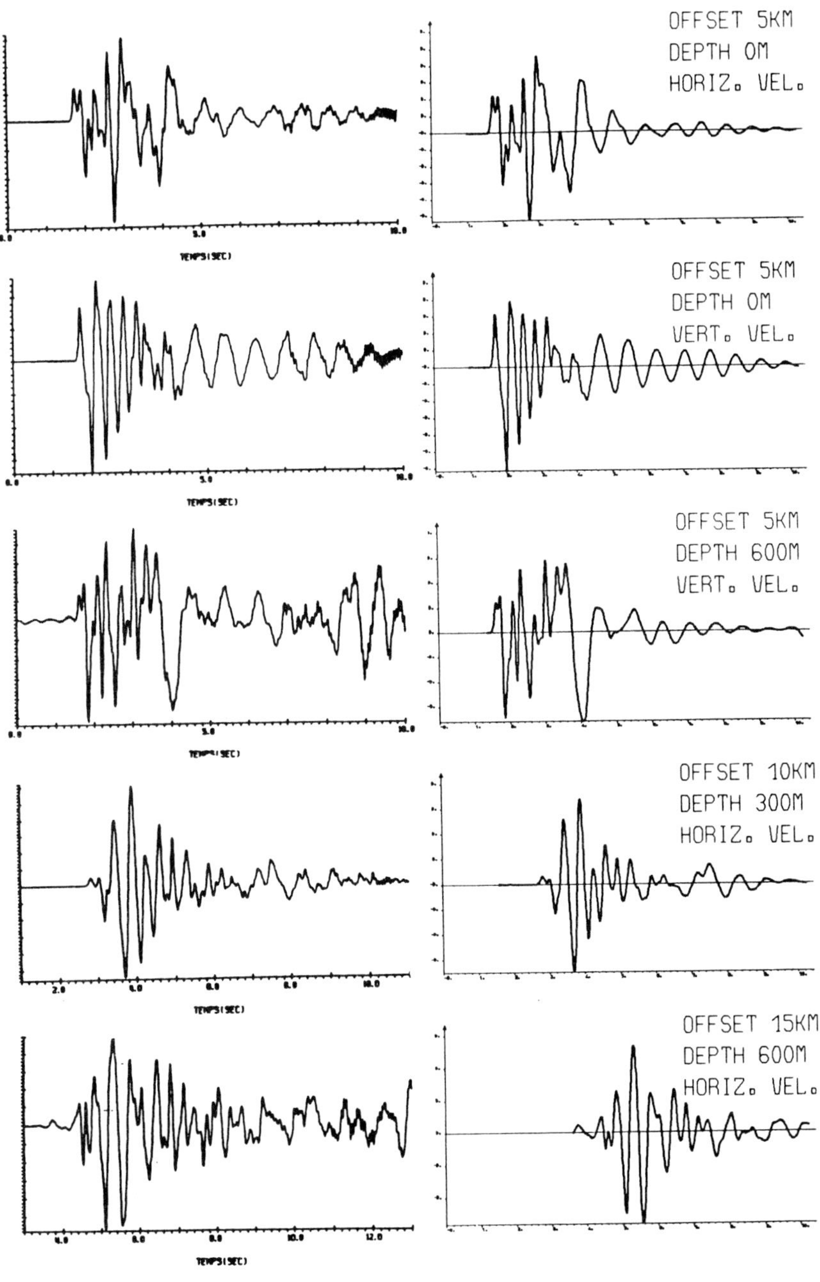

FIGURE 11: COMPARISON BETWEEN EXPERIMENTAL AND CALCULATED SEISMOGRAMS

different receiver points. We notice quite a good similarity of results. All effects are reproduced, even if high frequency disturbances are smoothed by calculation (figure 11). This may be due to internal damping introduced in the model, or to the shortage of high frequency components in the frequency domain.

FURTHER EXTENTS

The method displayed in this paper is the kernel of many programs used in civil engineering or geophysical analysis. An extension has been developed (Aubry et al[1,2], Tsakalidis[7]) to modelise soil-structure interaction with superficial foundations or deep foundations (piles). The near field soil displacement, compatible with the layered media is available, even if the incident wave propagation is not vertical. Green's function of a layered soil can be calculated and then, boundary integral method can be used to solve more complicated problems as local topographic effects on an earthquake, scattering over horizontal heterogeneities, triple interaction between soil, fluid and structure (for dam analysis), etc.

REFERENCES

1. Aubry D., Chapel F., Crepel J.M., (1984), Génie Parasismique, chapitre VI.2, Interaction sol-structure linéaire sur un sol hétérogène, Presse de l'ENPC.

2. Aubry D., Chapel F., Crepel J.M., (1984), Génie Parasismique, chapitre VI.6, Calcul sismique des fondations sur pieux, Presse de l'ENPC.

3. Dietrich M., (1983), Simulation numérique de profils sismiques verticaux, Thèse DDI, Université de Grenoble, France

4. Kennett, B.L.N., (1983), Seismic wave propagation in stratified media, Cambridge University Press

5. Smith E.A.L., (1962), Pile driving analysis by the wave equation, ASCE, paper 3306, vol. 127, part 1

6. Track, A., (1983), Calcul des fonctions de Green dynamiques par une méthode spectrale, Applications au génie parasismique, Thèse DDI, Ecole Centrale de Paris

7. Tsakalidis C. (1985), Diffraction d'ondes sismiques sur les structures sur pieux et fonctions de Green du sol multicouche, Thèse DDI, Ecole Centrale de Paris.

Effect of Layering on the Transmission of Ground Vibration

M. Petyt, D.V. Jones
Institute of Sound and Vibration Research, University of Southampton, Southampton, Hants, U.K.

ABSTRACT

The transmission of vibration over a distance of 25m of the surface of the ground is investigated theoretically. The mathematical model is an elastic layer of finite depth laying on an elastic half-space of different material properties. The model is a two-dimensional one and the surface is assumed to be subjected to a harmonic load acting on a strip of finite width. The solution is obtained using a dynamic stiffness matrix analysis. The attenuation of the amplitude of vibration is presented as a function of distance from the applied load for various frequencies.

To assist in the interpretation of the results, free wave propagation parallel to the surface is investigated by solving an appropriate eigenproblem. The results are presented in the form of dispersion curves.

1.0 INTRODUCTION

In the last decade there has been an increase in the number of complaints about vibration due to surface railways. The major type of complaint concerns the disturbing effects of the vibration, although there is fear that damage could also be caused to property.

The overall problem can be divided into four different aspects: (a) the dynamics of the vehicle-track system, (b) the propagation of vibration through the ground, (c) the response of buildings to the ground vibration and (d) the subjective response of the occupants of the buildings. This paper is concerned with the propagation of vibration through the ground. Of particular interest is the level of vibration within a distance of 25m of the track.

Early workers on ground vibration used analytical techniques to derive expressions for the level of vibration on the surface in the far field (see, for example, Miller and Pursey[1]). The reason for this was that the equations of motion were transformed into the wavenumber domain before solving. The inverse transforms of the solution could only be evaluated in the far field. Recent interest in machine foundations (Gazetas[2]) has led to the development of numerical methods for predicting the response of the ground immediately under the load. Such techniques have been found useful in predicting the level of vibration in the near field.

The model considered is an elastic layer of finite depth laying on an elastic half-space of different material properties. Both free wave propagation and forced response are investigated using the dynamic stiffness analysis method propsoed by Kausel and Roesset[3].

2.0 VIBRATION TRANSMISSION

The model considered is shown in Figure 1. It consists of a homogeneous, isotropic, elastic layer of depth d, overlaying a homogeneous, elastic half-space having different material properties. The surface of the ground is subjected to a uniformly distributed, harmonic load acting on a strip extending to infinity in the y direction, which is perpendicular to the x-z plane, and is of width 2a in the x direction. The magnitude of the load is P/2a per unit length in the x direction. Plane strain conditions are assumed to exist and, therefore, the model is essentially a two-dimensional one. The equations of motion of both the layer and half-space are of the form

$$(\lambda^* + \mu^*)\frac{\partial \Delta}{\partial x} + \mu^* \nabla^2 u = \rho \frac{\partial^2 u}{\partial t^2} \qquad (1)$$

and

$$(\lambda^* + \mu^*)\frac{\partial \Delta}{\partial z} + \mu^* \nabla^2 w = \rho \frac{\partial^2 w}{\partial t^2} \qquad (2)$$

where u,w are the components of displacement in the x and z directions and Δ the dilation which is given by

$$\Delta = \frac{\partial u}{\partial x} + \frac{\partial w}{\partial z} \qquad (3)$$

ρ is the density of the material and λ^*, μ^* are complex Lamé constants which are given by

$$\lambda^* = \frac{\nu E(1+i\eta)}{(1+\nu)(1-2\nu)}, \qquad \mu^* = \frac{E(1+i\eta)}{2(1+\nu)} \tag{4}$$

where E is Young's modulus, ν Poisson's ratio and η the loss factor of the material. Subscripts 1 and 2 in Figure 1 indicate the material properties of the layer and half-space respectively. Due to the uncertainty of the value of the loss factor, the same value was taken for both layer and half-space.

The boundary conditions for the problem consist of the following

(a) On z=0

$$\tau_{zz} = \begin{cases} -\frac{P}{2a}e^{i\omega t} & \text{for} \quad |x| < a \\ 0 & \text{for} \quad |x| > a \end{cases} \tag{5}$$

$$\tau_{zx} = 0 \tag{6}$$

where τ_{zz}, τ_{zx} are the normal and shear stresses, and ω is the frequency of the applied force.

(b) At z = d, the displacements u,w and stresses τ_{zz}, τ_{zx} are continuous across the boundary between the layer and half-space.

(c) At $z = \infty$, there are no reflections. This means that all waves travel in the positive z direction.

The stresses at any point are given by

$$\tau_{zz} = \lambda^* \frac{\partial u}{\partial x} + (\lambda^* + 2\mu^*) \frac{\partial w}{\partial z} \tag{7}$$

$$\tau_{xz} = \mu^* \left(\frac{\partial u}{\partial z} + \frac{\partial w}{\partial x}\right) \tag{8}$$

In principle, the problem posed by equations (1) and (2) subject to the boundary conditions (a), (b) and (c) can be solved by the following procedure:

(i) Introduce potentials, Φ, H such that

$$u = \frac{\partial \Phi}{\partial x} - \frac{\partial H}{\partial z}, \qquad w = \frac{\partial \Phi}{\partial z} + \frac{\partial H}{\partial x} \tag{9}$$

(ii) assume harmonic motion, and so

$$\Phi = \Phi(x,z)e^{i\omega t}, \qquad H = H(x,z)e^{i\omega t} \tag{10}$$

(iii) Take the Fourier Transform with respect to x of the resulting equations and boundary conditions. The Fourier transform of a function f(x) is given by

$$\bar{f}(k) = \frac{1}{2\pi} \int_{-\infty}^{+\infty} f(x)e^{-ikx}\, dx \tag{11}$$

where k represents wavenumber. This process converts the problem to the wavenumber domain.

(iv) Solve the transformed equations subject to transformed boundary conditions,

and (v) Perform the inverse transform

$$f(x) = \int_{-\infty}^{+\infty} \bar{f}(k)e^{ikx}\, dk \tag{12}$$

to return to the physical plane.

In practice, this technique suffers from numerical difficulties. The reason for this, is that the solution of the transformed equations involves exponential terms of the form $e^{\alpha z}$ where α is of similar magnitude to the wavenumber k. Evaluating these terms for z=d and large wavenumbers produces computational overflow.

This difficulty can be overcome by using the dynamic stiffness approach suggested by Kausel and Roesset[3]. This involves dividing the finite layer into a number of sub-layers, each of depth h, as shown in Figure 2. The general solution of the transformed equations are obtained for an individual sub-layer. The constants of integration are obtained by evaluating the solution for the transformed displacements at z=0 and z=h. The stresses are then calculated at z=0 and z=h. This results in the relationship

$$\{\bar{\tau}\} = [k]\{\bar{u}\} \tag{13}$$

for the layer, where

$$\{\bar{\tau}\}^T = \lfloor -i\,\bar{\tau}_{zz}(0) \;\; -\bar{\tau}_{zx}(0) \;\; i\bar{\tau}_{zz}(h) \;\; \bar{\tau}_{zx}(h) \rfloor \tag{14}$$

are the transformed components of stress,

$$\{\bar{u}\}^T = \lfloor\ i\ \bar{w}(0) \quad \bar{u}(0) \quad i\bar{w}(h) \quad \bar{u}(h)\ \rfloor \tag{15}$$

and [k] is a dynamic stiffness matrix, the elements of which are given by Kausel and Roesset[3], which they derived by a different procedure. The complex quantity $i = \sqrt{-1}$ has been introduced into (14) and (15) to ensure that the dynamic stiffness matrix is symmetric.

The dynamic stiffness matrix for the half-space takes the form (13) where

$$\{\bar{\tau}\}^T = \lfloor\ i\ \bar{\tau}_{zz}(0) \quad -\ \bar{\tau}_{zx}(0)\ \rfloor \tag{16}$$

and $$\{\bar{u}\}^T = \lfloor\ i\ \bar{w}(0) \quad \bar{u}(0)\ \rfloor \tag{17}$$

The sublayer dynamic stiffness matrices can be assembled using the same procedure that is used in finite element anlaysis (see Cook[4]) to give

$$\{\bar{T}\} = [K]\ \{\bar{U}\} \tag{18}$$

where $\{\bar{U}\}$, $\{\bar{T}\}$ contain the transformed components of displacement and applied stress at the interface between sub-layers and [K] the assembled dynamic stiffness matrix. Now the applied stress is zero at every level except the surface of the ground. Transforming equations (5) and (6) gives

$$\bar{\tau}_{zz}(0) = -\frac{P}{2\pi}\ \frac{\sin ka}{ka} \tag{19}$$

$$\bar{\tau}_{zx}(0) = 0 \tag{20}$$

Equation (18) is now solved for a given frequency and wavenumber to give the displacements at the interfaces between sublayers. Of particular interest are the components of displacement at the surface of the ground. This process is repeated for a range of wavenumbers. Finally, the inverse transform (12) is evaluated numerically. This is often carried out using discrete Fourier transforms. However, in the present case it was found that the functions to be inverse transformed were well behaved and that a quadrature method was more efficient. The method chosen is based upon the method of Piessens and Branders[5].

TABLE 1: MATERIAL PROPERTIES OF MODEL

Property	Layer	Half-space
$E(MN/m^2)$	269	20426
η	0.1	0.1
ν	0.257	0.179
$\rho(Kg/m^3)$	1550	2450
$c_1(m/s)$	459.4	3007.3
$c_2(m/s)$	262.7	1880.3
$c_R(m/s)$	241.9	1707.2

The material properties of the layer, which is 7m deep, and half-space are given in Table 1. Also included in this Table are the compression wave speed c_1, shear wave speed c_2 and Rayleigh wave speed c_R. In the case of the half-space the Rayleigh wave speed is that which would exist in the absence of the layer. The values of Young's modulus and Poisson's ratio are measured values for a particular site and represent average values for both the layer and half-space. The density was estimated using an empirical relationship between density and compression wavespeed. The loss factor is a typical value suggested by a survey of published measured values. The other parameters used are a=0.75m and $P=2\pi$ N/m^2.

Calculations were performed using two sublayers of equal depth. The means that the stiffness matrix in Equation (18) is of order (6x6). A typical plot of the modulus of the transform of the vertical displacement on the surface, $|w|$, as a function of wavenumber, is shown in Figure 3 for a frequency of 32Hz, in non-dimensional form. Within the range considered there are four peaks of differing magnitudes. The largest peak is the fourth one which occurs close to the layer Rayleigh wave value of ka=0.623. The second largest is the first peak which occurs close to the half-space shear wave value of ka=0.080. The two smaller peaks do not occur at wavenumbers which correspond to any of the other waves referred to in Table 1.

In the case of a layer on a rigid foundation, it is easy to see that standing waves could occur between the surface and the base. These standing waves can propagate to the left and right. It could be expected that these waves would give rise to peaks in a wavenumber plot. These facts have been demonstrated to be true by Jones[6]. The next section investigates whether such horizontally propagating waves exist in the present case, in an attempt to explain the presence of the additional peaks in Figure 3.

The inverse Fourier transform of the wavenumber plot gives the spatial variation of the displacements. Figure 4 shows the attenuation of the vertical component of displacement with distance for four frequencies 4,8,16 and 32Hz. In general the attenuation increases with distance and frequency. The curves corresponding to 4,8 and 16Hz exhibit less waviness than the one for 32Hz. This could be due to either fewer propagating modes or longer wavelengths. This is investigated in the next section.

3. FREE WAVE PROPAGATION

The equation of motion of free wave propagation can be obtained by putting $\eta=0$ and P=0 in equation (18). This gives

$$[K]\{\bar{U}\} = 0 \tag{21}$$

The stiffness matrix is a function of both frequency and wavenumber. In general, wavenumbers representing motion in a waveguide can be either real or complex. Only real wavenumbers are of interest since they represent modes which can propagate. Modes with complex wavenumbers do not propagate energy. Fortunately, in the present case, modes with complex wavenumbers only exist at wavespeeds greater than the shear wavespeed in the half-space, which is much higher than the body wavespeeds in the layer (see Table 1). The eigenproblem (21) can be solved for real wavenumbers by assuming a value for wavenumber and solving the eigenproblem for the frequencies which satisfy

$$|K| = 0 \tag{22}$$

This equation has been solved by the method developed by Wittrick and Williams[7]. The method uses the fact that the number of frequencies which satisfy equation (22) which are less than a fixed chosen frequency ω^*, $J(\omega^*)$ is given by

$$J(\omega^*) = J_O(\omega^*) + S([K(\omega^*)]) \tag{23}$$

where $J_O(\omega^*)$ is the number of frequencies of the system below ω^* when constraints are applied at the layer interfaces to prevent motion there. $S([K(\omega^*)])$ is known as the sign count function. It is evaluated by applying Gauss elimination to the matrix $[K(\omega^*)]$ to convert it to upper triangular form and then counting the number of negative elements on the main diagonal. If the layer is divided into sublayers of equal depth, $J_O(\omega^*)$ can be evaluate using a single sublayer having zero displacements on its two boundaries. In the present application the maximum frequency of interest is 64Hz. If the depth of a single layer is made sufficiently small, then $J_0(\omega^*)$ can be made zero for all $\omega^* \leq 128\pi$ rad/sec. Jones[6] has shown that for the application considered here, dividing the layer into four sublayers was sufficient. This means that equation (23) reduces to

$$J(\omega^*) = S([K(\omega^*)]) \qquad (24)$$

Having established a method of predicting the number of frequencies below a given frequency, ω^*, the values of the frequencies can be determined by a method of bisection. The mode shapes corresponding to these frequencies could be obtained by solving Equation (21) after evaluating [K] for the wavenumber-frequency pair under consideration. However, it was found to be more convenient to calculate the mode shapes from an analytical solution of the problem. Details are given by Jones[6].

As a check on the method, calculations were performed using the same data as Newlands[8] who produced results which are valid for large wavenumbers. It was found that for kd > 25 the results differed from Newlands' by less than 1%.

Calculations have also been performed using the parameters given in Section 2. The variation of wavenumber with frequency for the first six modes, for wavenumbers greater than the shear wavenumber in the half-space, is shown in Figure 5. For reference purposes, three additional lines have been included, to show the variation of wavenumber with frequency of the compression and shear waves in the layer and the shear wave in the half-space. The first mode is coincident with the line for the Rayleigh wave of the layer above 30Hz. Therefore, in this range the first mode propagates at the same speed as the Rayleigh wave. It can also be seen that at high frequencies the second mode propagates at the speed of the layer shear wave. Extending the figure further would indicate that the other modes do the same. The effect of the compression wave can be seen for modes 2,3 and 4. The curves for these modes exhibit an inflection as they cross the compression wave line.

Figure 5 shows that at 32Hz four modes with different wavenumbers can propagate. The first mode has a wavenumber close to the Rayleigh wavenumber for the layer and the fourth a wavenumber close to the shear wavenumber for the half-space. These produce the fourth and first peaks in Figure 3. The wavenumbers of the second and third modes correspond to the wavenumbers of the third and second peaks in Figure 3.

Figure 5 also shows that at 4,8 and 16Hz only two modes propagate, whilst at 32Hz four modes are propagating. This, together with the fact that wavelengths decrease with increase in frequency, explains the difference in the appearance of the 32Hz curve in Figure 4 from the 4,8 and 16Hz curves.

4. CONCLUSIONS

A two-dimensional model has been developed for investigating the propagation of surface vibration. The model consists of a finite layer overlaying a half-space having different material properties. Attenuation curves have been produced for various frequencies. These show that the attenuation increases with distance from the source and also frequency of the applied force. Free wave propagation analysis shows that at low frequencies only two modes propagate whilst at higher frequencies many more modes can propagate.

5. ACKNOWLEDGEMENTS

The authors would like to thank British Rail Research Department for their support and guidance and also for permission to publish the paper.

6. REFERENCES

1. Miller G.F. and Pursey H. (1954). The field and radiation impedance of mechanical radiators on the free surface of a semi-infinite isotropic solid. Proc.Roy.Soc.Lond., A223, pp521-541.

2. Gazetas G. (1983). Analysis of machine foundation vibrations: state of the art. International Journal of Soil Dynamics and Earthquake Engineering. Vol 2, pp2-42.

3. Kansel E. and Roesset J.M. (1981). Stiffness matrices for layered soils, Bulletin of the Seismological Society of America, Vol 71, pp1743-1761.

4. Cook R.D. (1981). Concepts and Applications of Finite Element Analysis. Second edition, John Wiley and Sons, New York.

5. Piessens R. and Branders M. (1975). Computation of oscillating integrals. Journal of Computational Applied Mathematics. Vol I, pp153-164.

6. Jones D.V. (1987). The Surface Propagation of Ground vibration, Ph.D. Thesis, University of Southampton.

7. Wittrick W.H. and Williams F.W. (1971). A general algorithm for computing natural frequencies of elastic structures. Quarterly Journal of Mechanics and Applied Mathematics, Vol 24, pp263-284.

8. Newlands M. (1952). The disturbance due to a line source in a semi-infinite elastic medium with a single surface layer. Philosophical Transactions, Vol 245, pp213-308.

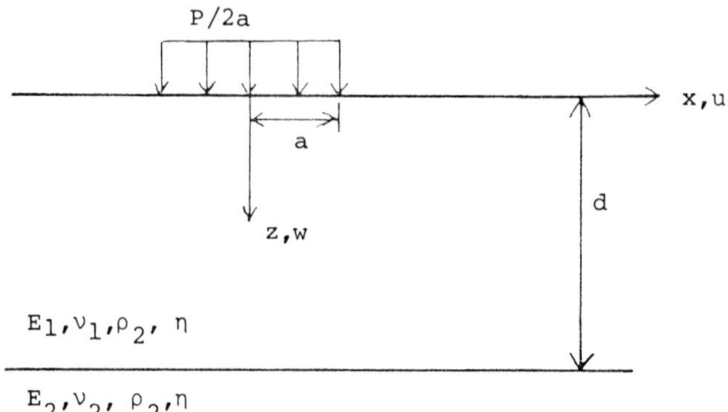

FIGURE 1: Two-dimensional model of layered ground

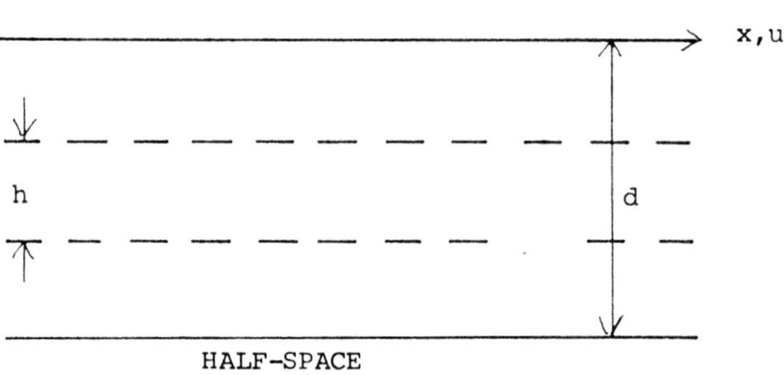

FIGURE 2: Division of finite layer into sub-layers

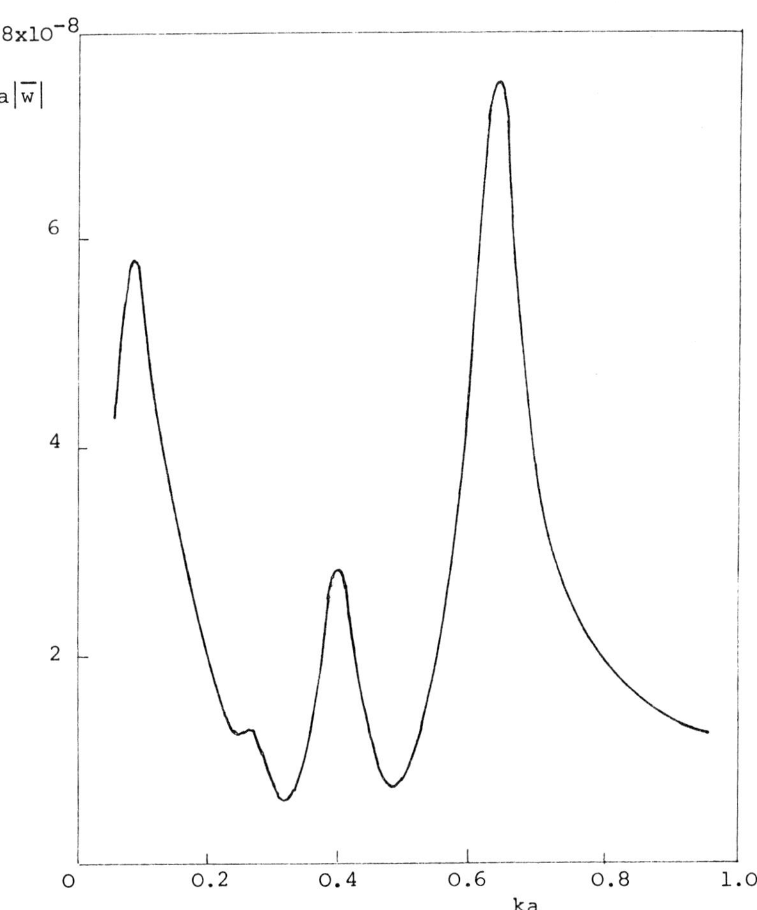

FIGURE 3: Modulus of the transform of vertical motion on the surface against wavenumber

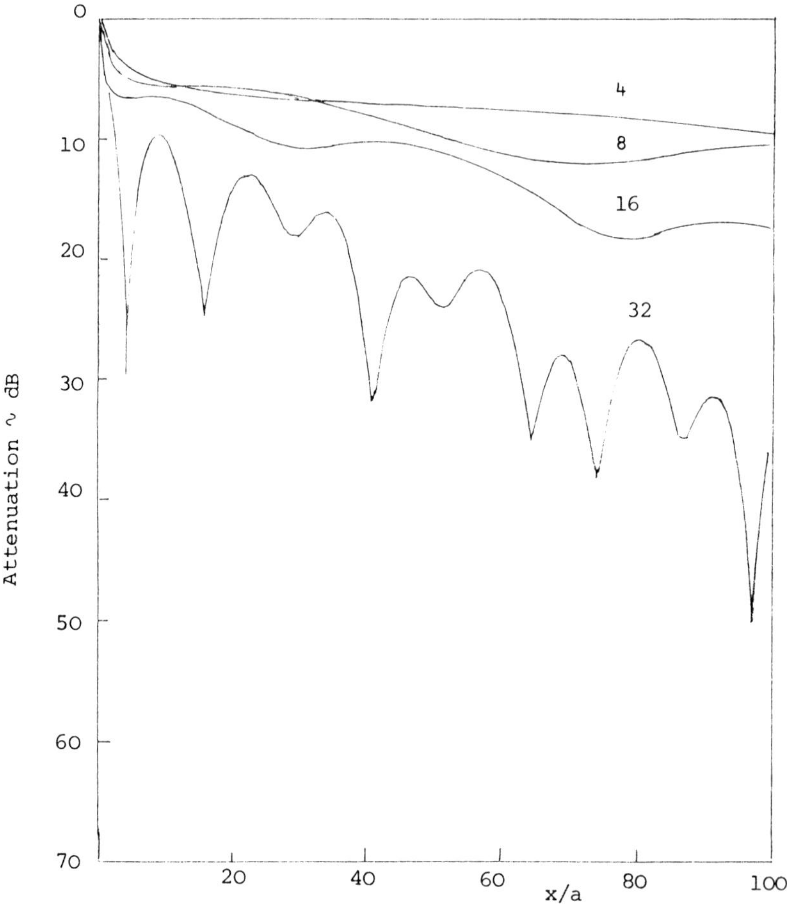

FIGURE 4: Attenuation with distance at 4,8,16,32Hz

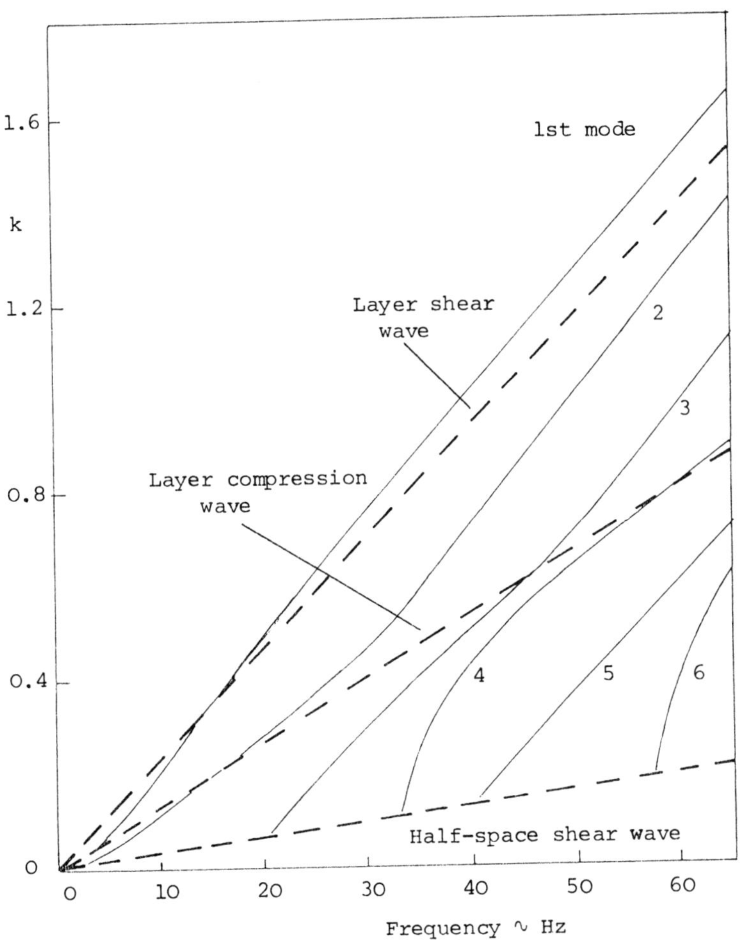

FIGURE 5: Dispersion curves for free wave propagation

Behaviors of Alluvial Plain with Irregular Topography by the Wave Propagation

H. Yokoyama
Technical Research Institute, Konoike Construction Co., Ltd., Osaka, Japan
I. Toriumi
The Faculty of Technology, Fukui University, Fukui, Japan

INTRODUCTION

One of the authors proposed a dynamic analysis of soil-foundation system, namely the numerical integral expression of the wave equations to determine dynamic response analysis which considers freely the side pressure and is applicable for any depth of embedment[1]. This analysis method[2] is a modification of the method of Takahashi and Honma[3] which was used for solving the heat conduction equations and the wave equations. The potentials ϕ and ψ of body waves (dilatational wave and distortional wave) at time $(t + \tau)$ in the medium through which wave spreads can be expressed by using the potentials ϕ and ψ of time (t) and $(t - \tau)$. This method allows separation of displacement potential of body waves. Namely it is step-by-step explicit analysis. It is useful for explaining and representation of wave motion.

The research group researched also the mechanics of surface wave which enters sideways into the alluvial plain[4]. In this research we analyzed two types of numerical analysis model which simplify the topographic irregularity of alluvial plain in the conditions of horizontal input from rigid bedrock and examined the behaviors of alluvial plain caused by irregularity. One of the models relates to the case where the input bedrock is horizontal and even but a homogeneous rectangular protrusion exists on the alluvial plain. Other model relates to the case where the free surface is horizontal and even but a rectangular dent exists on the input bedrock.

OUTLINE OF THE NUMERICAL INTEGRAL EXPRESSIONS FOR THE WAVE EQUATIONS

The numerical integral expressions for the wave equations used in this paper are roughly as follows. The detailed expressions are recorded in reference 1.

The potential functions ϕ and ψ of the wave equations can be expressed as follows.

$$\frac{\partial^2\phi}{\partial t^2} = \alpha^2\left(\frac{\partial^2}{\partial x^2} + \frac{\partial^2}{\partial z^2}\right)\phi \tag{1}$$

$$\frac{\alpha^2\psi}{\partial t^2} = \beta^2\left(\frac{\partial^2}{\partial x^2} + \frac{\partial^2}{\partial z^2}\right)\psi \tag{2}$$

where

ϕ and ψ are the potential functions of dilatational and distortional waves.

α is the velocity of dilatational waves, and β is the velocity of distortional waves.

$$\alpha^2 = \frac{\lambda + 2\mu}{\rho}, \quad \beta^2 = \frac{\mu}{\rho}$$

λ and μ are the Lame constants.
ρ is the density of the medium.

The wave equations (1), (2) can be written in the forms of numerical double integral expressions by Honma, Takahashi method respectively. The above double integral can be replaced by single integral and differential terms, with the aid of Tayler expansion. Applying Simpsons 1/3 rule for single integral terms and numerical differention by Collatz for differential terms, and adding an originated technique to the above expressions, ϕ (x, z, t + τ) and (x, z, t + τ) potential functions are expressed with the integral (grid) interval λs as follows finally.

$$\begin{aligned}\phi(x, z, t+\tau) &= \frac{1}{p}\{\phi(x, z-s, t) + \phi(x-s, z, t) - 4\phi(x, z, t) \\ &\quad + \phi(x+s, z, t) + \phi(x, z+s, t)\} + 2\phi(x, z, t) \\ &\quad - \phi(x, z, t-\tau) + o(s^4) + o(\tau^4)\end{aligned} \tag{3}$$

$$\begin{aligned}\psi(x, z, t+\tau) &= \frac{1}{p\gamma}\{\psi(x, z-s, t) + \psi(x-s, z, t) - 4\psi(x, z, t) \\ &\quad + \psi(x, z+s, t) + \psi(x+s, z, t)\} + 2\psi(x, z, t) \\ &\quad - \psi(x, z, t-\tau) + o(s^4) + o(\tau^4)\end{aligned} \tag{4}$$

where

τ is the numerical time interval, and s is the numerical integral (grid) interval.

$$S = \sqrt{P}\alpha\tau, \quad \gamma = \left(\frac{\alpha}{\beta}\right)^2$$

And then displacements u, w and stresses σx, σz, τxz are described as follows.

$$u(x, z, t) = \frac{1}{2s}\{-\phi(x-s, z, t) + \phi(x+s, z, t) - \psi(x, z-s, t) + \psi(x, z+s, t)\} + o(s^2) \quad (5)$$

$$w(x, z, t) = \frac{1}{2s}\{\phi(x, z-s, t) - \phi(x, z+s, t) - \psi(x-s, z, t) + \psi(x+s, z, t)\} + o(s^2) \quad (6)$$

$$\sigma x(x, z, t) = \frac{\lambda}{s^2}\{\phi(x, z-s, t) + \phi(x-s, z, t) - 4\phi(x, z, t) + \phi(x, z+s, t) + \phi(x+s, z, t)\} + \frac{\mu}{2s^2}\{4\phi(x-s, z, t) - 8\phi(x, z, t) + 4\phi(x+s, z, t) + \psi(x-s, z-s, t) - \psi(x-s, z+s, t) - \psi(x+s, z-s, t) + \psi(x+s, z+s, t)\} + o(s^2) \quad (7)$$

$$\sigma z(x, z, t) = \frac{\lambda}{s^2}\{\phi(x, z-s, t) + \phi(x-s, z, t) - 4\phi(x, z, t) + \phi(x, z+s, t) + \phi(x+s, z, t)\} + \frac{\mu}{2s^2}\{4\phi(x, z-s, t) - 8\phi(x, z, t) + 4\phi(x, z+s, t) - \psi(x-s, z-s, t) + \psi(x-s, z+s, t) + \psi(x+s, z-s, t) - \psi(x+s, z+s, t)\} + o(s^2) \quad (8)$$

$$\tau xz(x, z, t) = \frac{\mu}{2s^2}\{-\phi(x-s, z-s, t) + \phi(x-s, z+s, t) + \phi(x+s, z-s, t) - \phi(x+s, z+s, t) + 2\{-\psi(x, z-s, t) + \psi(x-s, z, t) - \psi(x, z+s, t) + \psi(x+s, z, t)\} + o(s^2) \quad (9)$$

For an example of the numerical calculations, when one cycle sine wave is excited at a point on surface of semi-infinite medium, the displacement-distribution is shown Fig. -1 [5].

BEHAVIORS OF ALLUVIAL PLAIN HAVING A HOMOGENEOUS RECTANGULAR PROTRUSION

If the plain bottom and free surface are horizontal and even, their motion caused by the same phase horizontal input is similar to shear vibrational mode (see Fig. 2-a)[6]. In order to reveal the behavior when a homogeneous protrusion exists on the plain surface we examined by horizontally inputting the sine wave to the plain at the same time, assuming height and width of protrusion and thickness of alluvial plain as variables [7].

Thickness of plain : HG = 1.0 λs, width of protrusion : B = 0.5 λs are taken as constant. Height of protrusion : Hp = 0.5 λs, 0.75 λs, 1.0 λs and 1.5 λs (where λs is distortional wave length) is varied. Fig. 2 b to e show the distribution of displacement in plain in the condition where the alluvial plain thickness and protrusion width are constant and protrusion height is varied as stated above. The distribution of protrusion displacement changes significantly depending on value of Hp. When protrusion height is Hp = 0.5 λs, one apparent vortex is formed, vortex center being center of figure of protrusion. Once the vortex is formed at the protrusion, it appears as if vortex moves only in the protrusion. If Hp = 0.75 λs, an apparent vortex grows more than at Hp = 0.5 λs, and spreads to the upper part, and below it the horizontal displacement appears significantly. But the rotational component of protrusion bottom is few. In the anti-symmetrical diagram the distribution of protrusion displacement appears as secondary mode. When Hp = 1.0 λs, two vortexes are formed, and moreover the rotational component of protrusion bottom grows. When Hp = 1.5 λs, one more vortex is being formed, and the rotational component of protrusion bottom also appears although it is not so distinctive as at Hp = 1.0 λs. The alluvial plain under the protrusion is significantly affected by the protrusion. The plain is most significantly disturbed when Hp = 1.0 λs where the rotational component of protrusion is most significant. The influence of protrusion propagates to the distance equal to approx. 5-fold width of protrusion.

Fig. 2-c, f, and g indicate the distribution of plain displacement when the protrusion width is changed, namely B = 0.5 λs, 1.0 λs and 1.5 λs, in the conditions that the alluvial plain thickness (HG = 1.0 λs) and protrusion height (Hp = 0.75 λs) are constant. The protrusion in case of B = 0.5 λs features significant bending deformation. The bending deformation reduces as B grows. After B = 1.0 λs the displacement distribution changes significantly, and the shape of vortex changes concurrently with this. The alluvial plain is also remarkably disturbed when B = 1.0 λs and 1.5 λs where the rotational component of protrusion bottom is significant as in case of above-stated change of Hp (Fig. 2-b to e).

Fig. 2-d, h, i and j show the distribution of displacement in plain when the shape of protrusion is kept unchanged (Hp = 1.0 λs, B = 0.5λs) and the alluvial plain thickness is changed HG = 0.5 λs, 1.0 λs, 1.5 λs and 2.0 λs. Although the distribution of protrusion displacement does not change significantly, vortex flow at the protrusion bottom varies significantly. Entry into the plain is most notable when plain thickness is small, i.e. HG = 0.5 λs. If HG = 1.0 λs and 1.5 λs, the state is almost similar to that in case of Hg = 0.5 λs, but if HG = 2.0 λs, entry into plain is less. Consequently, the alluvial plain under protrusion is disturbed. Fig. 2-k indicates an example where the input wavelength is twice as long as input wavelength shown in Fig. 2-b to j. If wavelength is long, the depth of plain to which influence of the protrusion spreads increases.

Thus, if a protrusion exists, the displacement distribution of the alluvial plain varies complicatedly depending on input wavelength, alluvial plain thickness, protrusion height and width. There appears fluid vortex which can not be treated as shear vibration.

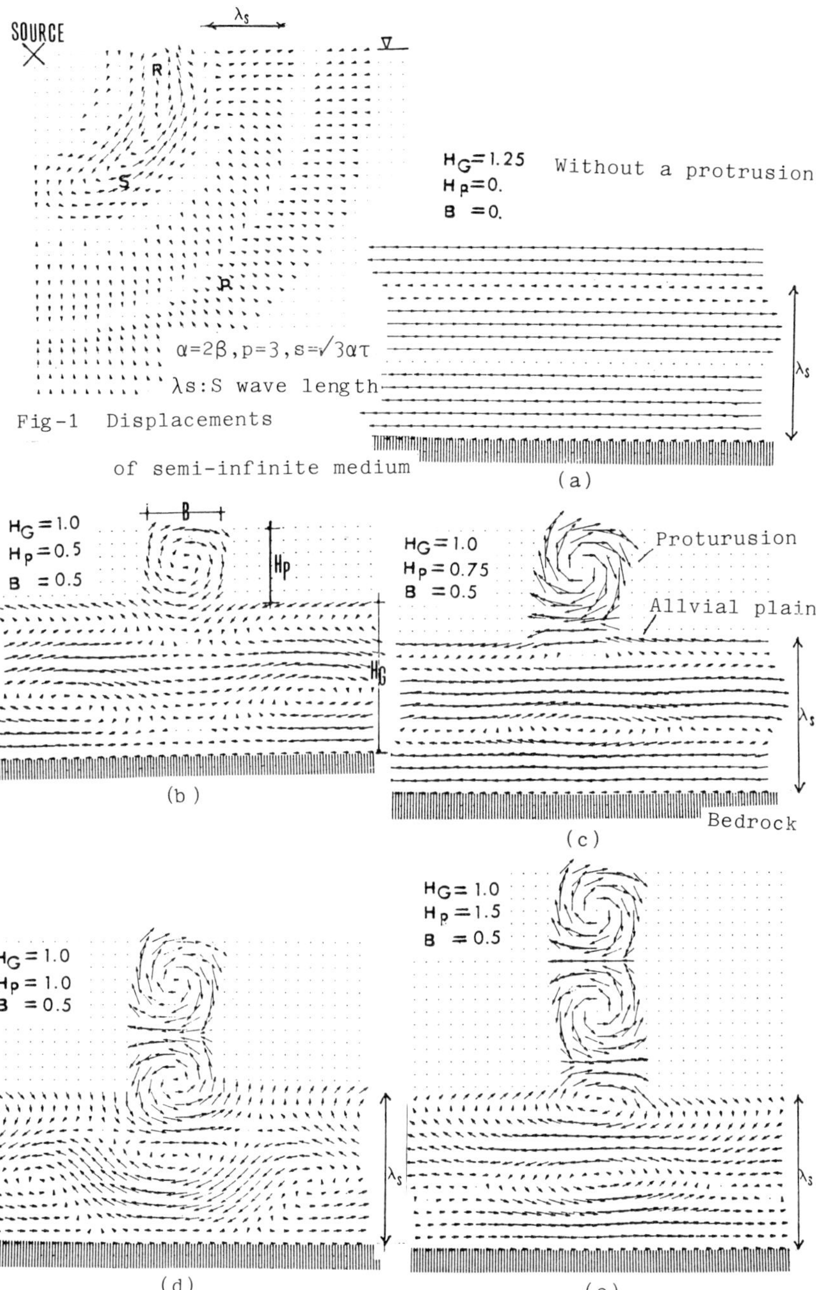

Fig-1 Displacements of semi-infinite medium

Fig-2(a)~(e) Displacements of alluvial plain with a protrusion. ($\alpha=2\beta, p=3, s=\sqrt{3}\alpha\tau$, λs: S wave length, Unit of HG, HP and B is λs)

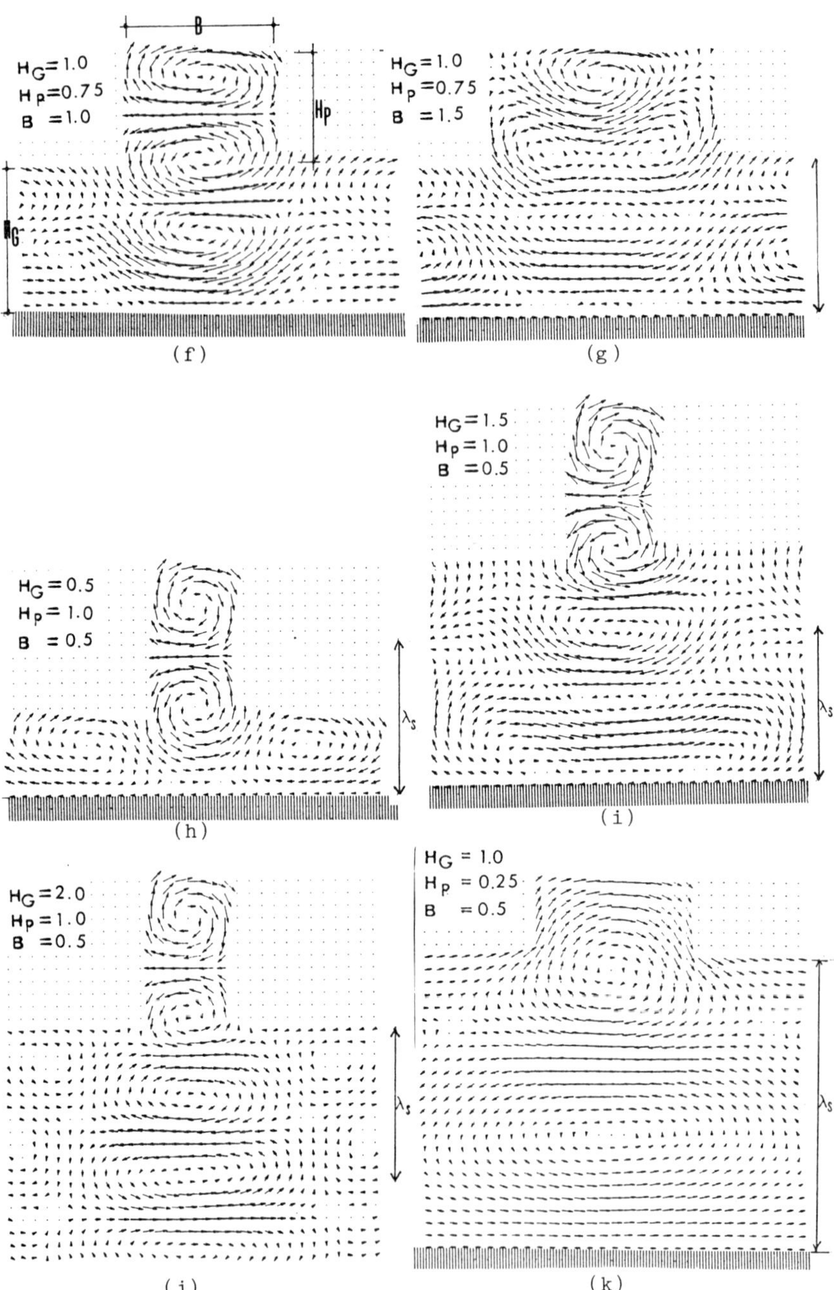

Fig-2(f)~(k) Displacements of alluvial plain with a protrusion. ($\alpha=2\beta$, p=3, $s=\sqrt{3}\alpha\tau$, λs: S wave length, Unit of HG, HP, and B is λs)

BEHAVIORS OF ALLUVIAL PLAIN HAVING A HOMOGENEOUS RECTANGULAR DENT ON BEDROCK

If the bedrock has non-uniform shape, naturally the behavior of alluvial plain is also affected significantly. Assuming that HG1 is layer thickness on dent, HG2 is layer thickness of other parts (hereinafter referred to as other layer) and Bw is width of dent, we examined two models (basic models), inputting horizontally the sine wave to the alluvial plain as stated in section above at the same time.

One of them is model 1 where the free surface of another layer becomes node when the free surface of dent upper layer becomes loop of vibration in case of same phase input as shown in Fig. 3, left (lower). With this model, the dent layer thickness is HG1 = 5/4 λs, thickness of other layer is HG2 = 1.0 λs. The other is model 2 (Fig. 3, right, upper) featuring inverse phase of motions of dent layer surface and surface of other layer where HG1 = 5/4 λs, HG2 = 3/4 λs. Fig. 3 represents free surface horizontal displacement (Us) of points when the dent width Bw is changed for models 1 and 2 (since the figure is anti-symmetrical, the left half is omitted. The same is valid for Fig. 4 and Fig. 5. The delta mark indicates the position of end of dent.) As a result it has been proved that [1] with model 1 the influence of dent upon the alluvial plain is less if dent width is smaller than Bw = 0.5 λs, [2] there is a peak of free surface horizontal displacement at a distance of about 3/4 λs to dent from the edge of dent except when Bw is small (0.5 λs, 0.67 λs), and [3] in case of Bw = 3.5 λs there are two peaks of displacement, and a valley is formed between these peaks, and if Bw = 2.5 λs it takes a shape resembling trapezoid, but if Bw = 1.5 λs one peak appears in the center, [4] the surface horizontal displacement is maximum if Bw = 1.5 λs. With the model 2, [1] the surface horizontal displacement of dent layer and other layer give motions of inverse phase if Bw is 1.67 λs or more (namely influence of dent is remarkable), [2] the influence of dent reduces, resulting in the motion of the same phase as that of other layer if Bw is less that 1.0 λs.

Fig. 4 represents distribution of alluvial plain displacement for models 1 and 2. The upper of figure is for model 1 where the dent width Bw is 0.5 λs, 1.67 λs and 2.5 λs whereas the lower is for model 2 where Bw is 0.5 λs, 1.5 λs and 2.5 λs. With models 1 and 2, the displacement distribution in dent layer differs remarkably form that in other layer owing to change of Bw. With model 1, significant motion occurs in the dent layer except when Bw = 0.5 λs where the influence of dent is less, and apparent vortex appears as with fluid. But motion of other layer is small. With model 2, the phase of dent layer motion is almost inverse to that of other layer motion except when Bw = 0.5 λs. Moreover vortex occurs. With models 1 and 2, the evident vertical components appear in the

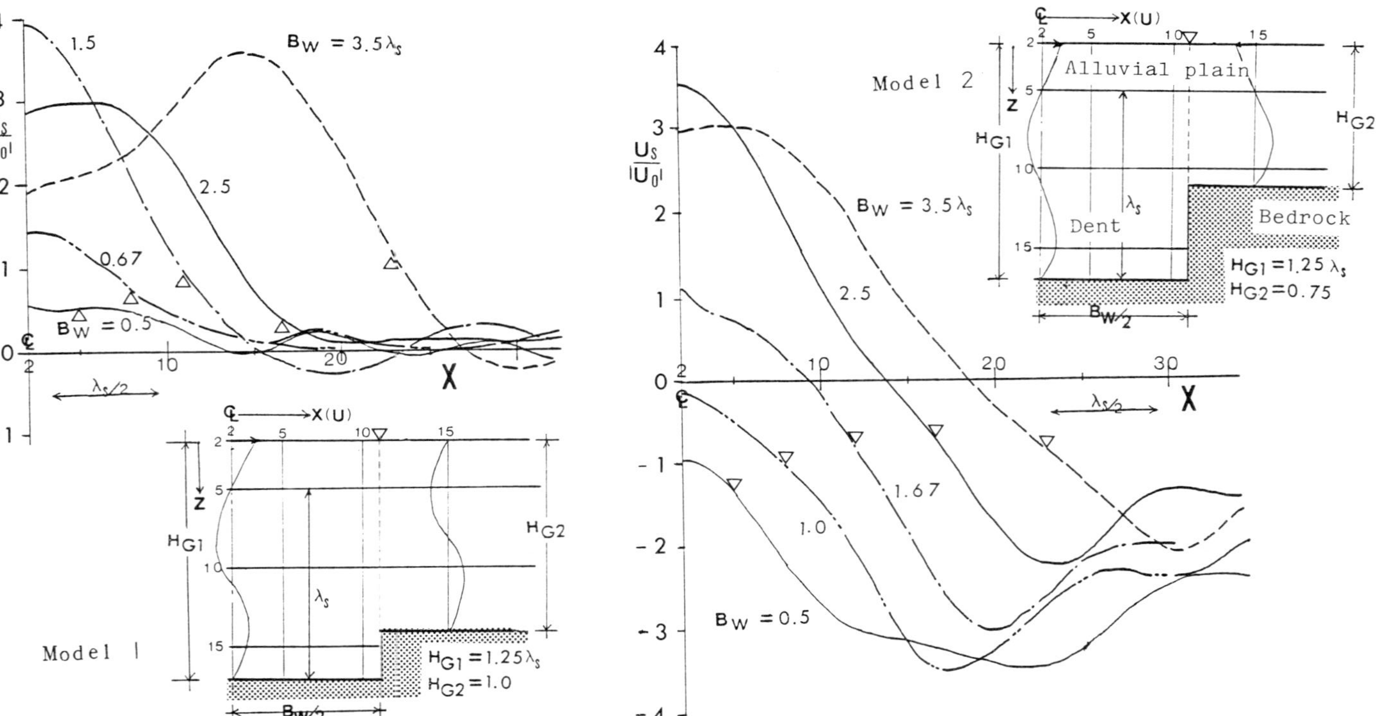

Fig-3 Horizontal displacement-amplitude curves (Us) on surface of alluvial plain with a dent. (The left of figures is for model 1 and the right is for model 2, $\alpha=2\beta$, p=3, $s=\sqrt{3}\alpha\tau$, Uo: Amplitude of input wave, Δ : Edge of dent, Unit of HG1, HG2 and Bw is λs)

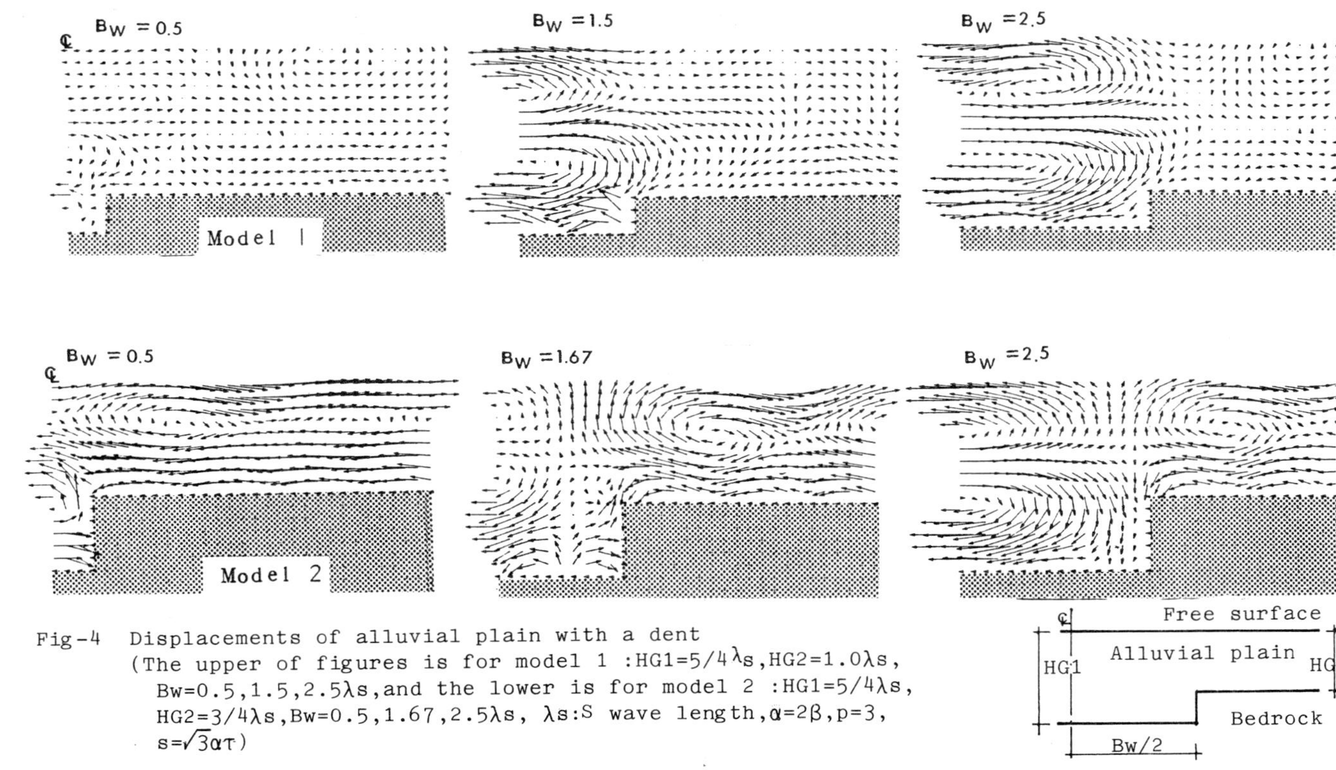

Fig-4 Displacements of alluvial plain with a dent
(The upper of figures is for model 1 :HG1=5/4λs,HG2=1.0λs, Bw=0.5,1.5,2.5λs,and the lower is for model 2 :HG1=5/4λs, HG2=3/4λs,Bw=0.5,1.67,2.5λs, λs:S wave length,$\alpha=2\beta$,p=3, $s=\sqrt{3}\alpha\tau$)

boundary zone between dent layer and other layer. This occurrence is especially intensive in case of model 2. The change of surface horizontal displacement (Us) shown in Fig. 3 is more understandable in Fig. 4.

Fig. 5 represents the stress distribution for models 1 and 2 when Bw = 2.5 λs. The upper of figure is for model 1 whereas the lower is for model 2. Each of them represents σz, σx and τxz. The size of circle shown in the figure corresponds to quantity. The circle with a center dot indicates negative value whereas the circle without dot indicates positive value. As is evident from the figure, σz appears remarkably near the boundary between dent layer and other layer (model 1) and from the boundary to the dent layer (model 2), except the boundary between bedrock and plain. The range of occurrence of model 2 is wider and larger than that of model 1. σx appears remarkably near the boundary surface between dent layer and other layer. Especially with the model 2, occurrence is more notable. It spreads remarkably to vicinity of center of dent. τxz approaches to shear stress distribution of plain without irregular topography as the distance to dent layer, especially to the center, reduces (model 1). The dent layer of model 2 represents similar, but not identical, situation as with model 1. Except for the vicinity of boundary where σz and σx occur remarkably, distribution is similar to that in plain without irregular topography in other layer although symbol is inverse to that in dent layer.

If the input bedrock surface has a dent, the dent layer swings significantly, moves with inverse phase opposite to the layer not having dent, or vertical motion occurs forcibly near the boundary between dent layer and other layer, depending on input wavelength, layer thickness and size of dent. The stress condition differs significantly from that of plain without irregular topography. Intensive stress σx occurs near the boundary. Actually there is a possibility of occurrence of tension crack on the free surface of plain. Intensive stress σz may affect also in the alluvial plain near the boundary. Hence it is important to consider the seismic problem near the boundary between dent layer and other layer. As this method allows us to trace successively the behavior of plain after occurrence of crack, we will examine it further.

CONCLUSION

We made a numerical calculation of simple models. The alluvial plain with irregular topography having a protrusion on the plain surface or a dent on the bedrock gives fairly complicated behavior as fluid which flows as vortex, differing remarkably from the alluvial plain without irregular topography even when horizontal sin waves input in same phase is given to the plain. If the topographic irregularity of the plain containing a

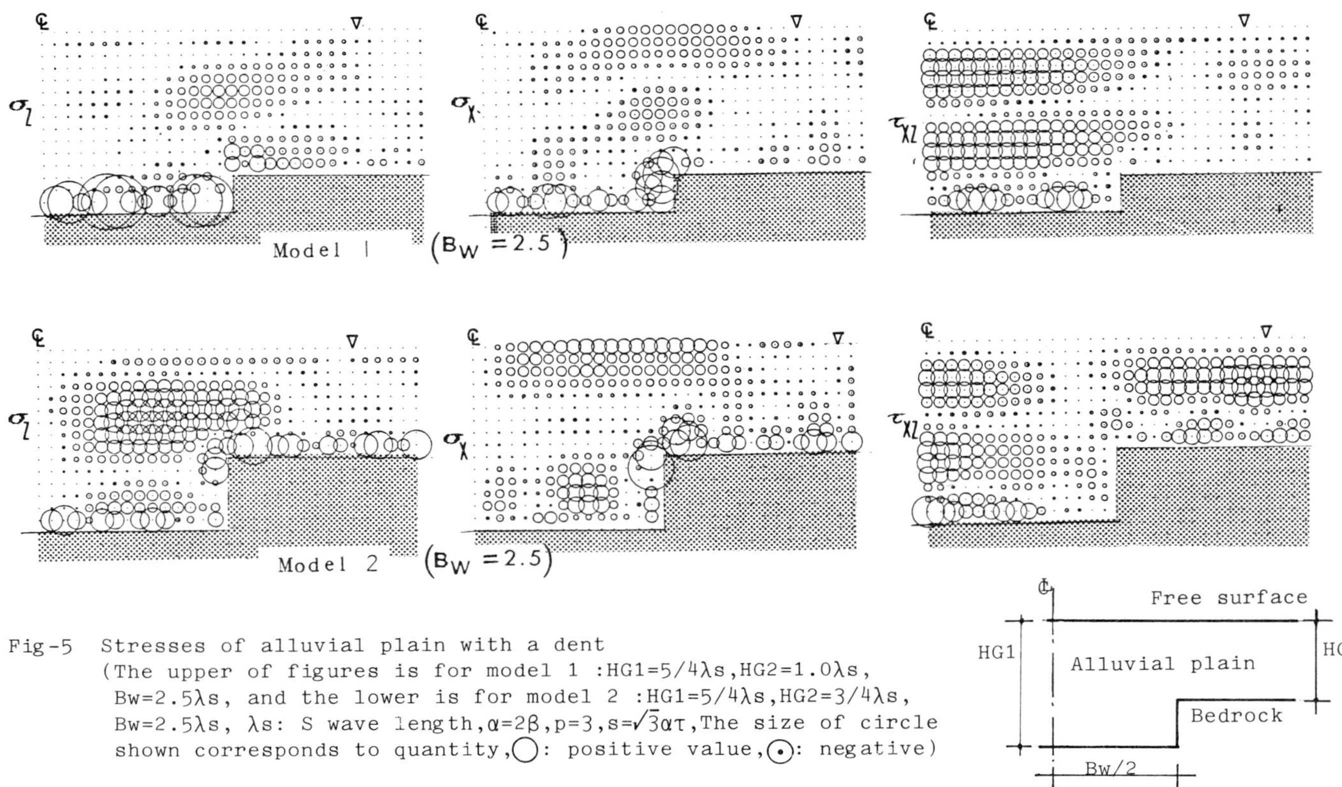

Fig-5 Stresses of alluvial plain with a dent
(The upper of figures is for model 1 : $HG1=5/4\lambda s$, $HG2=1.0\lambda s$, $Bw=2.5\lambda s$, and the lower is for model 2 : $HG1=5/4\lambda s$, $HG2=3/4\lambda s$, $Bw=2.5\lambda s$, λs: S wave length, $\alpha=2\beta$, $p=3$, $s=\sqrt{3}\alpha\tau$, The size of circle shown corresponds to quantity, ◯: positive value, ⊙: negative)

protrusion and a dent and input conditions, non-uniformity of geologic configuration are taken into consideration, the alluvial plain seems to transmit wave complicatedly and momently when earthquake occurs, but the fundamental phenomenon can be recognized from the simple models.

We intend to study the wave propagation problems of alluvial plain having irregular topography when input wave incidents from lateral side and dynamic embedded structure-soil system.

REFERENCES

1 Toriumi I. (1980), Research on the Dynamical Characteristics of the Foundation by the Use of Numerical Integral Expression for the Wave Equations, (Ed. Kowada A., Hisatoku T., Segawa T.), Vol. 5 pp 157 to 164, Proceedings of the 7th World Conf. on Earthquake Engineering, Istanbul, Turkey, 1980

2 Toriumi I. (1983) , An Improvement on a Numerical Integral Method for the Wave Equations, pp 729 to 780, conf. on Archi. Insti. of Japan, Hokuriku Area, Japan, 1983

3 Hidaka K. (1943), Numerical Integration, Iwanami Shoten, Tokyo, (in Japanese)

4 Toriumi I. (1986), Mechanics of Surface Wave in Alluvial Plain, pp 69 to 74, the 14th sympo. of Ground Motion, Tokyo, Japan, 1986

5 Sato Y. (1978), Theory of Elastic Wave Propagations, Iwanami Shoten, Tokyo, (in Japanese)

6 Yokoyama H. (1985), A Study on Dynamical Analysis of Soil-Structure System by the Wave Propagation, pp 207 to 208, Conf. on Archi. Insti. of Japan, Tokai Area, Japan, 1985

7 Yokoyama H. (1986), A Study on Dynamical Analysis of Elastic Stratum with a Projection over Rigid-Base by the Wave Propagations, pp 491 to 492, Conf. on Archi. Insti. of Japan, Hokkaido Area, Japan, 1986

SECTION 3: ENGINEERING SEISMOLOGY

Calculational Procedures for Seismic Hazard Analysis and its Uncertainty in the Eastern United States

G.R. Toro, R.K. McGuire
Risk Engineering, Inc., 5255 Pine Ridge Rd., Golden, CO 80403, U.S.A.

INTRODUCTION

Seismic hazard estimation in the eastern and central United States (EUS) is a challenging problem. Large, destructive, earthquakes have occurred in places like New Madrid, Missouri, and Charleston, South Carolina, but the annual rate of moderate and large earthquakes (m_b>5.5) in the entire EUS is relatively low. Because empirical information is scarce, there are multiple alternative explanations for the causes of earthquakes in EUS. Thus, there is considerable uncertainty about seismic hazard in EUS.

A study conducted by the Electric Power Research Institute (EPRI)[1] has developed and applied techniques to quantify and document the sources of uncertainty using a Bayesian framework, and to propagate these uncertainties through the seismic-hazard calculations. The result is a clear description of the hazard, its uncertainty, and the importance of the different contributors to that uncertainty. In this paper, we describe the sources of uncertainty, the calculation of seismic hazard in the EUS, and the characterization of uncertainty in hazard.

RANDOMNESS AND UNCERTAINTY

It is common in seismic hazard analysis to distinguish between randomness and uncertainty. Randomness is probabilistic variability that is inherent to the unpredictable nature of future events. Randomness can not be reduced by collecting additional information. Uncertainty represents our incomplete knowledge about the physical mechanisms that control the random phenomenon. Uncertainty can be reduced by collecting additional information.

Randomness in seismic hazard is described by the stochastic point process of earthquake occurrences and by distributions of earthquake magnitude, location, and amplitude of ground shaking (given magnitude and distance), for all active seismic sources in the vicinity of the site. Seismic hazard analysis (Cornell[2], McGuire[3,4]) integrates over these distributions to produce a hazard curve(i.e., probability of exceedance as a function of ground-motion amplitude). Because we are uncertain about the parameters of the point process and of the magnitude and location distributions, we consider several alternate models and parameter values and assign probabilities (in the Bayesian sense) to the alternatives according to their scientific credibilities. Each set of models or parameters leads to a different hazard curve, resulting in a large family of hazard curves, each member of the family having an associated probability

SOURCES OF UNCERTAINTY

Seismic Sources

Because there are multiple hypotheses about the causes of earthquakes in EUS, it is not known which of several potentially active tectonic features in the vicinity of a site, are actually seismic sources capable of generating destructive earthquakes. To capture this uncertainty, we consider all possible combinations of states (active or inactive) of all potential seismic sources that may affect the site, and associate a probability to each source combination.

The probability associated with a source combination is calculated from the marginal probability P^a of each source being tectonically active and from any dependency relationship among the sources. P^a is calculated from the physical characteristics of the source (which may themselves be uncertain) and the probability that a sources with a given set of characteristics is active.

Seismicity Parameters

These are the annual rate of earthquakes capable of structural damage (say, m_b>5) and the parameter of the assumed exponential magnitude distribution. These parameters are estimated using statistical methods; they are uncertain because a finite sample is available. Because potential seismic sources in EUS are often large and the spatial distribution of historic earthquakes is sparse, uncertainty arises about the spatial variation of seismicity within a seismic source. (Most seismic hazard studies ignore this important source of uncertainty.)

Maximum Magnitude

The maximum magnitude that a given seismic source can generate is an important, but difficult to estimate, parameter in seismic hazard analysis. The largest historic magnitude is a weak indicator of maximum magnitude because the historic record is short relative to the mean recurrence interval of large earthquakes (McGuire[5]); the largest historic magnitude provides only a lower-bound value for the maximum magnitude. Other criteria for estimating maximum magnitudes are physical constraints related to the size of the source and analogies with other regions (EPRI[1]). To characterize uncertainty in maximum magnitude, one can specify multiple values for each source, with subjectively assigned probabilities.

Ground-Motion Models

There is uncertainty about the mean amplitude of ground shaking produced by EUS earthquakes, given the magnitude of the earthquake and its distance to the site. This uncertainty is large because few strong-motion recordings from EUS earthquakes are available and no recordings of strong shaking have been obtained from large earthquakes. Ground-motion uncertainty is captured by considering alternate ground motion models, with weights assigned according to their credibilities.

LOGIC-TREE REPRESENTATION OF UNCERTAINTY

All the above contributors to uncertainty can be represented in a logic-tree format (Figure 1). Each level of the tree represents one source of uncertainty; each terminal node represents one "state of nature". Corresponding to each terminal node, there is a hazard curve. The probability associated with a terminal node (and with the corresponding hazard curve) is the product of the probabilities associated with all intermediate branches in the path from the root to the terminal node.

SEISMIC HAZARD CALCULATIONS

In the EPRI approach, we calculate hazard curves for all terminal branches of the logic tree. This is possible because of efficient procedures to group end branches and perform multiple, simultaneous calculations.

There are five conceptual steps in these calculations: estimation of seismicity parameters, calculation of hazards from individual sources, specifications of source combinations, combinations of source hazards, and calculation of summary statistics and sensitivity results.

Estimation of Seismicity Parameters
A procedure developed by Veneziano and Van Dyck[1,6,7] estimates seismicity parameters for each source using the catalog of historic earthquakes in the region. The following are three significant innovations of this methodology: seismicity parameters vary within seismic sources; it the earthquake catalog is corrected for incompleteness; and bootstrapping (Efron[8]) produces multiple sets of seismicity parameters for each source, to represent statistical uncertainty. The analyst controls the spatial variability of seismicity parameters within each source; he can choose multiple smoothing options if he is uncertain about this parameter.

Calculation of Hazards from Individual Sources
The contribution of each source to the seismic hazard at the site is calculated from the distributions of magnitude, distance, and ground-motion amplitude given magnitude and distance. This calculation takes the form:

$$\nu(a) = \sum \nu_i \iint P[A>a|m,r]\ f_{M(i)}(m)\ f_{R(i)}(r)\ dm\ dr \qquad (1)$$

in which the summation is performed over all sub-sources that comprise the source (one distribution of magnitude applies within each sub-source), ν_i is the mean annual rate of damaging earthquakes in sub-source i, $P[A>a|m,r]$ represents the ground-motion model, and $f_{M(i)}(m)$ and $f_{R(i)}(r)$ are the probability-density functions of magnitude and distance corresponding to sub-source i. This calculation is performed for several values of a, thus obtaining a hazard curve. The hazard defined in Equation 1 represents the annual rate at which ground-motion amplitude a is exceeded at the site; because it is much smaller than unity, this rate can be interpreted as the probability that ground-motion amplitude a be exceeded in any one year.

The above calculation is performed for all possible combinations of maximum magnitude, seismicity parameters, and ground-motion models. Calculations are performed simultaneously for all these combinations, in order to avoid repetition of time-consuming geometry calculations, but results are kept separate.

Specification of Source Combinations
In principle, the analyst should specify all possible source combinations and include them in the logic tree. If the number of potential sources is n, the number of source combinations is 2^n. The probability associated with each source combination is calculated from the activity probabilities of the sources and from any dependence

relationship among sources.

Because the analyst knows the contribution of each source to the hazard at the site, he can eliminate a source from source combinations in which it makes a negligible contribution. This operation may substantially reduce the number of terminal branches in the logic tree--without significant loss of accuracy--because some branches at the level of source combinations are merged and, more importantly, because the number of branches stemming from one source-combination branch is greatly reduced when the number of active sources in a source combination is reduced.

<u>Combination of Source Hazards</u>
Using the source hazard results and the source combinations, combined hazard results are generated for all end branches of the logic tree. For each source combination and each ground motion model, the algorithm calculates combined hazards and end-node probabilities for each combination of source parameters (seismicity parameters and maximum magnitude).

As an illustration of the number of combinations that arise in a typical application, consider the following example. If a source combination contains 4 active sources and each source has 3 possible sets of seismicity parameters and 2 possible values of maximum magnitude, the number of end branches for that source combination and one ground-motion model is equal to $(3 \times 2)^4 = 1296$. If there are 5 potential sources near the site and 3 possible ground-motion models, the total number of end branches is 50,421.

<u>Calculation of Summary Statistics and Sensitivity Results</u>
The family of hazard curves and their associated probabilities, corresponding to all end branches of the logic tree, contains all the information about seismic hazard at the site, its uncertainty, and the different contributors to that uncertainty. Given the very large number of hazard curves, and the possibility that their associated probabilities may differ considerably, it is not useful to display all the curves.

To represent the hazard at the site and its overall uncertainty, fractiles are calculated for each value of ground-motion amplitude and fractile curves are constructed for 0.15, 0.50 (median), and 0.85. The median curve is a measure of central value, whereas the spread between the 0.15 and 0.85 curves is a measure of overall uncertainty. If the number of hazard curves is large, fractiles are calculated using "bins" instead of sorting.

Because uncertainty in seismic hazard is large, it is important to quantify the contribution of the different parameters to the total uncertainty. The contribution of a given parameter to the total uncertainty depends on the sensitivity of the hazard results to that parameter and on the parameter's uncertainty.

Typical sensitivity analyses fix all but one of the parameters to their base-case (or best-estimate) values and then measure the effect of varying that one parameter by an amount proportional to its uncertainty. The EPRI methodology does not use this approach because, in general, there are no base-case values for non-scalar parameters (such as source combination) and because this approach ignores interaction effects.

Instead, the EPRI methodology follows an approach in which we measure the effect of varying one parameter, allowing all other parameters to remain random (i.e., to take all their possible values). Calculation of sensitivity to source combinations or ground-motion models is straightforward. To calculate sensitivity to source combinations, the median conditional hazard, given a source combination, is calculated for each source combination. If the median hazards corresponding to source combinations with substantial probability content vary substantially among themselves, there high sensitivity of seismic hazard to source combination. Results are presented as the set of conditional-median curves (because the number of these curves is small).

The analysis of sensitivity to seismicity parameters and maximum magnitude is slightly more complicated because these are source properties. Consider sensitivity to maximum magnitude for five sources in the vicinity of the site, with each source having two possible values of maximum magnitude (i.e., there 2^5 maximum-magnitude combinations). To calculate sensitivity to maximum magnitude, the geometric-mean conditional hazard is calculated for each maximum-magnitude combination. This calculation is not straightforward because most combined-hazard results are not disaggregated by maximum-magnitude combination. In fact, each hazard result corresponding to a source combination in which at least one source is not active must be proportionally allocated among various maximum-magnitude combinations. The final step is to calculate fractiles of these geometric-mean hazards and display them as fractile hazard curves. The spread between the 0.15 and 0.85 fractiles is a measure of sensitivity to maximum magnitude. Sensitivity to seismicity parameters is calculated in a similar manner.

Results from several expert teams--with each team specifying its own sources, activity probabilities, seismicity parameters and maximum magnitudes--can also be processed simultaneously to produce overall results, sensitivity to ground-motion model, and sensitivity to team. As a general rule, the teams are given equal weights.

Figures 2 through 4 depict results for a site in EUS. Figure 2 shows sensitivity to source combinations, Figure 3 shows sensitivity to ground-motion models (for several expert teams), and Figure 4 shows sensitivity to maximum magnitude.

Additional sensitivity results depict the contributions of different sources to the total hazard. 3-D plots depict the geographic distribution of contributions to hazard (for a given ground-motion amplitude $\underline{a}$) at the site and helps identify the sources, or portions thereof, that produce the most significant contributions to the hazard. Also, plots depicting the contribution of each source to the total hazard at the site as a function of ground-motion amplitude, indicate which sources dominate the hazard at different amplitudes. These results are important because they give an indication of the frequency content and duration of the ground motions associated with different hazard levels; these ground-motion characteristics significantly affect the performance of structures and high-frequency equipment (Kennedy et al. , Toro et al.).

<u>Organization of the Computer Calculations</u>

Three computer programs[9,1] (available from EPRI) perform the seismic-hazard calculations described above. Program EQPARAM generates multiple sets of seismicity parameters using as input the catalog of historic main earthquakes, assumptions about the regional variation of catalog completeness, and various alternate assumptions about the spatial variability of seismicity (<u>smoothing options</u>). Program EQHAZ calculates each source's contribution to the hazard at the site, for each possible combination of source parameters. The generation of source combinations is currently performed by the analyst but can be automated. Program EQPOST calculates total hazards for all source combinations and all combinations of parameters, and produces all statistical summaries and sensitivity analyses (to source combination, ground-motion model, maximum magnitudes, seismicity parameters, and teams).

By structuring the computations so that hazards are calculated separately for each source and then combined, we gain two significant computational advantages. First, source-hazard calculations are performed once, instead of being performed once for each source combination in which that source appears. Second, because the analyst already knows the source-hazard results when setting up the source combinations, he can exclude sources from source combinations in which they make negligible contributions, thus reducing the number of end branches in the logic tree. This added computational efficiency makes it feasible to perform an analysis by enumeration (i.e., considering all possible branches of the logic tree, instead of following a limited number of branches as in Monte Carlo methods).

The use of enumeration, and the structure of the calculations in EQPOST, make it possible to generate all overall and sensitivity results in one pass, without having to perform multiple runs with different input parameters.

CONCLUSIONS

The EPRI methodology for seismic hazard analysis in the eastern and central United States provides a framework for the quantification of uncertainties in input parameters and for the efficient propagation of these uncertainties through the hazard analysis. The result is a clear representation of seismic hazard, its total uncertainty, and the various contributions to this uncertainty.

ACKNOWLEDGMENTS

The work reported here was sponsored by EPRI-SOG projects P101-16 and P101-46; Carl Stepp and Jerry King were project managers. Daniele Veneziano, Josef Van Dyck, and Laurel Drake made substantial contributions to this work.

REFERENCES

1. Electric Power Research Institute (1986). Seismic Hazard Methodology for the Central and Eastern United States. Volume 1: Methodology, Palo Alto, Calif., EPRI report NP4726, Vol. 1.

2. Cornell, C.A. (1968). Engineering Seismic Risk Analysis, Bulletin, Seismological Society of America, Vol. 58, pp. 1583-1606.

3. McGuire, R.K. (1976). Fortran Computer Program for Seismic Risk Analysis, Reston, Virginia, U.S.G.S. Open-File Report 76-67.

4. McGuire, R.K. (1978). FRISK: Computer Program for Seismic Risk Analysis using Faults as Earthquake Sources, Reston, Virginia, U.S.G.S. Open-File Report 78-1007.

5. McGuire, R.K. (1977). Effects of Uncertainty in Seismicity on Estimates of Seismic Hazard for the East Coast of the United States, Bulletin, Seismological Society of America, Vol. 67, pp. 827-848.

6. Veneziano, D. and Van Dyck, J.F.M. (1986). Seismic Hazard Methodology for Nuclear Facilities in the Eastern United States: Appendix B, Golden, Colo., Risk Engineering, Inc., prepared for the Electric Power Research Institute.

7. Van Dyck, J.F.M. (1985). Statistical Analysis of Earthquake Catalogs, Cambridge, Mass., Massachussets Institute of Technology, PhD. Thesis, Department of Civil engineering.

8. Efron, B. (1982) The Jacknife, the Bootstrap and other Resampling Plans, Philadelphia, Penna., CBMS-NSF Regional Conference Series in Applied Mathematics.

9. Electric Power Research Institute (1986). Seismic Hazard Methodology for the Central and Eastern United States. Volume 2: EQHAZARD Programmer's Manual, Palo Alto, Calif., EPRI report NP4726, Vol. 2.

10. Electric Power Research Institute (1986). Seismic Hazard Methodology for the Central and Eastern United States. Volume 3: EQHAZARD User's Manual, Palo Alto, Calif., EPRI report NP4726, Vol. 3.

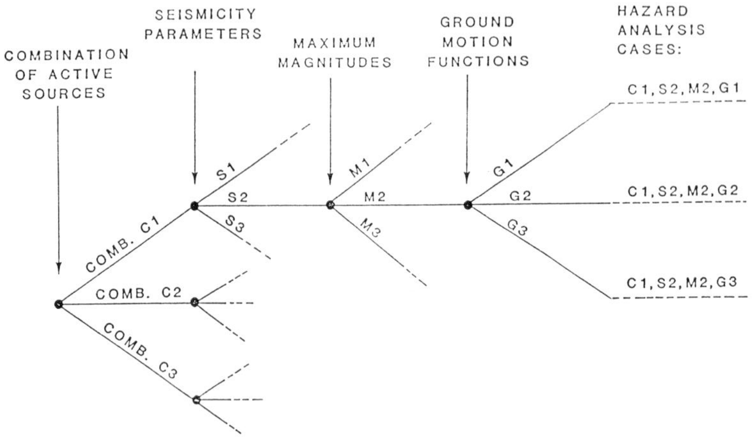

Figure 1. Logic-tree representation of uncertain parameters in seismic hazard analysis.

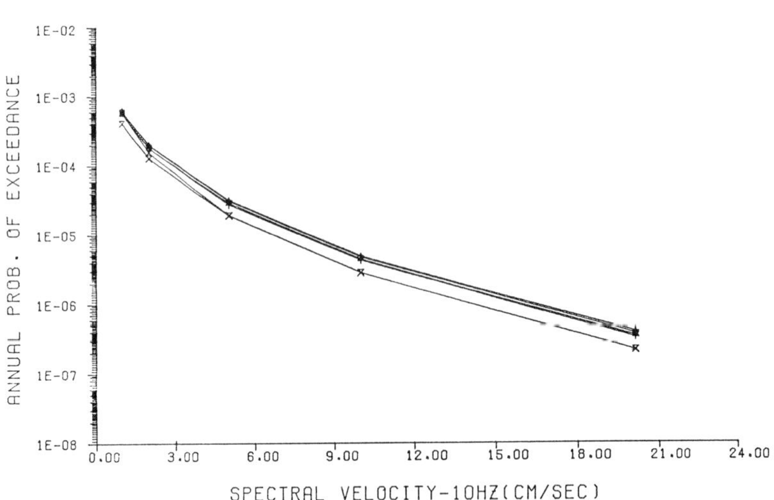

Figure 2. Sensitivity of Seismic Hazard to Source Combinations.

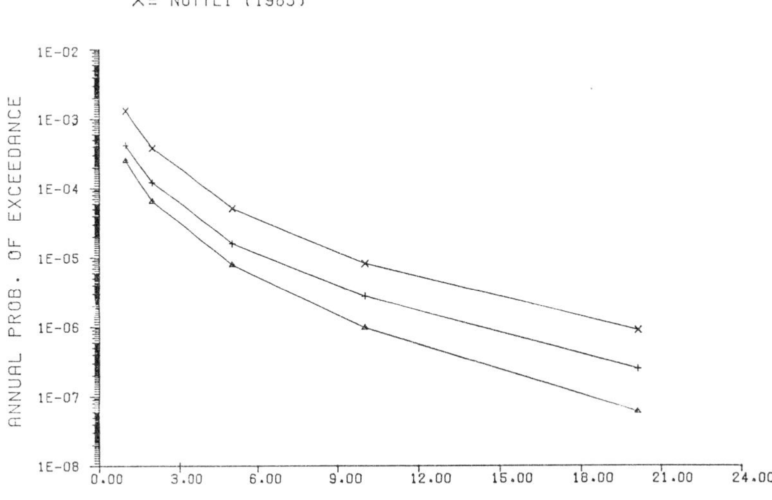

Figure 3. Sensitivity of Seismic Hazard to ground-motion models.

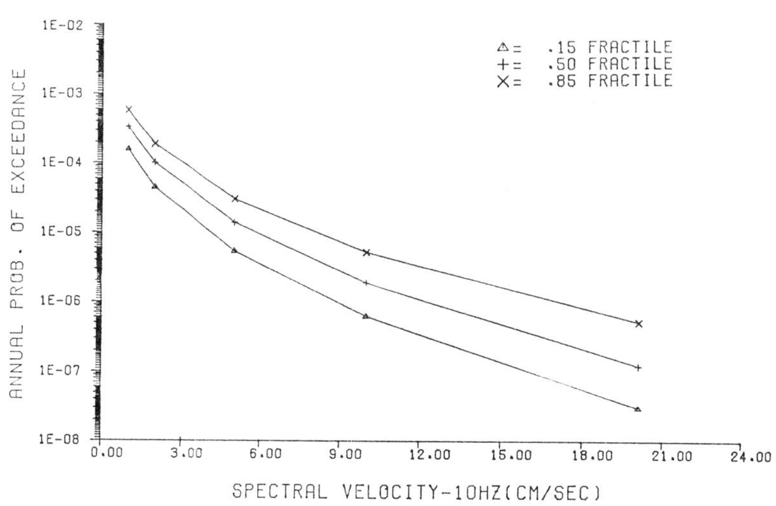

Figure 4. Sensitivity of Seismic Hazard to maximum magnitudes.

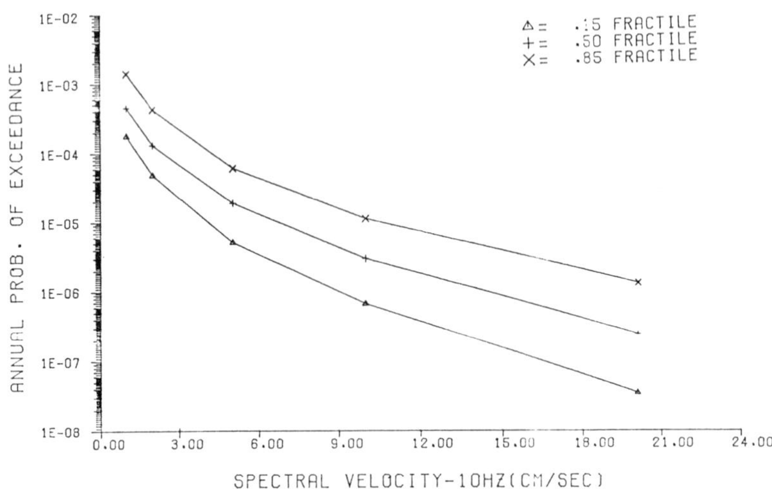

Figure 5. Total Uncertainty in seismic Hazard including variability among expert teams.

Tectonic Framework, Seismic Source Zones and Seismicity Parameters for the Eastern United States: An Application of the EPRI Methodology

C.T. Statton, R.C. Quittmeyer
Woodward-Clyde Consultants, P.O. Box 290, Wayne, NJ 07470, U.S.A.
T. Engelder
Department of Geology and Geophysics, the Pennsylvania University, State College, PA 16802, U.S.A.
T. Turcotte
Woodward-Clyde Consultants, Fox Plaza, San Francisco, CA 94102, U.S.A.
J. Kelleher
National Science Foundation, Washington, D.C. 20550, U.S.A.

INTRODUCTION

The Electric Power Research Institute (EPRI) has developed a probabilistic methodology for assessing seismic hazard in the central and eastern United States. The methodology is structured such that assumptions and thought processes are formally stated, but at the same time is flexible enough to accommodate different approaches to the problem. Six teams were employed to evaluate in parallel the seismotectonic framework of the study area. Teams rather than individuals were used because of the many diverse types of scientific information that are necessary to derive an estimate of seismic hazard. No one person can have expertise in all the necessary fields. Woodward-Clyde Consultants participated in this program as one of the teams.

The goal of the EPRI program was to capture a current assessment of seismic hazard. One characteristic of today's assessment is uncertainty in the input parameters. There is uncertainty in hypotheses to explain the contemporary seismicity, in the models used to predict the average rate of future earthquake occurrence, in the size of the largest earthquakes to be expected, in the ground motions caused by those events, and in the attenuation of those motions from their source to any particular site. The EPRI methodology incorporates these uncertainties directly in the evaluation. Thus, uncertainties must be estimated as well as the parameters themselves.

The methodology consists of a number of steps. First, tectonic features with any likelihood of generating seismic activity are compiled and described. These features include primarily geologic structures and geophysical anomalies thought to represent structures at depth. Second, for each feature its probability of activity is estimated. Next, source zones are defined on the basis of the identified features. Seismicity parameters to describe future earthquake occurrence are then determined for each source zone. Multiple hypotheses, both competing and alternatives, are accommodated for any area or volume of crust. For such cases, dependencies among such source zones are evaluated. These steps were carried out by each team. Subsequently, the team results were used as input to the hazard calculation.

This paper summarizes one team's approach to applying the EPRI methodology. While a complete description of what we did is beyond the scope of this presentation, we will attempt to provide enough details and examples so that the essentials of the application are conveyed. A more complete summary of our work can be found in EPRI[1].

DATA

The EPRI program included a large effort to compile available data that are pertinent to seismic hazard evaluation. This effort was carried out to increase the likelihood that differences between teams resulted from differences in interpretation rather than differences in data available for consideration. Geologic information, geophysical data, and an earthquake catalog were all made available to each team. Entries in the earthquake catalog, which was prepared by compiling previous catalogs, were examined individually to eliminate errors, duplicates, and mis-identified earthquakes. Thus, the data base available for this program was more complete, on average, than that used in any previous attempts to assess seismic hazard in the central and eastern United States.

TECTONIC FRAMEWORK

The first task in the EPRI methodology is to identify tectonic features with a potential for generating earthquakes. All features with a finite probability of producing such earthquakes are to be included; there need not be certainty of such an association. The probability that the feature is seismically active will be assessed in the next task. This collection of features focused on potential sources of larger events ($m_b > 5$), and included features supported by a variety of different working hypotheses.

Approach

Given the uncertainty in the interpreted causes of seismicity in the central and eastern United States, and the lack of detailed knowledge on specific tectonic features, it is difficult to define criteria by which any tectonic feature would be removed from consideration. The only criterion that we applied is one of size. Although data on rupture zone size in the eastern United States are scarce, the characteristic rupture radius for earthquakes of about magnitude 5 appears to be approximately one to several kilometers. Hence, features with lengths of 10 to 20 km are considered of sufficient size to support magnitude 5 earthquakes (albeit, perhaps with a very low probability). It is, therefore, assumed that significant earthquakes will rupture only a fraction of the total length of the faults with which they are associated.

We also adopt as a working hypothesis that the tectonic features that concentrate stress, and those features along which stress is released during an earthquake, need not be the same feature. For example, stress may be concentrated along a crustal block boundary, but the boundary may be inappropriately oriented in the current stress field to slip during an earthquake. Other faults oriented oblique to the boundary may release the stress through earthquakes instead. Thus, we include two types of features in our tectonic framework: stress concentrators, which will be associated spatially, but not causally with earthquake and stress relievers, which will be the actual sites of significant earthquakes.

The potential small size of features generating m_b 5 earthquakes makes it very difficult to identify all such features. They are probably within the geologic/tectonic "noise" of a study with a regional focus. For this reason, the concept of background zones is coupled to the tectonic framework. These zones, which may have a tectonic basis themselves, are broad areas over which it is believed that magnitude 5 (m_b) or greater earthquakes can occur, but for which specific fault structures cannot presently be identified. In part, the concept of stress concentrators also address this problem. Volumes of crust concentrating crustal stresses are used to accommodate seismically active regions whose traditional characteristics, such as mapped surface faults, provide no clear insight into earthquake occurrence.

Interpretation

Our interpretation of the tectonic framework for the central and eastern United States comprises over 70 features or classes of features. The list includes long linear geomorphic trends (e.g., the Hudson River trend), geophysical trends (e.g., the Mid-continent geophysical anomaly), well developed fault systems (e.g. the Ramapo fault system), intersections of geological and geophysical trends (e.g., the Anna, Ohio intersection), inferred rift structures (e.g., the New Madrid rift complex), basin structures (e.g., the Richmond, VA basin), intrusive rock bodies (e.g., the White Mountain intrusives), crustal blocks (e.g., the Tennessee crustal block), and regions of uplift (e.g., the central Kansas uplift). These features, and an assessment of their seismic potential, are summarized in Table 1. For more detail see Appendix A of EPRI[1].

TABLE 1
Features of the Tectonic Framework for the Central and Eastern United States

	Feature	Assessed Pa
1	Continental Shelf Edge	0.193
2	Cont inental Shelf Edge Intersections	0.147
3	Grand Banks Intersection	0.900
5	New Brunswick/Fundy/Nova Scotia Block	0.344
6	Norumbega Fault Zone	0.120
7	Passamaquoddy Bay Feature	0.281
8	White Mountain Intrusives	0.227
9	Central NH Crustal Saddle	0.227
11	SE New England Fault System	0.049
12	Charlevoix-LaMalbaie Structure	0.894
14	St. Lawrence Rift	0.250
17	Timiskaming Graben	0.275
18	Adirondack Dome/Uplift	0.335
19	Western Quebec Crustal Block	0.611
20	Mohawk River Trend	0.102
21	New Jersey Isostatic Gravity Anomaly	0.227
22	Newark Basin	0.082
23	Newark Basin Perimeter	0.394
24	Ramapo Fault System	0.133
25	Hudson River Trend	0.147
26	Central Virginia Gravity Saddle	0.488
27	State Farm Complex	0.149
28	Richmond Basin	0.096
29	South Carolina Gravity Saddle	0.486
30	Ashley River/Woodstock Faults	0.466
31	Eastern Tennessee Crustal Block	0.243
31A	Southern Appalachian Gravity/Crustal Block	0.227
32	Clarendon-Linden Fault System	0.139
33	Western New York - Southern Ontario Feature	0.462
34	Attica, NY Intersection Feature	0.509
35	Northeastern Ohio Gravity High	0.197
36	Michigan/Ohio Gravity Alignment	0.110
37	Bowling Green Fault System	0.073
38	Champaign-Anna Fault	0.067
39	Anna Geophysical Intersection	0.721
40	Central Reelfoot Rift	0.921
41	Reelfoot Rift	0.489
42	St. Louis Arm of the New Madrid Rift	0.728
43	Southern Indiana Arm of the New Madrid Rift	0.448
45	Nemaha Ridge/Humboldt Fault	0.201
46	Extended Southern Oklahoma Aulacogen	0.182
47	KS/NB Offset of MCG Anomaly	0.155
48	Southern Oklahoma Gravity Linear	0.263
49	Meers Fault	0.085
50	SD/MO Isostatic Gravity Low	0.136
51	Central Kansas Uplift	0.087
54	Great Plains Crustal Block	0.122
55	Plum River Fault, Illinois	0.011
56	Sandwich Fault, Illinois	0.061
57	Narragansett Basin	0.150
58	Ossipee Pluton	0.203
61	Tyrone-Mt. Union Lineament	0.055
63	Pittsburgh-Washington Lineament	0.055
70	Scranton Gravity High	0.062
71	New York/Alabama Lineament	0.081
72	Clingman Lineament	0.060
73	St. Lawrence Rift (Massena to Charlevoix)	0.180
74	Monteregian Hills	0.033
75	Boston Basin	0.052
76	Connecticut River Triassic Basin	0.212
77	Green Mountain/Berkshires Massif	0.060
78	Wabash River Valley Fault Zone	0.493
79	Western KY Arm, New Madrid Rift	0.045
80	Rome Trough	0.020
81	Anadarko Basin	0.143
82	Mid-continent Geophysical Anomaly	0.064
83	Stafford Fault Zone	0.014
84	Gettysburg Basin	0.049
85	Pascola Arch	0.098
86	Forest City Basin	0.057
87	Kentucky River Fault System	0.025
88	Precambrian/Cambrian Normal Faults	0.027
89	Brevard Zone	0.024
91	S. Carolina 250-km Low-Pass Gravity Saddle	0.579
92	Arkoma Basin	0.137
93	New Madrid 250 km Low-Pass Gravity Saddle	0.705
94	Pickens-Gilbertown Fault Zone	0.010
95	Ottawa-Bonnechere Graben	0.038
96	Ouachita Orogen	0.203
97	New England 250-km Low-Pass Gravity Saddle	0.327

SEISMIC POTENTIAL OF TECTONIC FEATURES

To accommodate the multiple hypotheses proposed to explain seismicity in the central and eastern United States, it is necessary to assign each hypothesis a weight according to its perceived probability of generating earthquakes. In this manner the uncertainty in describing the sources of future earthquakes can be incorporated in the hazard analysis. This is especially important in the study region because of the current lack of understanding of earthquake sources.

A systematic procedure was developed by the EPRI program methodologists for applying subjective judgement in assessing the potential for seismic activity of identified tectonic features. The methodology attempts to unravel the various threads of evidence and logic that form the basis of judgements, and to present them in a manner that is transparent to others. The methodology is designed to capture explicitly the uncertainties in assessing seismic potential, and to separate uncertainties that derive from different sources.

This process utilizes two matrices that encompass, respectively, the model and data components of seismic activity assessment. The first matrix- the generic matrix - describes the probability that a generic feature is seismically active given various combinations of physical characteristic states. This draws upon models that describe or accommodate earthquake generation, and identifies key characteristics used in model descriptions. Given the states of various physical characteristics relative to a generic tectonic feature, we assess the probability that such a feature based on the is seismically active. The second matrix- the data matrix - describes the degree to which existing data support or fulfill a specific state of the physical characteristics for a particular feature. The generic matrix describes our model uncertainties and is evaluated once. The data matrices must be evaluated for each tectonic feature based on the subset of data local and relevant to evaluation of the feature.

The physical characteristics used to model the potential for seismic activity represent a compromise between those characteristics that theoretically are thought to control earthquake generation, and those for which data are readily available. For example, we may wish to know the shear stress on a fault plane relative to the fault's strength, but it is more likely that we have only a general idea of the orientation of the regional tectonic stress field.

We reject an assessment methodology that depends solely on tectonic feature type. In other words, seismic potential is not related solely to the history or genesis of a geologic structure. For example, while some faults associated with Mesozoic rift basins may be related to contemporary seismicity, the observation that one such feature shows this association does not automatically require all such features to be seismically active. Rather the tectonic conditions and evidence must be considered independently for each such feature.

ORIENTATION IN STRESS FIELD / DEEPER CRUSTAL EXPRESSION	SPATIAL ASSOCIATION WITH SEISMICITY			
	YES		NO	
	FAVORABLE	UNFAVORABLE	FAVORABLE	UNFAVORABLE
WITH PROXIMITY TO STRUCTURAL INTERSECTIONS	0.99	0.90	0.15	0.035
WITHOUT PROXIMITY TO STRUCTURAL INTERSECTIONS	0.95	0.86	0.10	0.025
NO EXPRESSION	0.80	0.60	0.01	0.005

Figure 1. Generic matrix of physical-characteristics and associated probabilities of seismic activity.

As an example consider a specific feature, the Adirondack dome. The date matrix assessment for this feature is as follows:

Spatial Association with Seismicity	0.20
No Spatial Association with Earthquakes	0.80
Favorable Stress Orientation	0.50
Unfavorable Stress Orientation	0.50
Deep Crustal Expression & Intersections	0.30
Deep Crustal Expression Only	0.40
No Deep Crustal Expression	0.30

Thus there is a 0.03 probability (0.2 x 0.5 x 0.3) that the Adirondacks are characterized by spatial association with seismicity, favorable stress orientation, and deep crustal expression near intersections. The probability that a feature or crustal volume with this combination of physical characteristics is seismically active is obtained from the generic matrix to be 0.99. Hence, the contribution to the probability of activity for this feature from this combination is 0.0297 (0.03 x 0.99). By summing over the other combinations and their probabilities, the total probability of activity for the Adirondacks is calculated to be 0.335. This represents the team's assessment that the tectonic description of seismogenic processes and conditions in the Adirondack dome represent those that can and will support earthquake generation in the future. Probabilities of activity for all features of the tectonic framework are summarized in Table 1.

Three characteristics are identified as useful in assessing the seismic potential of tectonic features:

1. Spatial association with seismicity
2. Orientation relative to the tectonic stress field
3. Feature expression in the deeper crust.

Spatial association of earthquakes with a tectonic feature or crustal volume is considered a strong indicator of earthquake potential. Features spatially associated with earthquakes in the past are expected to be associated with events in the future. Further, such features may represent classes of seismogenic features thought to be active throughout the crust, independent of historical seismicity association. The orientation of a feature relative to the stress field is also important in assessing seismic potential. The closer a feature's orientation is to the expected plane of maximum resolved shear stress, the greater our expectation is for seismic activity. Finally, expression of a feature in the deeper crust (greater than 10 to 20 km) implies a dimension and tectonic significance that is thought to be conducive to earthquake generation. Here, inferred stress concentrations in the crust may be accommodated as important seismogenic phenomena independent of more traditional tectonic indicators. A slightly higher probability of activity is given to features if the deep crustal expression is also characterized by intersections. These intersections may serve as points at which stress concentrates.

The matrix of probabilities associated with the various combinations of physical characteristics is summarized in Figure 1. Spatial association with seismicity is the dominant factor in determining the probability of activity. Orientation relative to the stress field and deeper crustal expression influence the assessed probabilities, but not to as great a degree as seismicity.

Once probabilities have been assigned to the various cells of the generic matrix it is necessary to assess the degree to which a given tectonic feature is characterized by each combination of physical characteristic states corresponding to a cell of the matrix. For each feature, probabilities are subjectively assessed that each possible state accurately reflects the true state of the tectonic feature with respect to that characteristic. For example, given seismicity proximate to a feature, a confidence level is assigned describing the degree to which such data support clear association, or no association with the feature being assessed. The products of the probability that each combination of physical characteristics represents the feature's actual condition and the probability that the combination indicates a seismically active feature are summed to yield the probability of activity (P_a) for the feature.

SEISMIC SOURCE ZONES

To evaluate seismic hazard in the eastern United States it is necessary to define zones encompassing the sources of future earthquakes. Ideally, each source zone is based on a tectonic feature for which there is an explanation or working hypothesis of why the feature generates earthquakes. The assessed seismic potential of each zone reflects the certainty with which the tectonic explanation for the feature/source zone is believed to be capable of generating earthquakes.

In some cases, documented seismicity exists for which tectonic explanations are lacking. Source zones can also be defined for these cases. Boundaries are drawn around the regions of seismicity that are thought to have a similar seismic potential and maximum magnitude.

The focus of this study is on seismic hazard resulting from the occurrence of moderate-to-large earthquakes (m_b 5 or greater). Given this focus, seismic source zones are derived consistent with the hypothesis that moderate-to-large earthquakes occur as a manifestation of stress concentration within large volumes of crust that extend to depths of 10 to 20 km or greater. The near absence of both line sources and source zones based primarily on shallow geologic structures reflects a conscious bias that fault rupture during large earthquakes may occur along "faults of opportunity." That is, the fundamental cause of such earthquakes is unrelated to any particular shallow structure, but rather derives from tectonic processes or features that concentrate stress in the deeper crust (stress concentrators). The most favorably situated feature within the shallow crust, although not necessarily genetically related to the stress concentration, can serve as a focus for stress release (stress relievers).

For regions in which moderate-to-large earthquakes have occurred in the past, we require that the aggregate probability of activity for nearby features sum to at least 1.0. This corresponds to an assumption that if earthquakes have occurred in the past, they will occur again in the future (stress is renewable). In such regions, if the sum of the probabilities of activity do not sum to 1.0, a "default" or "none-of-the-above" zone is defined. These zones represent a belief that earthquakes will definitely occur in a region, but the actual feature causing the past and future events is unknown.

Future moderate-to-large earthquakes will also undoubtedly occur in locations where they have not previously been observed. In addition to such events associated with identified tectonic features, we also define a background zone to account for events in new locations, but for which a feature is unknown.

Dependencies among source zones that overlap are an important consideration in evaluating seismic hazard. We assume that, in general, only one tectonic feature is the cause of moderate-to-large earthquakes in a given region or volume. Thus, source zones are dependent and mutually exclusive. If one source zone is known with certainty to be active and the cause of seismicity, the other source zones are inactive. In a few cases, we allow more than one source zone to be active in an area. In these cases, the sources are independent of each other.

SOURCE ZONE SEISMICITY PARAMETERS

For each source zone, parameters describing the rate and magnitude distribution of seismicity must be estimated. Two seismicity distribution models were considered: the Poisson model and the characteristic earthquake model. For the Poisson model, the average rate of different size earthquakes is described by the relation:

$$\log (N) = a - bm \tag{1}$$

in which N is the number of earthquake greater than or equal to magnitude m, and a and b are constants. For the characteristic earthquake model, each fault is associated with a characteristic earthquake that recurs in a fairly regular fashion. The recurrence time between characteristic earthquakes may be large relative to the length of the historical record of seismicity, but it may be less than that predicted from a Poisson model. While the characteristic earthquake model has proved useful in understanding some seismicity in the western United States and along major plate boundaries, it is difficult to apply in the central and eastern United States because little information is available regarding recurrence times of large earthquakes.

The choice between the Poisson and characteristic earthquake hypotheses is important from a scientific point of view, but is not critical for hazard estimation in areas for which the recurrence times of damaging earthquakes are much longer than the lifespan of engineered structures (EPRI[2]). Since this is the case for most of the central and eastern United States, the Poisson model is used to estimate seismicity parameters in this study. The model is simple and data are available on which to estimate model parameters. Such data are not available for the characteristic earthquake model.

Two methods were used to estimate the values of a and b in equation (1). The first is the method of Weichert[3]. This method uses a maximum-likelihood approach to fit observed numbers of events in various magnitude classes to the model. The period of record completeness is allowed to vary with magnitude class, and a maximum magnitude cut-off is employed. Input values for this procedure were based on the EPRI earthquake catalog. Periods of completeness for each magnitude interval were estimated from "decade plots" in which the number of events per decade is used to assess the number of years that earthquake detection has been complete. This method assumes that earthquake occurrence is stationary as a function of time (i.e., the true rate of activity is not time dependent). Uncertainty in magnitude estimation and completeness is not incorporated in the Weichert method. The computed rates of seismicity are assumed to apply uniformly to the entire source zone.

A second method based on the maximum likelihood principle was developed as part of the EPRI program (EPRI[1]). This method includes several innovative features:

1. A model of earthquake detection is included so that data from periods of incomplete detection can be used in estimating recurrence parameters.
2. Uncertainty in magnitude estimates is incorporated in determination of a and b.
3. Data in different magnitude classes can be given different weights in the regression.
4. Seismicity recurrence parameters can vary spatially on a half-degree grid within a source zone. Different degrees of spatial smoothing can be chosen.
5. Default values of a and b can be prescribed and weighted in the estimation procedure.

These innovations provide great flexibility in estimating seismicity parameters, and allow much of the uncertainty in such estimates to be incorporated within the estimation procedure.

The EPRI methodology for estimating seismicity recurrence parameters was used for all of the source zones. In addition, for some source zones the Weichert method was also used. Typically, our options under the EPRI method included equal weighting of magnitude classes, high spatial smoothing of a and b values (little variation of a and b within a source zone), and default values for b of 0.8, 0.9, and 1.0. The weight for the default values was moderate so that in cases when little data were available, the default value dominates, but for source zones with many earthquakes, the data dominate. For our final selection of seismicity parameters, several combinations of these options were chosen and assigned weights to represent our uncertainty in recurrence parameter estimation.

Maximum Magnitude Estimates

Selection of the upper bound magnitude for individual sources in the central and eastern United States has little theoretical or empirical precedent. In this region, no moderate-to-large earthquake in the historical record has produced a documented surface rupture. Thus, methods tailored to feature dimensions or actual rupture plane properties are not useful. Rather, intuition and methods based on seismicity are adopted by default.

We examined a number of approaches and based our estimates on their combined predictions. These approaches were based on:

1. The maximum magnitude observed in the historical record.
2. Feature expression in the deep crust.
3. Characteristic dimension of the feature/source zone.
4. Seismic flux of the source zone.
5. Predicted 1000-year earthquake.

For each approach we derived a probability distribution for maximum magnitude. For example, the approach based on maximum observed magnitude has a distribution that ranges from 1/4 magnitude unit below to 3/4 magnitude units above the maximum observed. The distribution is slightly skewed to values above the maximum observed. The final distribution for the source zone is determined by summing and normalizing the component distributions. By using a distribution rather than a single value, we are able to express the large uncertainty associated with this parameter. This methodology for calculation of maximum magnitude distribution is systematic, transparent, and includes most factors that implicitly influence our judgement when producing a distribution based on "expert opinion".

CONCLUSIONS

A major question to be addressed by seismic hazard assessments for the central and eastern United States is the likelihood of large earthquakes at locations where they have not been documented in the past. For example, an earthquake with a magnitude of about 6-3/4 occurred near Charleston, South Carolina in 1866. To date, no geologic structure or set of tectonic conditions has been found in the epicentral region that would explain the source of this event. Hence, the question arises as to whether similar earthquakes can be expected at other sites in the study region.

Our assessment of the seismic source zone framework, seismicity parameters and maximum magnitude distributions suggest that large earthquakes can occur over much of the central and eastern United States. The probability of such occurrences however is low. For example, only two of the seismic source zones have no probability that their associated maximum magnitude event is greater than 6.75 (m_b). Over half have more than 10 percent probability of an upper bound magnitude greater than or equal to 6.75. On the other hand, the probability of activity is low for many of these zones. Thus, while we believe that many features can generate a large magnitude earthquake, the likelihood at any particular site is generally low.

A second issue to be addressed is the implication of the Meers fault. The Meers fault is located in Oklahoma and exhibits evidence of recent fault movement (i.e., a young fault scarp is visible at the surface). Currently, however, the region is seismically quiet. This observation suggests that other faults may exist in the central and eastern United States that show no present evidence of seismic activity, but may become active in the future. Inspection of our generic matrix shows that, for the three characteristics considered, if a feature is seismically quiet, but other characteristics are favorable, then there is approximately a 10 to 15% probability that such a feature is active. Even if no characteristics are favorable, we still believe that 5 out of every 1000 such features is seismically active.

The EPRI methodology for probabilistically assessing seismic hazard has proved to be useful in dealing with hazard in the central and eastern United States. It is extremely flexible and can handle many different approaches to the same problem. It requires that assumptions and reasoning be spelled out such that they are open to critical examination. Uncertainties are incorporated directly in the analysis to the extent that they can be quantified. It represents a structured and logical way of dealing with a complex problem.

REFERENCES

1. Electric Power Research Institute (1986). Seismic Hazard Methodology for the Central and Eastern United States, NP-4726, Research Project, P101-22, Volumes 1-10, EPRI, Palo Alto, CA 94304.

2. Electric Power Research Institute (1986). Applicability of the Poisson Earthquake-Occurrence Model, NP-4770, Research Project, P101-38, Final Report, prepared by Cygna Corporation (C.A. Cornell, Principal Investigator), EPRI, Palo Alto, CA 94304.

3. Weichert, D. (1980). Estimation of the Earthquake Recurrence Parameters for Unequal Observation Periods for Different Magnitudes, Bulletin of the Seismological Society of America, Vol. 70, pp. 1337-1346.

Comparison of Seismic Hazard Estimates Obtained by Using Alternative Seismic Hazard Methodologies *

D.L. Bernreuter, J.B. Savy and R.W. Mensing
University of California, Lawrence Livermore National Laboratory, P.O. Box 808, L-196, Livermore, CA 94550, U.S.A.

INTRODUCTION

Probabilistic seismic hazard analysis (PSHA) has almost come of age. For example, at a recent symposium on Seismic and Geologic Siting Criteria for Nuclear Power Plants[1], it was generally agreed that PSHA is an important element in the assessment of the seismic hazard at nuclear plant sites. Much of the credit for the increasing acceptance of PSHA is due in a large part to ongoing PSHA studies funded by the US Geological survey (Algermissen et al.[2]), by the U.S. Nuclear Regulatory Commission (Bernreuter et al. [3]), the Electric Power Research Institute study [4] and several utilities studies [5,6].

One problem that has arisen is the fact that the estimates of the seismic hazard at the same site from various studies are often significantly different. This is illustrated in Fig. 1 where the estimated annual probability of exceeding any given peak ground acceleration for two different studies are compared. One set of curves are taken from the Bernreuter et al. study [3] (hereafter referred to as the LLNL study) and the other set are the estimates obtained as part of the Electric Power Research Institute study [4] (hereafter referred to as the EPRI study) for the same site located in northern Illinois.

---

*This work was supported by the United States Nuclear Regulatory Commission under a Memorandum of Understanding with the United States Department of Energy.

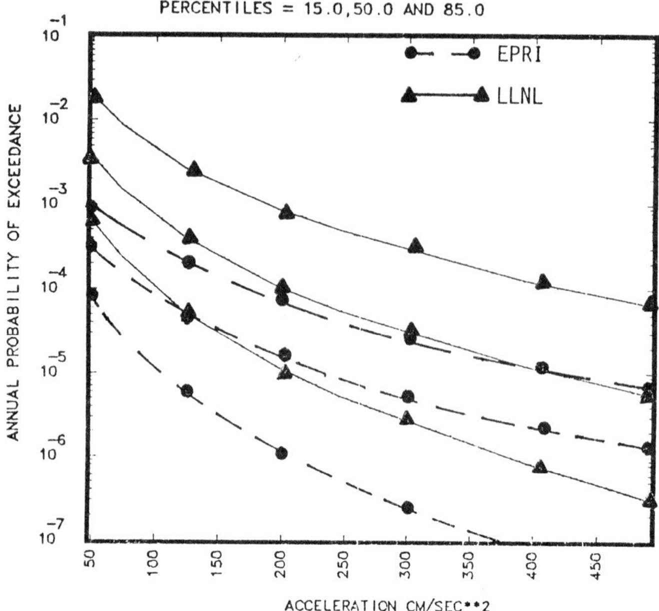

Figure 1. Comparison between the 15th, 50th and 85th constant percentile hazard curves (CPHC) for the LLNL and EPRI studies for the Braidwood site in Northern Ill.

It is seen from Fig. 1 that the LLNL and EPRI median hazard curves appear to differ by over a factor of 5. Such differences have led to the perception that the results of PSHA are too variable to be useful other than as a way of assessing the relative significance of alternative hypotheses. This perception of the great variability between PSHA studies arises in part because the large uncertainties surrounding all of the inputs into a typical PSHA indeed allow for various studies to arrive at different estimates and, in part, because of systematic differences between studies. That is, the various studies may not be solving the same problem or necessarily useing the same methodology.

In this paper a broad definition is being used for the term "seismic hazard estimation methodology". Specifically in this paper the term includes:

- The process used to develop and describe the seismicity and ground motion models and their uncertainties.
- The basic algorithm for computing the annual probability of exceedance of specific ground motion values.

- o The process used to estimate the uncertainty in the seismic hazard given the uncertainties in the inputs.

The purpose of this paper is to examine the differences between a number of studies. Clearly, it is not possible to put all the studies on the same common footing as this would require each study to redo their analysis with certain modifications. Instead we have made the necessary modifications to the our codes so that meaningful comparisons can be made between the LLNL study, the EPRI study[4], the Algermissen et al. study [2] and the utility sponsored Yankee Atomic Electric Company study[5]. In addition comparisons are made to seismic hazard estimates obtained using a nonparametric historical methodology.

OVERVIEW OF METHODOLOGIES

LLNL Seismic Hazard Methodology

The LLNL methodology[3] is similar in many ways to the methodology used in the other studies.[2,4,5,6] These studies are all based on the following four elements:

- o Identification of source zones affecting the site.
- o Description of the seismicity of a source zone using a recurrence model.
- o Identification of an appropriate ground motion model.
- o Estimation of the hazard by a probabilistic model.

It is assumed that the region affecting the ground motion at a site can be divided into discrete areas, referred to as source zones, of uniform seismicity characteristics. The seismicity of each source zone is described by the recurrence model which expresses the expected number of earthquakes, per year, exceeding a given magnitude. For the LLNL methodology, the zonation and estimation of the parameters (a,b) of the recurrence model are performed by seismicity experts. Numerous ground motion models exist for describing the attenuation of ground motion between the source and the site. The models are based on the opinion of ground motion experts.

Two panels of experts (Panel S for seismicity and G for ground motions) were formed. The S-Panel, included 11 experts knowledgeable about seismicity and zonation. The G-Panel, consisted of 5 experts knowledgeable about ground motion prediction. The individuality of the opinions of the experts was preserved by having them formulate their descriptions of zonation and seismicity without prior interaction and by encouraging them to use their own tectonic and seismicity information and data bases.The information was elicited from the panel members through a series of written questionnaires. This was followed up by joint feedback meetings in which each expert was provided with calibration hazard results based on his own input and during which the panel members were encouraged to interact. The experts' final opinions were elicited through feedback written questionnaires.

Retention of the diversity of opinions between experts is an important consideration in the LLNL methodology. A hazard estimate, i.e., calculation of a hazard curve, requires input from a member of each of the two panels. Thus, the hazard is estimated based on the inputs for every pair of experts, i.e., one S-expert and one G-expert. The variation in the hazard estimates between the 55 (11 x 5) pairs is representative of the diversity of opinions between experts.

The LLNL methodology makes a clear distinction between: (1) the random variation inherent in the occurrence of earthquakes affecting a site and the propagation of the related ground motion at the source to ground motion at the site, and (2) the uncertainties, due to diverse opinions between experts and the limited data base of past events, associated with the estimation process.

The uncertainties in the estimation process, referred to as modeling uncertainties, form the bases for the experts to describe their state of knowledge and level of confidence in the information used in formulating their opinions. Modeling uncertainties were introduced into the hazard analysis by having the experts provide alternative zonations and/or models as well as ranges of values for the seismicity parameters, e.g., (a,b) in the recurrence model. These uncertainties were themselves modeled in terms of probability distributions which were sampled, using Monte Carlo simulation, to describe the resulting uncertainty in the estimation of the seismic hazard. The LLNL study included the contribution of all earthquakes larger than 3.75 to the seismic hazard.

EPRI Seismic Hazard Methodology

It should be noted that description given here and results are from early draft interim reports describing the EPRI methodology and seismic hazard estimates[4] obtained at nine test sites. A number of changes have since been incorporated into their methodology as a result of peer review. However, updated results have not been released. Even though the final EPRI results are unavailable, for our purpose here, it is worthwhile to make comparisons to the interim version of their methodology and results.

While there are a number of similarities between the LLNL and EPRI methodologies there are some important differences. Both studies rely primarily on experts opinion to develop the inputs necessary and both studies attempt to carefully model the uncertainty. Both studies separate the random and modeling uncertainties.

EPRI formed six seismicity teams. Although the input from each team was kept separate there was extensive interaction among teams and consensus data sets were developed and used. No attempt was made to model the uncertainty in the ground motion and only few equally weighted ground motion models were used to develop representative interim results. This is very important because the differences in ground motion models was one of the major contributors to the differences between the LLNL and EPRI results shown in Fig. 1. EPRI's method for eliciting uncertainty is also different than LLNL's. The EPRI teams were asked to provide alternative discrete models for the seismicity. Each team provided at most two sets of seismicity parameters for each zone and in many cases only one model was provided. This is in contrast to the LLNL methodology where each S-Expert provided a best estimate and the 95th percentile confidence bounds for each input parameter. The LLNL approach generally leads to a much wider variation in the uncertainty bounds as modeled, by say the 15th and 85th percentile hazard curves than the EPRI approach. Because of the discrete nature of all of the inputs EPRI used a logic tree approach to model the uncertainty.

It should be noted that the differences between the logic tree and Monte Carlo approaches to model uncertainty are not important if the same distributions are modeled. The point made above was that there is significant differences in the distributions for the uncertainty in the input seismicity parameters between the LLNL and EPRI studies.

It should also be noted that the EPRI study only included the contribution to the seismic hazard for earthquakes greater than or equal to magnitude 5.

YANKEE Atomic Electric Company Methodology

The Yankee Atomic Electric Company[5] (YAEC) methodology is typical of a number studies performed by utilities to develop seismic hazard estimates for particular sites. The input was developed by the analyst team. A logic tree approach was employed to model the uncertainty. Thus only a few discrete alternative models were used for the seismicity parameters similar to the input provided by a single EPRI team. Only a few best estimate ground motion models were used. Only the contribution from earthquakes of magnitude 4 and greater were included in the hazard estimate.

Algermissen et al. Methodology

The Algermissen et al. methodology[2] is in sharp contrast to both the LLNL and EPRI methodologies. Input was solicited from regional teams of experts, however the final selection of all models was performed by the analyst team. Uncertainty was not directly modeled. Only a single best estimate model and a single best estimate seismic hazard estimate was developed. Random uncertainty was not included in the single ground motion model used. Finally, only earthquakes of intensity V and larger were included in estimation of the seismic hazard.

Nonparametric Historical Methodology

The methodologies discussed above can be characterized as a seismic-source approach to the calculation of the seismic hazard. They are based on utilizing statistical and geological evidence to define geographical regions (seismogenic zones) with homogeneous Poisson activity uniformly throughout the zone. This activity is described by a recurrence relationship. Calculation of the seismic hazard from the source geometry and seismic activity then depends on specification of a ground motion model, which expresses the decay with distance of the median value of a ground motion parameter, e.g., peak ground acceleration (PGA) for different magnitudes.

Given our lack of understanding of the process causing earthquakes in eastern North America it is natural to attempt to partiality validate such models. One method for validating the basic models inherent to the seismic-source approach is to compare the estimates with historical data for which no or minimal modeling is assumed. Such a nonparametric method was used in the LLNL methodology[3], to partially validate the models developed. In the nonparametric method[3,7] each historical earthquake in a catalog is attenuated to the site and the average exceedance rate for a given level of ground motion includes the observed random variability of the observed ground motion at similar sites between earthquakes of the same magnitude and located at the same distance from the site. The nonparametric methodology can also include a correction for catalog incompleteness. It produces a hazard estimate without quantifying the uncertainty.

The hazard curve, based on the historical data, of course, is only an estimate of the true historic hazard curve as no ground motion measurements were taken at any of the sites. The estimated hazard depends on the ground motion model and the model for the random variation in the ground motion parameter used to transform the ground motion from the source to the site.In our application of the historic method we chose not to apply a correction for incompleteness of the catalog because in this paper this would require an additional model assumption. The philosophy of the historic analysis is to make as few assumptions as possible.

COMPARISONS AND DISCUSSION

As noted in the introduction the necessary changes were introduced in the LLNL input and results were computed so that appropriate comparisons with other studies can be made. It is worthwhile to note that for given a set of inputs e.g., a single trial in the LLNL Monte Carlo approach or for a single branch in the EPRI or YAEC logic tree approach or the single model for the Algermissen et al. approach that if the same set of inputs are used each computer code would obtain very similar hazard curves. The primary differences in the results obtained using the various methodologies lies in how the input was developed and how the uncertainties were modeled.

LLNL AND EPRI

To compare LLNL's and EPRI's results on the same footing the LLNL results were recomputed for a lower bound magnitude increased from 3.75 to 5. In addition because the EPRI study (at this stage) did not attempt to fully model the uncertainty in the ground motion then the results should be compared using the same ground motion models. As EPRI provided results based on a single ground motion model the new set of LLNL results were obtained using only the input provided by LLNL's seismicity experts and the same ground motion model used by EPRI. Figure 2 gives a comparison of the 15th, 50th and 85th constant percentile hazard curves (CPHC) resulting from LLNL's and EPRI's methodologies when the same ground motion model and the same lower bound magnitude are used at the same site as in Fig. 1. The median hazard curves are now in much better agreement. One can also see the influence of the differences in methodology in the way the experts/teams were asked to model uncertainty on the 15th and 85th CPHC. Because of the differences in the way the uncertainty was modeled the mean hazard, rather than median, curves would not be in good agreement.

We made similar comparisons at the other eight sites included in the EPRI study. We found overall that there was

- o Excellent agreement of the median hazard curves at four sites.
- o Reasonable agreement at four site, i.e., the EPRI median hazard curves were only about a factor of 2 to 3 lower than the LLNL median hazard curves.
- o Only at one site did the EPRI median hazard curve fall below the LLNL 15th CPHC.

After carefully comparing EPRI's expert teams modeling to the LLNL's seismicity experts models we concluded that for the case when the same ground motion models are used, the primary reason for the difference between the median hazard curves are due to differences in zonation. That is, around the sites where the median hazard curves did not agree we found that the EPRI teams tended to have smaller more discrete zones as compared to the LLNL experts. At the four sites where there is excellent agreement between the two studies we found that

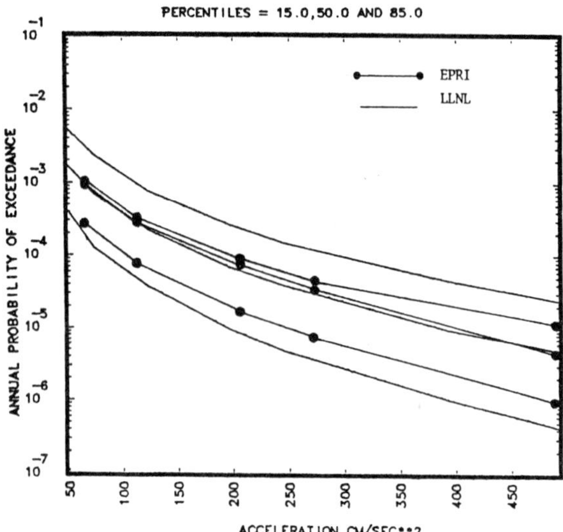

Figure 2. Comparison between the 15th, 50th and 85th CPHC for the LLNL and EPRI studies for the Braidwood site. The same ground motion model and lower bound magnitude was used to obtain both the LLNL and EPRI curves.

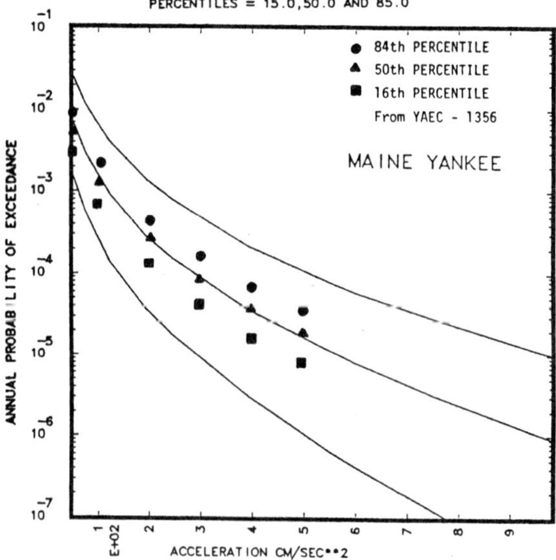

Figure 3. Comparison of CPHC obtained by using all the LLNL S-Experts and a set of ground motion models similar to ground motion models used by YAEC and the CPHC from YAEC (1985).[6] The same lower bound magnitude was used for both results.

the zonation was very similar,i.e., both the LLNL experts and the EPRI teams tended to define larger source zones.

LLNL and YAEC

In Fig. 3 we compare the results of YAEC's PSHA for the Maine Yankee site to those obtained by using the seismicity models provided by the LLNL S-Experts with a set of ground motion models similar to the ground motion models used in the YAEC study. It is observed from Fig. 3 that the median hazard curves are in excellent agreement, however, YAEC's uncertainty bounds, as represented by the 16th and 84th CPHC, are much smaller than LLNL's. This difference in uncertainty bounds is due in part to the fact that in the LLNL study eleven independent experts were used as contrasted to the single team for YAEC and in part to the way the uncertainty was modeled.

LLNL and Algermissen et al.

Algermissen et al. provided us with their zonation and seismicity parameters for each zone. We used this as input into our PSHA program along with the set of best estimate ground motion models provided by our G-Experts to compute a best estimate hazard curve. As noted earlier, because Algermissen et al. did not provide a model for uncertainty only a single best estimate hazard can be computed. In Fig. 4 we compare the combined over all LLNL S-Experts hazard curve to the hazard curve obtained using the Algermissen et al. seismicity model at the same site as for Figs. 1 and 2. It should be noted that for this comparison only earthquakes equal to or larger then modified Mercalli V are included in the PSHA. The agreement is excellent.

At other sites the Algermissen et al. curve is typically higher than the LLNL curve. However, the Algermissen et al. curve generally falls within the bounds of individual hazard curves for the LLNL S-Experts as shown in Fig. 5. Figure 5 also indicates the large differences between the various LLNL S-Experts indicating that one needs a wide sample of expert opinions.

LLNL and Historical

In Figs. 4 and 5 we also have plotted the resultant hazard curve obtained using the nonparametric historical approach discussed earlier. The curve shown in Fig. 4 is based on only the last 100 years of data. Better agreement is obtained if the last 200 years of data are used because of the 1811-12 New Madrid earthquakes. This illustrates one of the difficulties in using the nonparametric approaches; namely, that the return period of the largest event becomes the length of the catalog used. Nevertheless the nonparametric historical approach serves as a useful check. As expected, the agreement between the historical curve and one curve resulting from typical PSHA modeling is reasonably good in the 10^{-2} to 10^{-3} range of

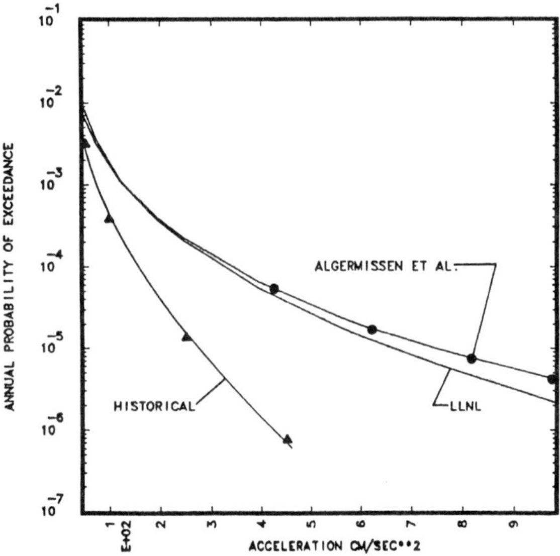

Figure 4. Comparison of the hazard curve obtained using Algermissen et al.'s seismicity model and the best estimate hazard curve combined over all LLNL's S-Experts for the Braidwood site. Also shown is the historical curve based on the last 100 years of data. The same ground motion models were used to generate all curves.

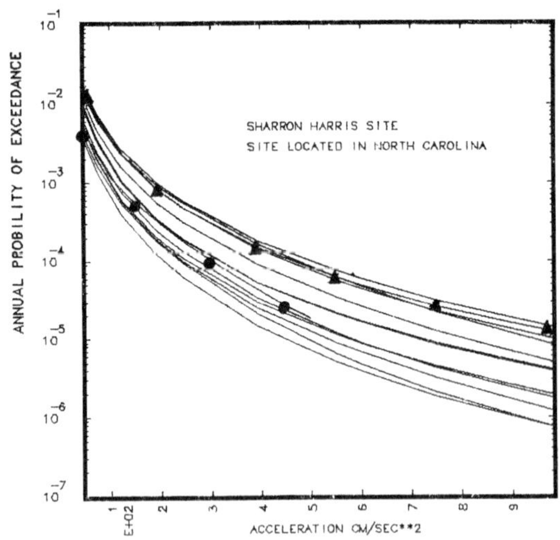

Figure 5. Comparison between the LLNL S-Experts' best estimate hazard curves (unlabeled curves) the historical hazard curve (curve labeled ●) and the Algermissen et al. hazard curve (curve labeled ▲). All hazard curves were obtained using the LLNL G-Experts' best estimate ground motion models.

annual probabilities of exceedance and much poorer at lower probability of exceedance.

CONCLUSIONS

In this paper we have tried to show that, because of the substantial uncertainty in the estimated hazard, care is required when comparing the results of various analyses. Often times the results of two analyses are substantially different because of systematic difference such as the lower bound magnitude used, the sets of ground motion models used, the range of expert judgement sampled and how uncertainty is modeled. Our studies have shown[3] that there are very significant difference in opinions among experts. Thus one must expect the possibility of substantial differences between studies which only have included a limited sample of expert judgement. On the other hand, it appears that the median hazard curves of reasonably complete studies which are based on a relatively large sample of expert judgement are in reasonable agreement. It should be noted that the mean hazard curve is governed by how the uncertainty analysis is performed. Thus when comparing mean hazard curves allowance should be made for how the uncertainty analyses were performed. For example, because of the differences in modeling the uncertainty between the LLNL and EPRI studies, one can often expect that the LLNL mean hazard curve will be higher than the EPRI mean even for the cases when the median hazard curves agree.

REFERENCES

1. Transcripts of the Symposium on Seismic and Geologic Siting Criteria for Nuclear Power Plants, Sponsored by the US Nuclear Regulatory Commission. October 7-9, 1986, Rockville, Maryland, Ann Riley Associates.

2. Algermissen, S.T., Perkins, D.M., Thenhaus, P.C., Hansen, S.L., and Bender, B.L., (1982) Probabilistic Estimates of Maximum Acceleration and Velocity in Rock in the Contiguous United States, U.S. Geological Survey Open-File Report 82-1033.

3. Bernreuter, D.L., Savy, J.B., Mensing R.W., Chen, H.C., Davis, B.C. (1985) Seismic Hazard Characterization of the Eastern United States: Vol. 1 and Vol. 2, LLNL Report UCID 20421.

4. EPRI, (1985) Seismic Hazard Methodology for Nuclear Facilities in the Eastern United States: Preliminary Seismic Hazard Test Computations for Parametric Analysis and Comparative Evaluations EPRI Research Project Number P101-29 (Draft).

5. Yankee Atomic Electric Company (1983), Maine Yankee Seismic Hazard Analysis, Report YAEC-1356.

6. Yankee Atomic Electric Company (1984), Review and Comments on NUREG/CR3756 Seismic Hazard Characterization of the Eastern United States: Methodology and Interim Results for Ten Sites, Report YAEC-1455.

7. Veneziano, D., Cornell, C.A. and O'Hara, T. (1984) Historical Method of Seismic Hazard Analyses, EPRI NP-3438, Project 1233-10, Interim Report, May 1984.

Ground Motion Relations for Eastern North American Earthquakes

O.W. Nuttli, R.B. Herrmann
Department of Earth and Atmospheric Sciences, Saint Louis University, St. Louis, MO 63103, U.S.A.

INTRODUCTION

There are at least four ways in which earthquakes in eastern North America differ from their more studied counterparts in western North America. The first is that earthquakes of a given magnitude occur approximately ten times more frequently west of the Rocky Mountains than east of them. The second is that the western earthquakes, even those of moderate magnitude, frequently cause the surface of the earth to rupture whereas, in historic times, no eastern earthquake has produced surface rupture. The third is that the anelastic attenuation of seismic waves at frequencies of 0.2 to 20 Hz, frequencies usually associated with damaging ground motion, is significantly smaller in the east than in the west. The fourth, which some of our seismological colleagues dispute, is that the scaling of ground motion with magnitude is different for large magnitude earthquakes in the two regions.

There is little to add to the first statement concerning the frequency of occurrence of earthquakes, except that the long recurrent times for eastern earthquakes make the general population complacent and often unwilling to invest in earthquake-resistant design and construction of structures. The federal government has played a leading role in requiring proper design of critical structures for which it is responsible, and in educating the engineers, scientists and political and industrial decision makers about the nature of the earthquake hazard.

EFFECTS OF EARTHQUAKE FOCAL DEPTH

With the exception of earthquakes in western Washington, western British Columbia, Alaska, the Aleutian Islands and Hawaii, the maximum focal depth of earthquakes in North America is about 25 km. Figure 1 shows the effects on near-field ground acceleration of variations in focal depth for a given epicentral distance, for earthquakes of 1 to 25 km in depth. The abscissa is horizontal distance measured on the earth's surface, either from the epicenter or from the surface projection of the nearest asperity point on the fault surface. For purposes of illustration a body-wave magnitude, m_b, of 6 is assumed. Body-wave magnitude is a measure of the strength of 1-Hz wave excitation at or near the earthquake source region.

From Figure 1 it can be seen that there are substantial differences in ground acceleration at horizontal distances of 1 to 10 km or more for focal depths typical of eastern North American earthquakes. There is almost an order of magnitude difference in acceleration at a distance of 1 km for focal depths of 1 and 25 km.

For eastern North American earthquakes, assuming a vertical fault plane, the minimum depth for an earthquake of $m_b = 6$ is 5 km if there is no rupture of the earth's surface. For western North American earthquakes, which often rupture the earth's surface, the depth could be as small as 1 km. These depth values are, of course, minimum values. In both regions earthquakes can occur as deep as 15 to 25 km. When they occur at these larger depths the near-field ground motion is almost an order of magnitude less whan when they occur at shallow depths.

EFFECTS OF ANELASTIC ATTENUATION

The two most important types of seismic wave attenuation are those caused by geometric spreading of the wave energy and by anelasticity. The former is dominant in the near-field region, whereas the latter becomes more important as the distance increases.

At distances as large as 1000 km, the geometric spreading term for body-wave amplitudes can be approximated by R^{-1}, where R is horizontal distance. For surface waves, in the frequency domain, the corresponding term is $R^{-\frac{1}{2}}$. In the time domain there is an additional effect due to the dispersion of the surface-wave train, so that the overall elastic attenuation is $R^{-5/6}$ for Airy phases (associated with group velocity minimum or maximum values) and R^{-1} for ordinary surface waves[1].

Unlike geometric-spreading attenuation, anelastic attenuation is frequency dependent. For body waves the effect can be approximated by the term exp(-ks), where k is the frequency-dependent coefficient of anelastic attenuation and s is distance measured along the ray path. For surface waves the corresponding term is exp(-kR), where R is the horizontal distance from the epicenter.

The term k is given by $\pi f/QV$, where f is the wave

frequency, Q is the specific quality factor and V is the velocity of the wave. V is independent of frequency for body waves, but frequency dependent for surface waves. Q, in general, is frequency dependent. Experimental evidence indicates that it can be represented by

$$Q(f) = Q_o f^n \tag{1}$$

where Q_o is the 1-Hz value of Q. Therefore,

$$k(f) = \pi f^{1-n}/Q_o V. \tag{2}$$

Q_o values for shear and surface waves were experimentally determined by Singh and Herrmann[2] for the United States. In general, the values are smaller in the United States west of the Rocky Mountains (150 to 400) and larger in the area to the east (400 to 1300). The values of n, on the other hand, are larger in the west (0.5 to 0.7) and smaller in the east (0.2 to 0.4). The overall effect of attenuation is illustrated in Figure 2, which shows ground displacement values for surface waves in the time domain for Q_o values of 150 (typical of coastal California) and 1000 (typical of central United States) and frequencies of 1, 5 and 20 Hz. For the purposes of this example a 1 km focal depth is assumed. A larger depth would decrease the ground displacement at distances out to about 10 km, as can be seen from Figure 1. Figure 2 shows that there is over an order of magnitude difference in 1-Hz ground displacements at 500 km for the central United States and coastal California curves. For 5-Hz waves at 500 km distance the difference is slightly less than two orders of magnitude, and for 20-Hz waves it is four orders of magnitude. At the lesser distance of 50 km the difference for 1-Hz waves is 1.3 times, for 5-Hz waves 1.5 times and for 20-Hz waves 1.9 times. Thus we can conclude that near-field ground-motion attenuation is not greatly different in the western and eastern United States, but that at distances greater than 100 km the effects of anelasticity become important and lead to large differences in ground-motion attenuation for different parts of the United States.

The Q_o and n values can be determined from the data of microearthquakes as well as from the data of damaging earthquakes. Therefore there are sufficient data to determine their values over most of North America.

EFFECTS OF MAGNITUDE SCALING

Magnitude scaling refers to the manner in which the ground motion, at a given distance, increases as the magnitude increases. The scaling relations are different for peak ground acceleration than for peak velocity and for peak displacement. The scaling relations themselves depend upon the manner in which the earthquake wave spectrum near the source changes as the earthquake magnitude changes.

The only region of North America for which there are adequate accelerogram data to determine scaling relations empirically is coastal California. For example, Joyner and Boore[3] obtained empirical relations for peak ground acceleration and velocity as a function of moment magnitude (a magnitude that measures the low-frequency seismic wave excitation).

In eastern North America there are no accelerogram data for earthquakes of m_b greater than about 5. Therefore, if the magnitude scaling relations are different for the eastern and western regions, the empirically determined western scaling relation for moderate to large earthquakes would not apply to the eastern region. Haar *et al*[4] concluded that for small to moderate earthquakes the scaling relations are the same in the two regions, and inferred that this would also be true for large earthquakes. Nuttli[5] concluded that the scaling relations for large earthquakes are different for the two regions. Iio[6], using data of Japanese earthquakes with a big range in magnitude, found that for small to moderate earthquakes his relation was similar to that of Nuttli[5] rather than to that of Haar *et al*[4]. We may tentatively conclude that the scaling relation for small to moderate earthquakes is the same for all regions, and corresponds to that found by Nuttli[5] and Iio[6]. This same scaling relations appears to hold also for large earthquakes in eastern North America, but not for large California earthquakes.

STRONG GROUND MOTION ATTENUATION AND SCALING RELATIONS FOR EASTERN NORTH AMERICA

The wave frequencies associated with peak ground acceleration differ from those for peak ground velocity and peak ground displacement. Furthermore, the wave frequencies for each of the three measures of strong ground motion decrease as the epicentral distance increases. Therefore, it is better to use empirical data, as done by Joyner and Boore[3], in the development of strong ground motion attenuation and scaling relations. However, when such data are unavailable, we have to proceed by making certain assumptions.

Initially we shall assume for eastern North America that at all distances the peak ground acceleration occurs at a frequency of 5 Hz. This will cause the peak ground acceleration to be slightly overestimated at distances of less than about 20 km and to be underestimated at distances beyond 20 km. For peak ground velocity the wave frequency initially will be assumed to be 1.5 Hz and for ground displacement 0.5 Hz. Using observed strong-motion data for earthquakes of m_b approximately equal to 5[7-10] to set the level of the curves, and taking into account that most of the accelerograph sites were on soft soil, the resulting equations for peak horizontal ground motion at soft soil sites in the near-field region (distances no greater than about 20 km) are

$$\log a_h = 0.57 + 0.50m_b - 0.83 \log (R^2 + h^2)^{\frac{1}{2}} - 0.00120 R \quad (3)$$

$$\log v_h = -3.60 + 1.00m_b - 0.83 \log (R^2 + h^2)^{\frac{1}{2}} - 0.00052 R \quad (4)$$

$$\log d_h = -6.81 + 1.50m_b - 0.83 \log (R^2 + h^2)^{\frac{1}{2}} - 0.00024 R \quad (5)$$

Figure 3 shows the behavior of peak horizontal acceleration (arithmetic average of the peak accelerations on the two horizontal accelerograms) versus epicentral distance, with body-wave magnitude as parameter. A focal depth of 15 km was assumed for all earthquake magnitudes. For large earthquakes, the so-called "Richter magnitude" often is synonymous with surface-wave magnitude, M_S, which is a measure of the 20-sec period source excitation. For eastern North America, an m_b of 5.0 is equal to an M_S of 4.4, an m_b of 6.0 to an M_S of 6.4, and an m_b of 7.0 to an M_S of 8.3. The largest historic earthquake in eastern North America, that in the New Madrid fault zone on February 7, 1812, had an estimated m_b of 7.4, corresponding to an M_S of 8.8[5].

At a distance of 200 km the a_h values calculated from Equation 3 need to be multiplied by about a factor of 2 because the dominant wave frequency will be closed to 1 Hz than to 5 Hz. At distances of 500 km the values calculated from Equation 3 need to be multiplied by about a factor of 4 because the dominant wave frequency will be closer to 0.3 Hz. In Figure 3 the dashed-line curves show the values calculated from Equation 3, assuming 5-Hz waves, and the solid-line curves show the effect of decreasing dominant frequency as epicentral distance increases.

THE MEXICO CITY SYNDROME FOR EASTERN NORTH AMERICAN EARTHQUAKES

The September 19, 1985 Mexican earthquake of M_S = 8.1 caused collapse of a number of structures of 8 to 20 stories height that were built on the old lake bed area of Mexico City. The epicentral distance to Mexico City was 400 km. At points of the city away from the lake bed the recorded peak ground acceleration was about 40 cm/sec^2, with dominant period of about 2 sec and a duration of about 30 sec[11]. At points on the lake bed the recorded peak acceleration was about 170 cm/sec^2, with a period of 2 to 3 sec and a duration of about 40 sec[12]. The soft sediments of the old lake bed caused the ground motion to be amplified, whereas the long duration of shaking and the dominant periods of 2 to 3 sec set the 8 to 20 storey buildings into resonance, with subsequent failure of the upper levels and sometimes of the entire building.

The metropolitan areas of eastern North America are not

built on old lake beds, but nevertheless high-rise structures are capable of suffering at least non-structural damage from large magnitude distant earthquakes[13]. Figure 3 indicates that at a distance of 400 km an M_S = 8.1 (m_b = 6.9) earthquake has a peak acceleration of 75 cm/sec^2. An M_S = 8.8 (m_b = 7.4) earthquake, the size of the February 7, 1812 New Madrid earthquake, would have a peak acceleration of 135 cm/sec^2 at 400 km. At 300 km earthquakes of M_S = 8.1 and 8.8 would produce peak horizontal accelerations of 110 cm/sec^2 and 200 cm/sec^2, respectively. The latter value equals or exceeds that measured at sites on the old lake bed of Mexico City.

Figure 4 illustrates that the ground motions caused by eastern North American earthquakes at large distances are of long duration, sinusoidal, and of frequency 1 to 0.3 Hz. The seismogram is from a vertical-component instrument that has a near constant magnification over the frequency range of interest. The earthquake occurred in northeastern Ohio on January 31, 1986 and had an m_b value of 5.0. The instrument is located at Saint Louis University, at a distance of 828 km from the epicenter. The seismogram shows two prominent trains of surface waves, one consisting of 1-Hz waves of approximately 40-sec duration that begin about 2 minutes after the first motion. The second begins about 3 minutes after the first motion and consists of waves of 0.3 to 0.5 Hz frequency which continue for approximately 60 sec.

During historic times, large magnitude earthquakes have occurred near Quebec City, Canada, in 1663 (M_S of about 8), the New Madrid fault region in the central United States in the winter of 1811 and 1812 (five earthquakes with M_S values of 8.0 to 8.8), and near Charleston, South Carolina in 1886 (M_S of about 7.7). The latter earthquake caused plaster to fall from the ceilings of the second and third floors of buildings in Chicago, at a distance of over 1200 km[14]. All of these major earthquakes occurred before the appearance of high-rise structures. Future large earthquakes in these or other regions of eastern North America will affect tall buildings in a number of metropolitan areas.

EFFECTS OF SMALL ANELASTIC ATTENUATION ON LOW-RISE STRUCTURES

Figure 2 shows that 5-Hz and 20-Hz wave amplitudes are much larger in the eastern than in the western United States for earthquakes of similar magnitude, particularly at distances of 100 km and larger. As a result, in the East damage to low-rise structures can extend to at least several hundred kilometers for moderate and large magnitude earthquakes, of m_b equal to 6 and larger. At distances of hundreds of kilometers the most common damage is to chimneys, followed by cracks in walls and foundation, and damage to plaster walls and ceilings. Damage of this type has resulted in millions of dollars of repairs for earthquakes as small as m_b equal to 5.

SUMMARY

Earthquakes in eastern North America cause damage over a wide area, but have a low frequency of occurrence. Of special concern is the potential for damage to high-rise structures at large epicentral distances, a result of low anelastic attenuation for waves of 0.2 to 1 Hz frequency and of surface-wave dispersion, which causes the duration of a sinusoidal wave train to increase with increasing epicentral distance.

REFERENCES

1. Ewing W.M., Jardetzky W.S. and Press F. (1957), Elastic Waves in Lavered Media, McGraw-Hill. New York.
2. Singh S. and Herrmann R.B. (1983), Regionalization of Crustal Coda Q in the Continental United States, Journal of Geophysical Research, Vol. 88, pp. 527-538.
3. Joyner W.B. and Boore D.M. (1981), Peak Horizontal Acceleration and Velocity from Strong-Motion Records Including Records from the 1979 Imperial Valley, California, Earthquake, Bulletin of the Seismological Society of America, Vol. 71, pp. 2011-2038.
4. Haar L.C., Mueller C.S., Fletcher J.B. and Boore, D.M. (1986), Comments on "Some Recent Lg Phase Displacement Spectral Densities and their Implications with Respect to Prediction of Ground Motions in Eastern North America" by R. Street, Bulletin of the Seismological Society of America, Vol. 76, pp. 291-295.
5. Nuttli O.W. (1983), Average Seismic Source-Parameter Relations for Mid-Plate Earthquakes, Bulletin of the Seismological Society of America, Vol. 73, pp. 519-535.
6. Iio Y. (1986), Scaling Relation between Earthquake Size and Duration of Faulting for Shallow Earthquakes in Seismic Moment between 10^{10} and 10^{25} Dyne-Cm, Journal of the Physics of the Earth, Vol. 34, pp. 127-169.
7. Chang F.K. (1983), Analysis of Strong-Motion Data from the New Hampshire Earthquake of 18 January 1982, Geotechnical Laboratory, U.S. Army Corps of Engineers. Vicksburg, Mississippi.
8. Herrmann R.B. and Nuttli O.W. (1980), Strong-Motion Investigations in the Central United States, Proceedings of the Seventh World Conference on Earthquake Engineering, Part II: Geoscience Aspects, Istanbul, Turkey, 1980.
9. Street, R. (1982), Ground Motion Values Obtained for the 27 July 1980 Sharpsburg, Kentucky Earthquake, Bulletin of the Seismological Society of America, Vol. 72, pp. 1295-1307.
10. Weichert D.H., Pomeroy P.W., Munro P.S. and Mork P.N. (1982), Strong Motion Records from Miramichi, New Brunswick, 1982 Aftershocks, Earth Physics Branch Open File Report 82-31. Ottawa, Canada.
11. Prince A. et al (1985), Acelerogramas en Ciudad Universitaria del Sismo del 19 de Septiembre de 1985, Informe

IPS-10A, Instituto de Ingenieria UNAM. Mexico City.
12. Mena E. et al (1985), Acelerograma en el Centro Scop de la Secretaria de Comunicaciones y Transportes, Sismo del 19 de Septiembre de 1985, Informe IPS-10B, Instituto de Ingenieria UNAM. Mexico City.
13. Malik L.E. and Nuttli O. (1986), Vulnerability of Tall Buildings in Atlanta, Chicago, Kansas City and Dallas to a Major Earthquake on the New Madrid Fault, Proceedings of the Third U.S. National Conference on Earthquake Engineering, Vol. II, pp. 859-870. Charleston, South Carolina.
14. Dutton C.E. (1889), The Charleston Earthquake of August 31, 1886, U.S. Geological Survey Ninth Annual Report, 1887-1888, pp. 203-528.

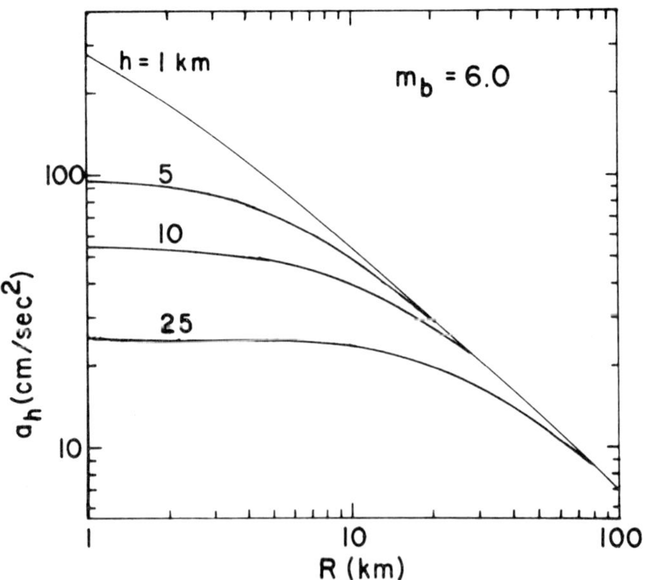

Figure 1. Variation of peak ground acceleration with earthquake depth and epicentral distance.

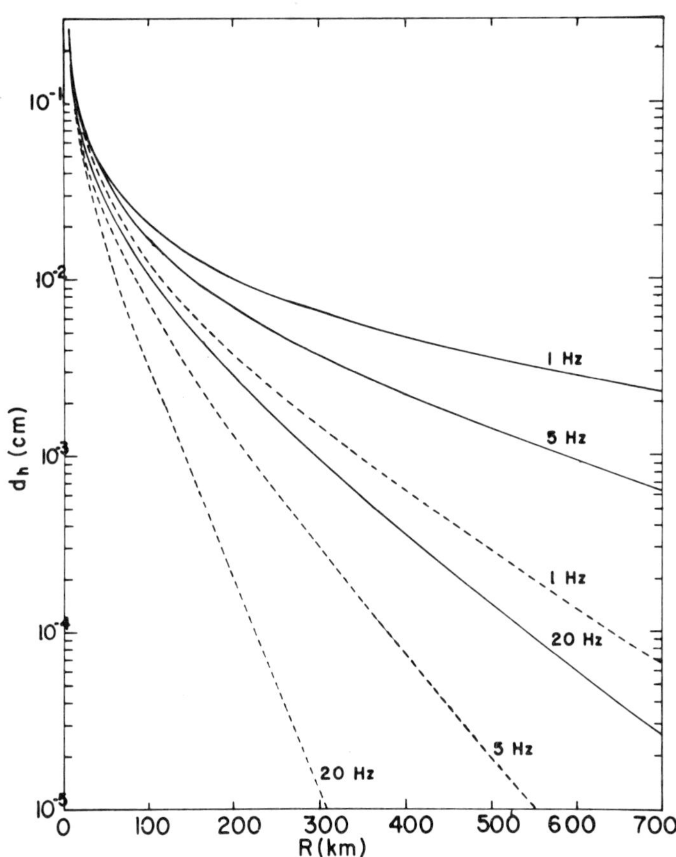

Figure 2. Variation of peak ground displacement with wave frequency and epicentral distance. The solid-line curves are for the central United States, and the dashed-line curves are for coastal California.

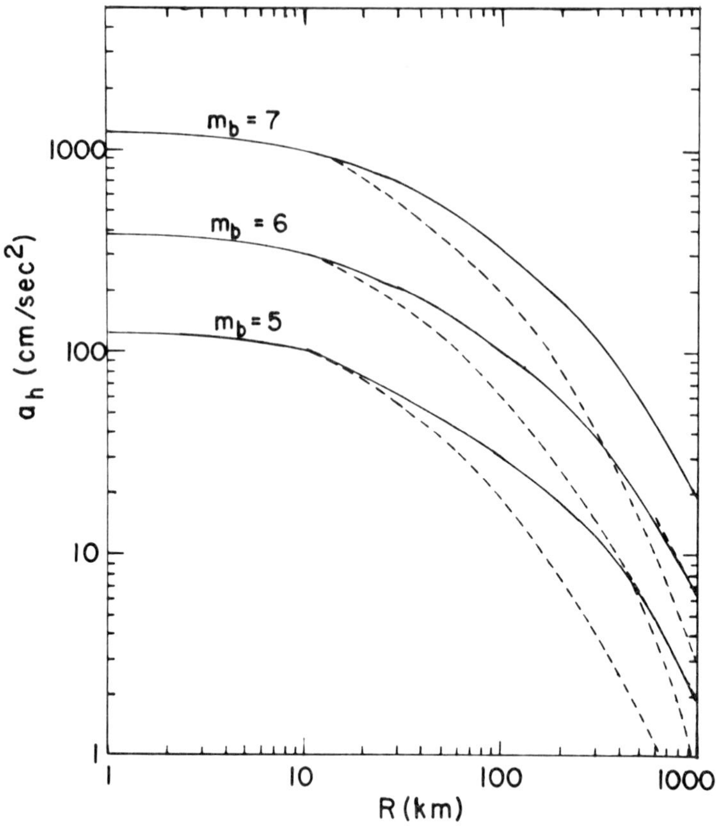

Figure 3. Variation of peak ground acceleration with earthquake magnitude and epicentral distance. The dashed-line curves are for a constant wave frequency of 5 Hz. The solid-line curves take account of a decrease in wave frequency with increasing distance.

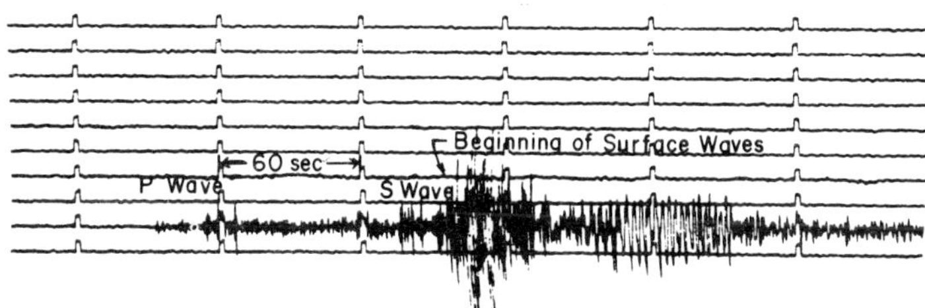

Figure 4. Vertical-component seismogram recorded at an epicentral distance of 828 km. There are two sinusoidal, long-duration trains of low frequency surface waves.

A Methodology to Correct for Effect of the Local Site Characteristics in Seismic Hazard Analyses*

J.B. Savy, D.L. Bernreuter and J.C. Chen
University of California, Lawrence Livermore National Laboratory, Livermore, California 94550, U.S.A.

1. INTRODUCTION

The complexities of the seismic ground motion observed at a particular site can be seen as the product of several independent phenomena. The seismic energy is released in the earth's crust, in general in dislocations distributed on a surface, at the earthquake source. An instrument standing at a site is subjected to the various types of seismic waves created by the source. These waves arrive at the site at different times and use a multiplicity of paths, they can carry vastly different proportions of the total energy and they also have very different frequency contents.

This paper is concerned with the transformation of the seismic waves as they travel the last few hundred feet before reaching the site. More specifically, we present a method to account for those local site effects for use in our seismic hazard analysis including the uncertainties in describing the site characteristics.

The purpose of the overall project[1] was to develop a methodology to estimate the seismic hazard at any point east of the Rocky Mountains. Because the strong ground motion data available is not sufficient to definitely chose between the various experts' interpretations, we use an array of ground motion models selected independently for the different regions of the eastern United States (EUS) by 5 ground motion experts.

*This work was supported by the U.S. Nuclear Regulatory Commission under a Memorandum of Understanding with the U.S. Department of Energy.

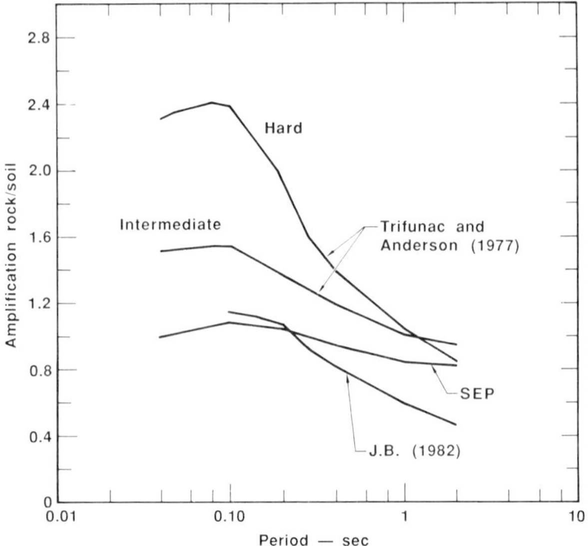

Figure 1. Simple correction factors obtained by regression analysis of WUS data.

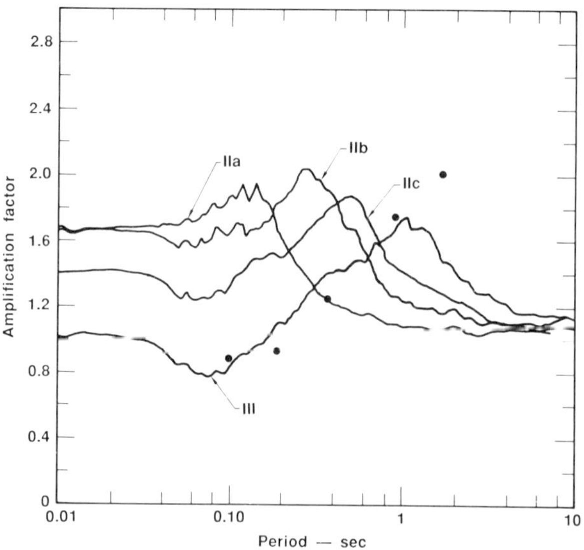

Figure 2. Comparison of the 7% median damping spectral amplification factors for sand–like sites for Categories IIa, IIb and III relative to rock. The solid dots show the results from J.B. (1982), Regression Analysis.

2. OVERVIEW OF THE METHOD

2.1 Introduction

One of the main objectives of the EUS project was to assess the uncertainty in the hazard estimates by including modeling as well as random uncertainty.

The experts' opinions were elicited via questionnaires, meetings and feedback questionnaires. They were asked to assign degrees of belief that each of the following methods of site correction is the appropriate one to use:

a. No correction.

b. Use only a simple soil or rock classification if available -- otherwise, no correction.

c. Develop correction factors for each ground motion model based on several generic site classifications, 1-D analysis data, and judgment.

d. Do a site specific analysis.

Each of these are discussed in more detail in the following sections:

2.2 No Correction

The expert might believe that his knowledge of EUS ground motion is so poor and the methods we have to assess local site effects so uncertain, that it would be better to use the generic ground motion models. All site types are treated the same using, for example, Nuttli's or Campbell's models which also fall into the "generic soil" category as they were developed using only generic soil data.

2.3 Simple Rock/Soil Correction

For this case, the site types are put into two (rock or soil) or (stiff or soft) or three categories (soft, stiff, basement rock) and a simple constant correction factor is applied (for each category).

Figure 1 shows typical correction factors going from soil to rock as a function of period found from WUS data. The curve labeled 1 is based on Joyner-Boore[2] regression analysis, the curves labeled 2 and 3 are based on Trifunac and Anderson[3] and the curve labeled 4 is based on the SEP results[4]. For Trifunac's model, curve 2 is between soft alluvium and hard sedimentary rock and curve 3 is between soft alluvium and basement or crystalline rock. Both Joyner-Boore and the SEP use only rock/soil categories.

Table 1 shows that there is considerable variation between the ratios rock/soil found in different studies.

In addition to possible deficiencies in the form of the mathematical models used for the regression analyses, all of the regression analyses were performed using less than perfect data sets. All the data sets suffer from the use of poor criteria for the identification of rock sites. Recently, more site boring data have become available to assist in properly sorting the data into categories. The Joyner-Boore[5] data set is the best in this regard, but it contains a number of questionable sites identified as rock sites. In contrast with the Joyner-Boore data set, most data sets contain data recorded in large buildings and/or in basements. These data sets were used to obtain the results plotted on Fig. 1 and given in Table 1. Joyner and Boore, Campbell[6,7] and others have shown that building type and location of the recorder in a sub-basement can have a significant effect on both the PGA and PGV.

TABLE 1

Model	Ratio PGA	PGV
Trifunac (1976a)[14] Basement Rock	1.93	1.07
(Intensity Based) Sedimentary Rock	1.40	1.03
Joyner-Boore (1982)	1.00	0.68
SEP	1.00	0.87
McGuire (1978)	1.22	0.93
Trifunac (1976b)[15] Crystalline Rock	0.76	0.55
Sedimentary Rock	0.87	0.74
Campbell (1981)	1.00	----

2.4 Generic Correction Factors

2.4.1 Overview of Approach The simple correction model might be adequate if enough categories were used; however, the data base is too sparse to define many categories. The approach described in this section, consists in supplementing the empirical data set with analysis. Our procedure is as follows:

a. Use available soil/rock pairs (soil and rock stations in close proximity to each other that have recorded the same earthquakes) to compute observed amplification factors. This provides a measure of the range of realistic amplification and the uncertainty introduced by source, travel path and rheological effects. These results are used to calibrate analytic results.

b. We defined eight categories based on three soil depth categories and two soil type categories plus a rock category and a deep soil category. Two basic soil types were chosen: (1) primarily a sandy type soil column and (2) primarily a "till like" column. Granted, most soil columns are mixed, but defining too many categories becomes pointless and a site specific approach should be used. Each category contains several different soil columns which are based on actual soil columns at nuclear power plant sites. We selected a set of time histories recorded at rock sites with a range of magnitudes and distances to incorporate the uncertainty from the source and travel path effects in the analysis. These time histories were used as input to the SHAKE computer program and the PGA and amplification factors were computed for each category and for each time history. Then the median and standard deviation of the correction factor were estimated for each category.

Given the lack of data recorded at true rock sites and the possible complexity of the systematic differences between rock and soil sites, it is not clear that even with added analysis we could develop an acceptable ground motion model for even weathered rock sites. For this reason we included a "generic" soil category in our analyses. Amplification factors were computed relative to the generic soil category as well as rock.

2.4.2 Selection of Time Histories Ideally, we would like to have a set of time histories recorded on unweathered hard rock from earthquakes with m_b magnitudes ranging from 4.0 to 7.0 and distances ranging from 2-1/2 km to several hundred kilometers. With such a set, we could examine the dependence on magnitude and distance. Unfortunately, the available set of time histories does not match the ideal set. Most records are recorded on weathered rock--and in fact, the shear wave velocity of the "rocks" at many sites is closer to a "soil" than a rock.

Table 2 gives a list of the records selected. We restricted our choice to recordings made either in the free-field or in small buildings.

Table 2
Rock Records Used in the Analysis

Station	Earthquake	Mag. M_L	Dist. R	Accel g's
Helena Fed. Bld.	Helena, Mont. 10/31/35	6.	8	0.15
Golden Gate Park	Daly City 3/22/57	5.3	8	0.13
Temblor	Parkfield 6/21/67	5.5	11	0.41
Pacoima Dam	San Fernando 2/9/72	6.4	3	1.20
Pacoima Dam	After Shock	5.4	12	0.11
Cal. Tech. Seism. Lab	San Fernando 2/9/72	6.4	18	0.19
Griffith Park Obs.	San Fernando 2/9/72	6.4	17	0.18
Cape Mendocino	Cape Mendocino 6/7/75	5.3	25	0.20
Oroville Seism. Sta.	Oroville 8/1/75	5.7	8	0.11
Gilroy Array No. 1	Coyote Lake 8/6/79	5.9	9	0.13
Gilroy Array No. 6	Coyote Lake 8/6/79	5.9	4	0.42
Superstition Mt.	Imperial Valley 10/15/79	6.6	25	0.21
Cerro Prieto	Imperial Valley 10/15/79	6.6	24	0.17
Superstition Mt.	Westmoreland 4/26/81	5.7	13	0.11
Rocca	Ancona, Italy 6/14/72	4.7	6	0.55
Rocca	Ancona, Italy 6/14/72	4.2	6	0.45
San Rocco	Friuli, Italy 9/15/76	6.1	9	0.12
San Rocco	Friuli, Italy 9/15/76	6.0	19	0.23
Bagnoli	Campunia Lucania 11/23/80	6.7	12	0.18
Sturno	Campania Lucania 11/23/80	6.7	18	0.23

2.4.3 Definition of Site Categories In order to define site categories, site data for more than 60 nuclear power plant sites throughout the United States were reviewed. Such site data includes geologic profile (layering and depth to rock), soil parameters (soil type, shear wave velocity, compressional wave velocity, density, shear modulus and damping ratio at high strain levels) and bedrock properties (shear wave velocity, compression wave velocity and density). Like most site classification systems, soil depth to the bedrock and soil type are the primary site parameters used to define site categories. Based on our review of the available data from FSAR and PSAR of U.S. nuclear plants sites we have defined the following categories based on the range of the thickness of the soils above bedrock and primary soil type:

Site Class I: Rock Sites. Sites with exposed bedrock including plutonic, igneous, metamorphic, crystalline and sedimentary rock. Sites where the thickness of the soil is less than 25 feet are also assumed to fall in this category and the surface material is neglected as it is generally removed. The mean shear wave velocity (Vs) from 60 sites is 6200 fps with a coefficient of variation (COV) of 40%.

Site Class II: Intermediate thickness soil sites. Sites having soil layering thickness ranging from 25 to 300 ft over bedrock. Based on the samples distribution of available sites, we further divide this site class into three subclasses IIa, IIb and IIc.

IIa: Soil deposit of 25 to 80 ft over rock. The mean V_s of the deposit is 1500 fps with a COV of 40%. The mean thickness is 48 ft with a COV of 30%. The mean and COV of bedrock V_s are 6000 fps and 30% respectively.

IIb: Soil thickness of 80 to 180 feet over rock. The mean V_s and the COV of sites are 1550 fps and 40% respectively. The mean soil thickness is about 120 feet with a COV of 40% and the V_s of bedrock is 6400 fps with a COV of 40%.

IIc: For soil depth of 180 to 300 feet over rock. Only four sites fell into this category. The mean V_s of soil and rock are 2000 fps and 9350 fps respectively. The mean soil thickness is 250 ft. No COVs were computed due to insufficient site data.

Site Class III: Thick soil sites. This is our generic site category and includes those sites having soil deposit more than 300 feet over the bedrock. The mean and COVs of the V_s among a set of 14 sites are 2115 fps and 26% respectively. The median value of soil thickness was found to be 650 feet with a COV of 40%. The mean V_s of the bedrock is 5700 fps with a COV of 45%.

2.4.4 Analysis Procedure The site response was calculated by assuming one-dimensional vertically propagating SH waves. The sites were modeled as a system of horizontal layers of infinite extent. A viscoelastic material model for each layer was assumed (i.e., shear modulus, density, Poisson's ratio, and material damping).

Site response calculation should account for the uncertainty contributed by the variation of depth of the soil model, dynamic soil properties, and the impedance ratio between soil and bedrock. All of these factors contribute significantly to the uncertainty of the calculated response. In addition, the seismic input motions are also an important contributor to the uncertainty. In our analysis to account for the above sources of uncertainty, we performed repeated deterministic analysis, each analysis simulating an earthquake occurrence. By performing many such analyses and by varying the values of the above input parameters in a Monte Carlo simulation, a mean response and its coefficient of variation are obtained. Variability in the seismic input is included by sampling one of the twenty time histories listed in Table 1 to obtain a different earthquake time history for each simulation. Variability in the dynamic modeling is introduced by sets of input parameters (mainly shear wave velocities of soil and rock, damping ratio of soil and the depth of soil deposit) from a log normal probability distribution for each simulation.

Table 3 shows the mean and the COV of four of the input parameters used in the simulation for site response analysis. The influence of non-linear soil behavior on site amplification is still an open research area. Currently, very little field data has been obtained to address this question. Tucker and King[8] show that the observed site amplifications are not much different between groups of strong and weak motions.

TABLE 3

The Means and COVs of the Input Parameters Used for Numerical Simulation for Site Response Analysis

	Class IIa		Class IIb		Class III	
	Mean	COV	Mean	COV	Mean	COV
H (ft; soil	48.	0.25	120.	0.25	650.	0.40
Vs (fps; soil)	1500.	0.40	1550.	0.40	2115.	0.26
D (%; soil)	7.	0.60	7.	0.60	7.	0.60
Vs (fps; rock)	6000.	0.40	6400.	0.40	6200.	0.40

H = Layer Thickness
V_s = Shear Wave Velocity
D = Damping Ratio

Practical non-linear soil constitutive models are not yet available. The non-linear behavior of soil materials cannot be fully described by constant elastic moduli and damping coefficients. However, a good approximation of the effects of soil non-linearities on the response can be obtained by the use of constant strain compatible moduli and damping ratios in a sequence of linear analyses. This method is known as the equivalent linear method[9].

Both linear and equivalent linear analyses were performed for site classes IIa, IIb and IIc. The results of our analysis are discussed in the following section.

2.4.5 Computed Correction Factors Our analysis of both actual and simulated data indicates that there is considerable variability in the correction factors from earthquake to earthquake. This variability can easily be accounted for in our hazard analysis by including it in the simulation process. The main problem, in our opinion, is in defining the ground motion levels and frequency content for the generic (base) case.

We have taken the "generic soil site" to be represented by a deep sand-like (shear-wave velocity function of the depth) site with linear viscoelastic properties, i.e. site Class III defined in Section 2.4.3.

For the purpose of this paper we have summarized the results relative to "rock"; i.e., relative to the set of time histories/spectra given in Table 2. Figure 2 shows a comparison of the median amplification factors relative to the rock site category computed for sand-like sites for categories IIa (25'-80'), IIb (80'-180'), IIc (80'-300') and III (deep). Also shown are the amplification factors computed by Joyner-Boore[2]. The match between Category III and Joyner and Boore's results is good. Note that our modeling results give a peak acceleration amplification factors of unity, in agreement with the results obtained by Joyner and Boore and Campbell. There is considerable departure at longer periods (greater than 1.5 seconds). This might be due, in part, to the fact that some of the rock records were not all base line corrected so that they contain some long period noise. It also might be due to the fact that, in our analysis, the damping is not a function of frequency; hence, the long period motion has the same damping as the short period high frequency motion.

A sensitivity analysis including a median damping of 2% with a COV of 40%, 7% with COV of 40% and an equivalent linear case with a best estimate curve based on available data showed that the damping of the soil is one of the most important parameters.

As noted earlier, because many of the ground motion models are generic soil models, we also give the amplification factors relative to Category III on Fig. 3 for sand-like sites and Fig. 4 for till-like sites. Also shown is the amplification factors relative to Category III for the rock set.

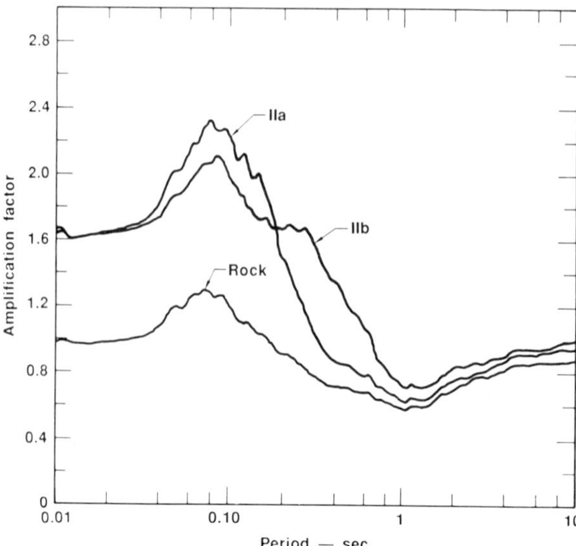

Figure 3. 7% median soil damping spectral amplification factors for sand–like soils relative to Category III. The amplification of the rock category relative to Category III is also shown for reference.

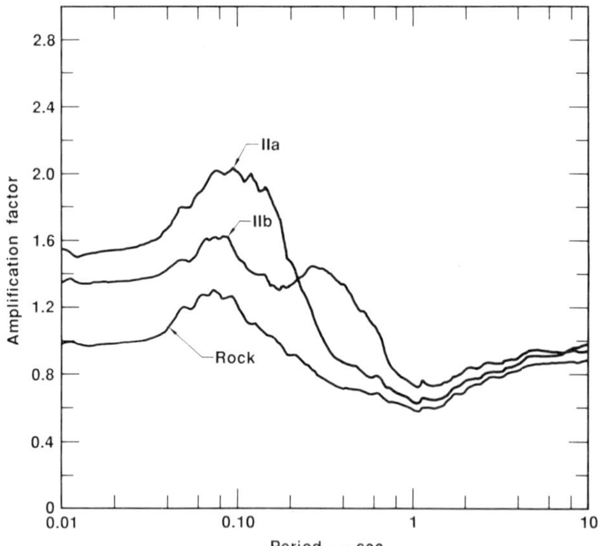

Figure 4. Comparison of the 7% median soil damping spectral amplification factors for till–like soils relative to Category III. The rock category relative to Category III is also shown for reference.

3. DISCUSSION

Interaction with the panel of experts contributing to the project identified several areas needing discussion. In particular it was suggested that surface waves and other non-vertically incident waves could be important. Also focusing and defocusing of rays are not considered.

To include the above considerations would require a very detailed site specific analysis. For a western U.S. (WUS) site where the configuration of major nearby active faults is known, it would be possible to perform such complex studies and examine them. However, such studies are almost beyond the current state-of-the-art and few even limited studies have been performed to address them. For the EUS these questions are even more difficult to assess because it is assumed that the earthquakes occur randomly around the site.

Our proposed approach evolved from the following observations: It is very difficult to separate out the wave type in the strong motion accelerograms. In part because strong motion accelerograms are generally recorded within 100 km of the source. Our analysis shows that much of the hazard is contributed by earthquakes located within 100 km of the site. It must also be kept in mind that we are primarily interested in the high frequency end of the ground motion spectrum (i.e. for frequencies greater than 1 or 2Hz.). Even for distant sources, in particular site amplification observed at sites from underground nuclear explosions[10,11], it has been found that the simple linear theory similar to our proposed approach, has been adequate to explain the important feature of the observed ground motion.

Also to address this issue as part of NRC funded SSMRP[12], we funded an analysis using earthquake source modeling to (in part) characterize the type and direction of incoming seismic waves at a typical EUS site as compared to WUS site[13]. This analysis performed for a soil site falling into our IIb category with a soil depth of about 100' over a bedrock with a shear wave velocity of 3.3 km/sec found that all of the energy emerged almost vertically at all frequencies. For a deeper WUS soil site the results were much different with waves generally emerging at angles of 20° or more relative to the vertical. It is assumed that the results for a deep soil EUS site would be similar.

The issue of ray focusing is very difficult to deal with given the random nature of seismic activity around any particular site. Except for a few very special potential earthquake locations around a few sites, it is not evident how to even approach this question.

REFERENCES

1. Bernreuter, D.L., Savy, J.B., Mensing, R.W., Chen, J.C. and Davis, B.C. Seismic Hazard Characterization of the Eastern United States, UCID 20421, 2 Volumes, (April, 1985).

2. Joyner, W.B. and Boore, D.M. Prediction of Earthquake Response Spectra, USGS open-file report 82-977, (1982).

3. Trifunac, M.D. and Anderson, J.G., "Preliminary Empirical Models for Scaling Absolute Acceleration Spectra", University of Southern California Dept. of Civil Engineering Report, No. CE77-03, (1977).

4. Bernreuter, D.L., Seismic Hazard Analysis: Application of Methodology, Results and Sensitivity Studies, NUREG/CR 1582, UCRL-53030 (1981).

5. Joyner, W.B. and Boore, D.M. "Peak Horizontal Accelerations and Velocity from Strong Motion Records Including Records from the 1979 Imperial Valley, California Earthquake", Bull. Seism. Soc. Am., 71, No.6, pp 2011-2038 (December 1981).

6. Campbell, K.W., "Near-Source Attenuation of Peak Horizontal Acceleration", Bull. Seism. Soc. Am., 71, No. 6 pp. 2039-2070, (December 1981)

7. Campbell, K.W. "The effects of Site Characteristics on Near-Source Recordings of Strong-Ground Motion", Proceedings of Conference XXII on Site Specific Effects of Soil and Rock on Ground Motion and the Implications for Earthquake-Resistant Design. USGS open-file report, 83-845, Walt Hays ed., Santa Fe, NM, (July 25-27, 1983).

8. Tucker, B.E. and King, J.L., "Dependence of Sediment-filled Valley Response on Input Amplitude and Valley Properties", Bull. Seism. Soc. Am., 74, No.1, pp 153-166, (February 1984).

9. Seed, H.B. and Idriss, I.M., "Influence of Soil Conditions on Ground Motions During Earthquakes", Journal of the Soil Mechanics and Foundation Division, Proceedings of the A.S.C.E., pp 99-137, (January 1969).

10. Hays, W.W, Procedure for Estimating Earthquake Ground Motions, USGS Professional Paper 1114, (1980).

11. Murphy, J.R., Davis, A.H. and Weaver, N.L., "Amplification of Seismic Body Waves by Low-Velocity Surface Layers", Bull. Seism. Soc. Am., 61, No.1, pp 109-145, (February 1971).

12. Bohn, M.P., Shieh L.L., Wells, J.E., Cover, L.C., Bernreuter, D.L., Chen, J.C., Johnson, J.J., Bumpus, S.E., Mensing, R.W., O'Connell, J., and Lappa, D.A., Application of the SSMRP Methodology to the Seismic Risk at the Zion Nuclear Power Plant, NUREG/CR-3428, UCRL-53483, (1984).

13. Apsel, R.J., Frazier G.A., Jurkevis, A., and Fried, J.C. Application of Earthquake Source Modeling to Assess the Relative Differences Between Seismic Ground Motion in the Eastern and Western Regions of the United States, and to Characterize the Type and Direction of Incoming Seismic Waves: Final Report. Del Mar Technical Associates, Del Mar, CA, DELTA-R-043, (September, 1980).

14. Trifunac, M.D., (1976a),"Preliminary Analysis of the Peaks of Strong Earthquake Ground Motion-Dependence of Peaks of Earthquake Magnitude, Epicentral Distance, and Recording Site Conditions", Bull, Seism. Soc. Am., 66, No. 1, pp 189-219, (February 1976).

15. Trifunac, M.D., (1976b),"Preliminary Empirical Model for Scaling Fourier Amplitude Spectra of Strong Ground Acceleration in Terms of Earthquake Magnitude, Source-to-Station Distance, and Recording Site Conditions", Bull. Seism. Soc. Am., 66, No. 4, pp 1343-1374, (August 1976).

SECTION 4: DYNAMIC METHODS IN SOIL AND ROCK MECHANICS

Some Aspects of the Dynamic Subsoil-Coupling Between Circular and Rectangular Foundations

Th. Triantafyllidis
Institute of Soil Mechanics and Rock Mechanics, University of Karlsruhe, Federal Republic of Germany

SUMMARY

The dynamic coupling between rigid foundations via the subsoil due to a harmonic excitation is considered. The mixed boundary-value problem is formulated by means of a set of Fredholm integral equations. The unknown stress distributions in the contact area between foundation and subsoil are uniform approximated using series expansions of orthogonal polynomials. The stiffness functions of the entire system can be determined rigorously using the orthogonality relations of the polynomials. In this paper the range of the dynamic subsoil-coupling between rigid square foundations as well as for rigid circular foundations is determined. Using this method and varying the inertia properties of the foundations as well as the geometry of the system problems of passive screening of vibrations as well as the dynamic behaviour of a foundation system due to a far-field excitation can be solved rigorously.

INTRODUCTION

Using the term "dynamic subsoil coupling" between adjacent foundations the manifold interdependence of the dynamic behaviour for each of the elements "foundation" within the system "foundations and subsoil" due to the presence of all the other foundations is mentioned. If one of the foundations is dynamically excited energy in form of waves is radiated in the halfspace. If the waves arrive at the non externally excited foundations, they start to move. Due to the inertia properties of the foundations stresses are generated at the contact area between the foundations and the subsoil. The generated stresses introduce new waves, which propagate in the subsoil and influence the dynamic behaviour of the first excited foundation.

Although for the solution of dynamic interaction problems of a single foundation a great amount of papers exist, some investigators considered dynamic subsoil coupling between two or more foundations. To the author's best knowledge the first consideration on dynamic subsoil coupling between foundations has been investigated by Warburton et al.[1], who analysed the dynamic behaviour of two circular rigid masses resting on the surface of a linear-elastic isotropic and homogeneous halfspace, one of them being excited by harmonic loads. The resulting stress distribution at the contact area between the free of external loads mass and the halfspace was determined numerically from the energy induced by the rigid displacement in comparison to the free-field displacement condition. Using this method the influence of the free of external loads mass upon the vibration behaviour of the externally excited mass was neglected. Due to a subdivision of the contact areas between rectangular foundations and the halfspace into rectangular elements with constant stress distribution the dynamic subsoil coupling between two rigid foundations was determined by Savidis et al.[2] for a linear-elastic, isotropic and homogeneous halfspace, by Gaul[3] for a linear visco-elastic homogeneous halfspace and by Wong et al.[4] for a linear visco-elastic, layered halfspace. Using this method a superposition of the displacements due to the stresses at the rectangular elements yields the stress distribution to satisfy the required displacement boundary conditions only at the centers of the elements in the contact areas between foundations and subsoil. Roesset et al.[5] considered two square, rigid, embedded foundations on a linear-elastic layer and determined their dynamic behaviour using the Finite-Element Method. Bielak et al.[6] calculated the dynamic behaviour of two square foundations resting on the surface of an elastic halfspace due to a harmonic seismic excitation by the Boundary Element Method.

While most of the investigators used relaxed boundary conditions at the interface between foundations and halfspace, within this paper a perfect bond is considered. The author has extended an analytical method of solution for soil structure interaction problems first presented by Oien[7], to solve dynamic subsoil coupling problems between adjacent, rigid strip foundations[8], rectangular[9,10] and circular[11] foundations, as well as soil structure interaction problems for circular[12] and rectangular[13] foundations.

In the analysis of subsoil coupling between adjacent foundations phenomena of wave propagation in the halfspace and soil structure interaction are included. A considaration of subsoil coupling between foundations is important only if the foundations are closely spaced and hence can be regarded as a more or less local effect. The material damping for such problems can not play an important rule in comparison to geometric damping caused by wave radiation. The opposite is true in wave propagation problems where material damping is important because of the distances the waves have to travel. Neglecting material damping leads to slight overestimation of vibration amplitudes of the foundations.

Within this paper the subsoil is presented as a linear-elastic, homogeneous and isotropic halfspace and the material damping is assumed to be zero. In the method presented below the boundary displacement conditions in the contact areas between subsoil and foundations are satisfied exactly. Assuming a perfect bond between foundations and halfspace the stress distributions in the contact areas are uniform approximated using orthogonal polynomials.

MATHEMATICAL FORMULATION AND METHOD OF SOLUTION

Without loss of generality two foundations which initially are regarded as massless, rigid plates resting on the surface of a halfspace are considered. Fig. 1 shows the geometry of the problem for circular and rectangular foundations, the choice of global and local co-ordinate systems and the main components of forces, moments, displacements and rotations. The local co-ordinate system for each foundation is marked by the index in brackets. The distance between the center of foundation i and the origin of the global co-ordinate system, located at the center of the foundation 1, is marked by the co-ordinates $x_1^{(0i)}$ and $x_2^{(0i)}$.

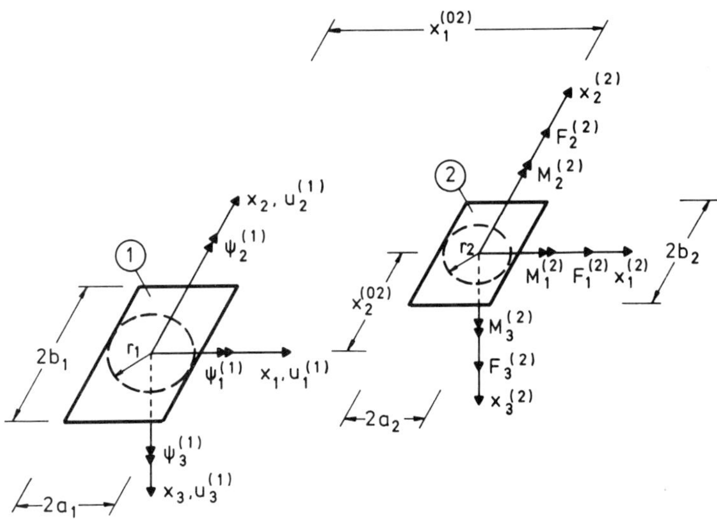

Figure 1. Two foundations (rectangular or circular) on the surface of the halfspace with the chosen global- and local co-ordinate systems and vectors of forces and moments.

The equation of motion for a linear-elastic, homogeneous and isotropic material is given by:

$$\mu \triangle \tilde{u} + (\lambda + \mu) grad(div \tilde{u}) = \rho \frac{\partial^2 \tilde{u}}{\partial t^2} \tag{1}$$

where λ, μ are the Lamé constants, G the dynamic shear modulus ($\mu :=$ G, $\lambda := \dfrac{2G\nu}{1-2\nu}$, ν = Poisson's ratio), ρ the material density, t the time, $\triangle$ the Laplace operator and $\tilde{u}$ the displacement vector. The boundary conditions of the mixed boundary value problem (the harmonic term $e^{i\omega t}$ is eliminated) are:

a) Stress boundary conditions:

$$\sigma_{3k}(x_1, x_2, 0) = 0; \qquad (k = 1, 2, 3) \tag{2}$$

for $|x_1 - x_1^{(0j)}| > a_j$ and $|x_2 - x_2^{(0j)}| > b_j$;(rectangular foundations)
(i,j=1,2)
$|x_i - x_i^{(0j)}| > r_j$; (circular foundations)

This condition requires a surface of the halfspace free from stresses outside of the foundations areas. In addition due to the boundedness of the solution for the boundary value problem the following relation has to be satisfied:

$$\sigma_{3k}(x_1, x_2, \infty) = 0. \tag{3}$$

b) Displacement boundary conditions:

$$\begin{aligned} u_1^{(j)}(x_1, x_2) &= u_1(x_1^{(0j)}, x_2^{(0j)}) - \psi_3^{(j)} x_2^{(j)} \\ u_2^{(j)}(x_1, x_2) &= u_2(x_1^{(0j)}, x_2^{(0j)}) + \psi_3^{(j)} x_1^{(j)} \\ u_3^{(j)}(x_1, x_2) &= u_3(x_1^{(0j)}, x_2^{(0j)}) - \psi_1^{(j)} x_2^{(j)} + \psi_2^{(j)} x_1^{(j)} \end{aligned} \tag{4}$$

for
$|x_1 - x_1^{(0j)}| \leq a_j$ and $|x_2 - x_2^{(0j)}| \leq b_j$; (rectangular foundations)
$|x_i - x_i^{(0j)}| \leq r_j$; (circular foundations)

The Eq.(4) describes the rigid body motion of the foundations. Because of small displacements and small deformation gradients, linearized kinematics are assumed.

To solve the mixed-boundary value problem the method of influence functions is used. A single mode of motion of one foundation is specified whilst all other modes of the foundation excited as well as of all the other foundations not excited are restrained. In all modes of motion specified stresses are generated in the contact area of the foundations, shear- and normal stresses contributing to the mode specified. The stress distributions at foundation i due to the dispacement condition (ε) (e.g. $\varepsilon = u_1$: horizontal unity displacement in the x_1-direction) on foundation j are noticed by:

$$\sigma_{3\ell}^{(i),(j\varepsilon)}(x_1^{(i)}, x_2^{(i)}), \qquad (\ell = 1, 2, 3).$$

From the stress distributions forces and moments in the contact area of the foundation result which can be determinated as follows:

$$\int_{A_i} \sigma_{3\ell}^{(i),(j\varepsilon)}(x_1^{(i)}, x_2^{(i)}) dA_i = F_\ell^{(i),(j\varepsilon)} \qquad (\ell = 1, 2, 3)$$

$$\int_{A_i} \sigma_{33}^{(i),(j\varepsilon)}(x_1^{(i)}, x_2^{(i)}) x_2^{(i)} dA_i = -M_1^{(i),(j\varepsilon)}$$

$$\int_{A_i} \sigma_{33}^{(i),(j\varepsilon)}(x_1^{(i)}, x_2^{(i)}) x_1^{(i)} dA_i = M_2^{(i),(j\varepsilon)} \tag{5}$$

$$\int_{A_i} \left\{ \sigma_{31}^{(i),(j\varepsilon)}(x_1^{(i)}, x_2^{(i)}) x_2^{(i)} - \sigma_{32}^{(i),(j\varepsilon)}(x_1^{(i)}, x_2^{(i)}) x_1^{(i)} \right\} dA_i = M_3^{(i),(j\varepsilon)}$$

where A_i : area of foundation i.
These forces and moments represent the generalized reaction of the half-space due to the unity modes of motion enforced and are called influence functions. They depend not only on the frequency of excitation but also on the geometry of the foundations, their mutual distance and the input parameters of the subsoil shear modulus G , mass density ρ and Poisson's ratio ν. If the forces and moments for all unit vibration modes of all the foundations have been calculated, the resultant reactions of the subsoil due to a genaral state of deformation of the foundations can be described as follows:

$$\begin{pmatrix} W^{(11)} & W^{(12)} \\ W^{(21)} & W^{(22)} \end{pmatrix} \begin{pmatrix} \tilde{u}^{(1)} \\ \tilde{u}^{(2)} \end{pmatrix} = \begin{pmatrix} \tilde{F}^{(1)} \\ \tilde{F}^{(2)} \end{pmatrix} \tag{6}$$

or in short form:

$$[W]\{\tilde{u}\} = \{F\} \tag{6a}$$

where $\tilde{F}^{(i)}$ represents the resulting reaction vector (3 forces and 3 moments resp.) of the halfspace on the foundation (i) due to the deformation vector $\tilde{u}$ of all the foundations, $\tilde{u}^{(i)}$ is the vector of displacements and rotations (6 degrees of freedom) of the foundation i and $W^{(ij)}$ is the influence matrix including as elements the reaction forces and moments (see Eq. 5) of the halfspace acting on the foundation i due to all unity vibration modes of the foundation j.

The elements of the influence matrix W can determined if the stress distributions for each unity vibration mode of the foundations are known. Using Green's functions, which yield the surface displacement of a linear-elastic, isotropic and homogeneous halfspace due to unit harmonic force, it is possible to calculate the surface displacements due to arbitrary stress distributions on the surface of the halfspace. In order to satisfy a given deformation condition within the area of the foundations superposition integrals (see Achenbach[14]) over the stress distributions and the respective Green's functions are used. These superposition integrals form a system of integral equations for the displacements required at the contact areas between foundations and subsoil ,which can be written in a short form as follows:

$$\sum_{i=1}^{n} \iint_{A_i} \sum_{\ell=1}^{3} \sigma_{3\ell}^{(i),(j\varepsilon)}(\xi_1^{(i)}, \xi_2^{(i)}) U_{k\ell}(x_1^{(\lambda)} - \xi_1^{(i)}, x_2^{(\lambda)} - \xi_2^{(i)}) d\xi_1^{(i)} d\xi_2^{(i)} = u_k^{(\lambda),(j\varepsilon)}(x_1^{(\lambda)}, x_2^{(\lambda)}) \tag{7}$$

for $|x_1^{(\lambda)}| \leq a_\lambda$ and $|x_2^{(\lambda)}| \leq b_\lambda$; (rectangular foundations)

$|x_1^{(\lambda)}|^2 + |x_2^{(\lambda)}|^2 \leq r_\lambda^2$; (circular foundations)

$(i, j, \lambda = 1, 2, ..., n.$ n : number of foundations), $(k, \ell = 1, 2, 3)$

where $U_{k\ell}(x_1, x_2)$ is the Green's function for the x_k-component of the surface displacement of the halfspace due to a unit harmonic force in the x_ℓ-direction acting on the origin of the global co-ordinate system[9].

The unknowns of the system of integral equations, which is of Fredholm type and first kind, are the stress distributions in the contact areas between foundations and halfspace. If the resultant loads acting on the foundations are required a uniform approximation of the stress distributions over these areas is sufficient. Thus the stress distributions can be represented by series expansions of orthogonal polynomials. The choice of the respective system of orthogonal polynomials has a great influence on the accuracy and the rate by which the solution converges. In the static case a singularity in the contact area between a rigid foundation and the elastic halfspace exist for the stress distributions of the following kind:

$$\frac{1}{\sqrt{a^2 - x_1^2}\sqrt{b^2 - x_2^2}} \qquad \text{(rectangular foundation} \quad 2a \times 2b)$$

$$\frac{1}{\sqrt{r^2 - x_1^2 - x_2^2}} \qquad \text{(circular foundation)}$$

Using the above functions as a start point for the approximation of the stress distributions, the following series expansions are postulated:

a.) Rectangular foundations

$$\sigma_{3k}^{(i),(j\varepsilon)}(x_1^{(\ell)}, x_2^{(\ell)}) = G \sum_{N=0}^{N} \sum_{M=0} A_{(k),NM}^{(i),(j\varepsilon)} \frac{T_N(\chi_1^{(\ell)}) \cdot T_M(\chi_2^{(\ell)})}{\sqrt{1 - (\chi_1^{(\ell)})^2}\sqrt{1 - (\chi_2^{(\ell)})^2}} \tag{8a}$$

$$(k = 1, 2, 3); (i, j, \ell = 1, 2, ..., n)$$

where: $\chi_1^{(\ell)} = \dfrac{x_1^{(\ell)}}{a_\ell}$, $\chi_2^{(\ell)} = \dfrac{x_2^{(\ell)}}{b_\ell}$

and $T_N(\chi) := cos(N \cdot arccos(\chi))$ is the Chebychev polynomial.

b.) Circular foundations

$$\sigma_{3k}^{(i),(j\varepsilon)}(x_1^{(\ell)}, x_2^{(\ell)}) = G \sum_{N=0}^{N} \sum_{M=0} A_{(k),NM}^{(i),(j\varepsilon)} \frac{U_{N,M}^0(\tilde{\chi}^{(\ell)})}{\sqrt{1 - |\tilde{\chi}^{(\ell)}|^2}} \tag{8b}$$

$$(k = 1, 2, 3); (i, j, \ell = 1, 2, ..., n)$$

where: $\tilde{\chi}^{(\ell)} = (\chi_1^{(\ell)}, \chi_2^{(\ell)}) \qquad \chi_m^{(\ell)} = \dfrac{x_j^{(\ell)}}{r_\ell}$

The polynomials $U^0_{N,M}(\tilde{x})$ belong to a general class of polynomials $U^s_{M,N}(\tilde{x})$ defined over a circular area by:

$$U^s_{N,M}(\tilde{x}) = \frac{(-1)^m (s)_m}{2^m(\frac{s+1}{2})_m M!\,N!(1-|\tilde{x}|^2)^{\frac{s-1}{2}}} \frac{\partial^2}{\partial x_1^N \partial x_2^M}(1-|\tilde{x}|^2)^{m+\frac{s-1}{2}} \qquad (9)$$

with:

$$m = N + M, \quad (s)_m = \frac{\Gamma(s+m)}{\Gamma(m)}, \qquad \text{and} \qquad \Gamma(x): \text{ the Gamma-function.}$$

The polynomials chosen are orthogonal to each another within the respective foundation area with respect to the weight function appearing in the denominator of the Eqs. (8a,b) for the stress distributions. Introducing the Eqs. (8a,b) into the system of integral equations (Eq. 7) one can calculate the integrals analytically (see Triantafyllidis[9,10,11,12]). Applying the Bubnov-Galerkin method[15] the system of integral equations (7) can be transformed into a linear algebraic system with complex, frequency depending coeffients, which after an integration in the complex plane can be given in an analytical form (for the complex integration see Triantafyllidis[9]). The unknowns of the algebraic system are the coefficients $A^{(i),(j\varepsilon)}_{(k),NM}$ of the series expansions for the stress distributions (see Eq. 8a,b). Solving this system for each of the prescribed vibration mode $(j\varepsilon)$ the coefficients of the influence matrices (e.g. the reactions of the halfspace on the contact areas of the foundations due to the excitation mode) result directly[9,10,11,12] from the coefficients $A^{(i),(j\varepsilon)}_{(k),NM}$ due to the biorthogonality relations[16] of the polynomials. The coefficients of the influence matrices represent also the "lumped parameters" (stiffness functions) of a foundation system on the halfspace.

STATIONARY FAR-FIELD EXCITATION

Much of the destructive capacity of earthquakes is associeted with the resonant vibration of structures excited by low-frequency seismic motions of relatively long duration. Of particular interest is the dynamic behaviour of a system of foundations resting on the halfspace due to this kind of excitation. We now consider plane Rayleigh waves with an angle of insidency ϑ with respect to the chosen co-ordinate system (see Fig. 2). The vector of the displacement field on the surface of the halfspace $\tilde{u}^I(x_1, x_2) \cdot e^{i\omega t}$ is given as follows:

$$\tilde{u}^I(x_1, x_2) \cdot e^{i\omega t} = A(B \cdot \tilde{e}_\xi + \tilde{e}_3) \cdot e^{i(\omega t - k_I \xi)} \qquad (10)$$

where:

$$cos\vartheta = \tilde{e}_\xi \tilde{e}_1; \qquad B = i\frac{2qs - s^2 - 1}{q(1 - s^2)}$$

$$q = \sqrt{1 - \alpha^2 K^2}; \qquad s = \sqrt{1 - K^2}$$

$$K = \frac{c_R}{c_S}; \qquad \alpha = \sqrt{\frac{1 - 2\nu}{2 - 2\nu}}$$

and A is the amplitude of the vertical displacement of the incident Rayleigh waves, k_I the wave number of the incident waves, c_S the shear wave velocity, c_R the Rayleigh wave velocity and $\tilde{e}_\eta$ denotes the unity vector on the η- direction.

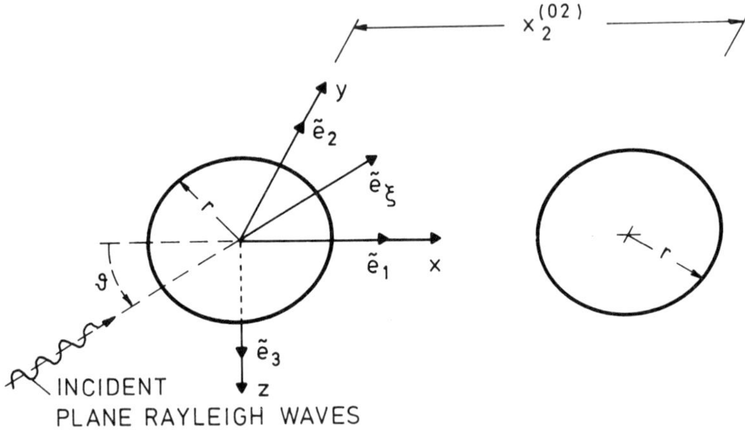

Figure 2. Plane Rayleigh waves with a angle of insidency ϑ acting on the system of two identical circular foundations.

Following a procedure suggested by Thau[17] the motion of a rigid foundation can be divided in two parts (ommiting the term $e^{i\omega t}$):

$$\tilde{u}_R = \tilde{u}^S + \tilde{u}^I \tag{11}$$

where $\tilde{u}_R$ denotes the rigid body motion of the foundation and $\tilde{u}^S$ the motion of the foundation due to the scattered waves. Thus the scattered displacement field within the area of a foundation can be represented as: $\tilde{u}^S = \tilde{u}^D + \tilde{u}_R$, where $\tilde{u}^D = -\tilde{u}^I$ denotes the wave field diffracted by the immobile foundations. Using the same system of integral equations (see Eq. 7) as for the determination of the halfspace reaction due to the rigid body motion (Eq. 6) one can determine the loads acting on the immobile foundations due to the far-field excitation $\tilde{u}^I$ replacing the right hand side of the Eqs. (7) by the displacement wave field $\tilde{u}^D$ within the areas of the foundations. Transforming the system of integral equations into a linear algebraic system by use of the Bubnov-Galerkin method one can determine the coefficients $A^{(i),(D)}_{(k),NM}$ (see Eqs. 8a,b) of the series expansions for the stress distributions for the case of the displacement field $\tilde{u}^D$. Finally using the biorthogonality relations of the polynomials the loads $\tilde{F}^{(i),D}$ acting on the immobile foundations can directly determined from the coefficients $A^{(i),(D)}_{(k),NM}$ by the Eq. (5).

CONSIDERATION OF INERTIA PROPERTIES

In the previous analysis the foundations are regarded as inertialess plates perfectly bonded to the halfspace. The reaction of the halfspace due to a harmonic excitation has been calculated at the contact area between foundation and halfspace. Considering a matrix $[M]$ which involve the inertia properties and the local vectors between the bases of the foundations and the gravity centers the dynamic response of a group of foundations (at the centers of gravity) can be determined by the Newton's low of motion:

$$\left\{ \begin{array}{c} Q^{*(1)} \\ Q^{*(2)} \end{array} \right\} e^{i\omega t} + \left\{ \begin{array}{c} F^{(1)} \\ F^{(2)} \end{array} \right\} e^{i\omega t} = \frac{d^2}{dt^2}\left([M] \left\{ \begin{array}{c} \tilde{u}^{(1)} \\ \tilde{u}^{(2)} \end{array} \right\} e^{i\omega t} \right) \qquad (12)$$

where $F^{(i)}$ is the reaction load vector of the halfspace upon the foundation i due to the rigid body motions of all the foundations and $Q^{*(i)}$ are the external loads upon the foundation i and/or the loads $F^{(i),D}$ acting at the immobile foundations due to the far-field excitation. Setting Eq. (6) into the Eq. (12) yields :

$$\left\{ \begin{array}{c} \tilde{u}^{(1)} \\ \tilde{u}^{(2)} \end{array} \right\} = -\left[W + \omega^2 M \right]^{-1} \cdot \left\{ \begin{array}{c} Q^{*(1)} \\ Q^{*(2)} \end{array} \right\} \qquad (13)$$

and thus the vibration behavior of the foundations can be determined.

NUMERICAL RESULTS

Using the analytical method presented a great number of problems involving subsoil coupling between foundations can rigorously be solved. An application of this method for problems of passive screening of vibrations has already been published (see Triantafyllidis et al.[10]). In this paper the change of the dynamic stiffness of one foundation due to the presence of a second identical one with respect to the distance of the centers of gravity and the excitation frequency is of particular interest. Changing the distance between the two foundations one can also determine the range in which the subsoil coupling between the two foundations is important. The range limits are obtained if the stiffness functions within the system of two foundations tend to the stiffness functions of a single foundation. An extensive study to determine this range using rectangular foundations of several length/wide ratio b/a as well as for circular foundations has already been treated (see Triantafyllidis et al.[18]). It has been seen that the vertical vibration mode of the excited foundation is the most sensitive one in comparison to the others within the system of two foundations. In this paper only the dynamic stiffness function for the vertical vibration mode of the excited foundation within the system of two identical square or circular foundations of the same area are presented. The dimensionless dynamic stiffness functions (real part := k_{zz}, imaginary part := r_{zz}) for the vertical vibration mode u_z ($z \hat{=} x_3$) are defined by:

$$RE(F_z) = Gru_z \cdot k_{zz},$$
$$IM(F_z) = \omega r^2 \sqrt{G\rho} u_z \cdot \frac{r_{zz}}{a_0}. \tag{14}$$

where F_z is the reaction in z- direction of the halfspace at the contact area between the foundation excited by u_z and the halfspace, a_0 is the dimensionless frequency ($a_0 := \omega r/c_S$) and r the radius of the excited circular foundation or the equivalent radius of a circular foundation of the same area if square foundations are considered.

Fig. 3a shows the real part k°_{zz} for the circular and $\bar{k}_{zz}$ for a square foundation with the same area of the dynamic dimensionless stiffness functions for the vertical vibration mode of the excited foundation with respect to the excitation frequency a_0 and the dimensionless distance $\bar{x}/r$ ($\bar{x} := x_1^{(02)}$) between two identical circular or square foundations with equivalent area resp. The centers of gravity of the foundations are located in the x_1-axis ($x_2^{(02)} = 0$, see also Fig. 1).

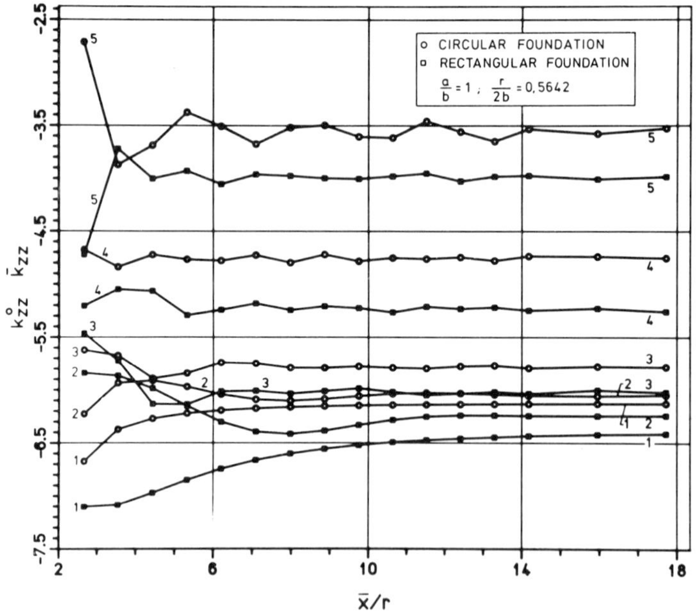

Figure 3a. Spring coefficient k_{zz} of the excited foundation for the vertical vibration mode within the system of two (circular, square) foundations (ν=1/3).

The results in the Figs. 3a, 3b for the circular foundations are marked with the symbol ○ and the respective results for the square foundations with the symbol □. The numbers on the curves indicate the chosen dimensionless excitation frequency a_0 and are defined as follows:
$1 : a_0 = 0.03989, \quad 2 : a_0 = 0.39894,$
$3 : a_0 = 0.79789, \quad 4 : a_0 = 1.59577, \quad 5 : a_0 = 2.39365.$

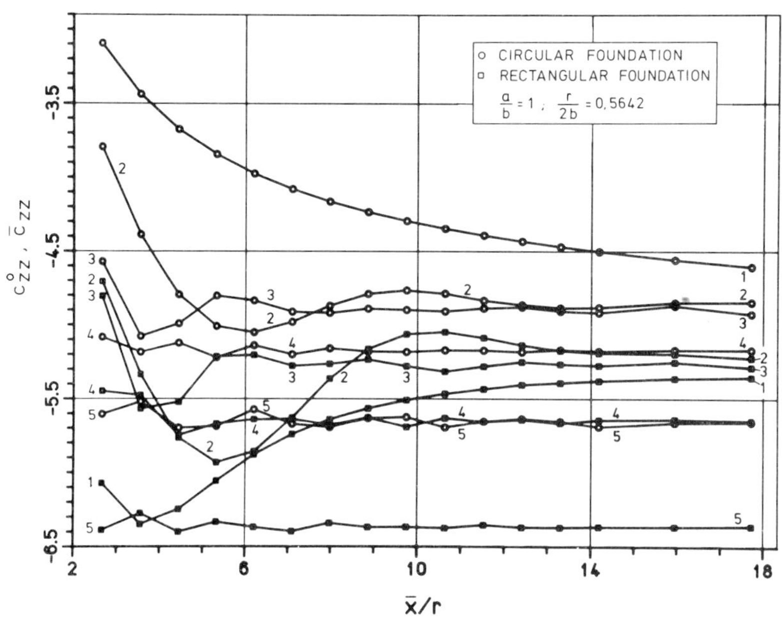

Figure 3b. Dashpot coefficient c_{zz} of the excited foundation for the vertical vibration mode within the system of two (circular, square) foundations (ν=1/3).

Fig. 3b shows the imaginary part c_{zz} ($c_{zz} := r_{zz}/a_0$ of the dynamic dimensionless stiffness functions (c° for the circular and $\bar{c}$ for the rectangular foundation) for the vertical vibration mode with respect to the dimensionless distance $\bar{x}/r$ between the two foundations and the excitation frequency a_0. Using a lumped parameter model the function k_{zz} represents the spring- and the function c_{zz} the dashpot coefficient for the vertical vibration mode of the excited foundation within the system of the two identical foundations. From the Figs. 2a,b one can see that the presence of the second foundation has a great influence upon the vibration behaviour of the excited foundation. This influence increases with decreasing foundation spacing.

An other very important phenomenon appeared if two foundations are excited by plane Rayleigh waves with a angle of incidency $\vartheta = 90°$(see Fig. 2). The greatest part of energy during a seismic excitation is associated with the

Rayleigh waves, and thus for a seismic excitation of relatively long duration one can regard harmonic Rayleigh waves as a seismic excitation. For the determination of the loads on the foundations due to the incident plane Rayleigh waves, which tend to force the foundations to move we regard the foundations as massless fixed circular plates. According to the Eq. (13) these loads can be regarded as external loads acting on movable foundations with several inertia properties. Due to the diffraction waves produced by both foundations not only forces in y- direction and moments about the x- axis are acting at the interfaces between the foundations and the subsoil but also simultaneously forces are acting in x-direction (perpendicular to the incidency direction of the Rayleigh waves) as well as moments about the y- and the z- axis. Each of the foundations can be interpreted as a new simple source of waves within the incident plane Rayleigh waves due to the principle of Huygens, which states that every point on a wave surface becomes in turn a source for a new disturbance. Due to the boundary conditions for the fixed circular plates and the superposition of the diffraction waves between the two foundations it is possible that forces can appear in a direction perpendicular to the incident Rayleigh waves. The same is also possible for moments about the y- and the z- axis. If the distance between the two foundations is very large (i.e. the subsoil coupling between the foundations is negligible), than the forces and moments produced by the diffraction wave field between the two foundations dissapear.

Because, to the author's best knowledge, such a study has not appeared in the literature, it is interesting to calculate e.g. all the forces and moments acting on two identical immobile foundations with respect to the excitation frequency. The distance between the centers of gravity of the two foundations is assummed to be $x_1^{(02)} = 8.0$ m (see Fig. 2), both radii $r = 1.5$ m and the amplitude of the vertical component of the incident Rayleigh waves 1 mm. A Poisson's ratio of the halfspace $\nu = 1/3$ is chosen. Figs. 4a-c show the diffraction loads acting at the interface between the immobile, massless foundations and the halfspace due to the incident Rayleigh waves with respect to the dimensionless excitation frequency a_0. In the Figs. 4a-c the solid curves represent the real part (RE) and the dashed curves the imaginary part (IM) of the respective loads. The forces F_y as well as the moments M_x acting on the two foundations are equal in magnitude and phase but the forces F_x as well as the moments M_y, M_z acting on the foundations 1 and 2 have the same magnitude but a phase difference of π. This means that the sum of the forces $F_x^{(1),(D)}$ and $F_x^{(2),(D)}$ as well as the sumes of the moments $M_y^{(1),(D)}$ and $M_y^{(2),(D)}$ and the moments $M_z^{(1),(D)}$ and $M_z^{(2),(D)}$ are all zero, thus the global system remains in equilibrium. If a structure (e.g. a frame or a bridge) is supported on the two foundations, in adddition to the forces $F_y^{(i),(D)}$ and the moments $M_x^{(i),(D)}$ the loads $F_x^{(i),(D)}$ as well as $M_y^{(i),(D)}$ and $M_z^{(i),(D)}$ have to be considered.

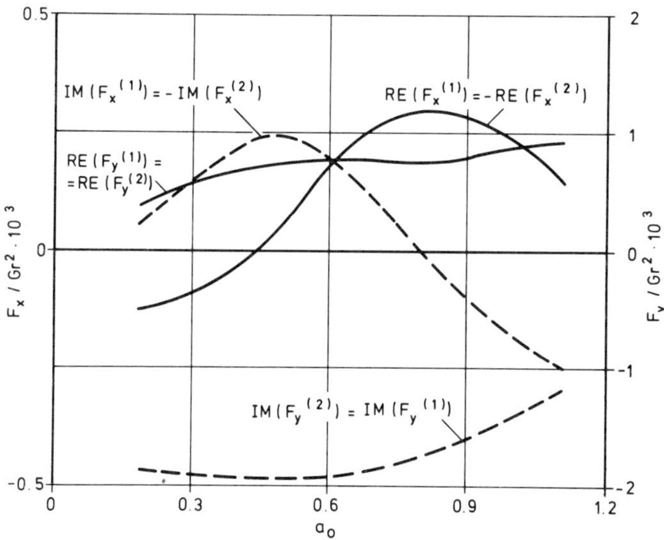

Figure 4a. Dimensionless horizontal forces acting on two massless, immobile foundations due to Rayleigh waves excitation with an angle of incidency of 90^0 (see Fig. 2).

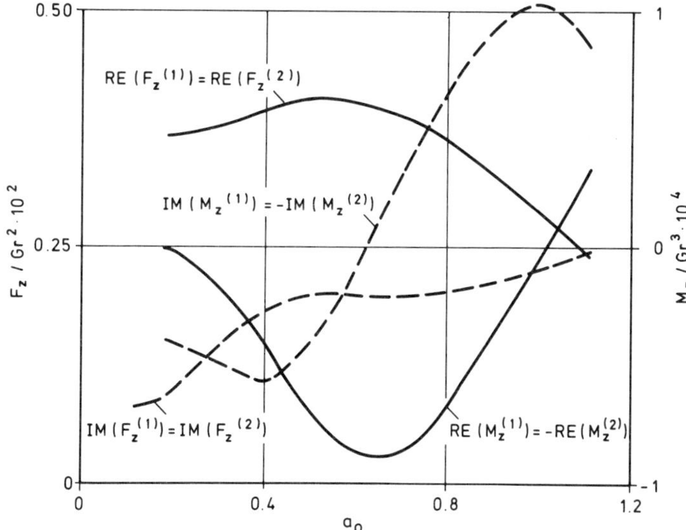

Figure 4b. Dimensionless vertical forces and moments about the z - axis acting on two massless, immobile foundations due to Rayleigh waves excitation with an angle of incidency of 90^0 (see Fig. 2).

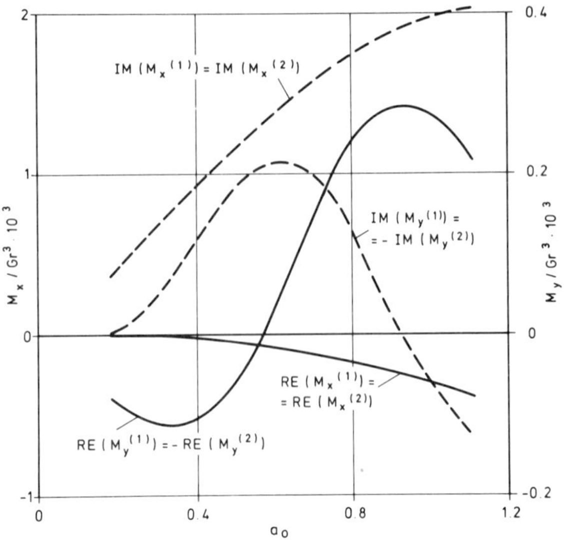

Figure 4c. Dimensionless moments about the x - and the y - axis acting on two massless, immobile foundations due to Rayleigh waves excitation with an angle of incidency of 90^0 (see Fig. 2).

The presentation of the magnitudes of the loads acting in perpendicular direction with respect to the incident waves form a better idea of the amount of these loads than the presentation of the real- and imaginary part of them. The magnitudes of these additional loads during an earthquake depend on the subsoil parameters, the geometry of the foundations and the excitation frequency. For the chosen geometry and the assumed vertical amplitude of 1 mm for the Rayleigh waves the magnitude of the force F_x acting on one of the foundations in relation to the magnitudes of the forces F_y and F_z as well as the magnitudes of the moments M_y and M_z in relation to the magnitude of the moment M_x are shown in Figs. 5a and 5b resp. depending on the excitation frequency.

Since these effects seem not to be negligible for designing the overlying structure it is nessecary to make an extensive study of this problem using a variation of the distance between the two foundations, the excitation frequency and the subsoil input parameters. A paper about the theoretical background and some results of these effects varying the distance between the two foundations and the wavelength of the incident Rayleigh waves is in preparation.

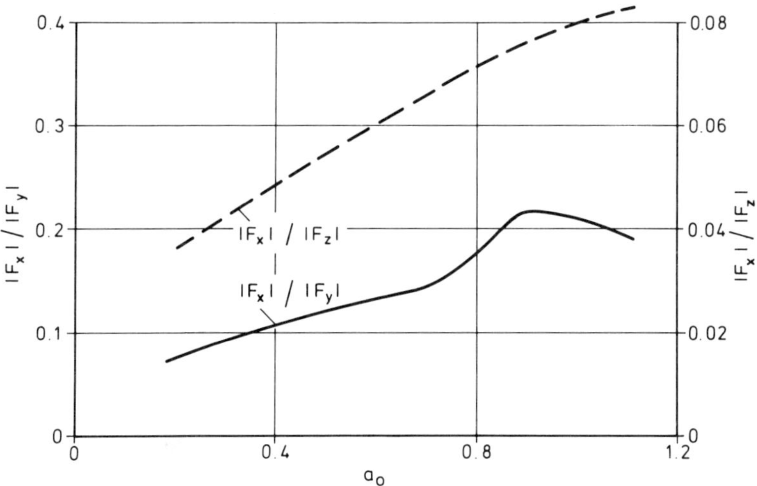

Figure 5a. Magnitudes ratio between the force F_x and the forces F_y, F_z due to Rayleigh waves excitation within the system of two foundations.

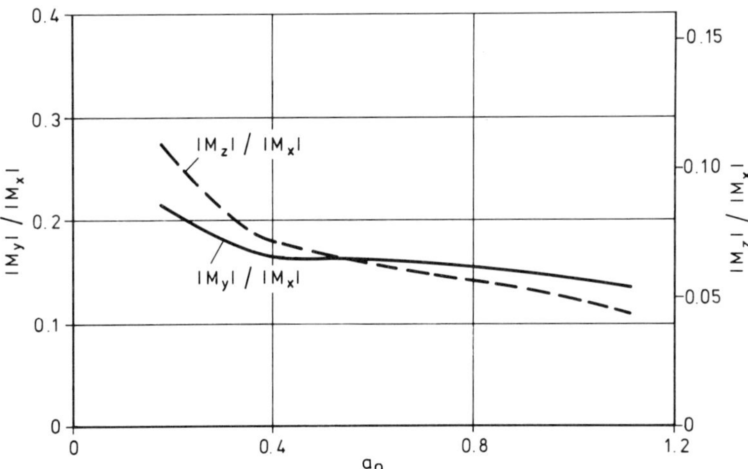

Figure 5b. Manitudes ratio between the moments M_y , M_z and the moment M_x due to Rayleigh waves excitation within the system of two foundations.

CONCLUSIONS

By means of the analytical method presented problems of subsoil coupling between foundations can be solved rigorously. The dynamic behaviour of a foundation within a system of foundations resting on the surface of a linear-elastic, homogeneous and isotropic halfspace differs in particular cases very much from the vibration behaviour of a single foundation. Some important effects showed up for the case of a seismic excitation of long duration (Rayleigh waves) acting upon a group of foundations giving a basis for further investigations on this topic with respect to the safety of structures supported by a finite number of foundations.

REFERENCES

1. Warburton G.B. Richardson H.D. and Webster J.J. (1972), Harmonic Response of Masses on an Elastic Half-Space, Journ. of Eng. for Ind., Trans. ASME, Vol. 194, pp. 193-200

2. Savidis S. and Richter T. (1977) Dynamic Interaction of Rigid Foundations, Vol. II, pp. 369-374, Proccedings of the IX. Int. Conf. on Soil Mechanics and Foundation Engineering, Tokio, Japan, 1977, Japanese Society of Soil Mechanics and Foundation Engineering.

3. Gaul L. (1980). Zur Dynamik der Wechselwirkung von Strukturen auf dem Baugrund, Habilitationsschrift, Institut für Mechanik der Universität Hannover

4. Wong H.L. and Luco J.E. (1986), Dynamic Interaction between Rigid Foundations in a Layered Half-Space, Soil Dyn. and Earthq. Eng., Vol. 5, No. 3, pp. 149-158.

5. Roesset J.J. and Gonzalez J.J. (1978) Dynamic Interaction between Adjacent Structures (Ed. Prange B.), Vol. 1, pp. 127-166, Proceedings of the Conf. on Dynamical Methods in Soil and Rock Mechanics, Kalrsruhe, W. Germany, 1977, Balkema. Rotterdam.

6. Bielak J. and Coronato J.A. (1981) Response of Multiple-Mass Systems to Nonvertically Incident Seismic Waves, Vol. II, pp. 801-804, Proceedings of Int. Conf. on Recent Advances in Geotechnical Earthquake Engineering and Soil Dynamics, St. Louis, USA, 1981, University of Missuri-Rolla.

7. Oien M.A. (1971), Steady Motion of a Rigid Strip Bonded to an Elastic Half Space, Journal of Applied Mechanics, Vol. 38, No. 2, Trans. of the ASME, Series E, June 1971, pp. 328-334.

8. Triantafyllidis Th. (1982), Ein analytisches Verfahren zur Berechnung der Untergrundkopplung von mehreren starren, auf der Halbraumoberfläche liegenden Streifenfundamenten bei harmonischer Erregung, Ing.-Archiv, Vol. 52, pp. 145-157.

9. Triantafyllidis Th. (1984). Analytische Lösung des Problems der dynamischen Untergrundkopplung starrer Fundamente, Dissertation, Veröffentlichungen des Instituts für Bodenmechanik und Felsmechanik der Universität Karlsruhe, Heft Nr. 97.

10. Triantafyllidis Th. and Prange B. (1987), Dynamic Subsoil Coupling between Rigid, Rectangular Foundations, Soil Dyn. and Earthq. Eng., (in print).

11. Triantafyllidis Th. and Prange B. (1987), Dynamic Subsoil Coupling between Rigid, Circular Foundations on the Halfspace, Soil Dyn. and Earthq. Eng., (to appear).

12. Triantafyllidis Th. and Prange B. (1987), Rigid Circular Foundation: Dynamic Effects of Coupling to the Half-Space, Soil Dyn. and Earthq. Eng., (in print).

13. Triantafyllidis Th. (1986), Dynamic Stiffness of Rigid Rectangular Foundations on the Half-Space, Earth. Eng. and Struct. Dyn. ,Vol. 14, pp. 391-411

14. Achenbach J.D. (1980). Wave Propagation in Elastic Solids, North-Holland Publ. Comp. Amsterdam.

15. Michlin S.G. and Smolitski K.L. (1957). Approximate Methods for Solution of Differential and Integral Equations, Elsevier Publ. Co. New York.

16. Erdelyi A. (1953). Higher Transcendental Functions, Vol. 2, McGraw-Hill. New York.

17. Thau S.A. (1967), Radiation and Scattering from a Rigid Inclusion in an Elastic Medium, Journal of Applied Mechanics, Vol. 34, No. 2, Trans. ASME, Vol. 89, Series E, pp. 509-511

18. Triantafyllidis Th. and Vrettos Ch. (1986). Entwicklung analytischer Lösungen für das Problem der dynamischen Untergrundkopplung zwischen benachbarten, dynamisch beanspruchten runden, starren Fundamenten. Arbeitsbericht zum DFG- Forschungsvorhaben: - Untergrundkopplung, Inhomogenität - , Kennzeichen II D4-Pr 102/7-2.

Surface Waves in a Layered Half-Space with Bending Stiffness

H.-B. Mühlhaus, Th. Triantafyllidis
Institute of Soil Mechanics and Rock Mechanics, University of Karlsruhe, Federal Republic of Germany

SUMMARY

The propagation of surface waves in a layered, incompressible half-space with periodically varying stiffness is investigated. The layers are assumed to have a finite thickness and are defined by linear elastic incompressible materials with different shear moduli. In the formulation of the propagation problem, the half-space, which would be materially non homogeneous in a classical continuum is represented as a materially homogeneous half-space of a Cosserat-continuum. Relationships between the elasticities of the orthotropic Cosserat-continuum and the elasticities of the individual layers are given. Due to the material nonhomogeneity the velocity of the surface waves is dispersive. Results of the analysis are represented and compared for special cases with classical analytical solutions.

1. INTRODUCTION

In this paper the propagation of surface waves in a layered medium is investigated. Such a medium is obtained by superposition of adhering layers which are alternately hard and soft. The hard and the soft material occupy resp. fractions α_1 and α_2 of the total thickness h. The investigation is restricted to the case of a medium which is incompressible in layers with shear moduli G_1 and G_2 and densities ρ_1 and ρ_2 of the hard and soft layers respectively. Furthermore it is assumed that the deformation is infinitesimal. Under certain conditions these restrictions will provide a drastic simplification of the algebra of the problem without altering the essential mechanical features of the results.

Because the half space is nonhomogeneous the complete solution of the propagation problem is still highly complex: Equilibrium has to be satisfied in layers, transition conditions have to be taken into account at the interfaces of the layers[1]. If, on the other hand, only wavelengths l with $l/h \gg 1$ are of interest, then the layered half space can be modeled by an homogeneous, nonisotropic medium.

In the following an alternative approach is proposed: The half space which is nonhomogeneous in the classical continuum will be modeled as a materially homogeneous half space of a Cosserat continuum (Cosserat[2], Günther[3,4], Schäfer[5], Mühlhaus et al.[6], Mühlhaus[7,8]). The mathematical effort required for the solution of the propagation problem is then comparable to the solution that the classical continuum theory offers for the long wave length limit.

In a Cosserat continuum each material point has three additional rotational degrees of freedom besides the three translational degrees of freedom which are considered in the classical continuum. In the present case we will use the rotational degrees of freedom for the representation of the bending effects in the layers.

The next section gives a brief introduction to the statics and kinematics of Cosserat continuum. In the third and fourth section constitutive relations are presented for an orthotropic, incompressible Cosserat continuum. For simplicity we restrict our considerations to the case of plane deformations. In the fifth section the surface wave problem is solved and the results are discussed.

2. FORMULATION

In Cosserat theory one distinguishes between two types of surface loads: The components of the one type of load are given by the physical components of the stress tensor, which is nonsymmetric in general. The components of the other type of load are defined by the components of the couple stress tensor. The assumption of such a continuum may be motivated as follows: Consider an element (ℓ_1, ℓ_2) of the classical continuum (Fig. 1a); σ_{12} and σ_{21} are the average values of shear stresses at the sides ℓ_1 and ℓ_2 respectively.

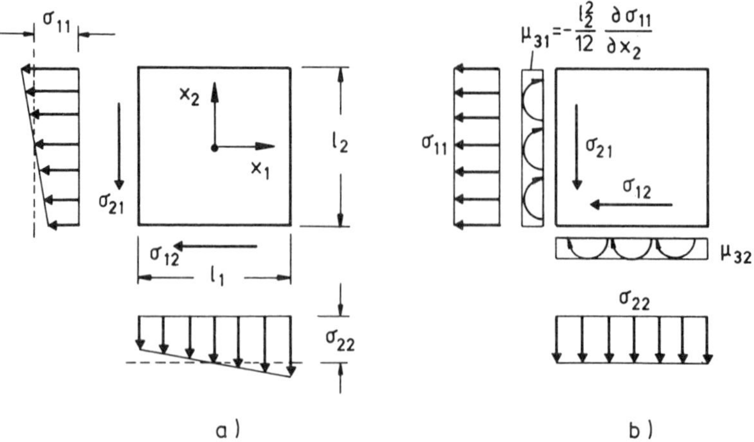

Figure 1. Stresses at an element (ℓ_1, ℓ_2) of the classical continuum.

The normal stresses along ℓ_1 and ℓ_2 are assumed to be linearily variable. Now it is well known from elementary statics, that a linearily variable normal stress is statically equivalent to a force and a moment or - because we have in view a field theory - by a force intensity (= stress) and a moment intensity (= couple stress) (Fig. 1b).

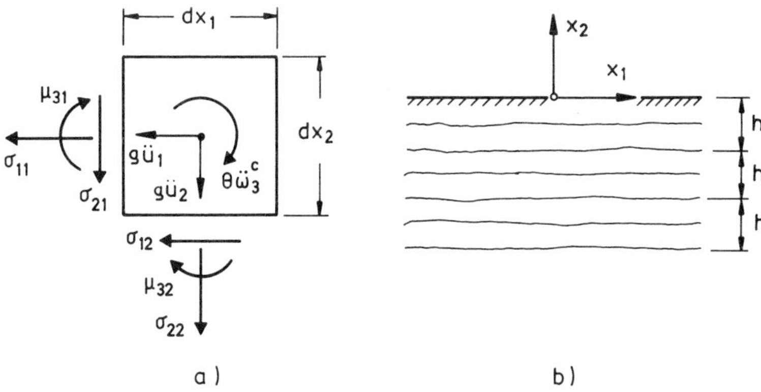

Figure 2. (a) Stresses and inertia loads acting at an infinitesimal element of a Cosserat continuum. (b) Geometry of the layered halfspace.

Fig. 2a shows the stresses and inertia loads acting on an infinitesimal element of the Cosserat continuum in plane deformations. The displacements are denoted as u_i and ω_3^c is the nonvanishing component of Cosserat rotation vector; Θ is the moment of inertia.

Force equilibrium and moment equilibrium give:

$$\sigma_{11,1} + \sigma_{12,2} = \rho\ddot{u}_1, \tag{1}$$

$$\sigma_{21,1} + \sigma_{22,2} = \rho\ddot{u}_2, \tag{2}$$

$$\mu_{31,1} + \mu_{32,2} + \sigma_{21} - \sigma_{12} = \Theta\ddot{\omega}_3^c, \tag{3}$$

where:

$$(\cdot)_{,i} = \frac{\partial}{\partial x_i}(\cdot) \tag{4}$$

and

$$\dot{(\cdot)} = \frac{d}{dt}(\cdot). \tag{5}$$

Eq. (1)-(3) are, together with the constitutive relations (comp. Sect. 3), the equations of motion of a Cosserat continuum in plane deformations. Next the deformation measures which are conjugate in energy to σ_{ij} and μ_{ij} are derived: Eq. (1) and Eq. (2) are multiplied with the virtual velocities $\delta\dot{u}_1$ and $\delta\dot{u}_2$ respectively; Eq. (3) is multiplied with the virtual Cosserat rotational velocity $\delta\dot{\omega}_3^c$. As usual it is assumed that $\delta\dot{u}_i$, $\delta\dot{\omega}_3^c$ vanish at those parts of the boundary where kinematic boundary conditions are prescribed. Now the resulting expressions are added and then integrated over the domain under consideration. Application of Gauss' theorem yields:

$$\int_B (\sigma_{ij}\delta\dot{\epsilon}_{ij} + \mu_{3i}\delta\dot{\kappa}_{3i})dV = \int_{\partial_1 B} t_i\delta\dot{u}_i dA + \int_{\partial_2 B} m_3\delta\dot{\omega}_3^c dA - \int_B (\rho\ddot{u}_i\delta\dot{u}_i + \Theta\ddot{\omega}_3^c\delta\dot{\omega}_3^c)dV, \tag{6}$$

where:

$$\sigma_{ij}n_j = t_i \quad \text{on} \quad \partial_1 B, \tag{7}$$

$$\mu_{3i}n_i = m_3 \quad \text{on} \quad \partial_2 B, \tag{8}$$

$$\epsilon_{ij} = E_{ij} + W_{ij} - W_{ij}^c, \tag{9}$$

$$\kappa_{3i} = \omega_{3,i}^c. \tag{10}$$

In Eq. (9) E_{ij} and W_{ij} denote the symmetric and the antimetric part of the displacement gradient respetively. W_{ij}^c denotes the antimetric tensor corresponding to the rotation vector ω_i^c. In the present case all components of W_{ij}^c vanish, except $W_{21}^c = -W_{12}^c = \omega_3^c$. ϵ_{ij} and κ_{ij} are the deformation measures of the Cosserat continuum in infinitesimal deformations (e.g. Günther[3]; Schaefer[5]). Eq. (6) is the virtual work equation of the Cosserat continuum in plane deformations. Eq. (6) was derived in such a way that each solution of the equations of motion which satisfies the boundary condition, does also satisfy the virtual work equation. On the other hand, in analogy to the situation in the classical continuum, all fields $(\sigma_{ij}, \mu_{ij}, \ddot{u}_i, \ddot{\omega}_3^c)$ which satisfy Eq. (6) for arbitrary $\delta\dot{u}_i, \delta\dot{\omega}_i^c$, do also satisfy Eqs. (1)-(3) and the stress boundary conditions (Eqs. 7,8).

In the solution of the surface wave problem it is assumed that the layers are parallel to the free surface of the half space and that the x_2- axis is oriented orthogonally to the layers (Fig. 2b). On the free surface the following boundary conditions have to be satisfied:

$$\sigma_{22} = 0, \quad \sigma_{12} = 0, \quad \mu_{32} = 0, \qquad \text{on} \quad (x_2 = 0, x_1). \tag{11}$$

3. CONSTITUTIVE EQUATIONS

In our analysis we will assume the linear-elastic constitutive relationships proposed by Mühlhaus[7] and Mühlhaus et al.[6]:

$$\sigma_{11} = 2T\epsilon_{11} + p \quad , \qquad \sigma_{22} = 2T\epsilon_{22} + p \quad , \tag{12}$$

$$\sigma_{12} = G(\epsilon_{12} + \epsilon_{21}) \quad , \qquad \sigma_{21} = \sigma_{12} + G^c\epsilon_{21} \quad , \tag{13}$$

$$\mu_{32} = 0 \quad , \qquad \mu_{31} = \frac{4T^c h^2}{12}\kappa_{31} \quad , \tag{14}$$

where:

$$\epsilon_{11} + \epsilon_{22} = 0 \qquad \text{and} \quad p = \frac{1}{2}(\sigma_{11} + \sigma_{22}). \tag{15}$$

The moduli T and G have the same physical significance as in an incompressible, orthotropic classical continuum (e.g. Biot[9]). The moduli which are specific for the Cosserat continuum are the additional shear modulus G^c and the bending modulus T^c.

If deformations with $\epsilon_{12} + \epsilon_{21} = 0$ are considered then Eqs. (12)-(15) reduce to the constitutive relations of Reissner's[10] plate theory. In this theory ω_3^c is the rotation of the cross sections of the plate and h is the thickness of the plate. In Mühlhaus[7] and Mühlhaus et al.[6] the following relations have been derived between the moduli G_1 and G_2 of the individual layers and the moduli T, G, G^c and T^c in Eqs. (12)-(15):

$$T = \alpha_1 G_1 + \alpha_2 G_2 \quad , \tag{16}$$

$$G = \frac{G_1 G_2}{\alpha_2 G_1 + \alpha_1 G_2} \quad , \tag{17}$$

$$T^c = \frac{T \alpha_1 \alpha_2}{\alpha_2 G_1 + \alpha_1 G_2}(T - G) \quad , \tag{18}$$

$$G^c = T - G. \tag{19}$$

4. AVERAGE DENSITY AND MOMENT OF INERTIA

In this section we want to express ρ and Θ by the corresponding quantities of the layered continuum. For evaluation of the average density ρ one proceeds as follows: The velocity field $\dot{u}_1 = cx_1$, $\dot{u}_2 = -cx_2$, $\omega_3^c = 0$ shall be prescribed in the Cosserat- and in the layered continuum. According to Eq. (16) the kinetic energy and the strain energy of both continua are equal if

$$\rho = \alpha_1 \rho_1 + \alpha_2 \rho_2. \tag{20}$$

Next consider two semi-infinite strips. One of the strips be defined in the Cosserat continuum and the other in the layered continuum (Fig. 3a,b). Assuming the lowest eigenfrequencies of both strips to be equal, one obtains a relationship between Θ and the quantities defined in layers. We will evaluate the frequencies by the Ritz' method.

First the strip of Cosserat material is considered: In the case of homogeneous kinematic boundary conditions one can put $\delta \dot{u}_i = \dot{u}_i$ and $\delta \dot{\omega}_3^c = \dot{\omega}_3^c$ in Eq. (6). The modes:

$$u_2 = U \sin(\pi \frac{x_1}{L}) \exp(i\omega t), \tag{21}$$

$$\omega_3^c = \Omega \sin(\pi \frac{x_1}{L}) \exp(i\omega t), \tag{22}$$

$$u_1 = 0 \quad ; \qquad i = \sqrt{-1}, \tag{23}$$

are inserted into Eqs. (12)-(14), (6) and one obtains:

$$\left[\left(\frac{\pi}{L}\right)^2 GU^2 + G^c\left(\frac{\pi}{L}U - \Omega\right)^2 + \frac{1}{3}T^c\left(\pi\frac{h}{L}\right)^2\Omega^2\right] - \omega^2\left[\rho U^2 + \Theta\Omega^2\right] = 0 \quad (24)$$

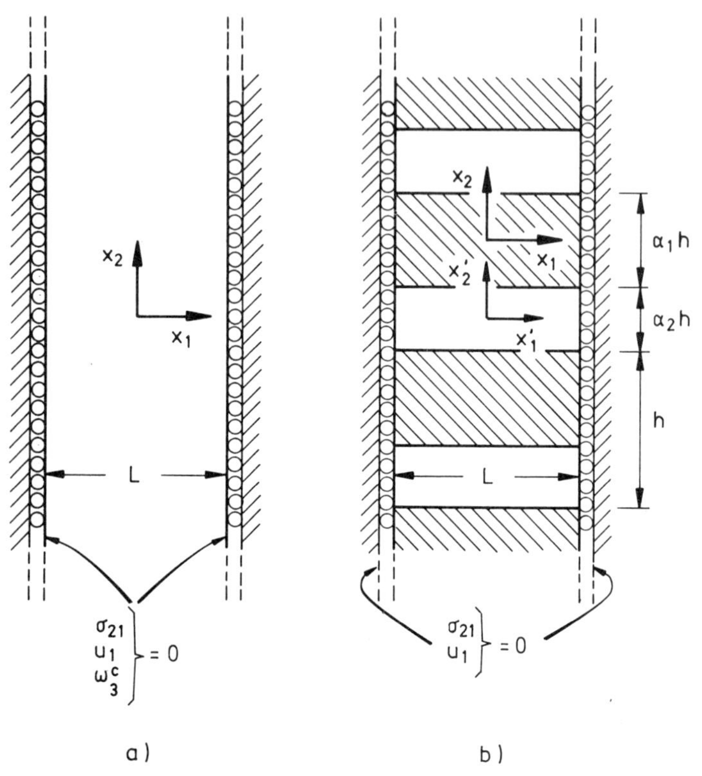

Figure 3. Goemetry and boundary conditions of semi infinite strips (a) in the Cosserat continuum and (b) in the layered classical continuum.

According to Ritz' method the gradients with respect to U and Ω of the left hand side of Eq. (24) have to vanish. The resulting homogeneous system of equations is nontrivially solvable if:

$$\left[1 + \frac{1}{3}\frac{T^c}{G^c}\left(\pi\frac{h}{L}\right)^2 - \frac{\Theta}{G^c}\omega^2\right]\left[1 + \frac{G}{G^c} - \frac{\rho L^2}{\pi^2 G^c}\omega^2\right] = 1. \quad (25)$$

For the identification of Θ we require a frequency which is independent of the wavelength L. We consider the limit of Eq. (25) for $L \to \infty$ and obtain:

$$\omega^2 = \frac{G^c}{\Theta}. \quad (26)$$

The frequency ω (Eq. 26) appears in connection with rotational vibrations of the cross sections of the layers.

We add a brief discussion of Eq. (25) before the layered strip (Fig. 3b) is considered: In case of an isotropic (i.e. $G_1 = G_2$) medium from Eqs. (16)-(19) it follows that $G^c = 0$ and $T^c = 0$. Eq. (25) reduces to:

$$\frac{L^2\omega^2}{\pi^2} = c_S^2 \quad , \qquad c_S = \sqrt{\frac{G}{\rho}}, \tag{27}$$

where c_S is the shear wave velocity of the isotropic medium. Now we consider the case $G_2 = 0$, $G_1 > 0$, $\alpha_1 = 1$. Then the strip decomposes into decoupled plates of thickness h and length L. According to Eqs. (16)-(19) one obtains $G = 0$ and $T^c = G^c = G_1$. For the special case $\Theta = 0$, Eq. (25) yields:

$$c_f^2 = \frac{1}{\rho}\frac{\frac{1}{3}G_1\left(\pi\frac{h}{L}\right)^2}{1+\frac{1}{3}\left(\pi\frac{h}{L}\right)^2} \quad ; \quad c_f = \frac{L\omega}{\pi}. \tag{28}$$

If Eq. (28) is expanded into powers of $(\pi h/L)^2$ one finds that the first term is identical with the flexural wave velocity of a thin incompressible plate (Achenbach[11]). For the short wave length limit one finds that $c_f \to c_S$.

Now the case of the layered strip (Fig. 3b) is considered: In both types of layers it is assumed that

$$u_2 = U_2 \sin(\pi\frac{x_1}{L})\exp(i\omega t). \tag{29}$$

Within the hard and the soft layers we assume that

$$u_1 = U_1\frac{2x_2}{\alpha_1 h}\cos(\pi\frac{x_1}{L})\exp(i\omega t), \tag{30}$$

and

$$u_1 = -U_1\frac{2x_2'}{\alpha_2 h}\cos(\pi\frac{x_1'}{L})\exp(i\omega t), \tag{31}$$

respectively (see Fig. 3b). The displacements (Eqs. (29)-(31)) are continuous at the layer interfaces and satisfy the kinematic boundary conditions $u_1 = 0$ if $|x_1| = L/2$ or $|x_1'| = L/2$. Now again Ritz' method is applied. An analogoue calculation as in the case of the strip of Cosserat material yields for the limiting case $L \to \infty$:

$$\omega^2 = \frac{12}{h^2}\frac{\alpha_1 G_1 + \alpha_2 G_2}{\rho_1 {\alpha_1}^3 + \rho_2 {\alpha_2}^3} \tag{32}$$

We assume that the frequency (Eq. 26) of the strip of Cosserat material is equal to the frequency (Eq. 32) of the layered strip and obtain:

$$\Theta = \frac{\rho_1\alpha_1^3 + \rho_2\alpha_2^3}{12}h^2\left(1 - \frac{T}{G}\right). \tag{33}$$

According to Eqs. (16)-(17) in the case of isotropy (i.e. $G_1 = G_2$) $T = G$, thus $\Theta = 0$. On the other hand if $G_2 = 0$ and $\alpha_1 = 1$ Eq. (33) reduces to $\Theta = \rho h^2/12$, where $\rho = \rho_1$.

5. SOLUTION OF THE SURFACE WAVE PROBLEM

In order to decouple the equations of motion the following procedure is used: Differentiation of the Eqs. (1)-(2) with respect to x_1 and x_2 resp. and substracting the results gives:

$$\sigma_{11,12} - \sigma_{22,21} = \rho(\ddot{u}_{1,2} - \ddot{u}_{2,1}) + \sigma_{21,11} - \sigma_{12,22} \tag{34}$$

Using the constitutive relations (Eq. 12), the left hand side of Eq. (34) can be written as follows:

$$\sigma_{11,12} - \sigma_{22,21} = 2T(\epsilon_{11,12} - \epsilon_{22,21}) \tag{35}$$

Differentiating twice the expressions for σ_{21} and σ_{12} of the constitutive relations (Eq. 13) with respect to x_1 and x_2 resp. and substituting Eq. (35) into Eq. (34), by means of the incompressibility condition: $\epsilon_{11} + \epsilon_{22} = 0$, one obtains:

$$4T\epsilon_{11,12} - 2G(E_{12,11} - E_{12,22}) - G^c\epsilon_{21,11} = \rho(\ddot{u}_{1,2} - \ddot{u}_{2,1}) \tag{36}$$

In a similar manner replacing the constitutive relations into Eq. (3) one obtains:

$$\alpha\omega^c_{,11} + G^c u_{2,1} - G^c\omega^c - \Theta\ddot{\omega}^c = 0, \tag{37}$$

where: $\alpha = 4\dfrac{T^c h^2}{12}$.

Incompressibility of the material under consideration permits the introduction of a stream function $\Psi(x_1, x_2, t)$ such that:

$$u_1 = \Psi_{,2} \quad , \qquad u_2 = -\Psi_{,1} \tag{38}$$

Thus Eqs. (36),(37) by means of Eqs. (38) can be expressed as follows:

$$(G+G^c)\frac{\partial^4\Psi}{\partial x_1^4} + (4T-2G)\frac{\partial^4\Psi}{\partial x_1^2\partial x_2^2} + G\frac{\partial^4\Psi}{\partial x_2^4} + G^c\frac{\partial^2\omega^c}{\partial x_1^2} = \rho\left(\frac{\partial^4\Psi}{\partial t^2\partial x_2^2} + \frac{\partial^4\Psi}{\partial t^2\partial x_1^2}\right) \tag{39}$$

$$\alpha\frac{\partial^2\omega^c}{\partial x_1^2} - G^c\frac{\partial^2\Psi}{\partial x_1^2} - G^c\omega^c - \Theta\frac{\partial^2\omega^c}{\partial t^2} = 0 \tag{40}$$

The Eqs. (39),(40) define a system of two partial differential equations with the Ψ- and ω^c functions as unknowns. Considering plane harmonic waves, which propagate in the x_1- direction with the wave number γ and the excitation frequency ω the following expressions for the functions $\Psi(x_1, x_2, t)$ and $\omega^c(x_1, x_2, t)$ can be assumed:

$$\Psi(x_1, x_2, t) = f(x_2)e^{i\gamma x_1}e^{i\omega t} \tag{41}$$

$$\omega^c(x_1,x_2,t) = \Omega(x_2)e^{i\gamma x_1}e^{i\omega t} \tag{42}$$

Using Eqs. (41) and (42) the system of coupled partial differential equations (Eqs. 39-40) can be reduced to a system of two ordinary differential equations for the unknown functions $f(x_2)$ and $\Omega(x_2)$, which is given as follows:

$$\Omega(x_2)(-\omega^2\Theta + G^c + \gamma^2\alpha) = G^c\gamma^2 f(x_2) \tag{43}$$

$$Gf''''(x_2) - (4T-2G)\gamma^2 f''(x_2) + T\gamma^4 f(x_2) - G^c\gamma^2\Omega(x_2) = \\ = \rho\omega^2(\gamma^2 f(x_2) - f''(x_2)) \tag{44}$$

where $(\cdot)'$ denotes differentiation with respect to x_2.

Substitution of Eq. (43) into Eq. (44) gives a fourth order differential equation for $f(x_2)$:

$$\frac{G}{T}f''''(x_2) + \hat{\alpha}(\gamma)f''(x_2) + \hat{\beta}(\gamma)f(x_2) = 0 \tag{45}$$

where:

$$\hat{\alpha}(\gamma) = \bar{k}_\beta^2 - 4\gamma^2 + 2(\frac{G}{T})\gamma^2$$

$$\hat{\beta}(\gamma) = \gamma^4\tilde{\beta}(\gamma) - \bar{k}_\beta^2\gamma^2$$

$$\tilde{\beta}(\gamma) = \frac{(\frac{G}{T}) + h^2\gamma^2 A_G - \bar{k}_\beta^2 h^2 A_\rho}{1 + h^2\gamma^2 A_G - \bar{k}_\beta^2 h^2 A_\rho}$$

$$A_\rho = \frac{\rho_1\alpha_1^3 + \rho_2\alpha_2^3}{12(\alpha_1\rho_1 + \alpha_2\rho_2)}$$

$$A_G = \frac{4\alpha_1\alpha_2(\alpha_1 G_1 + \alpha_2 G_2)}{12(\alpha_2 G_1 + \alpha_1 G_2)}$$

$$\bar{k}_\beta^2 = \frac{\omega^2\rho}{T}$$

The variable $\bar{k}_\beta$ used in the above expressions represents the wave number of a shear wave velocity in a homogeneous material having shear modulus and density equalling the average shear modulus and density of the two layers resp. Assuming the solutions of Eq. (45) to be of the form:

$$f(x_2) = Ce^{\lambda x_2} \tag{46}$$

one obtains (if $G \neq 0$):

$$\lambda = \pm\sqrt{\frac{-\hat{\alpha}(\gamma) \pm \sqrt{\hat{\alpha}^2(\gamma) - 4\frac{G}{T}\hat{\beta}(\gamma)}}{2\frac{G}{T}}} \tag{47}$$

In the case $G_1 = 0$ or $G_2 = 0$(i.e. $G = 0$) the differential equation with respect to the function $f(x_2)$ is of second order and the solution for λ is:

$$\lambda = \pm\sqrt{-\frac{\hat{\beta}(\gamma)}{\hat{\alpha}(\gamma)}} \tag{47.a}$$

The next step to find the solution for the surface waves of an alternately layered halfspace is to apply the stress boundary conditions. Considering surface waves whose displacements vanish in the limit as $x_2 \to \infty$ only two of the four solutions for λ, which have to be real and positive,are taken into account (boundedness of the solution). In the expressions for the boundary conditions (Eq. 11) the isotropic part p of the stress tensor σ_{ij} appears. By use of the Eq. (1), the constitutive relations and, in analogy to the stream Ψ and the Cosserat rotation ω_3^c, assuming that p is of the form:

$$p = q(x_2)e^{i\gamma x_1}e^{i\omega t}, \tag{48}$$

one can express the unknown function $q(x_2)$ in terms of the function $f(x_2)$ as follows:

$$q(x_2) = i\frac{T}{\gamma}\Big[\big(\frac{G}{T}\big)f'''(x_2) + \Big(\bar{k}_\beta^2 + \big(\frac{G}{T}\big)\gamma^2 - 2\gamma^2\Big)f'(x_2)\Big] \tag{49}$$

Thus the stress boundary conditions can be expressed as functions of $f(x_2)$. The results are:

$$\begin{aligned} \sigma_{22}\big|_{x_2=0} &= 0 = \big(\frac{G}{T}\big)f'''(x_2) + \Big(\bar{k}_\beta^2 + \big(\frac{G}{T}\big)\gamma^2 - 4\gamma^2\Big)f'(x_2) \\ \sigma_{12}\big|_{x_2=0} &= 0 = f''(x_2) + \gamma^2 f(x_2) \end{aligned} \tag{50}$$

Substituting the solution of the Eq. (45) with respect to the boundedness of the solution into the stress boundary conditions (Eqs. 50) the following algebraic system has to be solved:

$$\begin{aligned} C_1\lambda_1\Big[\big(\tfrac{G}{T}\big)\big(\lambda_1^2-\gamma^2\big)+\hat{\alpha}(\gamma)\Big] &+ C_2\lambda_2\Big[\big(\tfrac{G}{T}\big)\big(\lambda_2^2-\gamma^2\big)+\hat{\alpha}(\gamma)\Big] &= 0 \\ C_1\Big[\lambda_1^2+\gamma^2\Big] &+ C_2\Big[\lambda_2^2+\gamma^2\Big] &= 0 \end{aligned} \tag{51}$$

A non trivial solution of the above algebraic system exists if the determinant of the equation system (51) vanishes; thus:

$$\Big(\lambda_1 - \lambda_2\Big)\Big[\hat{\beta}(\gamma) - \gamma^4\big(\frac{G}{T}\big) - \lambda_1\lambda_2\big(\bar{k}_\beta^2 - 4\gamma^2\big)\Big] = 0 \tag{52}$$

From Eq. (52) and for a given frequency one can calculate the wave number γ of the surface waves and thus the dispersion relation for the given incompressible alternately layered halfspace can be determined. The Eq. (52) includes some special cases the solutions of them being well known. Considering a homogeneous, isotropic and incompressible halfspace (i.e. $G = T$) the solutions of the Eq. (45) (satisfying the boundedness of the solution) are:

$$\lambda_1 = \gamma \quad , \quad \lambda_2 = \sqrt{\gamma^2 - \bar{k}_\beta^2} \tag{53}$$

Substituting the Eqs. (53) into Eq. (52) one obtains:

$$\left(2\gamma^2 - \bar{k}_\beta^2\right)^2 - 4\gamma^3\sqrt{\gamma^2 - \bar{k}_\beta^2} = 0 \tag{54}$$

Eq. (54) represents the Rayleigh equation for the homogeneous, isotropic and incompressible halfspace.

Next we consider the special case $G_2 = 0$, $\alpha_2 = 0.5$, $\rho_2 = 0$. This case represents a halfspace consisting of plates of thickness $h^{(1)} = 0.5h$ and within the spacing between the plates material with zero stiffness is prescribed. The solution for this problem has to degenerate to the well known solution of the classical plate bending theory. Analytical expressions for the relations between frequencies and the wave numbers by consideration of long wave lengths (small wave numbers) are already given in the literature (see Achenbach[11]). For a plate of uniform thickness d consisting of an incompressible material the following relation holds:

$$\frac{\omega}{c_s} = \gamma^2 d \frac{1}{\sqrt{3}} \tag{55}$$

For $G_2 = 0$ it follows that $G = 0$. Now a solution of Eq. (45), where $f(x_2)$ is independent on x_2, exists if:

$$\hat{\beta}(\gamma) = 0 \tag{56}$$

The solution of Eq. (56) yields:

$$\bar{k}_\beta = \frac{\gamma^2 h}{\frac{5}{8} h \bar{k}_\beta \pm \sqrt{12}\sqrt{1 + \frac{3}{256} h^2 \bar{k}_\beta^2}} \tag{57}$$

If long wave lengths are of interest the limit for $h\bar{k}_\beta \to 0$ of the Eq. (57) has to be taken and the result is:

$$\bar{k}_\beta = \frac{\omega}{\sqrt{\frac{T}{\rho}}} = \frac{\omega}{c_s^{(1)}} = \gamma^2 \frac{h}{\sqrt{12}} = \gamma^2 \frac{h^{(1)}}{\sqrt{3}} \tag{58}$$

In this paper for the numerical calculation of the dispersion curves of a periodically interlayered halfspace with the period h only a variation of the shear moduli ratio G_1/G_2 of the two layers has been taken in to account. Furthermore it is assumed that: $\alpha_1/\alpha_2 = 1$ and $\rho_1/\rho_2 = 1$ and the G_1/G_2 ratios to be: 1, 0.5, 0.1, 0.01, 0. Fig 4 shows for the five cases under consideration the dispersion relations between the dimensionless surface wave velocity $c/\bar{c}_s$ ($\bar{c}_s = \omega h/\bar{k}_\beta$, average shear wave velocity) with respect to the dimensionless wavelength $2\pi h/L_R$ (L_R: wave length of the surface waves).

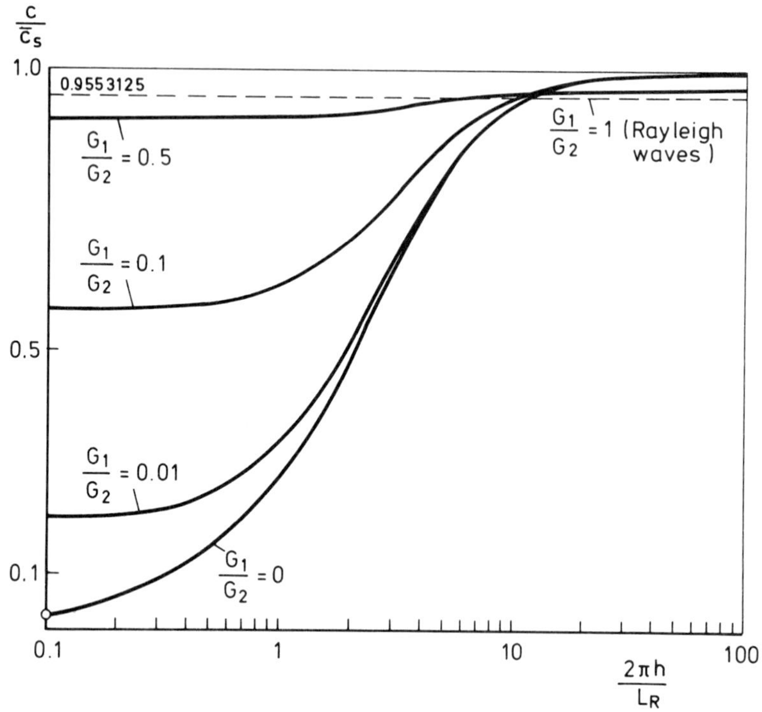

Figure 4. Dimensionless dispesion relations between the phase velocity c for the Rayleigh mode and the wave length L_R of a periodically interlayered halfspace with the period h.

The circle o in Fig. 4 represents the dispersion relation of a plate for the flexural vibration mode for long wave lengths (see Achenbach[11]). For higher frequencies the phase velocity of the surface waves tends to the average shear wave velocity of the two layers. In this paper only the so called "Rayleigh mode" is presented (λ_1, λ_2 real and positive). In general the Eq. (52) has also imaginary roots for λ, which are associated with waves propagated within the layers (normal modes) as well as complex roots for λ, which represents reflections at the interfaces between the layers (e.g. Bragg reflection[12]). A paper about these modes will be published as soon as possible.

CONCLUSIONS

In this paper the problem of propagation of surface waves in an incompressible periodically interlayered halfspace is solved. It has been shown that the originally nonhomogeneous halfspace can be maped into a homogeneous halfspace of a Cosserat continuum. The resulting mathematical problem then is relatively easy to solve. In case of compressible materials one can procede essentially in a similar manner as in the present paper to determine the dispersion relations. Of course, some modifications in the constitutive relations (Eqs. 12-20) have to be applied in this case. In more complicated initial- boundary value problems the virtual work equation (Eq. 6) can serve as a starting point for finite element formulations.

REFERENCES

1. Haskell N.A. (1953), The Dispersion of Surface Waves in Multilayered Media, Seism. Soc. America Bull., Vol. 43, pp. 17-34.

2. Cosserat E. and Cosserat F. (1909), Theorie des Corps Deformables, Hermann et Fils, Paris.

3. Günther W. (1958), Zur Statik und Kinematik des Cosserat'schen Kontinuums, Abh. Braunschweigische Wiss. Ges., Vol. 10, pp. 195-213.

4. Günther W. (1962), Schalenstatik im räumlichen Cosserat'schen Kontinuum, Miszellaneen der Angewandten Mechanik, Festschrift W. Tollmien, (Ed. Schäfer M.), Akademie Verlag, Berlin, pp. 67-77.

5. Schaefer H. (1967), Das Cosserat Kontinuum, Zeitschrift Angew. Math. Mech., Vol. 47, pp. 485-498.

6. Mühlhaus H.-B. and Vardoulakis I. (1986), Axially- Symmetric Buckling of the Surface of a Laminated Half Space with Bending Stiffness, Mech. of Materials, Vol. 5, pp. 109-120.

7. Mühlhaus H.-B. (1985), Surface Instability of a Half Space with Bending Stiffness (in German), Ing.- Archiv, Vol. 55, pp. 388-400.

8. Mühlhaus H.-B. (1986), Shear-band Analysis in Granular Materials by Cosserat Theory (in German), Ing.- Archiv, Vol. 56, pp. 389-399.

9. Biot M. (1965). Mechanics of Incremental Deformations, John Wiley & Sons Inc., New York, London, Sydney.

10. Reissner E. (1945), The Effect of Transverse Shear Deformation on the Bending of Elastic Plates, J. Appl. Mech., Vol. 12, pp. A 69-A 77.

11. Achenbach J. (1980), Wave Propagation in Elastic Solids. Chapter 6, Approximate Theories for Plates, pp. 254-257, North-Holland Publ. Co., Amsterdam.

12. Grant F.S. and West G.F. (1965), Interpretation Theory in Applied Goephysics. Chapter 3, Plane Seismic Waves in Layered Media, pp. 51-91, McGraw-Hill, Inc. New York, St. Luis, San Francisco, Toronto, London, Sydney.

Transient Behavior of Strip Foundations Resting on Different Soil Profiles by a Time Domain BEM

H. Antes, O. von Estorff

Institut für Mechanik, Ruhr-Universität Bochum, P.O.B. 102148, D-4630 Bochum, West Germany

INTRODUCTION

During recent years several numerical techniques, e.g. the finite element method (FEM), the half space method and the boundary element method (BEM), have been developed for the study of the behavior of dynamically loaded structures interacting with the surrounding media. In principle, the application of the FEM to soil-structure interaction problems appears to be very effective since it can easily handle complex structural geometries located on the surface as well as within homogeneous and nonhomogeneous soils. The FEM, however, has the disadvantage that a semi-infinite medium must be represented by a finite size model. To remedy this situation one may use either transmitting boundaries[1], special infinite elements[2], or hybrid techniques, i.e. combined halfspace/FE approaches[3].

The third of the first-mentioned methods, the BEM proved to be well suited for handling soil dynamic problems. It can easily be applied to problems involving infinite domains because, due to the presence of appropriate fundamental solutions, it automatically takes into account the effect of radiation damping. Moreover, the dynamic behavior of a structure resting on the soil mass is influenced to an important extent by the foundation geometry (shallow or deep foundation) and by the type of the soil model. Using BEM, it is a very easy matter to change these geometries.

While the BEM has undergone a rapid development relatively little work has been done on elastodynamics and, especially, on dynamic soil-foundation interaction problems. Moreover, most of the studies available to date have been carried out in the frequency domain[4-7]. This is a clear disadvantage because the approach cannot be extended to non-linear problems. BEM solutions in elastodynamics using time-stepping techniques, i.e. direct solutions in the time domain have been mainly performed for special problems. Among them is the work by Wolf et al.[8,9] who used appropriate Green's functions for the layered halfspace[8], by Karabalis and Beskos[10] considering 3-D rigid surface foundations, and by Spyrakos et al.[11,12] who examined rigid and flexible, surface and embedded strip footings. A time-dependent boundary element technique to treat general 2-D elastodynamic problems was developed by Mansur[13] for zero initial conditions, while Antes[14] derived a similar, more general formulation which includes the contribution of arbitrary initial conditions. The BEM procedure of Antes[14], which was compared to that of Spyrakos[11] in references[15,16], has been applied successfully to several interaction problems, e.g. the interaction between elastic structures or soils and compressible fluids[17-20,22]

This paper reports recent progress in the application of this method[14]. It provides information for assessing the damping effect of the embedment of a foundation, and it investigates the influence of a layer on rigid bedrock and of a stratum over a halfspace on the behavior of rigid strip foundations. In this context the possibility of wave focussing in the case of special rigid bedrock geometries is discussed.

BASIC DIFFERENTIAL AND INTEGRAL EQUATIONS

Under the assumption of small displacements the motion in a homogeneous, isotropic, linear elastic medium can be described in terms of the derivatives of the displacement components $u_i(\mathbf{x},t)$ as follows

$$(c_1^2 - c_2^2)\, u_{i,ij} + c_2^2 u_{j,ii} - \ddot{u}_j = -\frac{1}{\rho}\, b_j \quad , \; i,j=1,2 \; . \qquad (1)$$

In (1), a Cartesian reference frame has been used and the commas and dots indicate partial space and time derivatives, respectively. The propagation velocities of the dilatational and distortional waves for the case of plane strain are given by

$$c_1 = \sqrt{E(1-\nu)/\rho(1+\nu)(1-2\nu)} \quad , \quad c_2 = \sqrt{E/2\rho(1+\nu)} \quad , \qquad (2)$$

where E is Young's modulus and ν Poisson's ratio. Also, ρ stands for the mass density of the elastic medium and b_i for the body forces. The initial and boundary conditions are defined as follows

$$u_i(\mathbf{x},t) = \bar{u}_{io}(\mathbf{x}); \; \dot{u}_i(\mathbf{x},t) = \bar{v}_{io}(\mathbf{x}) \text{ for } t = 0 \text{ in } \Omega + \Gamma \quad , \qquad (3)$$

$$\begin{aligned} u_i(\mathbf{x},t) &= \bar{u}_i(\mathbf{x},t) && \text{for } t > 0 \text{ on } \Gamma_1 \\ T_i(\mathbf{x},t) &= \sigma_{ik} n_k = \bar{T}_i(\mathbf{x},t) && \text{for } t > 0 \text{ on } \Gamma_2; \; \Gamma = \Gamma_1 + \Gamma_2 \; . \end{aligned} \qquad (4)$$

The prescribed values are indicated by super bars. The traction components can be expressed as functions of displacement derivatives according to

$$T_i = \rho[\delta_{ik}(c_1^2 - 2c_2^2)\, u_{j,j} + c_2^2(u_{i,k} + u_{k,i})]\, n_k, \tag{5}$$

where n_k are the direction cosines of the outward normal to the boundary Γ of the elastic medium.

The basic idea of the proposed boundary element method is to describe this initial - boundary value problem by integral equations. In order to deduce these using Graffi's[21] elastodynamic reciprocal theorem, it is necessary to consider two states which satisfy equations (1), whereby one is subjected to the actual body force distribution b_j and the other to a unit impulse force in the direction x_i at time τ and point $\boldsymbol{\xi}$. Thus, one obtains[14]

$$\begin{aligned} c_{ij}u_j(\boldsymbol{\xi},t) = \int_0^t \{ & \oint_\Gamma [\overset{*}{u}{}_j^{(i)}(\mathbf{x},\boldsymbol{\xi},t-\tau)T_j(\mathbf{x},\tau) - \overset{*}{T}{}_j^{(i)}(\mathbf{x},\boldsymbol{\xi},t-\tau)u_j(\mathbf{x},\tau)]d\Gamma_{\mathbf{x}} \\ & + \int_\Omega \overset{*}{u}{}_j^{(i)}(\mathbf{x},\boldsymbol{\xi},t-\tau)b_j(\mathbf{x},\tau)d\Omega_{\mathbf{x}}\}d\tau \\ & + \rho\int_\Omega [\bar{v}_{jo}(\mathbf{x})\overset{*}{u}{}_j^{(i)}(\mathbf{x},\boldsymbol{\xi},t) - \bar{u}_{jo}(\mathbf{x})\overset{*}{v}{}_j^{(i)}(\mathbf{x},\boldsymbol{\xi},t)]\, d\Omega_{\mathbf{x}}, \end{aligned} \tag{6}$$

where $c_{ij} = 0.5\delta_{ij}$ for points $\boldsymbol{\xi}$ on a smooth boundary Γ, while $c_{ij} = \delta_{ij}$ holds for interior points $\boldsymbol{\xi}$. The appropriate impulse solution is given by the response function

$$\begin{aligned} \overset{*}{u}{}_j^{(i)}(\mathbf{x},\boldsymbol{\xi},t') = \frac{1}{2\pi\rho}\{ & \frac{H(c_1t'-r)}{c_1r^2}[\frac{2c_1^2t'^2-r^2}{R_1} r_{,i}r_{,j} - R_1\delta_{ij}] \\ & - \frac{H(c_2t'-r)}{c_2r^2}[\frac{2c_2^2t'^2-r^2}{R_2} r_{,i}r_{,j} - (R_2+\frac{r^2}{R_2})\delta_{ij}]\}, \end{aligned} \tag{7}$$

where

$$R_\alpha = (c_\alpha^2 t'^2 - r^2)^{1/2}; \quad \alpha=1,2 \; ; \quad t' = t-\tau , \tag{8}$$

and H is the Heaviside step function guaranteeing the causality of the motion.

The corresponding singular tractions $\overset{*}{T}{}_j^{(i)}$, determined by using relation (4), contain derivatives of the Heaviside function and, in order to eliminate these, integration by parts with respect to time is carried out. Assuming zero body forces and initial conditions, the result of this regularization procedure is the following system of integro - differential equations[14]

$$c_{ij}u_j(\boldsymbol{\xi},t) = \int_o^t\oint_\Gamma [\, \overset{*}{u}{}_j^{(i)}(\mathbf{x},\boldsymbol{\xi};t')T_j(\mathbf{x},\tau) + \overset{*}{P}{}_j^{(i)}(\mathbf{x},\boldsymbol{\xi},t')u_j(\mathbf{x},\tau) - \overset{*}{Q}{}_j^{(i)}(\mathbf{x},\boldsymbol{\xi},t')\dot{u}_j(\mathbf{x},\tau)]d\Gamma_{\mathbf{x}}d\tau, \quad (9)$$

where

$$\overset{*}{P}{}_j^{(i)}(\mathbf{x},\boldsymbol{\xi},t') = \sum_{\alpha=1}^{2}(-1)^{\alpha}\frac{H(c_\alpha t'-r)}{2\pi c_\alpha R_\alpha}\,[\,2\overset{o}{a}_{ij}(r)\frac{2c_\alpha^2 t'^2-r^2}{r^3} + \overset{\alpha}{a}_{ij}(r)\frac{r-c_\alpha t'}{R_\alpha^2} + (\overset{1}{a}_{ij}(r)+\overset{2}{a}_{ij}(r))\frac{c_\alpha t'}{r^2}] \quad (10)$$

$$\overset{*}{Q}{}_j^{(i)}(\mathbf{x},\boldsymbol{\xi},t') = \sum_{\alpha=1}^{2}(-1)^{\alpha}\frac{H(c_\alpha t'-r)}{2\pi c_\alpha^2}\,[\,(\overset{1}{a}_{ij}(r)+\overset{2}{a}_{ij}(r))\frac{R_\alpha}{r^2} + \overset{\alpha}{a}_{ij}(r)\frac{1}{R}\,], \quad (11)$$

with

$$\overset{o}{a}_{ij}(r) = c_2^2(n_i r_{,j} + n_j r_{,i} + \delta_{ij} r_{,n} - 4r_{,i}r_{,j}r_{,n})$$

$$\overset{1}{a}_{ij}(r) = 2c_2^2 r_{,i}r_{,j}r_{,n} + (c_1^2-2c_2^2)n_j r_{,i}$$

$$\overset{2}{a}_{ij}(r) = c_2^2(2r_{,i}r_{,j}r_{,n} - r_{,j}n_i - r_{,n}\delta_{ij}), \quad (12)$$

The only singularities in these integro - differential equations are those that occur when r and t' tend to zero simultaneously. When $r \to 0$ and $c_\alpha t' \neq 0$ only pseudo-singularities appear which vanish when contributions from similar terms referring to dilatational and shear waves are calculated simultaneously[13]. Thus, these time-dependent equations can be integrated without difficulty, provided that appropriate shape-functions are used.

NUMERICAL IMPLEMENTATION

Equation (9) can be used to determine time-dependent displacements in the interior of Ω as well as along the boundary Γ. However, one must first find values for the unkown boundary tractions **t** and displacements **u**. The treatment thus consists of[14,22]

(i) a discretization of the boundary in which displacements and tractions are assumed to be constant over each boundary element Γ_l.

(ii) a step-by-step integration in time where displacements and tractions are taken to be linear and constant over each time interval Δt, respectively.

In order to arrive at systems of algebraic equations, collocation is used at every node $\boldsymbol{\xi}_\lambda$ and at all time steps $t_m = m\Delta t$. Then, integrations over each time interval and over each boundary element have to be carried out, according to (i) and (ii), e.g.

$$\mathbf{U}^{nm} := \left(U_{j\ 1\lambda}^{(i)nm} \right) = \int_{(n-1)\Delta t}^{n\Delta t} \int_{\Gamma_1} U_j^{(i)}(\mathbf{x},\xi_\lambda,m\Delta t-\tau)d\Gamma_{\mathbf{x}}d\tau \ . \qquad (13)$$

The integrations with respect to time have been performed analytically; the results cannot be given here for lack of space, but they may be found in the paper of Antes[14]. The integrals over each boundary segment have been evaluated approximately using Gaussian quadrature. Finally, we obtain the discrete analogues of equations (9). They can be cast in the following form[14]

$$\hat{\mathbf{T}}\cdot\mathbf{u} = (0.5\mathbf{I} + \mathbf{T})\cdot\mathbf{u} := \sum_m (0.5\mathbf{I} + \mathbf{T}^m)\cdot\mathbf{u}^m = \sum_m \mathbf{U}^m\cdot\mathbf{t}^m =: \mathbf{U}\cdot\mathbf{t}. \qquad (14)$$

According to the causality condition the matrices $\mathbf{T}$ and $\mathbf{U}$ are lower triangular, where, if all time-steps Δt have the same duration, all blocks along the diagonals are equal (see Figure 1). Thus, for each time step t_m, it is necessary to determine only one extra blockmatrix $\mathbf{U}^m$ and $\mathbf{T}^m$.

$$\begin{bmatrix} U^{11} & & & \\ U^{21} & U^{22} & & \mathbf{0} \\ U^{31} & U^{32} & U^{33} & \\ \vdots & \vdots & \ddots & \\ U^{n1} & U^{n2} & \cdots & U^{nn} \end{bmatrix} = \begin{bmatrix} U^{1} & & & \\ U^{2} & U^{1} & & \mathbf{0} \\ U^{3} & U^{2} & U^{1} & \\ \vdots & \vdots & \ddots & \\ U^{n} & U^{n-1} & \cdots & U^{1} \end{bmatrix}$$

Figure 1. Structure of the U-matrix

COUPLING ALONG INTERFACES

In the analysis of the dynamic response of a foundation on a layered soil or on a single layer - halfspace system the equations of motion governing the coupled system must be solved simultaneously, and suitable conditions must be prescribed along the interfaces.

Assuming completely bonded contact, the following conditions have to be satisfied at each point of the interface Γ_I between the two deforming subregions Ω_A and Ω_B

(i) compatibility of the motions, i.e. the components of the motion of subregion Ω_A must coincide with those of the subregion Ω_B.

(ii) equilibrium of tractions, i.e. the components of the tractions acting on the subregion Ω_A must be opposite to those on the subregion Ω_B.

In order to incorporate these conditions, the systems of boundary element equations encompassing the subregions Ω_A and Ω_B have to be combined. After reordering the entire set of equations we obtain a coupled matrix form at each new time step, e.g. in the case of a single layer - halfspace (S - H) system with prescribed tractions on the surface (see Figure 10)

$$\begin{bmatrix} \hat{\mathbf{T}}^S_{oo} & \mathbf{T}^S_{oI} & -\mathbf{U}^S_{oI} \\ \mathbf{T}^S_{Io} & \hat{\mathbf{T}}^S_{II} & -\mathbf{U}^S_{II} \\ \mathbf{O} & \hat{\mathbf{T}}^H_{II} & \mathbf{U}^H_{II} \end{bmatrix} \cdot \begin{bmatrix} \mathbf{u}^S_o \\ \mathbf{u}^S_I \\ \mathbf{t}^S_I \end{bmatrix} = \begin{bmatrix} \mathbf{U}^S_{oo} \\ \mathbf{U}^S_{Io} \\ \mathbf{0} \end{bmatrix} \cdot \begin{bmatrix} \bar{\mathbf{t}}^S_o \end{bmatrix} \quad \begin{matrix} \text{for } \xi_\lambda \varepsilon \Gamma^S_o \\ \text{for } \xi_\lambda \varepsilon \Gamma^S_I \\ \text{for } \xi_\lambda \varepsilon \Gamma^H_I . \end{matrix} \quad (14)$$

In (14) $\Gamma^S_I = \Gamma^H_I$ denotes the interface and Γ^S_o stands for the non-common "outer" boundary, the surface of the single-layer.

DAMPING OF EMBEDDED FOUNDATIONS

Consider now a rigid, massless strip-foundation of width 2B = 2 m (6.56 ft) bonded to a supporting homogeneous soil medium with shear modulus $G = 10^5$ KN/m^2 (14.5$\cdot 10^3$ psi), mass density ρ = 2000 kg/m^3 (3.867 lb/sec^2/ft^4) and Poisson's ratio $\nu = 0.33$. Under plane strain conditions the wave propagation velocities are c_1 = 443.9 m/s (1456 ft/sec) and c_2 = 223.6 m/s (733 ft/sec).

The soil-foundation interface is discretized into four equal elements at the basemat and two or four elements at each vertical face (T/B=1 or T/B=2) while at both sides of the foundation, i.e. at the free soil surface, four elements of the same length are used (see Fig. 2).

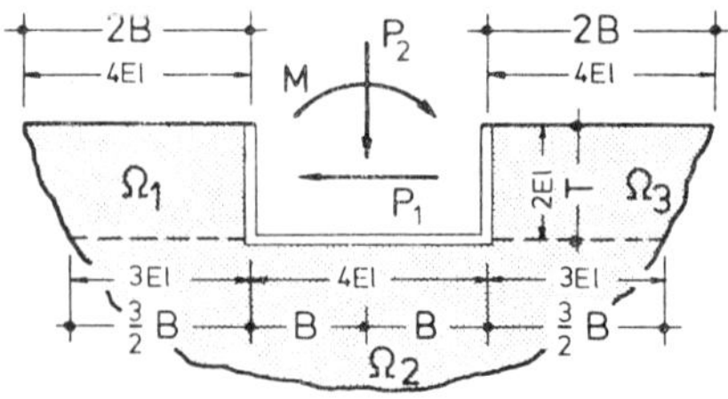

Figure 2. Geometry and discretization of an embedded foundation (T/B=1)

This foundation has been subjected to a horizontal, vertical and rocking rectangular impulse of intensity $P_h = P_v = 1776$ KN/m ($1.217 \cdot 10^5$ plf) and M = 1776 KNm/m (3.99^3 lb), respectively, and a time duration of $\Delta t = 5.63 \cdot 10^{-3}$ secs has been chosen. In Figures 3 to 5 the time history ($\tilde{t} = tc_2/B$) of the responses ($f_\alpha = Gu_\alpha$, $f_\varphi = GB^2 u_\varphi$) are plotted for the first 80 time-steps.

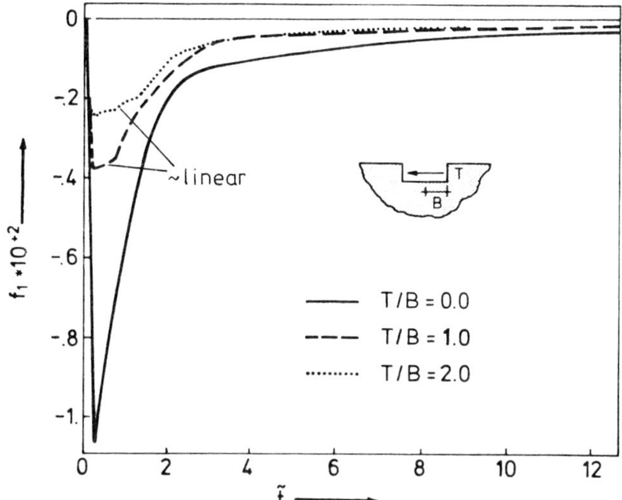

Figure 3. Dependence of horizontal impulse response on embedment depth

First, it is evident that, since the intensity of the acting impulse remains constant, the amplitudes at the beginning of the responses decrease with increasing contact area.

The second aspect to be noted is the influence of the embedment depth on the radiation damping, that is the energy transmitted into the soil and dissipated by waves propagating outward and downward. These waves are generated at every point on the soil-foundation interface so that, in general, the damping increases with increasing area of contact. Yet, the magnitude of this increase is dependent on the type of the impulse and on the geometry of the foundation. Thus, for surface foundations a horizontal impulse produces mainly shear waves whereas a vertical as well as a rocking impulse generates mainly compression - extension waves, and for swaying embedded foundations both types of waves appear.

Since - as we know - p-waves radiate only 7-8 percents of the energy while s-waves and Rayleigh waves radiate approximately 28 and 67 percents, respectively, the change in wave production by embedding the foundations must also alter the amount of decrease of the responses.

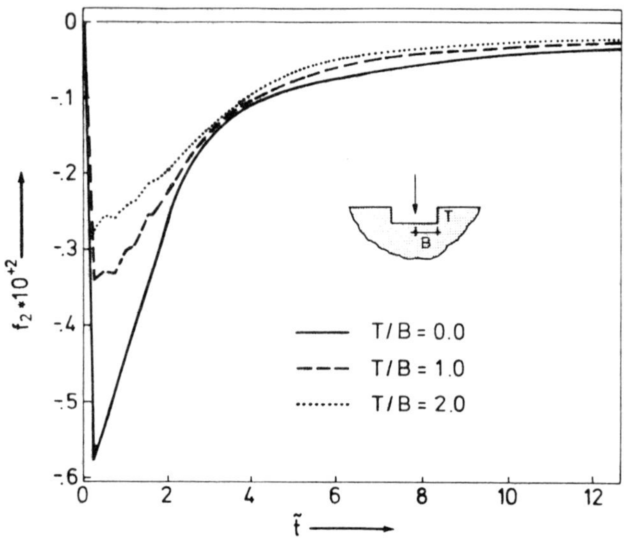

Figure 4. Dependence of vertical impulse response on embedment depth

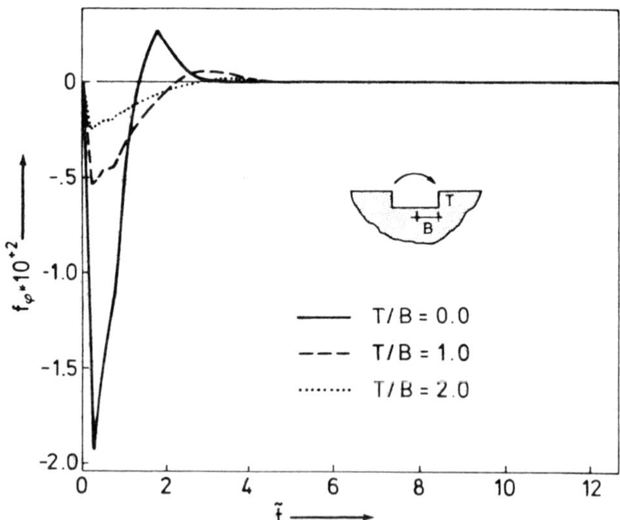

Figure 5. Dependence of rocking impulse response on embedment depth

This is obvious, for example, from the time history of the horizontal response due to a horizontal impulse (Fig. 3). While the response of the surface foundation decreases exponentially from the beginning due to the predominance of the shear waves, the response of an embedded foundation decreases only linearly during the first time steps because of the p-waves produced by the side walls. Later on, refractions on the free surface produce shear and Rayleigh waves so that the energy radiation causes an exponential decrease of the response in all cases.

The time history of the vertical response shows a similar behavior. Here the vertical impulse to the surface foundation produces mainly p-waves so that in the beginning the response of the surface foundation decreases only linearly.

REFLECTION AND REFRACTION IN LAYERED SOIL

Consider now a rigid, massless strip-foundation of width 2B = 2 m (6.56 ft) bonded to a homogeneous elastic soil-layer of thickness H over a rigid bedrock. Using the same material data as before, this surface foundation is subjected to horizontal and vertical impulsive forces and rocking moment of intensity 740 KN/m ($5.07 \cdot 10^4$ plf) and 740 KNm/m ($1.663 \cdot 10^5$ lb), respectively. The duration of one time step is taken to be $\Delta t = 1.3516 \cdot 10^{-3}$ secs.

The discretization consists of four equal elements for the layer-foundation interface, three elements of the same length at both sides of the foundation and ten equal elements for the bedrock (see Fig. 6).

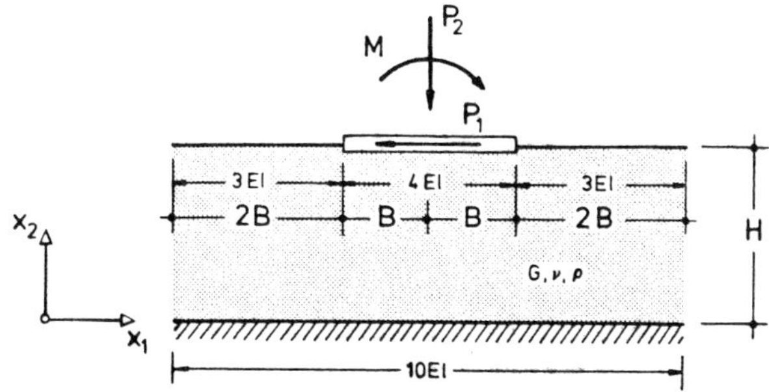

Figure 6. Geometry and discretization of a surface foundation on an elastic layer

In Fig. 7, the time-dependent horizontal, vertical and rocking responses are plotted for 60 time-steps, that is for the first 0.081 secs. Obviously, after the time required by the s-waves and p-waves to traverse the distance from the foundation to the bedrock and back, reflections become noticeable which induce vibrating motions of the rigid foundation. Also, radiation of energy to infinity in the horizontal direction damps the responses, but much more slowly than in the case of an elastic halfspace considered before.

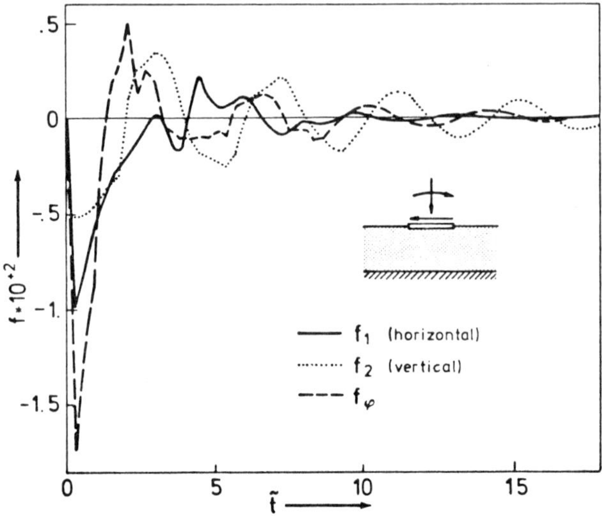

Figure 7. Influence of a rigid bedrock on time-dependent responses

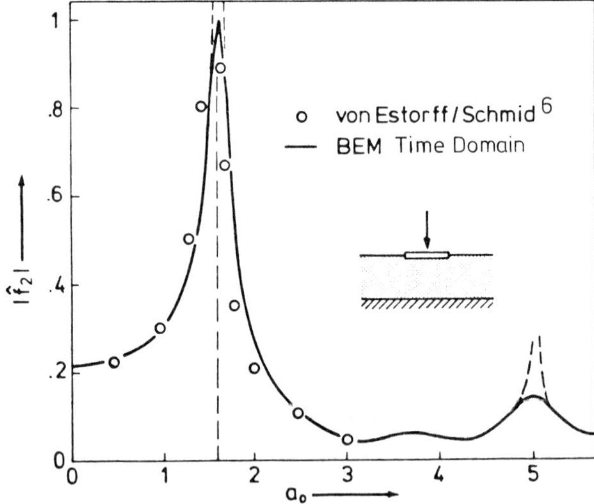

Figure 8. Amplitudes of vertical motion of a surface foundation on a soil layer (H/B = 2) versus frequency

In order to check the time-dependent responses, the time history of the vertical impulse solution has been transformed by Fast-Fourier transforms into the frequency domain. The result is presented in Figure 8. It shows the continuous variation of the nondimensionalized amplitude of the vertical motion over a frequency range of $0 < \omega < 1230$ rad/sec ($\omega = a_o c_2/B$) along with values for distinct frequencies (marked by circles) which were determined using a frequency domain integral equation[6]. The agreement of both approximations is excellent so that the time domain solution procedure is shown to produce a correct time history approximation.

Next, in a parametric study the influence of the layer thickness H has been examined. Figure 9 demonstrates clearly the expected and already mentioned fact: the thicker the layer, the more the response approaches the 'halfspace solution (H/B=∞)', and the later reflections due to the rigid bedrock give rise to vibrating motions.

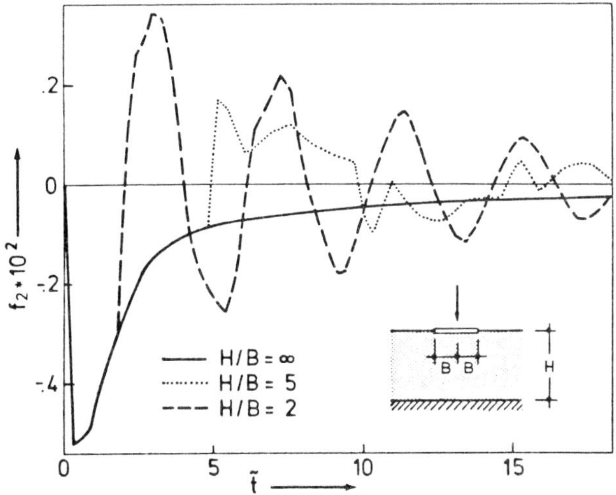

Figure 9. Influence of layer thickness to the vertical response of a surface foundation

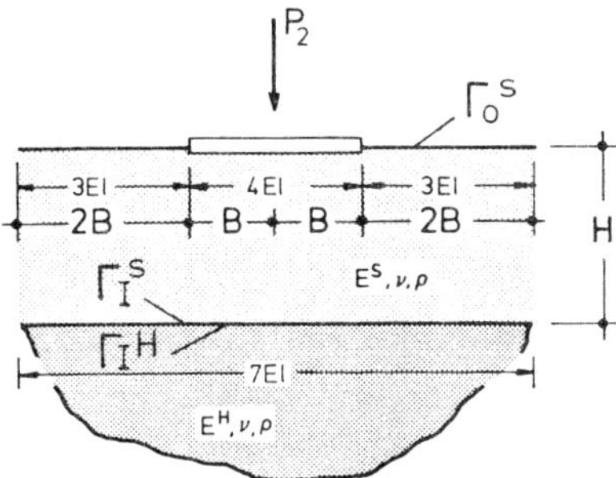

Figure 10. Geometry and discretization of a surface foundation on a layer over the halfspace

Of even more practical importance is the case when the surface foundation is bonded to a layer over an elastic halfspace, i.e. when the material data of the layer and of the halfspace are different (see Fig. 10). The interface between the elastic layer and the halfspace is discretized into seven equal elements while the same discretization as before has been used for the free surface and the foundation.

In Figure 11, the time history of the vertical displacement of the same impulsively excited surface foundation as in Figs. 9 and 10 is presented. The curves given there correspond to a range of ratios of Young's moduli for the layer $E^S = 2.66 \cdot 10^5$ KPa ($3.858 \cdot 10^4$ psi) and the halfspace E^H. The rest of the material data remain unchanged.

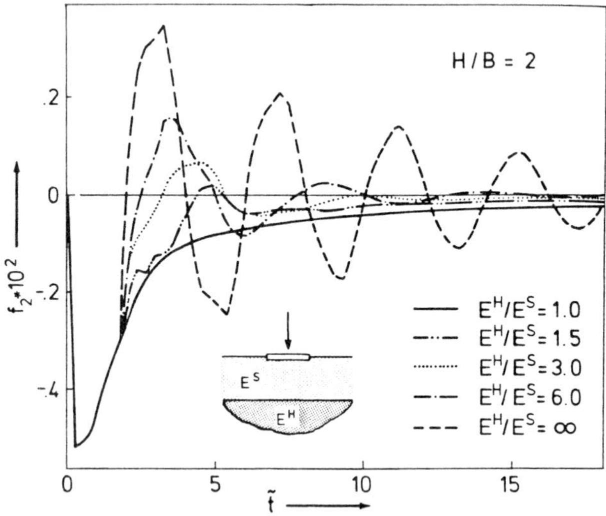

Figure 11. Influence of ratios of Young's moduli on the vertical response of an impulsively excited surface foundation

It is obvious that the reflection and refraction generates motions which vary between the response for the halfspace ($E^H/E^S = 1$) and that for the rigid bedrock ($E^H/E^S = \infty$). This study clearly demonstrates that the softer the halfspace under the layer, the greater is the damping.

WAVE FOCUSSING BY SPECIAL SOIL PROFILES

In special bedrock geometries it happens that the reflected waves will be focussed so that the responses of the foundations will be amplified. In order to demonstrate this behavior, we again consider the surface strip-foundation on a soil layer and merely change the geometry of the bedrock. While the layer thickness H at both open sides remains unchanged (H/B=2), the bedrock under the foundation has now a circle-shaped trough (radius r=3.625 B, see Fig. 12).

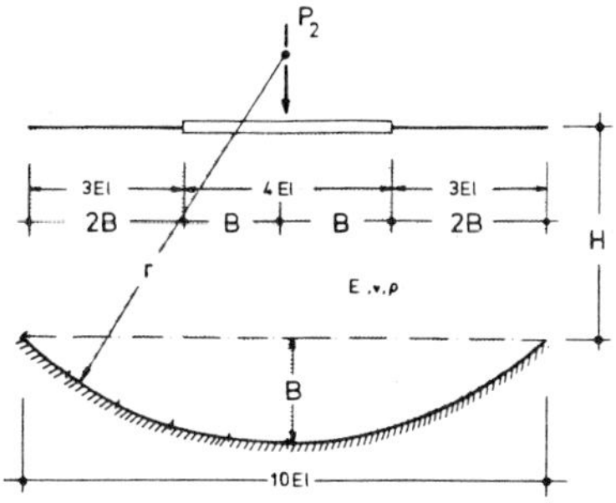

Figure 12. Geometry and discretization of a surface foundation on an elastic layer with a basin-shaped bedrock

In Figure 13, the comparison of the time history of the vertical response of the foundation, excited by a vertical rectangular impulse, for a layer with constant thickness and for the layer with a basin-shaped bedrock is plotted. It shows significant differences, although energy can radiate to infinity out of equal 'openings'. Thus, the only interpretation for this response amplification is wave focussing.

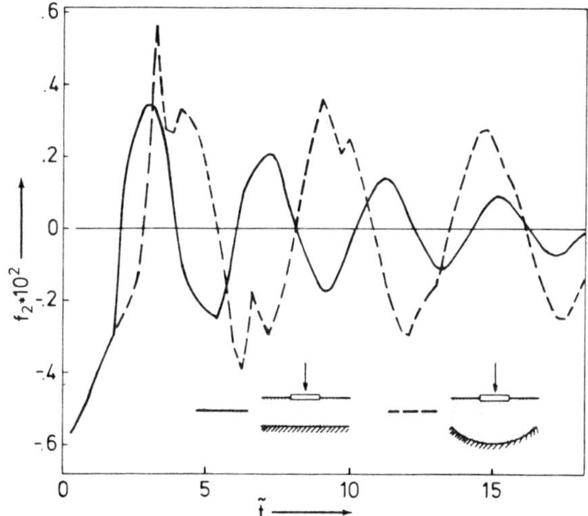

Figure 13. Influence of basin-shaped bedrock on the vertical response of an impulsively excited surface foundation

CONCLUSIONS

The direct time domain BEM has been used for the determination of the transient reponse of massless rigid strip-foundations under conditions of plane strain. Studies have been conducted for surface and embedded foundations for various types of soil profiles. In all cases it was observed that the damping of the motions is highly dependent on the soil layer and bedrock geometry.

ACKNOWLEDGEMENTS

This work was supported by the Deutsche Forschungsgemeinschaft (SFB 151, project C1).

REFERENCES

1. Kausel, E. and Tassoulas, J.L. (1981), Transmitting Boundaries: a Closed Form Comparison. Bull. Seism. Soc. Am., Vol. 71, pp. 143-159
2. Chow, Y.K. and Smith, I.M. (1982), Infinite Elements for Dynamic Foundation Analysis, in Numerical Methods in Geomechanics (Ed. Eisenstein, Z.) Vol. 1, pp. 15-22, Proc. of the Conf., Edmonton, Balkema, Rotterdam.
3. Dasgupta, G. (1982), A Finite Element Formulation for Unbounded Homogeneous Continua, J. Appl. Mech. ASME, Vol. 49, pp. 136-140.
4. Dominguez, J. (1978), Dynamic Stiffness of Rectangular Foundations, Publ. No. R78-20, Dept. Civil Engg., M.I.T.
5. Niwa, Y., Kobayashi, S. and Fukui, T. (1976), Applications of Integral Equation Method to Some Geomechanical Problems, in Numerical Methods in Geomechanics (Ed.: Desai, C.S.) pp. 120-131, ASCE, New York
6. Estorff, O. von and Schmid, G. (1984), Applications of the Boundary Element Method to the Analysis of the Vibration Behavior of Strip Foundations on a Soil Layer (Eds.: Beskos, D.E., Krauthammer, Th. and Vardoulakis, I.) pp. 11-17, Proc. Int. Symp. Dynamic Soil-Structure Interaction, Minneapolis, USA, Balkema, Rotterdam - Boston
7. Abascal, R. and Dominguez, J. (1985), Dynamic Response of Embedded Strip Foundations Subjected to Obliquely Incident Waves, in Boundary Elements VII (Eds.: Brebbia, C.A. and Maier, G.) pp. 6-63 to 6-69, Proc. of the 7th Int. Conf., Como, Italy, Springer Verlag, Berlin-New York
8. Wolf, J.P. and Obernhuber, P. (1985), Nonlinear Soil-Structure-Interaction Analysis using Green's Functions of Soil in the Time Domain, Earthqu. Engg. Struct. Dyn., Vol. 13, pp. 213-223
9. Wolf, J.P. and Darbre, G.R. (1986), Non-linear Soil-Structure Interaction Analysis based on the Boundary -Element Method in

Time Domain with Application to Embedded Foundation, Earthqu. Engg. Struct. Dyn., Vol. 14, pp. 83-101

10. Karabalis, D.L. and Beskos, D.E. (1984), Dynamic Response of 3-D Rigid Surface Foundations by Time Domain Boundary Element Method, Earthqu. Engg. Struct. Dyn., Vol. 12, pp. 73-93

11. Spyrakos, C.C. and Beskos, D.E. (1985), Dynamic Response of Rigid Strip-Foundations by Time Domain Boundary Element Method, Int. J. Num. Meth. Engng., Vol. 23, pp. 1547-1565

12. Spyrakos, C.C., Patel, P. and Beskos, D.E. (1986), Dynamic Analysis of Flexible Embedded Foundations: plane Strain Case, Proc. 3rd. Int. Conf. Compt. Meth. Exper. Meas., Porto Carras, Greece

13. Mansur, W.J. (1983), A Time-Stepping Technique to Solve Wave Propagation Problems Using the Boundary Element Method, Ph. D. Thesis, Southampton

14. Antes, H. (1985), A Boundary Element Procedure for Transient Wave Propagations in Two-Dimensional Isotropic Elastic Media, Finite Elem. Anal. Design, Vol. 1, pp. 313-322

15. Spyrakos, C.C. and Antes, H. (1986), Time Domain Boundary Element Method Approaches in Elastodynamics: A Comparative Study, Computers & Structures, Vol. 24, pp. 529-535

16. Antes, H. and Spyrakos, C.C. (1986), Boundary Element procedures for Transient Analysis of Soil-Structure Interaction Problems, (Ed.: Kounadis, A.N.) , Proc. 1st. Nat. Congr. Mechanics, Athens, Greece

17. Antes, H. and Estorff, O. von (1986), Dynamic Soil-Fluid Interaction Analysis by the Boundary Element Method, in BETECH 86 (Eds.: Brebbia, C.A. and Connor, J.J.), Proc. 2nd. Bound. Elem. Techn. Conf., M.I.T., Boston, CM Publ.

18. Antes, H. and Estorff, O. von (1987), Analysis of Absorption Effects on the Dynamic Response of Reservoir Systems by Boundary Element Methods, Earthqu. Engng. Struct. Dyn. (accepted)

19. Antes, H. and Estorff, O. von (1986), Dynamic Soil-Structure Interaction by BEM in the Time and Frequency Domain, Vol. 2, pp. 5.5-33 to 5.5-40, Proc. 8th. Europ. Conf. Earthqu. Engng., Lab. Nac. Eng. Civil, Lisbon, Portugal

20. Antes, H. and Estorff, O. von (1987), Dynamic Response Analysis of Rigid Foundations and Elastic Structures by Boundary Element Procedures, Soil Dyn. Earthqu. Engng. (submitted)

21. Graffi, D. (1947), Sul Teorema di Reciprocita nella Dinamica dei Corpi Elastici, Mem. Accad. Sci. Bologna, Vol. 18, pp. 103-109

22. Estorff, O. von (1986), Analysis of dynamic interaction between structures and surrounding media by time-dependent BEM (in German), TWM 86-10, Inst. Konstr. Ingenieurbau, Ruhr-Universität Bochum, Germany

Frequency Domain Analysis of Two-Dimensional Wave Propagation with Applications to Earthquake Engineering

P.K. Hadley, A. S. Çakmak
Department of Civil Engineering, Princeton University, U.S.A.
S. Altay
Ebasco Corporation, U.S.A.
A. Askar
Department of Mathematics, Bogazici University, Istanbul, Turkey

ABSTRACT

A frequency domain analysis is presented for the propagation of SH-waves in a homogeneous two-dimensional medium. Using previously presented boundary element techniques for computing steady-state, harmonic response[1] [2] [3] the admittance function for several points in various geometries is computed for a range of frequencies. These geometries can contain differing topographies and cavities such as hills and tunnels.

The incident earthquake is transformed into the frequency domain by a fast Fourier transform [7], convoluted with the admittance function, and rendered into the time domain.

Examples of the technique are given for several different geometries subjected to two different earthquake time histories. The plots presented for these cases show the reasonableness of the results and thereby imply the correctness of the procedure which generated them.

An effort is made in understanding the effect of the different topographies and tunnels on wave-scattering patterns.

INTRODUCTION

Results have been reported for the steady-state scattering of harmonic waves in a homogeneous medium[1] [2] [3] The ultimate purpose of these efforts is to conduct a frequency domain analysis of earthquake wave propagation. This paper describes the implementation of the frequency-domain analysis and presents several examples of its use. The problem of earthquake wave propagation is solved for several different profile geometries containing surface irregularities and cavities subjected to two incident time series--a Gaussian pulse and the San Fernando earthquake.

FORMULATION

Newton's second law, when applied to an infinitessimal of elastic material, becomes the familiar wave equation. By subjecting this to a Fourier transform, the time dependence of the equation is separated out and the remaining problem is to solve the reduced wave equation for the response of the medium to a steady-state harmonic input. When this response is computed for all frequencies, the results are convoluted with the transform of the actual incident wave and a solution in the time domain is achieved.

The wave equation for a homogeneous, elastic medium is

$$\nabla^2 \underset{\sim}{u} = \frac{1}{c^2}\frac{\partial^2 \underset{\sim}{u}}{\partial t^2} \tag{1}$$

where $\underset{\sim}{u}$ = displacement and c = wave velocity. In order to remove the dependence on time, a Fourier transform is applied

$$\int_{t=-\infty}^{\infty} \nabla^2 \underset{\sim}{u}\, e^{i\omega t} dt = \nabla^2 \underset{\sim}{\bar{u}}$$

and (2)

$$\int_{t=-\infty}^{\infty} \frac{1}{c^2}\frac{\partial^2 \underset{\sim}{u}}{\partial t^2} e^{i\omega t} dt = \frac{\omega^2}{c^2}\underset{\sim}{\bar{u}}$$

yielding the reduced wave equation

$$\nabla^2 \underset{\sim}{\bar{u}} + \xi^2 \underset{\sim}{\bar{u}} = 0 \tag{3}$$

where $\underset{\sim}{\bar{u}}$ = the steady-state displacement vector and $\xi = \frac{\omega}{c}$ = the wave number.

For the two-dimensional case, $\frac{\partial}{\partial z} = \frac{\partial}{\partial x_3} = 0$. If consideration is limited to antiplane strain (SH-waves), then $\bar{u}_1 = \bar{u}_2 = 0$. Thus, for SH-waves considered in a two-dimensional medium, equation (3) becomes

$$\nabla^2 \bar{w} + \xi^2 \bar{w} = 0 \tag{4}$$

where $\bar{w} = \bar{u}_3$ = the harmonic displacement in the z-direction.

The Green's function for this reduced anti-plane wave equation, given by Eringen and Suhubi (4) and Mow and Pao(5), is

$$g(\underset{\sim}{x}, \underset{\sim}{x}') = \frac{iH_0^{(1)}(\xi\bar{r})}{4} \tag{5}$$

where $\underset{\sim}{x}$ = the point of obversation, $\underset{\sim}{x}'$ = the point of integration, $H_0^{(1)}$ = the Hankel function of 0th order and the first kind, and $\bar{r} = |\underset{\sim}{x} - \underset{\sim}{x}'|$ = distance between $\underset{\sim}{x}$ and $\underset{\sim}{x}'$. Thus, the differential equation in equation (4) becomes the integral equation

$$\int\int_{V'}(g\frac{\partial\bar{w}}{\partial n'} - \bar{w}\frac{\partial g}{\partial n'})dV' = \begin{cases} \bar{w}(\underset{\sim}{x}) \text{ for } \underset{\sim}{x} \in V \\ 0 \quad \text{ for } \underset{\sim}{x} \notin V \end{cases} \tag{6}$$

By applying the divergence theorem to this equation, the following boundary integral expression is produced

$$\int_{S'}(g\frac{\partial\bar{w}}{\partial n'} - \bar{w}\frac{\partial g}{\partial n'})dS' = \begin{cases} \bar{w} \text{ for } \underset{\sim}{x} \in V \\ 0 \text{ for } \underset{\sim}{x} \notin V \end{cases} \tag{7}$$

which gives the displacement coefficient for the harmonic excitation for any point in V as a function of the displacement and outward normal derivatives along the boundary.

To solve the equation, $\underset{\sim}{x}$ is moved to the boundary so that in the numerical formulation, the same points can be both integration and observation points. This process causes the kernel to become singular as $\underset{\sim}{x}' \to \underset{\sim}{x}$. Fortunately, the singularity is integrable and is equal to $0.5\bar{w}$ (5). Thus, the final integral equation, derived by extracting this singularity and interpreting the integral in the Cauchy principal value sense, is

$$0.5\bar{w} + P\int_{S'}(g\frac{\partial\bar{w}}{\partial n'} - \bar{w}\frac{\partial g}{\partial n'})dS' = \begin{cases} \bar{w} \text{ for } \underset{\sim}{x} \in V \\ 0 \text{ for } \underset{\sim}{x} \notin V \end{cases} \tag{8}$$

The total wave field is comprised of the incident wave plus the scattered wave $\bar{w}^t = \bar{w}^i + \bar{w}^s$. Because the incident field comes from infinity, it is not singular while the scattered field is. Therefore, the left side of the above expression is zero for the incident field and $\bar{w}^s$ for the scattered field. Adding the equations for the scattered and incident fields and rearranging produces

$$0.5\bar{w}^t - P\int_{S'}(g\frac{\partial\bar{w}^t}{\partial n'} - \bar{w}^t\frac{\partial g}{\partial n'})dS' = \bar{w}^i \tag{9}$$

By discretizing the domain, this equation can be solved for $\bar{w}^t$ (1) (2) (7) (3).

The solution for the transform, $\bar{w}$, is then convoluted back to the time domain by the inverse Fourier transform

$$w(\underset{\sim}{x}, t) = \frac{1}{2\pi}\int_{\omega=0}^{\infty}\bar{w}(\underset{\sim}{x}, \omega)\bar{u}^i(\omega)e^{-i\omega t}d\omega \tag{10}$$

where $w(\underset{\sim}{x}, t)$ = displacement time history at point $\underset{\sim}{x}$ and $\bar{u}^i(\omega)$ = incident field transform.

Because the transfer function, $\bar{w}(\underset{\sim}{x}, \omega)$, contains the information of the response of the point, $\underset{\sim}{x}$, to a unit incident field at a given frequency and the forward transform of the earthquake $\bar{u}^i(\omega)$, contains the frequency content of the incident earthquake, the inverse transform contains the time history of the response of the particular point of the profile to the incident earthquake.

Because the matrices involved are large, they have the potential of requiring great amounts of computer time to solve. However, these equations seem amenable to solution by a process similar to the Gauss-Seidel iteration scheme. In computing the admittance function for a range of frequencies, the matrices are constructed and solved for an array of closely spaced frequencies. The solution at one frequency is a good first approximation for the solution at the next frequency. Thus, with one iteration a good update is made for the solution at successive frequencies and the whole admittance function is constructed quite efficiently. Using this procedure can also save significantly on storage requirements.

Example

Figure 1 shows the principal steps in computing the time history of point 3 of plot (a) subjected to a vertically incident earthquake as recorded in (b). The surface(s) of the profile are first divided into discreet portions by defining observation points which are marked by o's in plot (a). The resulting segments are then represented by their lengths and their two Gauss quadrature points which are marked by x's. The points (1 to 5) at which time histories are to be generated are then chosen.

After establishing the geometry of the profile, the next step is to create an admittance function for each of the points (1 to 5) for which a time history is to be computed. The admittance function is constructed frequency by frequency where it is necessary to compute the response of all the observation points to a unit incident harmonic wave at each frequency. This is accomplished through the discretization of the domain and the numerical analysis as described previously. These results are shown for two frequencies in plots (c) and (d).

When these values are computed, the responses at the points of interest are kept and the rest are discarded. After the results have been reserved for each of the evenly spaced frequencies from 0 to 157 rad/s ($=50\pi$rad/s$=50$ Hz), each point has an admittance function as shown in plot (e) where the steady-state response of the point is given for each of the harmonic incident frequencies.

These admittance functions (or their interpolates) are then multiplied by the Fourier transform of the incident earthquake, shown in (f), to obtain the transfer function (g) of the point in response to the incident wave. This transfer function is the forward Fourier transform of the response of the point.

Accordingly, the application of the inverse (or backward) transform to the transfer function results in the time history of the response of the point to the given incident earthquake. This response in the time domain is given in plot (h). Thus, within the

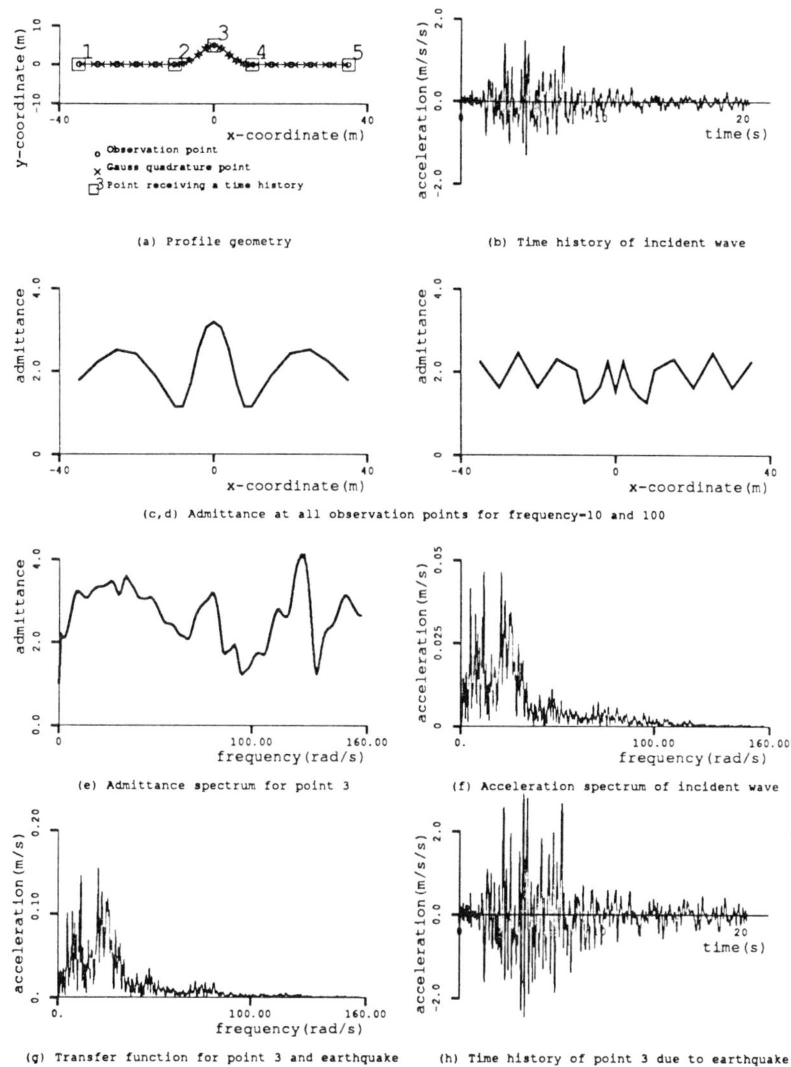

Figure 1: Example of the process

These plots show the major steps in producing a frequency-domain analysis of a point in the Herrera hill profile subjected to the San Fernando earthquake.

limits of the numerical processes, plot (h) contains the response of point 3 in plot (a) to the incident field of plot (b).

Profiles and time histories used in the examples

Figure 2 shows the geometry profiles used in the example problems.

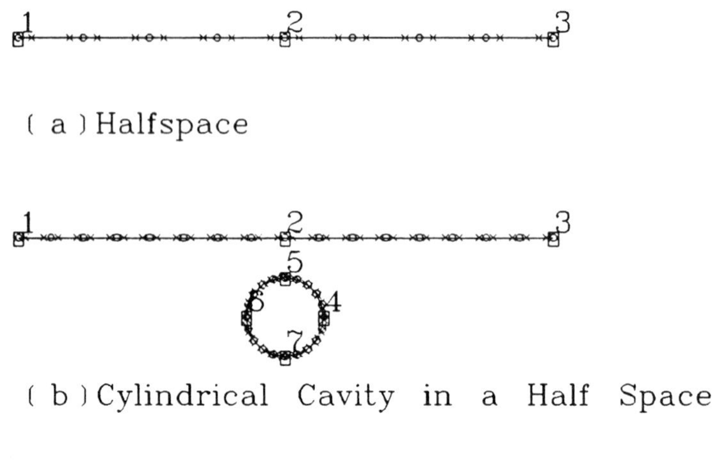

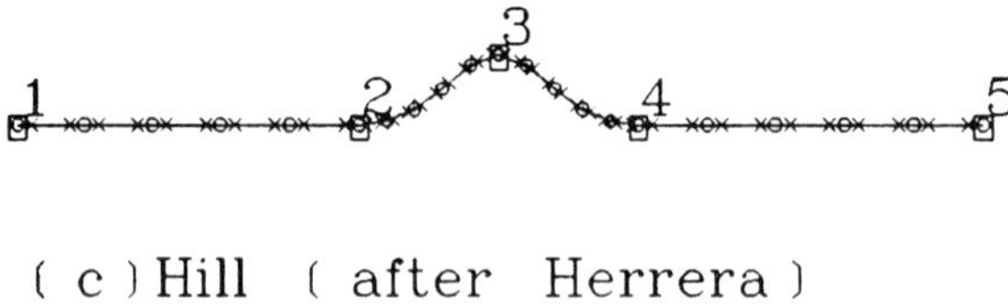

o – Observation points
× – Gauss quadrature point
□1 – Point of time history

Figure 2: Geometry Profiles

All profiles shown here have the surface of the half space truncated at (-35.0,0.0) and (35.0,0.0)

Figure 2(a) is a half-space which is included because of its simplicity. Because the free surface reflects any incident wave with a 180° phase shift, the response of any point in the half-space is easily computed exactly. Any point on the surface itself should have a response double the incident field. The surface of the half space is truncated at ±35m. Figure 2(b) represents a cylindrical cavity in a half-space which can be thought of as a pipeline or subway tunnel in a seismically prone region. The surface of the half space is truncated at ±35m, the circle has its center 10m below the surface and has a radius of 5m. Figure 2(c) is a hill of the type analyzed by Herrera *et al.* [6], and Figure 2(d) is a half space with a valley in the surface. For both the hill and the valley the half space is truncated at ±35m and the hill or valley extends from -10m to 10m and is 5m high (or deep).

All of the profiles are analyzed for a medium with a shear wave velocity of 50m/s. While this is a low number, it does represent a more difficult range of wave numbers in which to do calculations. These examples are chosen as a representative spectrum of the problems which are amenable to solution by the present method.

Figure 3 shows the time histories and Fourier spectra of the two earthquakes to which all of the profiles were subjected. Figure 3(a) is the time history of a Gaussian pulse with a wide Fourier spectrum as shown in Figure 3(b). This clean spike shows how a single pulse of energy would propagate through the profile and affect the several points of interest in it. Figure 3(c) and (d) contain the time history and Fourier spectrum of the east-west trace of the San Fernando (1971) earthquake. The units of both time histories can be considered to be in m/s^2. All of the admittance functions from the above profiles were convoluted with both of these records.

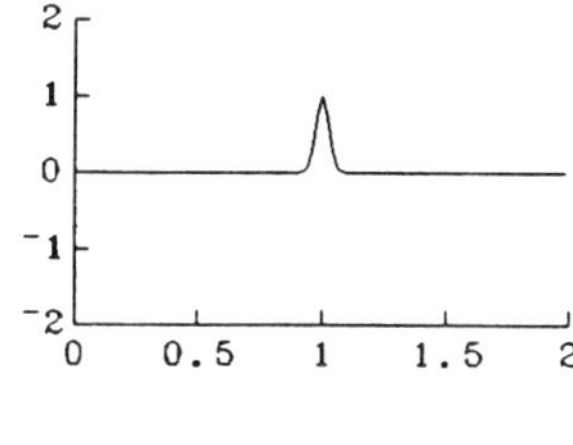

(a) Time history

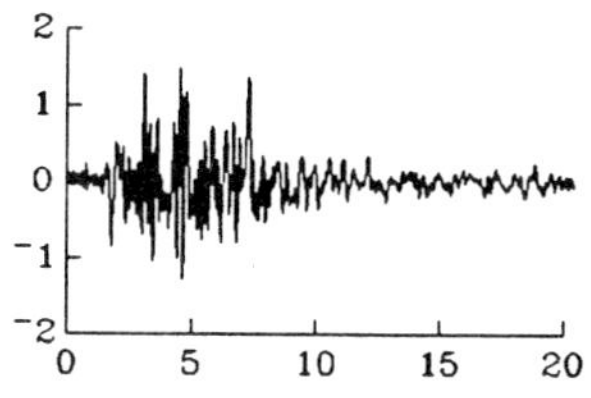

(c) Time history

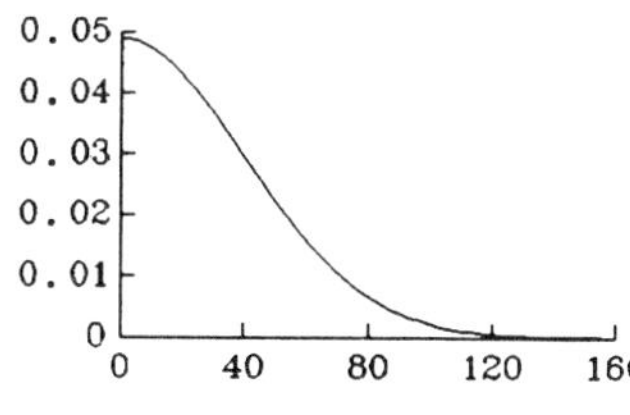

(b) Fourier spectrum

Gaussian Pulse

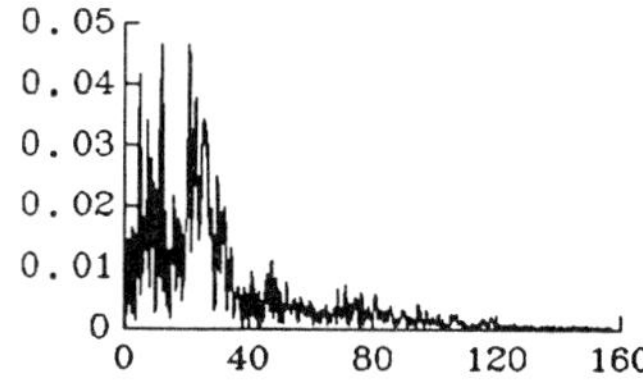

(d) Fourier spectrum

Sanfernando Earthquake

Figure 3: Time History Inputs

EXAMPLES

Figure 4 shows the responses of points on the half-space of Figure 2(a) to the Gaussian pulse of Figure 3(a). Analytical theory predicts that the response of the free surface is to double the amplitude of the response. This is, in fact, what the present techniques also arrive at. Thus, for this simple case where an analytic solution exists, the method is shown to work well. This response is also valuable for comparison to the response of modified geometries presented later.

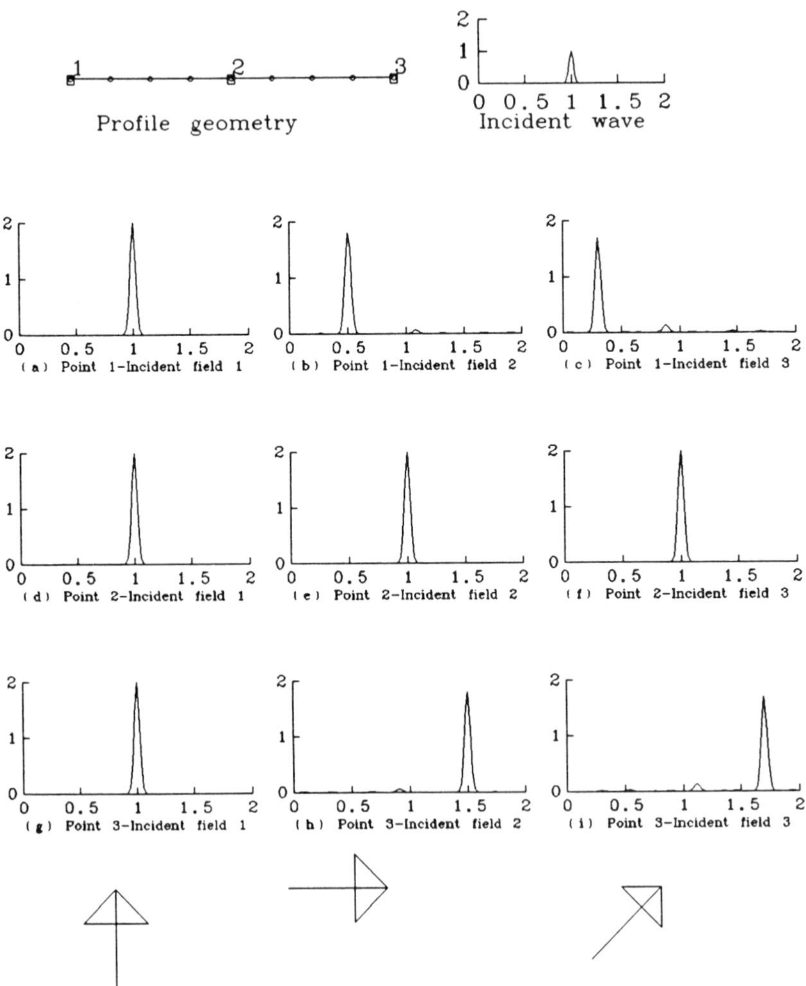

Figure 4: Halfspace of Figure 2(a) subjected to Gaussian pulse of Figure 3(a)

The rows contain the responses of the three points. The three columns represent response to the wave at vertical, 45°, and horizontal incidence.

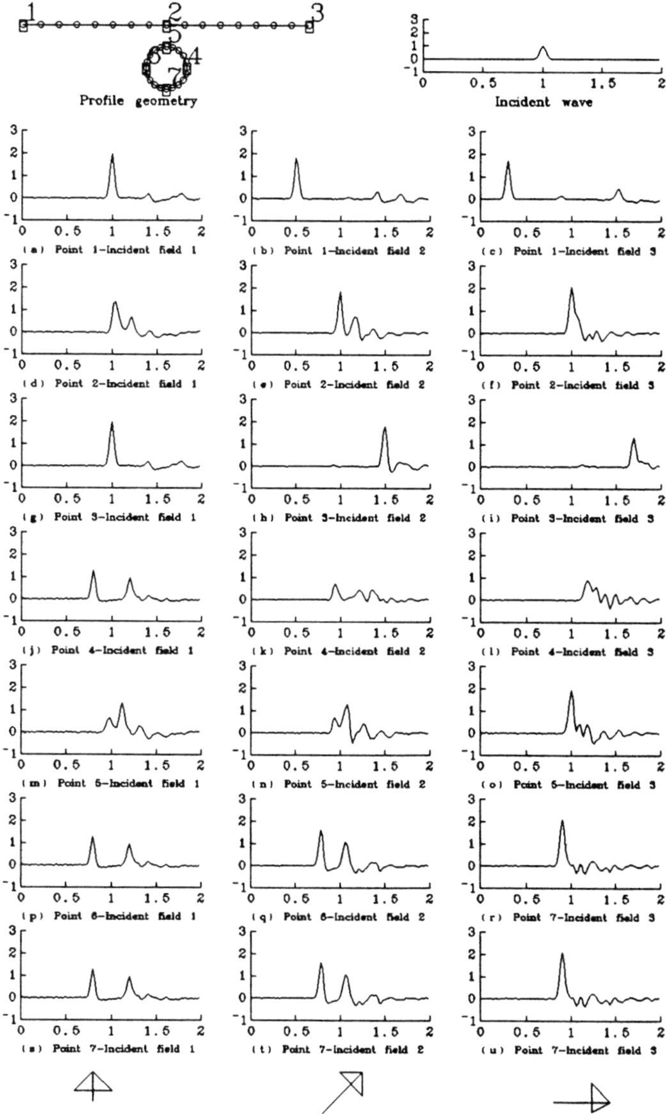

Figure 5: Circle in half-space (Figure 2(a)) subjected to Gaussian pulse (Figure 3(a))

The seven rows contain the responses of the seven points The three columns represent responses to the wave at vertical, 45°, and horizontal incidences.

Figure 5 shows the response of the seven points of interest shown in Figure 2(b) to the Gaussian pulse of Figure 3(a). The first row (a)-(c) shows the response of the right side of the cavity. (a) shows the pulse of energy both as it ascends and as it descends after reflection. (b) shows the response to a wave incident at a 45° angle where the point is shielded by the cavity and, because of the obliqueness of incidence, feels the reflected wave front shortly after the incident. (c), the response to horizontal incidence, shows that incident and reflected wave travel together.

The fifth row shows the responses of point 5 on the surface directly above the cavity. They all demonstrate the amplification of the displacement which one would expect at a flat free surface. (m) also has a slight dip after the peak which could be wave energy trapped between the cavity and the surface. Naturally, this phenonemon is not as pronounced for the two non-vertical incidences shown in (n) and (o).

Figure 6 shows the response of the five points of interest shown in Figure 2(c) to the Gaussian pulse of Figure 3(a). These fifteen plots show the way the presence of the hill modifies the response which one would get in a half-space (Figure 4). One can see in (a) that the response is initially quite similar to the clean response in a half-space but later is slightly perturbed by the energy focused by the hill. For the oblique incidences this energy is not focussed at the top. (d) and (g) show symmetric response to the vertical wave at the symmetrically placed base points of the hill. (e) and (h) show the difference caused by having an asymmetric incident wave and the more diffuse nature of the reflected wave as opposed to the first wave.

Figure 7 shows the response of the five points of interest shown in Figure 2(d) to the Gaussian pulse of Figure 3(a). These fifteen plots show the way the presence of the valley modifies the response which one would get in a half-space (Figure 4). One can see in (a) that the response is initially quite similar to the clean response in a half-space but later is slightly perturbed by the energy focused by the valley.

Figure 8 shows the response of the three points on a half-space Figure 2(a) to the Sanfernando earthquake of Figure 3(c). As can be expected, the effect of the free surface is to double the amplitude of the response. Thus, one can see that the techniques presented in this article work well for this simple case for which an analytic solution exists. This response is also valuable for comparison to the response of modified geometries.

Figure 9 shows the response of the six points of interest shown in Figure 2(b) to the San Fernando earthquake of Figure 3(c).

The second row shows the responses of point 2 on the surface directly above the cavity. They all demonstrate the amplification of the displacement which one would expect at a flat free surface. (d) also has a slight dip after the peak which could be wave energy trapped between the cavity and the surface. Naturally, this phenonemon is not as pronounced for the two non-vertical incidences shown in (e) and (f). The fifth row (m)-(o) shows the response of the right side of the circle. (m) shows the pulse of energy both as it ascends and as it descends after reflection. (n) shows the response to a wave incident at a 45°. angle where the point is shielded by the cavity and, because of the obliqueness of incidence, feels the reflected wave front shortly after the incident. (o), the response to horizontal incidence, shows that incident and reflected wave travel together.

Although this geometry does not possess an analytical solution, the reader should be able to convince himself at least of the reasonableness of the rest of the responses.

Figure 10 shows the response of the three points of interest on the hill shown in Figure 2(c) to the San Fernando earthquake of Figure 3(c). These fifteen plots show

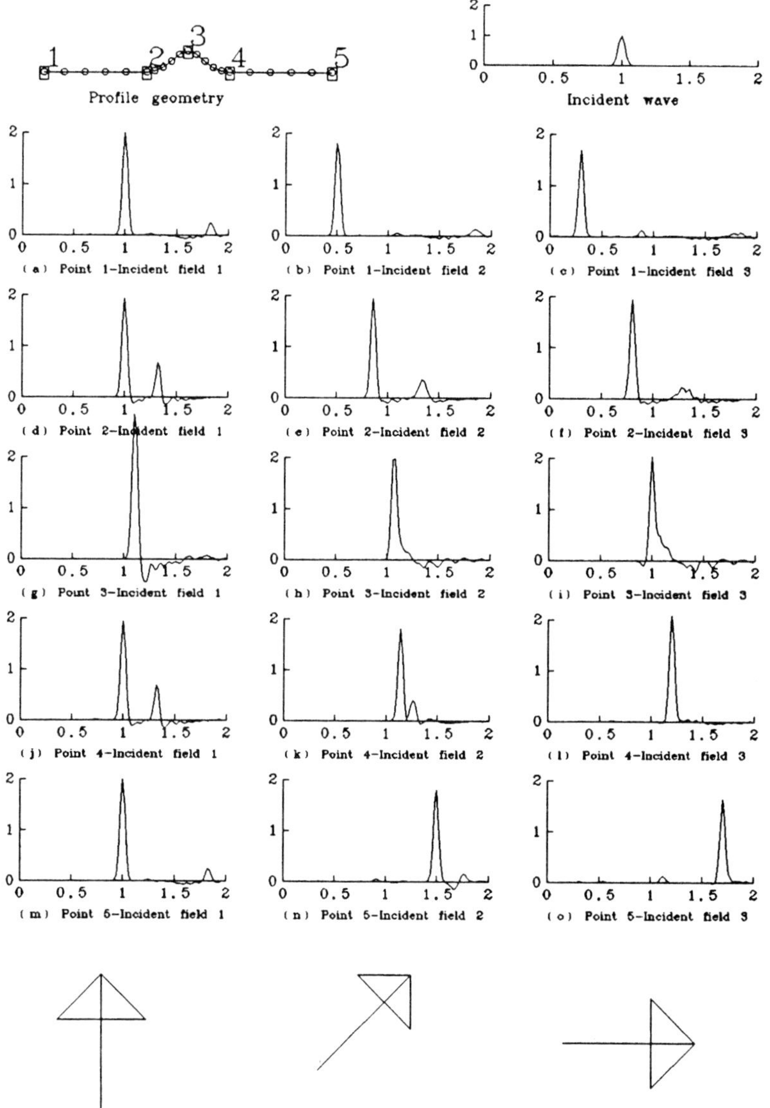

Figure 6: Hill (Figure 2(c)) subjected to Gaussian pulse (Figure 3(a))

The five rows show responses at the left endpoint, the left toe of the hill, the hill top, the right toe, and the right endpoint. The three columns represent response to waves at vertical, 45°, and horizontal incidences.

the way the presence of the hill modifies the response which one would get in a half-space (Figure 8). One can see in (g) that the response is initially quite similar to the clean response in a half-space but later is slightly perturbed by the energy focused

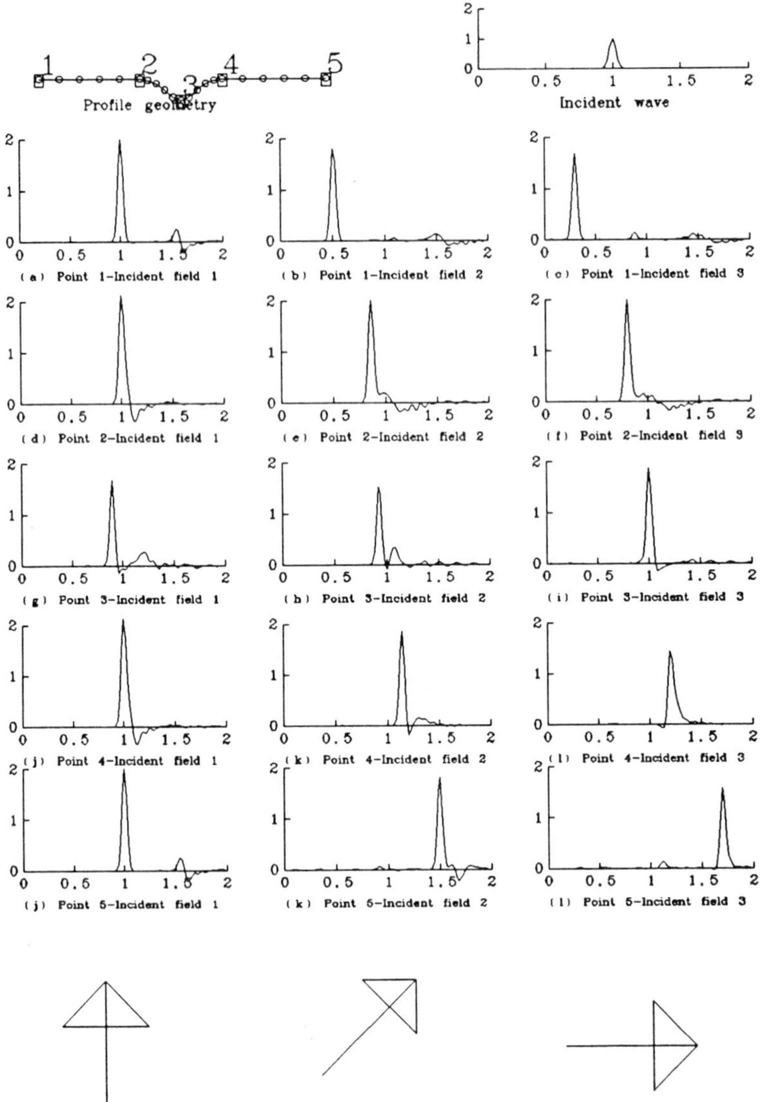

Figure 7: Valley (Figure 2(d)) subjected to Gaussian pulse (Figure 3(a))

The five rows contain the responses of the five points The three columns represent response to waves at vertical, 45°, and horizontal incidences.

by the hill. For the oblique incidences this energy is not focussed at the top. (d) and (j) show symmetric response to the vertical wave at the symmetrically placed base points of the hill. (e) and (k) show the difference caused by having an asymmetric

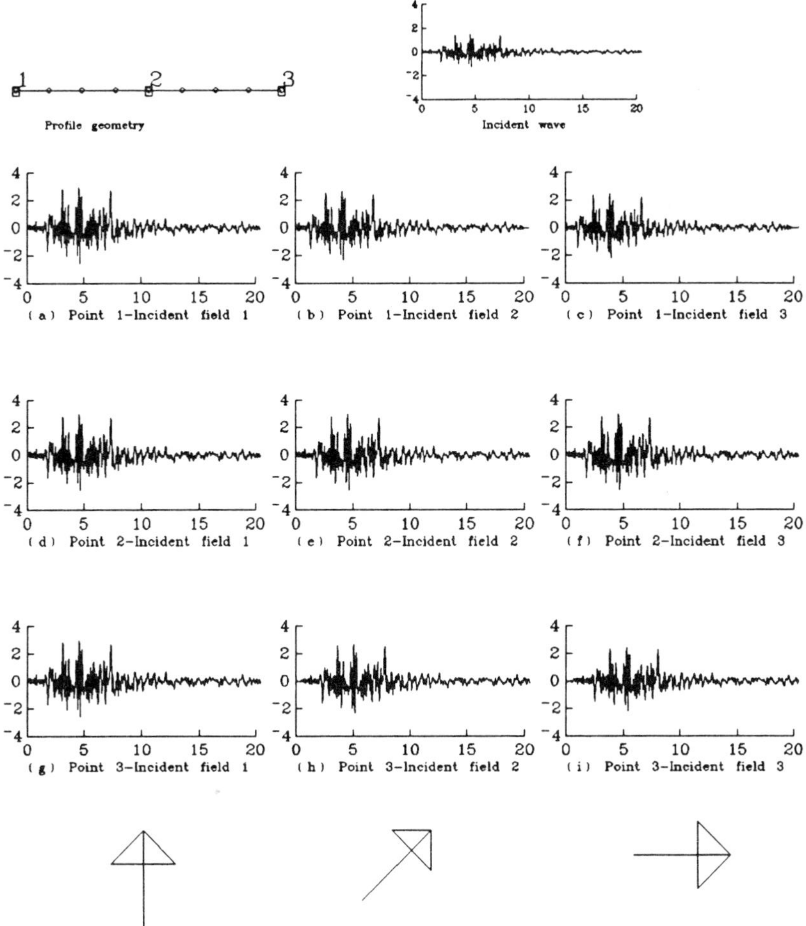

Figure 8: Halfspace (Figure 2(a)) subjected to Sanfernando earthquake (Figure 3(c))

The rows represent the responses of the three points The columns represent response to the wave at vertical, 45°, and horizontal incidences.

incident wave and the more diffuse nature of the reflected wave as opposed to the first wave.

Figure 11 shows the response of the three points of interest shown in Figure 2(d) to the San Fernando earthquake of Figure 3(c). These fifteen plots show the way the presence of the valley modifies the response which one would get in a half-space (Figure 8). One can see in (g) that the response is initially quite similar to the clean response in a half-space but later is slightly perturbed by the energy diffused by the valley.

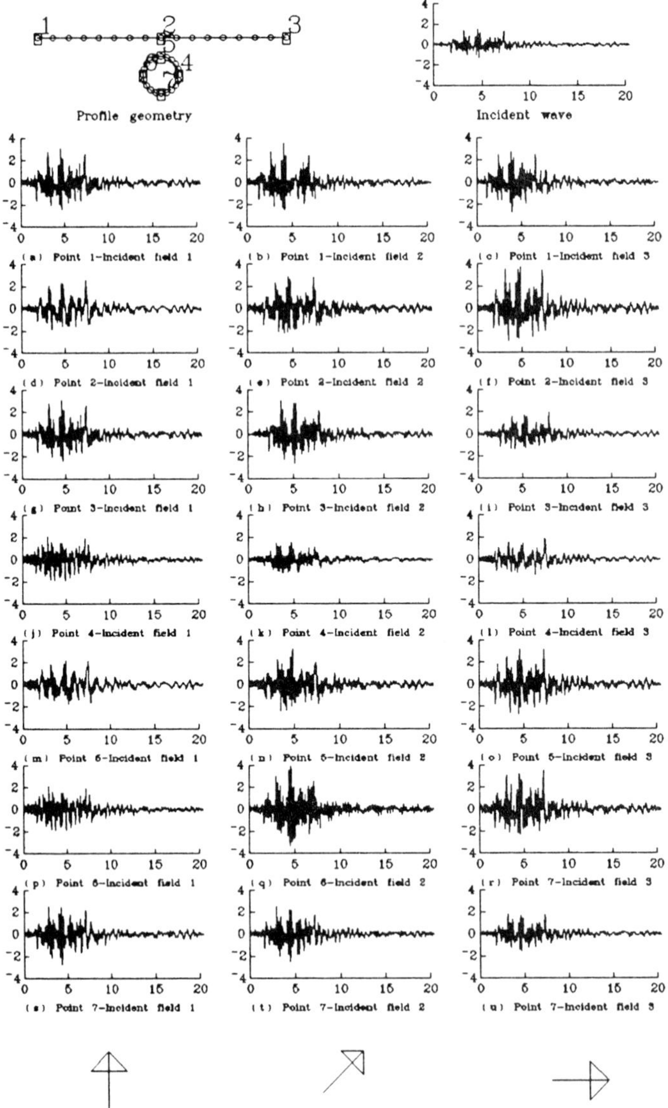

Figure 9: Circle in half-space (Figure 2(b)) subjected to San Fernando earthquake (Figure 3(c))

The seven rows contain the responses of the seven points. The three columns represent response to waves at vertical, 45°, and horizontal incidences.

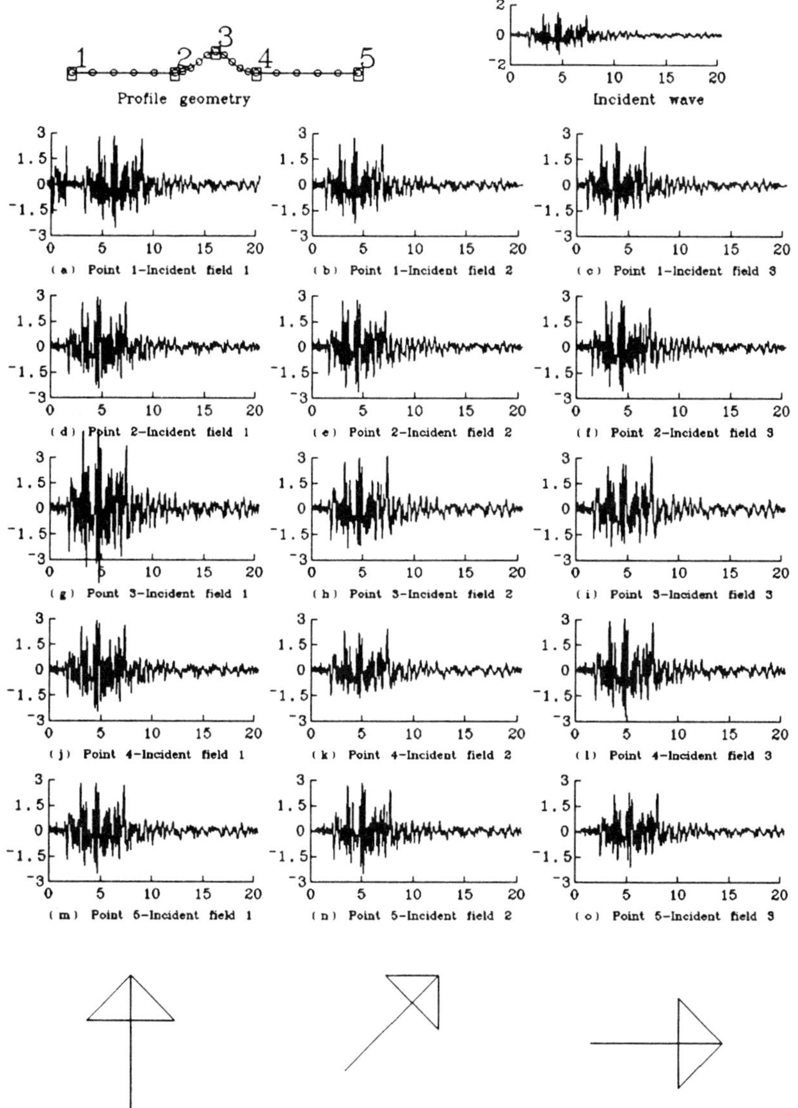

Figure 10: Hill (Figure 2(c)) subjected to San Fernando earthquake (Figure 3(c))

The five rows contain the responses of the five points
The three columns represent response to waves at vertical, 45°, and horizontal incidences.

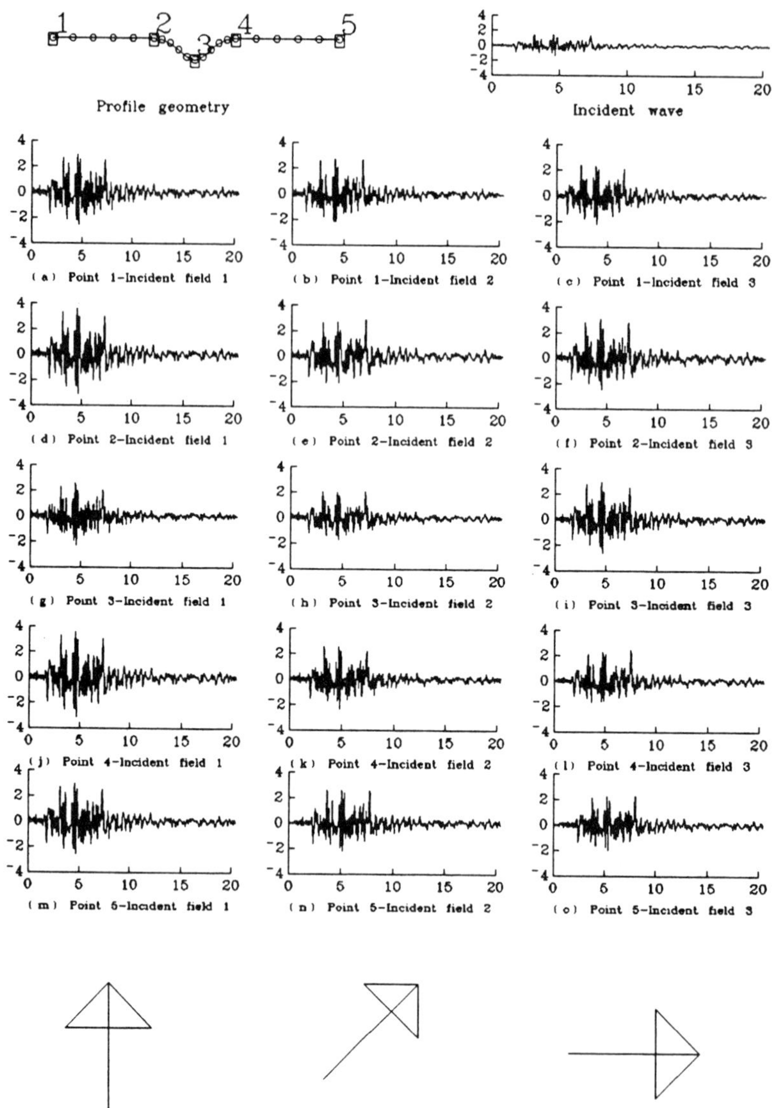

Figure 11: Valley Figure 2(d) subjected to San Fernando earthquake Figure 3(c)

The five rows contain the responses of the five points. The three columns contain the responses to waves at vertical, 45°, and horizontal incidences.

ACKNOWLEDGEMENT

This research was supported by a grant (ECE8417678) from the National Science Foundation; this support is gratefully acknowledged.

CONCLUSIONS

The boundary method is seen to be effective for investigating the response of two-dimensional profiles to incident SH-waves. For those few situations in which an exact solution is available, the method shows good agreement with theory. In those situations in which the computed response can be classified as reasonable or impossible, the method produces reasonable solutions. In those situations where the response cannot be predicted at all well, the method is shown to be applicable.

Therefore, the authors feel that the boundary element approach to a frequency-domain solution to the problem of earthquake response can be a valuable tool for research and design.

Bibliography

1. Askar, A., X. Zeng, and S. Altay, Explicit integration of boundary integral equations and scattering of seismic waves by underground structures, World Conference on Earthquake Engineering, San Francisco (San Francisco, 1984).

2. S. Altay, Explicit integration of boundary integral equations in the frequency domain for wave-scattering problems, Ph. D. dissertation, Princeton University (Princeton, 1986).

3. Askar, A., A. S. Cakmak, S. Altay, P. K. Hadley, Explicit integration of boundary integral equations in the frequency domain for wave-scattering problems, Recent Applications in Computational Mechanics (New Orleans, 1986).

4. Eringen, A. C. and E. S. Suhubi, *Elastodynamics--Vol.2: Linear Theory,* Academic Press, New York (1971).

5. Mow, C. C. and Y. H. Pao, *Diffraction of Elastic Waves and Dynamic Stress Concentrations,* Crane-Russak, New York (1972).

6. Sanchez-Sesma, F.J., I. Herrera, and J. Aviles, A boundary method for elastic wave diffraction: Application to scattering of SH waves by surface irregularities, **72** (Bull. Seism. Soc., 1985) p. 473. (1985).

7. IMSL, Standard Mathematical Subroutine Library,

Attenuation Analysis of High Frequency Seismic Waves in Randomly Heterogeneous Rock Media by Finite Difference Simulations

E. Faccioli, A. Tagliani

Department of Structural Engineering, Politecnico, P.za L. da Vinci 32, 20133 Milano, Italy

INTRODUCTION

Our interest in wave propagation in heterogeneous elastic media was stimulated by studies on the band-limited character of acceleration Fourier Amplitude Spectra (FAS) of real earthquakes and particularly on the causes of the frequency cutoff (known as f_{max}) observed in logarithmic plots of such spectra (Hanks[1], Aki[2]). In earthquake engineering applications, the choice of an appropriate FAS model for frequencies ≳10 Hz is critical for predicting the scaling of peak ground acceleration versus magnitude (Boore[3], Faccioli[4]). Moroever, unrealistic assessements of the high frequency content of expected seismic motions may result in overconservative design, e.g. of rigid structures sited on rock.

For characterizing the high-frequency seismic motions Anderson and Hough[5] proposed a simple form of the acceleration FAS, A(f), namely

$$A(f) = A_o \exp(-\pi k f) \qquad f>f_E \tag{1}$$

where f is frequency in Hz and f_E denotes the point beyond which this asymptotic form is valid. Based on the analysis of strong-motion data, the previous authors found that the attenuation parameter k increases with distance R, and can be approximately written in the form:

$$k(R) = k_o + mR \tag{2}$$

This suggests that the zero-distance intercept k_o, which varies between about 0.01 (rock sites) and 0.07 (deep soil sites), is caused by a "near-site" effect due to attenuation in the weathered layers close to the surface. If we restrict our attention to rock sites, and consider that the opening of existing cracks in rock is largest at the surface and should vanish at 2-3 km depth, it seems reasonable to associate near-site frequency attenuation with the scattering suffered by seismic waves in the region where the cracks are open. On the other hand, the distance-dependent term in (2) can be interpreted to describe frequency-independent, whole path attenuation. Herein, only k_o will be considered, and simply denoted as k in the following.

To quantify near-site attenuation, we analyze propagation of seismic signals through one-and two-dimensional models of heterogeneous rock media. The heterogeneity is introduced through the seismic shear wave velocity, based on the evidence of velocity logs from wells drilled in rock, which show strong small-scale fluctuations (Sato[6] ; Moos and Zoback[7]). These can be described by a spatial stochastic process, superimposed on a deterministic trend which is usually an increasing function of depth. Elastic wave propagation in randomly heterogeneous materials is performed by high order finite difference (FD) approximations. Our main purpose is to determine whether the spectral characteristics of seismic signals propagated for sufficient distances in random media of the previous type conform with the simple observational model of Eq. (1), and also whether the resulting values of the limiting frequency f_{max} and of the attenuation parameter k agree with those observed for rock sites. Although several numerical studies of attenuation in random elastic materials have been published,[8,9,10] to our knowledge none of them is specifically devoted to simulating near-site effects on surface ground motions.

NUMERICAL MODEL AND PROBLEM FORMULATION

We consider in general two-dimensional SH-wave propagation in a heterogeneous elastic medium with constant density ρ and variable shear modulus μ (x,z) or, equivalently, with variable shear wave velocity β (x,z). The horizontal displacement v in the y-direction satisfies the scalar wave equation

$$\frac{\partial^2 v}{\partial t^2} = \frac{\partial}{\partial x}\left[\beta^2(x,z)\frac{\partial v}{\partial x}\right] + \frac{\partial}{\partial z}\left[\beta^2(x,z)\frac{\partial v}{\partial z}\right] \qquad (3)$$

The velocity β (x,z) will be taken as the sum of a deterministic component, function of depth z only, and a stochastic fluctuation,

$$\beta(x,z) = \beta_o(z) + \tilde{\beta}(x,z) \qquad (4)$$

The spatial domain is discretized by a rectangular grid with step h both in x and z directions, and time is discretized by a constant step Δt. The time derivative in Eq. (3) is replaced by a second-order, centered FD approximation, while the space derivatives are replaced by fourth-order (i.e. five-point) approximations[11]. Minimization of the grid dispersion error and stability of the fourth-order scheme are achieved by using 5 grid steps for the shorter wavelengths and $\Delta t < \sqrt{3/8}\, h/\beta_{max}$, where β_{max} is the largest velocity in the grid (Alford et al.[12]). For $1.2 \lesssim \beta \lesssim 3.5$ km/sec, numerical tests showed that we can propagate frequencies up to 80-90 Hz without noticeable dispersion by using h=5m and Δt=0.5 msec; the results of the next section were all obtained with these values. Spurious reflections from the bottom edge of the grid are avoided by introducing the A2 and A1 paraxial approximations of Clayton and Engquist[13] at the base and at a fictitious line below it, respectively. Since the base motion is vertically incident, no strong parasite reflections are generated from the vertical edges of the grid in 2-D models, and simple symmetry conditions are used for them. The zero-stress (free-surface) condition developed by Ilan et al.[14] is imposed at the top edge, and a fictitious line with the same displacement values is introduced above it, giving accurate results.

The vertically incident plane wave, except when otherwise indicated, is an impulse of the form

$$v_o(t) = \begin{cases} C(t-t_o)e^{-\alpha(t-t_o)} & t \geq t_o \\ 0 & t < t_o \end{cases} \tag{5}$$

where C is a constant and α controls the duration of the signal. Its FAS has a corner frequency at $f_c = \alpha/2\pi$ and decays as f^{-2} for $f > f_c$. The corresponding acceleration spectrum is flat for $f > f_c$, and is well suited for analyzing high frequency attenuation effects within the band allowed by the FD scheme. If the excitation consists of a long duration earthquake signal, it is computationally efficient to solve the problem in two steps. First we compute at the desired receiver points the impulse-response of the model to a smoothed delta-function (Boore[15]), and subsequently we perform the convolution between impulse-response and earthquake signal by Fast Fourier Trasform techniques.

The method for generating the random fluctuations of β, both in one and two dimensions, is described by Frankel and Clayton[10]. We considered fluctuations given by Gaussian white-noise (referred to as "Gaussian"), white-noise with a Gaussian filter ("filtered Gaussian"), and self-similar. They are characterized by their correlation function (save for the Gaussian case), probability distribution, and standard deviation. In practice, for a given medium, the governing parameters are the correlation distance a, and the standard deviation σ of the fluctuations. The distance a controls the extent of the interaction between the incident wavelengths and the irregularities of the medium: small a generally implies that only the short wavelengths will feel the "roughness" of the medium and viceversa. However, one-dimensional media of the self-similar type tend to be insensitive to a , i.e. they exhibit roughness over a very broad range of length scales. A reported estimate of a for real rock media is the order of 20m [6]. On the other hand, σ controls the fluctuation amplitude relative to the mean value of the deterministic component. Except possibly for near-surface layers, realistic values of σ are probably of the order of 0.10 [9,10].

RESULTS

The results illustrated herein refer to 1-D (i.e. horizontally layered) models of random media, considered to be rapresentative when the excitation is vertically incident. Presentation of 2-D analyses will be made in a subsequent paper.

We consider first the response of a filtered Gaussian medium 1.5km thick, characterized by a =10m and σ =0.10, to a base input given by Eq. (5) with $f_c=\alpha/2\pi$ =12Hz, t_o=0. The random velocity fluctuations are superimposed on a constant value β_o=2.2km/sec, as shown in Fig.1a. To test the accuracy of the FD scheme, the displacement at the surface (lower trace in Fig.1b) is compared with that calculated by the Haskell matrix method[16] (upper trace in Fig.1b). The two synthetic seismograms agree quite well; the smoothed onset of the main pulse in the upper trace is related to frequency resolution limitations in implementing Haskell method. Also illustrated in Fig.1c are the acceleration FAS of the excitation (multiplied by 2) and of the surface response, computed over a 0.3 sec time window containing the main pulse,shown by the horizontal segment in Fig.1b. The present medium is clearly not causing high-frequency attenuation, because the mean wavelength associated with the corner frequency of the input spectrum($\lambda_c \simeq$180m) is much longer than the correlation distance a =20m, and the medium is quite "smooth" at scale lengths larger than a . Quite a different effect is produced by the Gaussian medium shown in Fig.2a, also having β_o=2.2km/sec. This medium is irregular at all scale lengths (compatible with the grid step size) and the difference in small scale "roughness" with respect to the previous case is especially evident. This causes a visible broadening of the main impulse at the surface (Fig.2b, upper trace), more clearly indicated by the sharp falloff in the acceleration FAS beyond $f_{max} \simeq$40Hz (Fig.2c).

A convenient estimate of f_{max} is provided by $f_{0.95}$, i.e. the frequency for which the cumulated squared spectral acceleration attains 95% of its value for $f \rightarrow \infty$ [17]. On the other hand, the value of k in Eq.(1) can be estimated from our numerical experiments by measuring the slope of the logarithm of the response/excitation spectral ratio versus frequency in linear scale, over a band spanning between f_E and a limiting frequency of 80-90 Hz. We did this on a set of 3 simulations, of which Fig.3 shows one example. In all simulations the deterministic component β_o increases monotonically from 1.5 km/sec at the surface to 2.8 km/sec at 3.0 km depth, and Gaussian fluctuations with σ =0.10 (relative to the mean value of β_o) are superimposed on it, as shown in Fig.3a. The response of the medium to the same impulse as in Fig.2 is calculated at points 1,2 and 3, located respectively at 2km and 1km depth and at the surface. Figure 3b illustrates the calculated acceleration FAS of the main impulse at these points, together with the excitation spectrum multiplied by 2. It is evident that f_{max} decreases and the rate of spectral falloff increases as the propagation path becomes longer. In fact, $f_{0.95}$ values are very similar for the 3 simulations, and their average equals 62Hz at point 2 and 42Hz at point 3 (the value at point 1 is close to 90Hz, and therefore not significant). Observed f_{max} values are typically 5-15Hz for strong-motion spectra [4], but can be 40-50Hz for microearthquakes digitally recorded on rock outcrops [18], in rough agreement with our results for a 3km propagation path. Note also the low-frequency amplification produced by the variable β_o in the upper spectrum of Fig.3b: the interaction between such amplification and the high-frequency falloff may introduce an apparent corner frequency which differs from the excitation corner frequency.

Estimates of the attenuation parameter k over a given propagation distance are influenced both by the considerable scatter of the spectra and, to a lesser extent, by the choice of the lower frequency limit f_E. The average k values for 3 simulations are 0.002±0.003, 0.006±0.003 and 0.016±0.004 for points 1,2,3, respectively, see Fig.3c. The value k=0.016 agrees well with the observed mean value (k=0.015) determined by De Natale et al. [19] from microearthquakes of 3-5km focal depth, digitally recorded at Campi Flegrei, Italy. Since k increases with distance, we are naturally led to compare the assumed spectral decay $\exp(-\pi kf)$ with approximations of the form $\exp(-\pi fR/\beta_{av}Q)$ where β_{av} is the mean velocity over the propagation distance R, and Q is the apparent quality factor of an equivalent homogeneous anelastic medium. The previous mean values of k correspond to Q=190,140 and 82 for distances of 1,2,3km, respectively. The variable Q-value can be explained by noting that, (a) for constant β_o the apparent Q^{-1} is roughly proportional to σ^2 (Richards and Menke [8]), and (b) for β_o decreasing with increasing path length and a constant r.m.s. fluctuation amplitude (0.22km/sec in our case), the ratio σ/β_{av} effectively increases with path length, thus giving rise to a lower Q.

Three additional simulations, not illustred here, were performed with media having the previous β_o values and self-similar fluctuations with a =20m and σ^o =0.10. The resulting k values are significantly smaller and the f_{max}

values larger than those of the Gaussian media (the average k and $f_{0.95}$ for point 3 are 0.01±0.0007 and 74±4 Hz). This shows that, for given β_o and σ, Gaussian fluctuations are the most effective in suppressing the high frequencies.

Lastly we studied propagation of a real acceleration signal, using first a medium with $\beta_o(0)$=1.2km/sec, $\beta_o(3km)$=2.6km/sec, and Gaussian fluctuations with σ =0.10. The input accelerogram was recorded in Friuli (Italy) at the Somplago-D SMA-1 station, in a power house about 300m underground in limestone rock, during the M5.1, 9/11/1976 16:35 GMT earthquake at a distance of 6-8km. The significant portion of the record, low-pass filtered at 25-27 Hz and baseline corrected, and its FAS are shown in Fig.4a and 4b. No f_{max} is visible below the filter cutoff at 25Hz ($f_{0.95}$=24Hz). The amplitude of the record was reduced to be compatible with the depth of the base of the model, based on β_o values. The impulse response at the surface of the medium was calculated using a smoothed delta-function of total duration 0.10sec, with a flat displacement spectrum up to a corner frequency of 30Hz. The accelerogram obtained at the surface and its FAS are illustrated in Figs.4c and d. The spectrum exhibits a minor depletion in relative high frequency content with respect to the input spectrum, indicated by a $f_{0.95}$ value of 22 Hz but this is rather a consequence of the low frequency amplification effect.

To investigate the influence of the propagating medium on the result, four additional simulations were carried out, with self-similar fluctuations having σ =0.10 and a=20, 50 and 200m, and with filtered Gaussian fluctuations having σ =0.10 and a =20m. In all cases the deterministic trend $\beta_o(z)$ was the same as in Fig.4, and the same seed was used to generate the initial sequence of random numbers. The resulting $f_{0.95}$ values range between 22 and 24Hz, and the surface accelerogram and its FAS have only minor differences with respect to that of Figs.4c and d. Unfortunately, no simultaneous recording at the rock surface is available for the event of Fig.4a to validate our estimates of f_{max}.

The heterogeneities and path length used in our experiments do not give rise to enough scattering to cause a spectral falloff in the frequency range where it is generally observed in strong-motion spectra (f_{max} < 15Hz). Using a larger σ and/or a longer propagation path would bring f_{max} closer to observations, but may imply arbitrary assumptions on the size and distribution of the small-scale fluctuations and on the extent of the region where they are present in real rock media. Our result, however limited and preliminary, does not preclude the possibility that f_{max}-values observed at short distances in strong earthquakes may be related to source effects [2]. Additional analyses with 2-D models using fluctuation parameters obtained from real data will help clarify this aspect.

CONCLUSIONS

Numerical simulations of 1-D seismic wave propagation in randomly heterogeneous rock media, performed with a high accuracy finite-difference model, show that:

1. The values of the spectral decay parameter k and of the limiting frequency f_{max} obtained with Gaussian white noise fluctuations of $\beta(\sigma=0.10)$ superimposed on a deterministic component increasing with depth, are in agreement with microearthquake data recorded on rock, provided the randomly heterogeneous path length is about 3 km;

2. The equivalent Q in a medium of the previous type decreases with increasing path length;

3. The simulated f_{max} values are consistently higher than those observed in acceleration spectra of strong earthquakes, and do not rule out the possibility that source effects are a controlling factor.

ACKNOWLEDGEMENTS

The work described in the paper was supported in part by ENEA of Italy under the Research Contract 86/0426/0001 997

REFERENCES

1. Hanks T. (1982), f_{max}, Bull. Seismol. Soc. Am., 72, pp.1867-1879.

2. Aki K. (1986), Physical Theory of Earthquakes, Proceedings of the Strarbourg, France, 1986 Summer School on "Seismic Hazard in the Mediterranean Region", in press

3. Boore D. (1983), Stochastic Simulation of High-frequency Ground Motion Based on Seismological Models of the Radiated Spectra, Bull. Seismol. Soc. Am., 73, pp.1865-1894.

4. Faccioli E. (1986), A Study of Strong Motions from Italy and Yugoslavia in Terms of Gross Source Properties, in Earthquake Source Mechanics (Eds. Das S. Boatwright J. and Scholz C.), pp. 297-409, Geophysical Monograph 37, Am. Geophys. Union, Washington.

5. Anderson J. and Hough S. (1984), A Model for the Shape of the Fourier Amplitude Spectrum of Acceleration at High Frequencies, Bull. Seismol. Soc. Am., 74, pp.1969-1994.

6. Sato H. (1984), Attenuation and Envelope Formation of 3-Component Seismograms of Small Local Earthquakes in Randomly Inhomogeneous Litosphere, J. Geophys. Res., 89

(B2), pp. 1221-1241

7. Moos D. and Zoback M. (1983), In Situ Studies of Velocity in Fractured Crystalline Rocks, J. Geophys. Res., 88(B3), pp. 2345-2358.

8. Richards P. and Menke W. (1983), The Apparent Attenuation of a Scattering Medium, Bull. Seismol. Soc. Am., 73, pp.1005-1022.

9. Frankel A. and Clayton R. (1984), A Finite-Difference Simulation of Wave Propagation in Two-Dimensional Random Media, Bull. Seismol. Soc. Am., 74, pp.2167-2186.

10. Frankel A. and Clayton R. (1986), Finite Difference Simulations of Seismic Scattering, J. Geophys. Res., 91(B6), pp.6465-6489.

11. Abramowitz M. and Stegun I. (Eds.) (1965), Handbook of Mathematical Functions, pp.883-884, Dover Publications, New York.

12. Alford R. Kelly K. and Boore D. (1974), Accuracy of Finite-Difference Modeling of the Acoustic Wave Equation, Geophysics, 39, pp.834-842.

13. Clayton R. and Engquist B. (1977), Absorbing Boundary Conditions for Acoustic and Elastic Wave Equations, Bull. Seismol. Soc. Am., 67, pp.1529-1540.

14. Ilan A. Ongar A. and Alterman Z. (1975), An improved Rapresentation of Boundary Conditions in Finite Difference Schemes for Seismological Problems, Geophys. J. Royal. Astr. Soc., 43, pp. 727-747

15. Boore D.(1972), Finite Difference Methods for Seismic Wave Propagation in Heterogeneous Materials, in Methods in Computational Physics (Eds. Alder B. Fernbach S. and Rotenberg M.), Vol.2, pp.17-19, Academic Press, New York.

16. Haskell N. (1960), Crustal Reflection of Plane SH Waves, J. Geophys. Res., 65, pp.4147-4150.

17. Anderson J.(1986), Implication of Attenuation for Studies of the Earthquake Source, in Earthquake Source Mechanics (Eds. Das S. Boatwright J. and Scholz C.), pp.311-418, Geophysical Monograph 37, Am. Geophys. Union, Washington.

18. Fletcher J. Haar L. Vernon F. Brune J. Hanks T. and Berger J. (1986), The Effects of Attenuation on the Scaling of Source Parameters for Earthquakes at Anza, California, in Earthquake Source Mechanics (Eds. Das S. Boatwright J. and Scholtz C.), pp.331-438, Geophysical Monograph 37, Am.

Geophys. Union, Washington.

19. De Natale G. Faccioli E. and Zollo A. (1987), Scaling of Peak Ground Motions from Digital Recordings of Small Earthquakes at Campi Flegrei, Southern Italy, Pure and Applied Geophysics, in press.

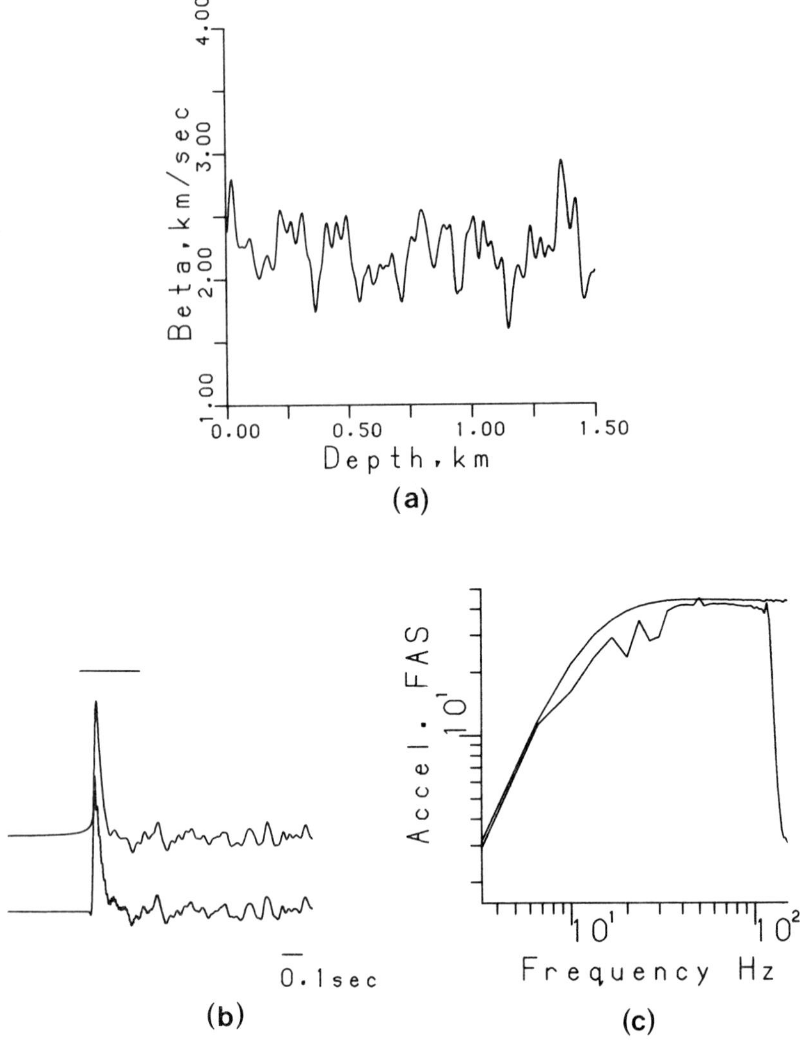

Figure.1. (a)Seismic velocity versus depth in 1-D, "filtered" Gaussian medium with a =20m, σ =0.10; (b) Surface displacement calculated by FD (lower trace) and Haskell method (upper trace); (c) FAS of surface acceleration compared with that of excitation (smooth curve).

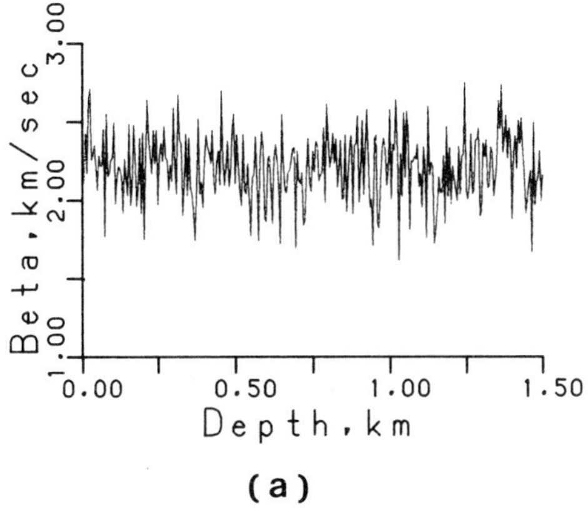

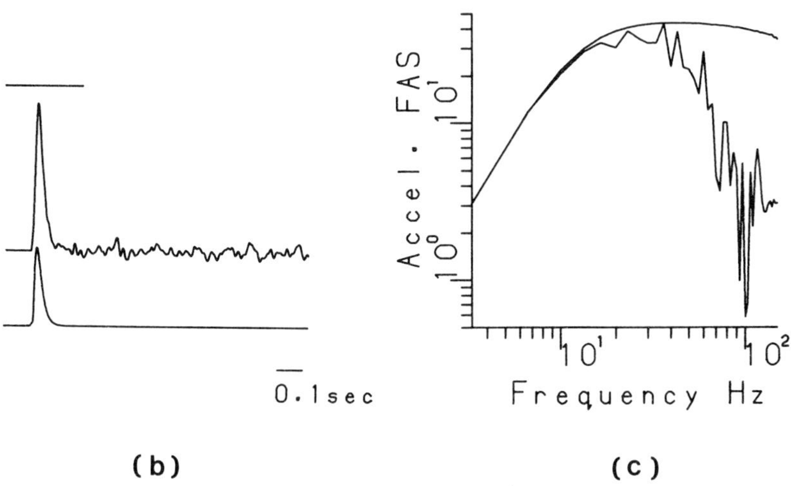

Figure.2. (a) Seismic velocity versus depth in 1-D white noise Gaussian medium with σ =0.10; (b) displacement impulse used for base excitation (lower trace) and surface response (upper trace); (c) FAS of surface acceleration compared with FAS of excitation (smooth curve).

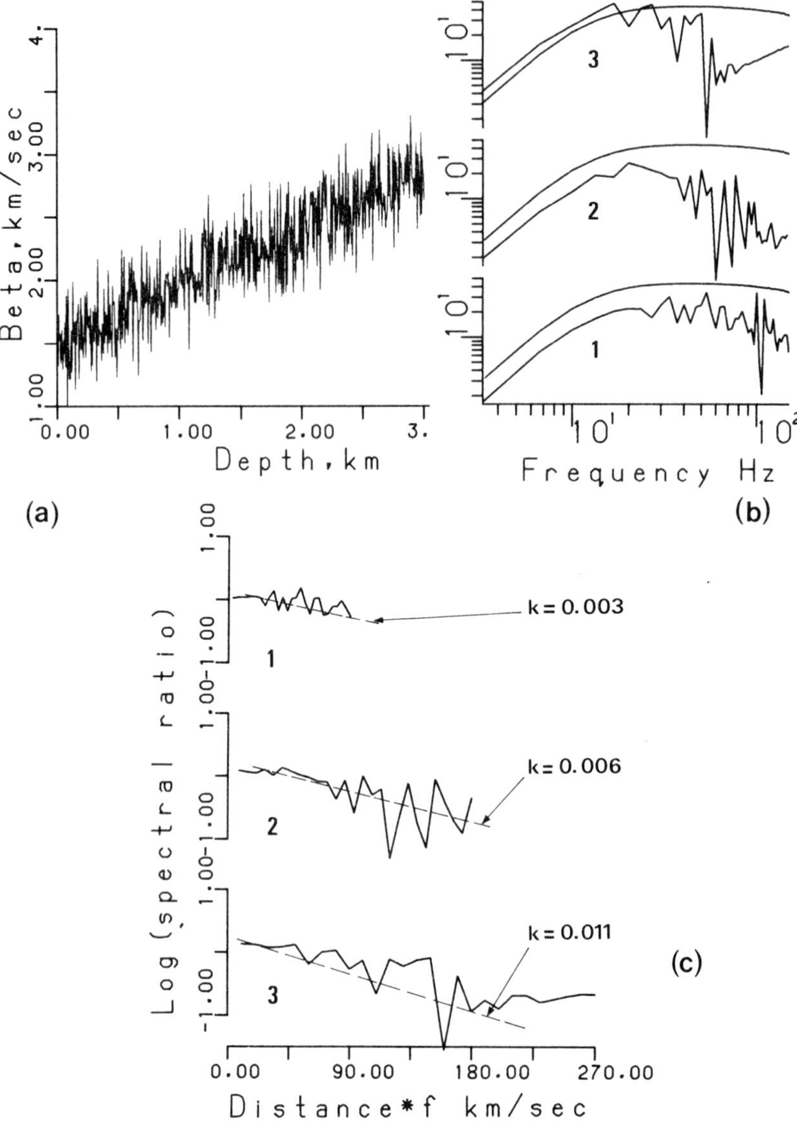

Figure.3. (a)Seismic velocity versus depth ; (b) Acceleration FAS at receiver points 1, 2, 3, compared with the input FAS (smooth curves); (c) Plots of the log (response/excitation) spectral ratios against the product of frequency and propagation distance, with least-squares straight lines giving the value of k.

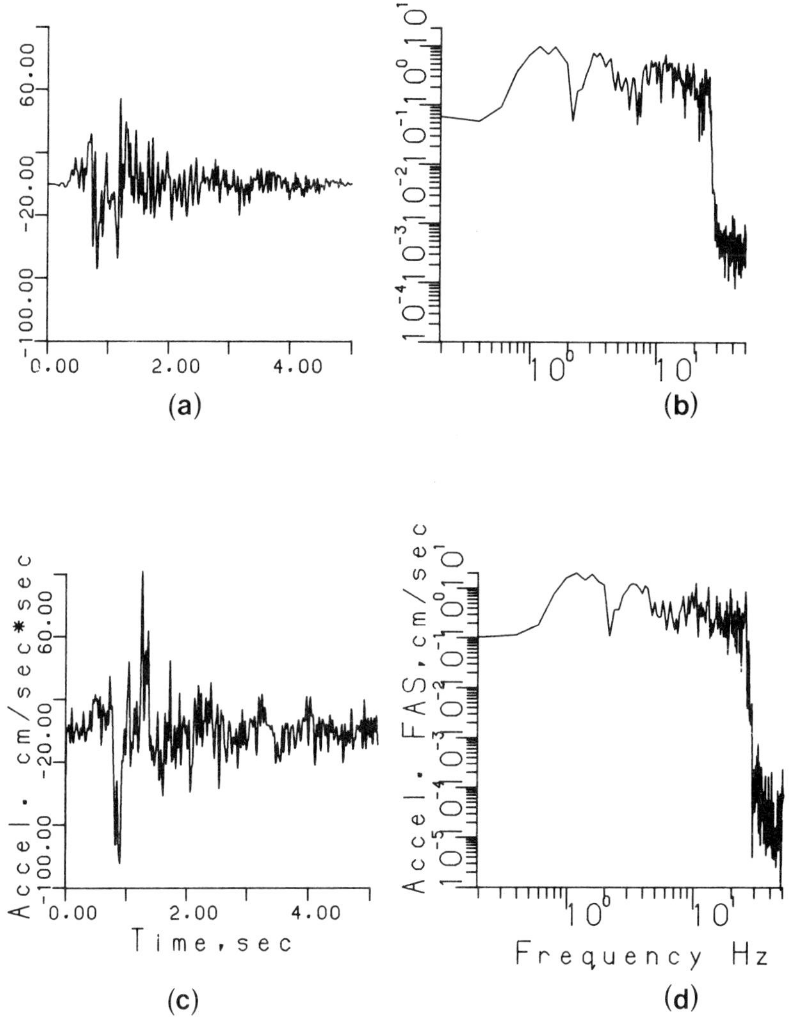

Figure.4.(a) Input acceleration time history (not scaled); (b) FAS of input acceleration; (c) Surface response acceleration of Gaussian medium 3km thick, with σ =0.10; (d) FAS of surface acceleration.

Green's Function for Layered Half-Space

A. Umek, A. Štrukelj
Department of Civil Engineering, Faculty of Technical Sciences, University of Maribor, 62000 Maribor, Yugoslavia

INTRODUCTION

The first step in dealing with the soil structure interaction problems by the method of integral equations e.g. Wong and Luco[1] or by the boundary element method, which are the two most promising techniques, is to obtain the appropriate Green's function. In the soil-dynamics problems one can consider the whole-space Green's function as an appropriate one. In this case the integration path has to be taken or boundary elements must be placed along all the surfaces separating the areas with different elasto-dynamic properties. Using this approach we are working with a relatively simple Green's function on one side and must cope with many boundaries some of which could be of infinite extent. The latter fact represents the major obstacle in both integral equations and boundary element methods. The just mentioned difficulties disappear entirely if we are able to incorporate all the mechanical and geometrical properties of the soil in to the Green's function. In this case the integration or the boundary elements cover only the area of the contact between the soil and the superstructure, which is in all cases finite. The main difficulty using this approach is in obtaining such a Green's function and that it has to be derived for each case separately. Therefore some simplifications concerning the geometrical and mechanical properties of the soil are desirable. To our knowlage the only work addressing the problem of this kind has been published by Kausel[2], in which the Green's function for n layers on a rigid foundation has been obtained. In our work we are modeling the soil as n layers with parallel contact surfaces on the top of an elastic half-space. The material properties within each layer and throughout the half-space are constant. The method of solution combines analytical and numerical techniques and is believed to be new in the concept and the realisation.

DESCRIPTION OF THE MODEL

Physical model

Our principal model consists of n layers, with the surfaces parallel to each other, supported by an elastic half-space (Figure 1). The material properties in each layer and in the half-space are considered to be elastic, homogeneous and isotropic,

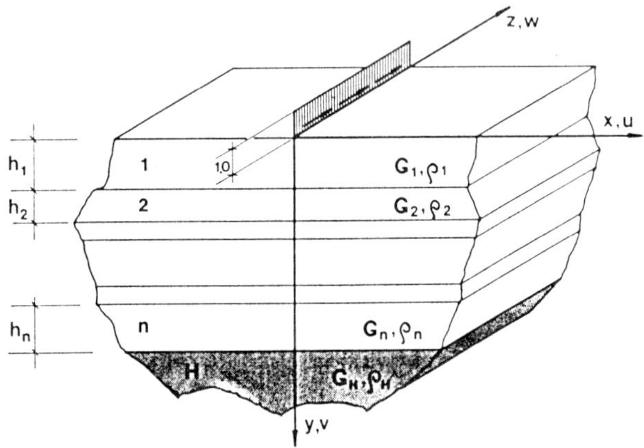

Figure 1. The layered half-space soil model considered.

so they differ from the layer to the layer only and are constant within the layers. The material damping is neglected in the case treated in this paper. The contact surfaces between the layers are defined with y=constant. Across these surfaces the continuity of all the displacements and stresses is assumed.

To obtain the Green´s function in the space the discussed half-space is loaded with appropriate point and line loads respectively. In the present paper we limit our attention to a unit, tangential line load with the harmonic time dependence, acting in the direction as shown in Figure 1.

Mathematical model

The just described physical model constitutes in the elasto-dynamic sense an antiplane problem, where all the displacements are in the direction of the z-axis only and are functions of x and y but not of z. They are governed by the following differential equations:

$$\nabla^2 w = - \frac{\omega^2}{C_{is}^2} w \qquad (1)$$

where C_{is} is the phase velocity of the shear wave in the i-th layer.

The continuity of the displacements and the stresses across the contact surfaces between the layers yields the subsequent conditions:

$$w_i(x, y_i, \omega) - w_{i+1}(x, y_i, \omega) = 0$$

$$G_i w_{i,y}(x, y_i, \omega) - G_{i+1} w_{i+1,y}(x, y_i, \omega) = 0 \qquad (2)$$

Here y_i is the defining coordinate of the contact surface between the layers i and i+1 and G_i is the shear modulus in the i-th layer. On the surface of the half-space the following boundary condition must be satisfied:

$$G_1 w_{1,y}(x, 0, \omega) = - \delta(x) \qquad (3)$$

In addition to the Equations (1), (2) and (3) the radiation condition in the half-space must be obeyed.

METHOD OF SOLUTION

The Fourier transform in the x-coordinate is applied to the Equations (1), (2) and (3). The Equations (1) become ordinary differential equations with the following solution:

$$\overline{w}_i = A_i(k, \omega) \exp(a_i y) + B_i(k, \omega) \exp(-a_i y) \qquad (4)$$

where k is the Fourier transform parameter, A_i and B_i are the integration constants to be determined and a_i are given by:

$$a_i = \left(k^2 - \frac{\omega^2}{C_{is}^2} \right)^{1/2} \qquad (5)$$

It can be seen that the solution given by equation (4) is not unique. It has several branch points in the complex k-plane, which are located at $a_i(k)=0$. To make the solution unique the branch cuts are introduced. Those belonging to the layers can be led arbitrarily. This is however not the case with the branch cuts in connection with the underlying half-space. They must be chosen in such a way that they at the same time ensure the uniqueness of the solution and together with the selection of one of the constants of integration A_H or B_H to be zero satisfy the radiation condition. It can be shown that the proper position of the half-space branch cuts corresponding to $A_H=0$ is from the particular branch point outward along the positive or negative real k-axis respectively. It is resonable to lead all the other branch cuts in the same way. The k-plane obtained in this way is shown on Figure 2. The integration constants A_i and B_i can now be uniquely determined by satisfying the Equations (2) and (3). This leads us to the system of algebraic equations, which can be solved for the pairs of the discrete values of the parameter k and ω. Limiting our attention to the surface of the half-space the Green´s function in the transformed domain is given by:

$$\bar{w}(k,0,\omega) = A_1(k,\omega) + B_1(k,\omega) \qquad (6)$$

Since we are given A_1 and B_1 for the discrete values of k and ω only the same goes for $\bar{w}(k,0,\omega)$ also. This makes the analytical Fourier inversion impossible and the numerical contour integration as the mean for the Fourier inversion seems to be the natural choice.

It could be seen from Figure 2 that the original Fourier

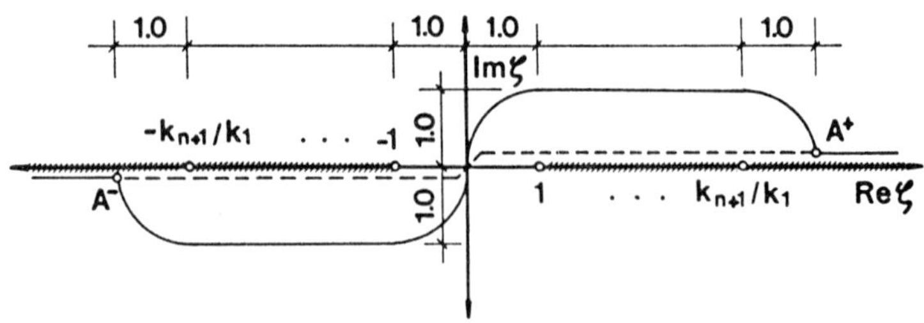

Figure 2. The Fourier inversion paths; a) – – – the analytical; b) ——— equivalent selected in our work for the numerical inversion.

inversion path goes right through all the singularities of the integrand. Therefore it is convenient to introduce an equivalent integration path, which would run sufficiently far from all the singularities of the integrand to make the numerical integration problemless. To achieve this the coordinate transformation $k=k_1\zeta$ is introduced, where k_1 gives the position of the singularity closest to the origin of the coordinate axes and the equivalent integration path shown in Figure 2 is chosen. The inversion integral of Equation (6) now becomes:

$$w(x,0,\omega) = I_1 + I_2 + I_3 + I_4 \qquad (7)$$

where are

$$I_1 = \frac{k_1}{2\pi}\int_{-\infty}^{A^-} (A_1(k_1\zeta,\omega) + B_1(k_1\zeta,\omega)) \exp(i\zeta k_1 x)\, d\zeta \qquad (8)$$

$$I_2 = \frac{k_1}{2\pi}\int_{A^-}^{0} (A_1(k_1\zeta,\omega) + B_1(k_1\zeta,\omega)) \exp(i\zeta k_1 x)\, d\zeta \qquad (9)$$

$$I_3 = \frac{k_1}{2\pi}\int_{0}^{A^+} (A_1(k_1\zeta,\omega) + B_1(k_1\zeta,\omega)) \exp(i\zeta k_1 x)\, d\zeta \qquad (10)$$

$$I_4 = \frac{k_1}{2\pi}\int_{A^+}^{\infty} (A_1(k_1\zeta,\omega) + B_1(k_1\zeta,\omega)) \exp(i\zeta k_1 x)\, d\zeta \qquad (11)$$

The integrals I_2 and I_3 are easy to evaluate using the Romberg procedure as given by Davis[3]. The semiinfinite integrals I_1 and I_4 are first reduced to the finite range by:

$$I_1 = \frac{k_1}{2\pi}\int_{A^- - \pi/(k_1 x)}^{A^-} S_1(\zeta,\omega) \exp(i\zeta k_1 x)\, d\zeta \qquad (12)$$

$$I_4 = \frac{k_1}{2\pi}\int_{A^+}^{A^+ + \pi/(k_1 x)} S_4(\zeta,\omega) \exp(i\zeta k_1 x)\, d\zeta \qquad (13)$$

where the sums S_1 and S_4 are given as:

$$S_{1,4}(\zeta,\omega) = \sum_{r=0}^{\infty} (-1)^r \left(A_1(k_1(\zeta \mp \frac{r\pi}{k_1 x}), \omega) + \right.$$

$$\left. B_1(k_1(\zeta \mp \frac{r\pi}{k_1 x}), \omega)\right) \qquad (14)$$

It was noticed however that the convergence of both infinite series given in equation (14) was extremly slow, so that several tenthousand terms were necessary to obtain the five digit accuracy. Since this computational effort seemed to us to be great, we looked for methods to speed up the convergence. It was found that the Shanks transformation[4] gives the best results. After its introduction ten to at most twenty terms are needed to evaluate the series in Equation (14) with the five digit accuracy. The results obtained so far were introduced into the Equations (12) and (13) and the integrals evaluated by the Romberg´s procedure again. This completes the computation of the Equation (7) and yields the desired Green´s function.

ILLUSTRATIVE EXAMPLE

As an example we take a layered half-space consisting from one layer and the suporting, homogeneous half-space. The problem is normalized in the following way. All the distances and the mass-densities are divided by the thickness and the mass-density of the layer respectively. The unit time is the time needed for a shear wave to transverse the thickness of the layer. So we obtain that $H_1=1.0$, $G_1=1.0$ and $C_{1s}=1.0$ in our example. For a realistic situation we assume that the shear modulus and the shear wave velocity in the half-space are biger than in the layer. In example given here they are taken $G_H=2.0$ and $C_{Hs}=2.0$. The antiplane Green´s function for the just described layered half-space and the driving frequency $\omega=10.0$ is computed and plotted in Figure 3 with the solid line. For comparison the antiplane Green´s function for the homogeneous half-space with the material characteristics of the layer is given by the broken line in Figure 3.

CONCLUSIONS

A method for computing the antiplane problem Green´s function for a layered half-space has been presented. It is characterized by very modest computational effort and good numerical

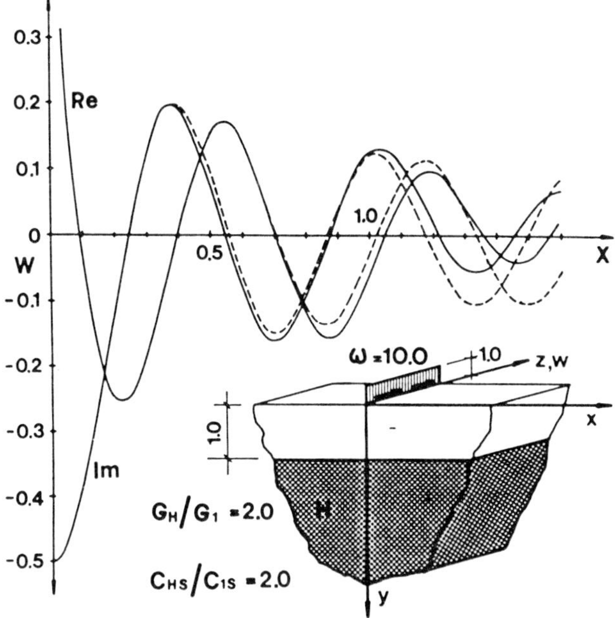

Figure 3. Antiplane problem Green´s functions; —— layered half-space; – – – homogeneous half-space.

stability in the range of low and moderate driving frequencies. It is believed and our experiences indicate that there should be no difficulties in applying the same technique to the plane and three-dimensional problems of the layered half-space.

REFERENCES

1. Wong H.L. and Luco J.E. (1978). Tables of Impedance Functions and Input Motions for Rectangular Foundations, Report No. CE78-15, University of Southern California.
2. Kausel E. (1981). An Explicit Solution for the Green Function for the Dynamic Loads in Layered Media, Research Report R81-13, Massachusetts Institute of Technology.
3. Davis P.J. and Rabinowich P. (1967). Numerical Integration, Blaisdell Publ. Co., London.
4. Hamming R.W. (1973). Numerical Method for Scientist and Engineers, McGraw-Hill, New York.

Modelling in Soil Dynamics by a Finite Domain with Respect to Transient Excitation

K.-H. Elmer, H.G. Natke and R. Thiede
Curt-Risch-Institut für Dynamik, Schall- und Meßtechnik, Universität Hannover, 3000 Hannover 1, Federal Republic of Germany

INTRODUCTION

If wave propagation from a machine hall through the soil to the residential building nearby has to be calculated, the soil is generally treated as an elastic half space. Because an analytical solution of the problem is mostly not possible it has to be gained by a numerical method. Here only a small section of the half space can be treated, i.e. artificial boundaries will be inserted. Waves striking such boundaries will be reflected (Fig. 1) and the geometrical damping of the halfspace will not be modelled. Various proposals exist on how to form such boundaries in order to avoid wave reflection. Unfortunately, most of these proposals can only be realized with special computer programmes. Only a few proposals are practicable if the solution has to be realized with a FEM programme (like ADINA) that is usually available in computer centres. One possibility often used is to arrange dashpots at the artificial boundaries (Lysmer and Kuhlemeyer[1]). Problems appear then at the lateral boundaries of the model, because frequency-dependent damping constants have to be chosen in order to damp out the striking Rayleigh waves. If the excitation is harmonic, the damping constants can easily be tuned to the excitation frequency. Problems appear if the excitation is transient. At this point the question has to be answered: for which frequency must the damping constants be chosen? Consideration for instance of the measured foot force spectra of different types of weaving machines shows that some excitations can be roughly regarded as harmonic ones because one frequency dominates (Fig. 2 a, b, c), so the damping constants can be tuned to this frequency. Other spectra, in contrast, clearly show different frequencies of nearly the same magnitude (Fig. 2 d), and thus no specific frequency can be quoted. In generalization the same statement holds true for transient excitation applying the FFT-technique.

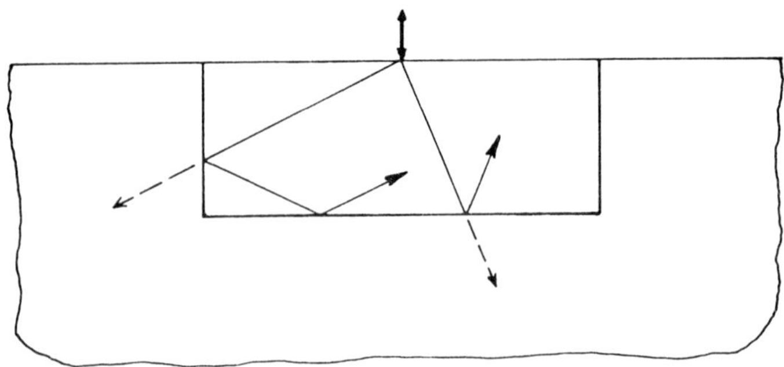

Figure 1. Finite domain with wave reflection.

In the next chapter proposals are made for a way of modelling the boundaries of the finite domain under transient excitation. The influences of the time step and the element size will be discussed in further chapters.

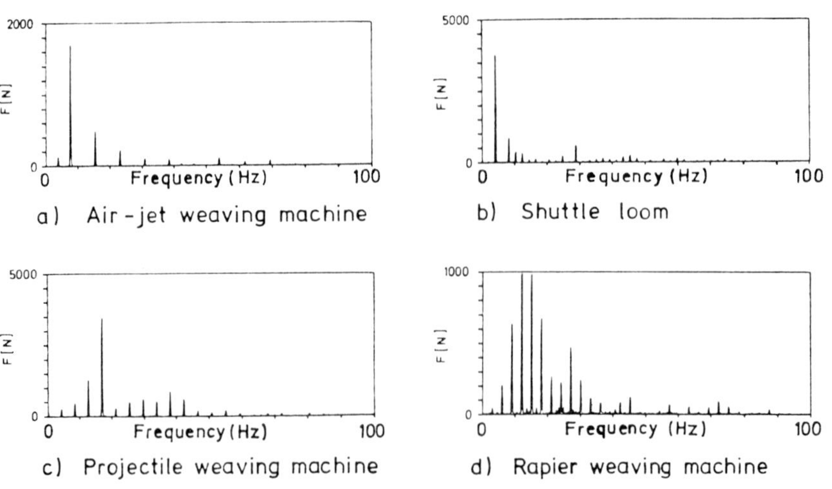

Figure 2. Spectra of the dynamic foot pressure F of different weaving machines.

BOUNDARY MODELLING

In order to deal with wave propagation problems with non harmonic excitation spectra like Fig. 2 d, both lateral boundaries of the finite domain are modelled by several dashpot rows placed side by side so that Rayleigh wave energy can be absorbed (Fig. 3). Each of these dashpot rows is designed for a different frequency of the excitation spectrum.

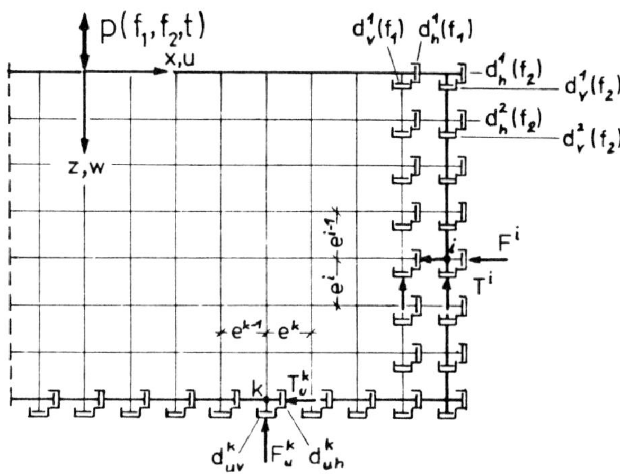

Figure 3. Modelling of the boundaries of the finite domain.

Following Lysmer and Kuhlemeyer[1] the following are the boundary damping forces for the node i

$$F^i = a^i \rho \, v_P \, \dot{u}^i \, \frac{e^{i-1} + e^i}{2} = d_h^i(f_n) \, \dot{u}^i \tag{1}$$

$$T^i = b^i \rho \, v_S \, \dot{w}^i \, \frac{e^{i-1} + e^i}{2} = d_v^i(f_n) \, \dot{w}^i \; . \tag{2}$$

The factors a^i and b^i can be calculated by the formulas

$$a^i = \frac{v_P}{v_R} \left[1 - \left(1 - 2 \frac{v_S^2}{v_P^2} \right) \frac{h'(k_R z^i)}{g(k_R z^i)} \right] \tag{3}$$

$$b^i = \frac{v_S}{v_R} \left[1 + \frac{g'(k_R z^i)}{h(k_R z^i)} \right] \; . \tag{4}$$

The horizontal and the vertical components u and w of the Rayleigh wave can be represented in the following form

$$u = g(k_R z)\sin(\omega t - k_R x) \tag{5}$$

$$w = h(k_R z)\cos(\omega t - k_R x) \quad . \tag{6}$$

Here

$$g(k_R z) = D\left[e^{-qz} - \frac{2sq}{s^2 + k_R^2} e^{-sz}\right] \tag{7}$$

$$h(k_R z) = D\left[-\frac{q}{k_R} e^{-qz} + \frac{2qk_R}{s^2 + k_R^2} e^{-sz}\right] \tag{8}$$

with $q = \sqrt{k_R^2 - k_P^2} \quad , \quad s = \sqrt{k_R^2 - k_S^2}$

The symbols used in the preceding formulas are:

V_R Rayleigh wave velocity
V_P compression wave velocity
V_S shear wave velocity

$k_R = \frac{\omega}{V_R}$, $k_P = \frac{\omega}{V_P}$, $k_S = \frac{\omega}{V_S}$ wave numbers

$\omega = 2\pi f$ circular frequency of excitation
f_n n-th frequency of excitation
e^{i-1}, e^{i} length of the elements neighbouring the boundary node i
ρ density
D constant

$$(\ldots)' = \frac{d(\ldots)}{d(k_R z)} \quad , \quad (\dot{\ldots}) = \frac{d(\ldots)}{dt} \quad .$$

As expressions (3) and (4) show, the factors a and b are dependent on depth and frequency, and therefore the damping constants d_h and d_v are also dependent on depth and frequency. For given frequencies of excitation f_n these damping constants can be calculated for particular nodes i with the help of the preceding relations. It does not matter whether this node i lies at the boundary or in the interior of the finite domain of the half space, because the horizontal distance x does not appear in the calculation of the damping constants.

At the lower boundary of the finite domain dashpots have to be designed only in the boundary nodes, even if the excitation is non harmonic, because the damping constants are independent of frequency, as is shown by Lysmer and Kuhlemeyer[1]. The damping forces perpendicular and parallel to the boundary (Fig. 3) result in

$$F_u^k = a^k \rho\, v_p\, \dot{w}^k\, \frac{e^{k-1}+e^k}{2} = d_{uv}^k\, \dot{w}^k \tag{9}$$

$$T_u^k = b^k \rho\, v_s\, \dot{u}^k\, \frac{e^{k-1}+e^k}{2} = d_{uh}^k\, \dot{u}^k \quad . \tag{10}$$

With sufficient accuracy the factors a and b can be chosen equal to 1. It follows that for all the nodes k of the lower boundary, damping constants can be stated independent of depth and frequency:

$$d_{uv}^k = \rho\, v_p\, \frac{e^{k-1}+e^k}{2} \tag{11}$$

$$d_{uh}^k = \rho\, v_s\, \frac{e^{k-1}+e^k}{2} \quad . \tag{12}$$

A precondition for modelling of this kind is that the lower boundary of the finite domain is placed so deep that the action of the Rayleigh wave can be neglected when compared with the compression wave and the shear wave.

Some results of calculated examples are presented here in order to demonstrate the effectiveness and usefulness of the suggested possibility of solving wave propagation problems in the soil. All the calculations are carried out with the model of the finite domain shown in Fig. 4.

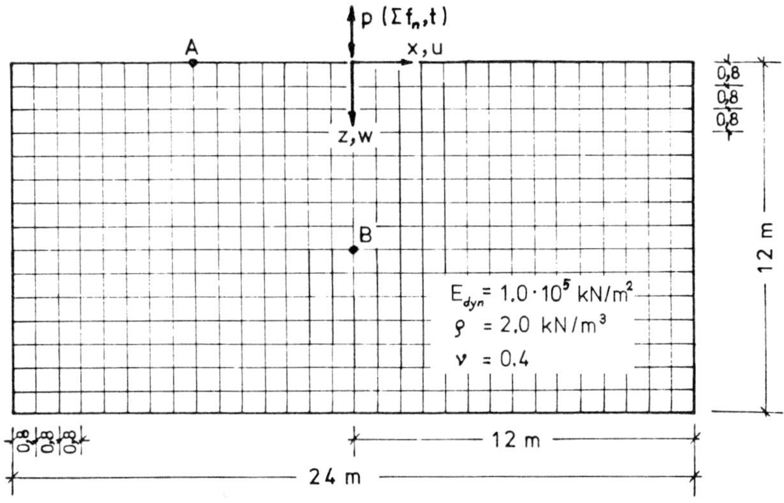

Figure 4. Finite element model for example calculation.

Example 1: The excitation on the surface of the finite domain includes two frequencies of 17.5 Hz and 22.5 Hz (Fig. 5 b), the vertical displacements in the excitation point are plotted in Fig. 5 a. Corresponding to the excitation, the lateral boundaries are modelled with two dashpot rows. The damping constants in the boundary nodes are tuned to 22.5 Hz, and those in the node row beside the boundary are tuned to 17.5 Hz. Fig. 6 shows the dynamic vertical displacement and the spectrum in the observation point A on the surface, and Fig. 7 the corresponding quantities in point B in the interior of the domain. The time plots of the displacement in all three points show stationary behaviour. If there were any perceptible reflections the displacement amplitudes would reveal them with advancing time.

Example 2: The model is the same as in Example 1, the only difference being in the excitation. Now only one frequency of 20 Hz is included (Fig. 8 b). This produces the peculiarity that, with respect to the frequency, the damping constants and the excitation do not correspond. Nevertheless, the results plotted in Figs. 8, 9, and 10 also show stationary behaviour.

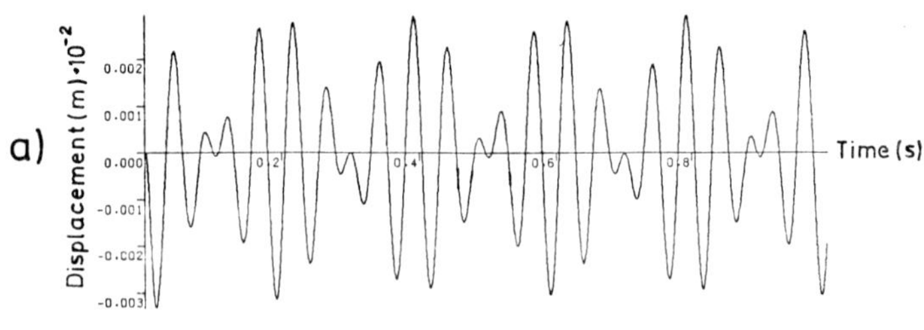

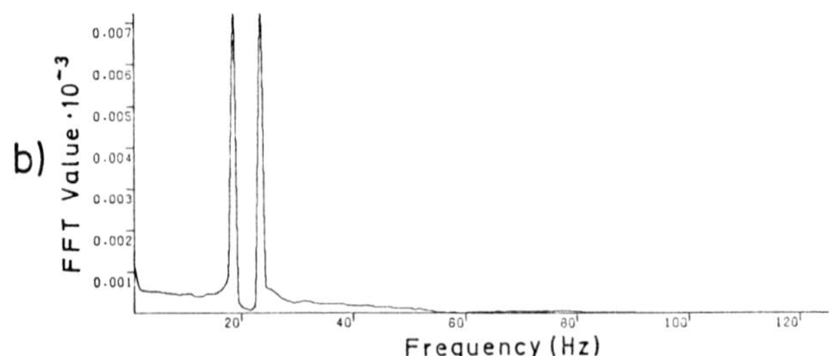

Figure 5. Dynamic vertical displacement and spectrum in the excitation point. Damping constants tuned to 17.5 Hz and 22.5 Hz.

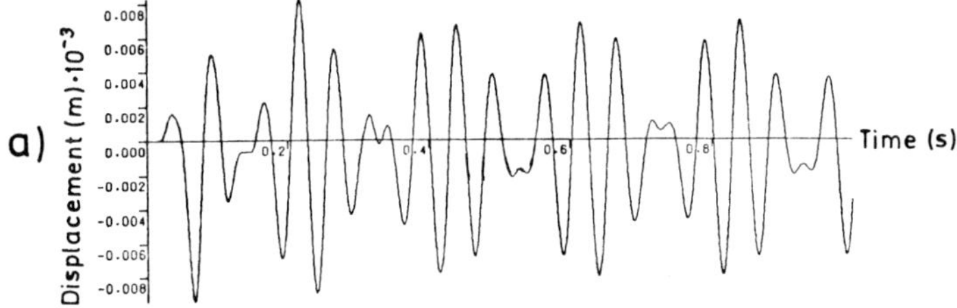

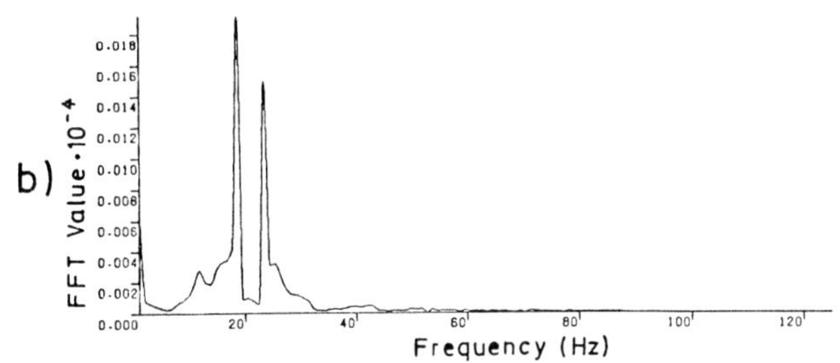

Figure 6. Dynamic vertical displacement and spectrum in point A. Damping constants tuned to 17.5 Hz and 22.5 Hz.

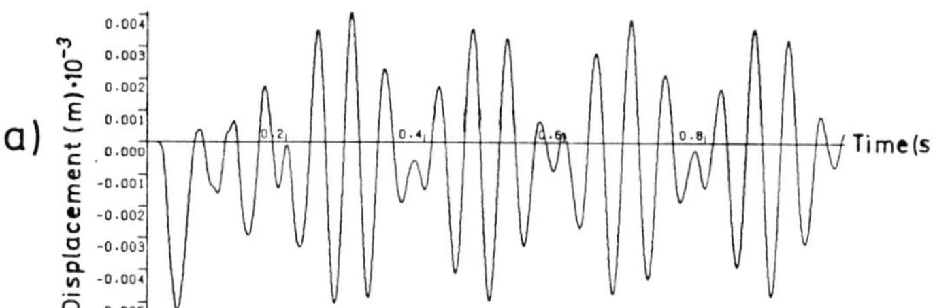

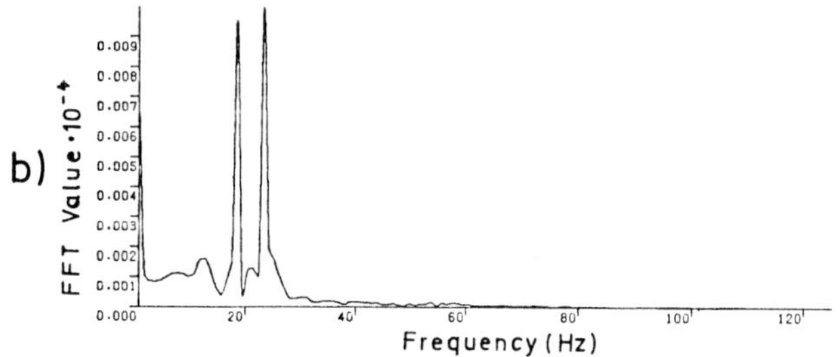

Figure 7. Dynamic vertical displacement and spectrum in point B. Damping constants tuned to 17.5 Hz and 22.5 Hz.

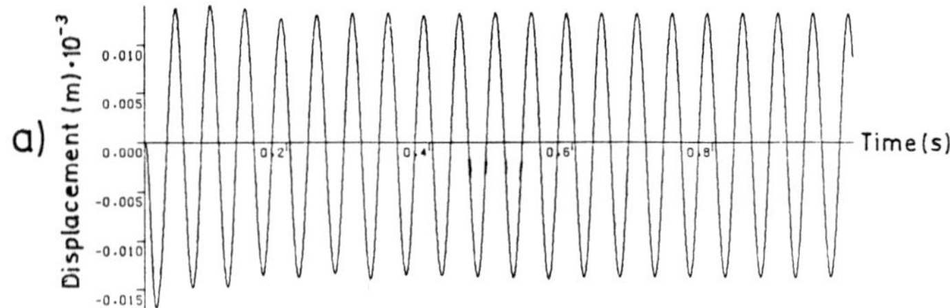

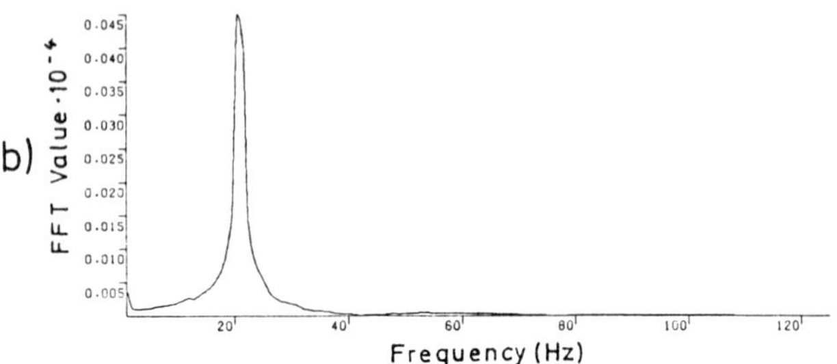

Figure 8. Dynamic vertical displacement and spectrum in the excitation point. Damping constants tuned to 17.5 Hz and 22.5 Hz.

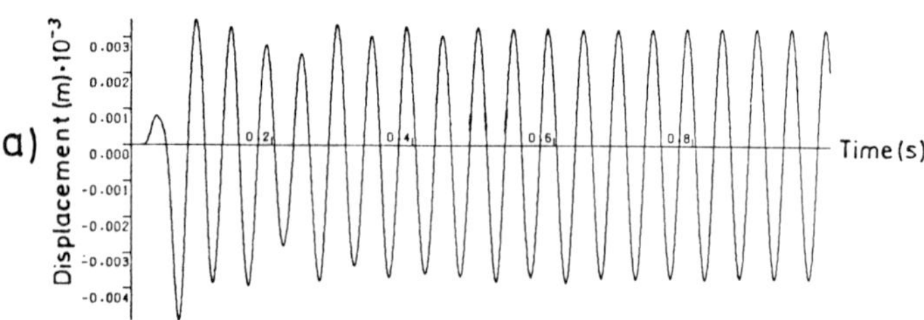

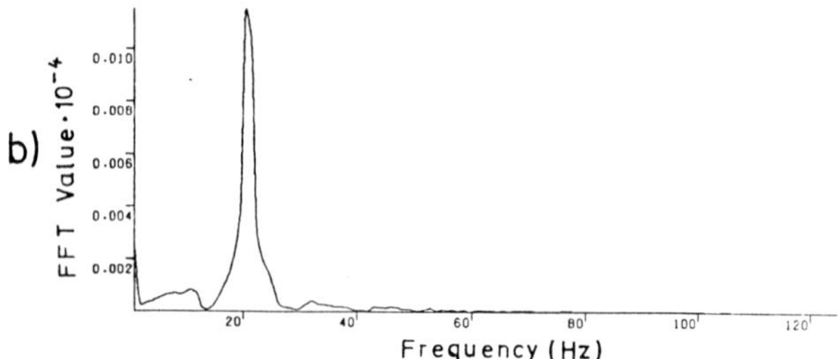

Figure 9. Dynamic vertical displacement and spectrum in point A. Damping constants tuned to 17.5 and 22.5 Hz.

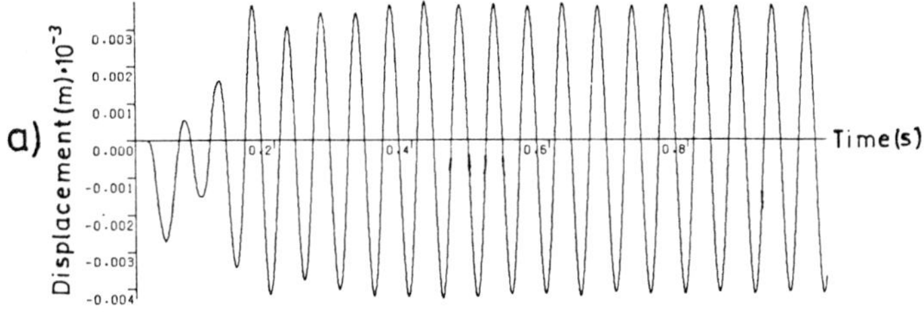

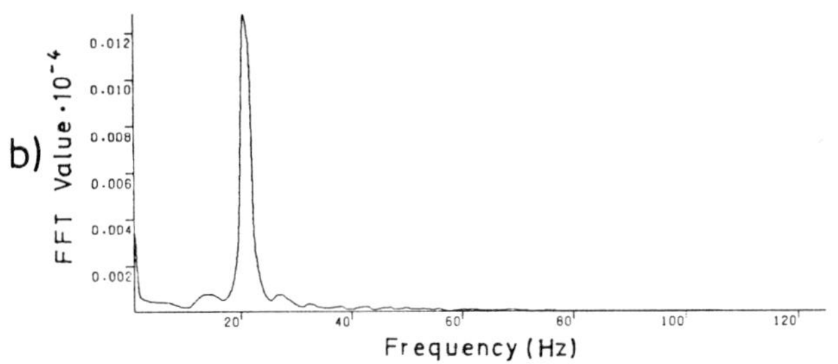

Figure 10. Dynamic vertical displacement and spectrum in point B. Damping constants tuned to 17.5 and 22.5 Hz.

Example 3: Example 3 corresponds to Example 2, but now the damping constants are changed. The damping constants in the lateral boundary nodes are calculated for 25 Hz, and those in the node row beside the boundary are tuned to 15 Hz. Although these damping constants differ from the frequency of excitation (20 Hz) about 25%, the results in Figs. 11, 12, and 13 hardly show differences from the corresponding results in Figs. 8, 9, and 10. Despite the doubled calculation time, the time plots of the displacements still show roughly stationary behaviour, i.e. perceptible reflections cannot be identified.

The results of Examples 2 and 3 indicate that it is not necessary to tune the damping constants to the excitation with great accuracy. But these are first results, and it is not possible to generalize them at the moment. More investigations and more detailed ones must be carried out in order to be able to do that.

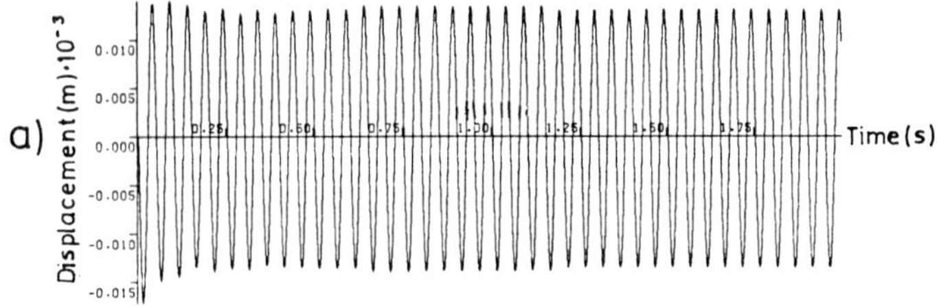

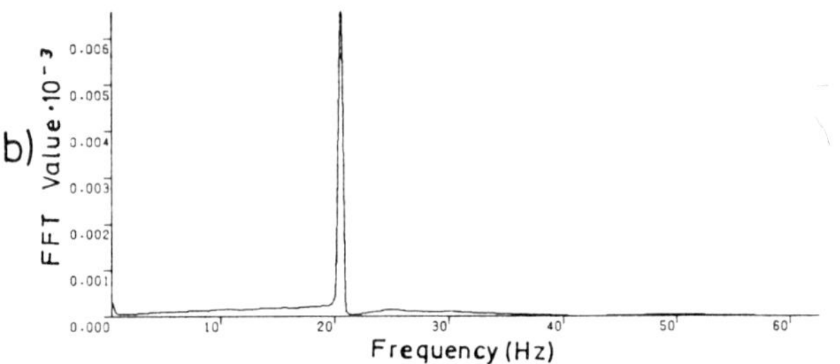

Figure 11. Dynamic vertical displacement and spectrum in the excitation point. Damping constants tuned to 15 Hz and 25 Hz.

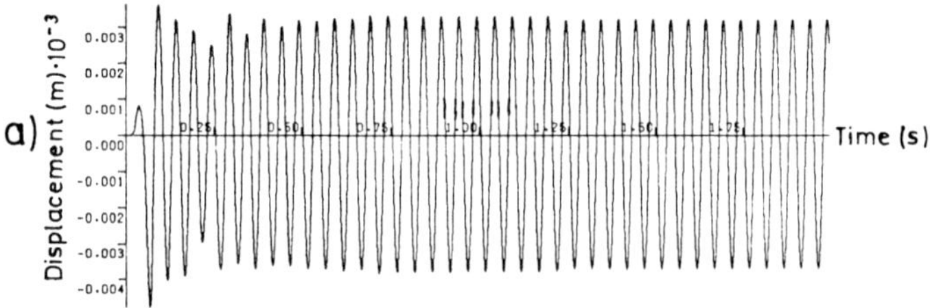

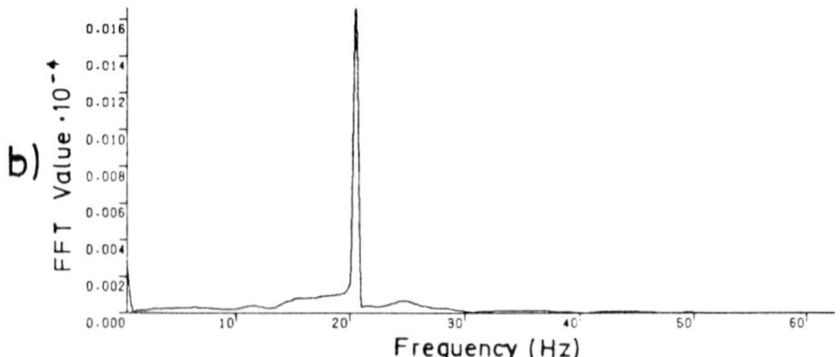

Figure 12. Dynamic vertical displacement and spectrum in point A. Damping constants tuned to 15 Hz and 25 Hz.

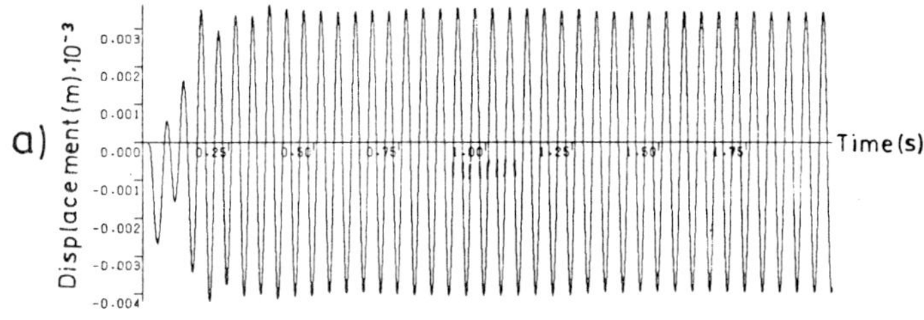

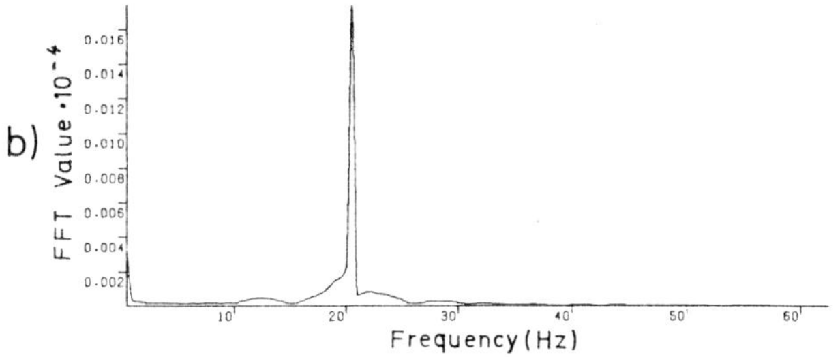

Figure 13. Dynamic vertical displacement and spectrum in point B. Damping constants tuned to 15 Hz and 25 Hz.

Example 4: In Example 4 we have three dominating frequencies of excitation with 12.5 Hz, 15.5 Hz, and 18.5 Hz. Three rows of corresponding dampers are modelled at the lateral boundaries. The results in Figs. 14, 15, and 16 do not show noticeable disturbances caused by reflections at the boundaries.

THE INFLUENCE OF THE TIME STEP

In numerical solutions to wave propagation problems in the soil special attention has to be paid to temporal discretisation. It depends on the choice of the right time step whether a special frequency is included, whether wave propagation can be calculated at all in connexion with the given spatial discretisation, and whether the process of calculation advances stably or not. In order to include a frequency with accompanying period T_n, experience shows that the time step Δt should be chosen as

$$\Delta t \leq \frac{T_n}{20} \qquad (13)$$

(see Bathe[2] chap. 9.4.4 and Levy and Wilkinson[3] chap. 1.7).

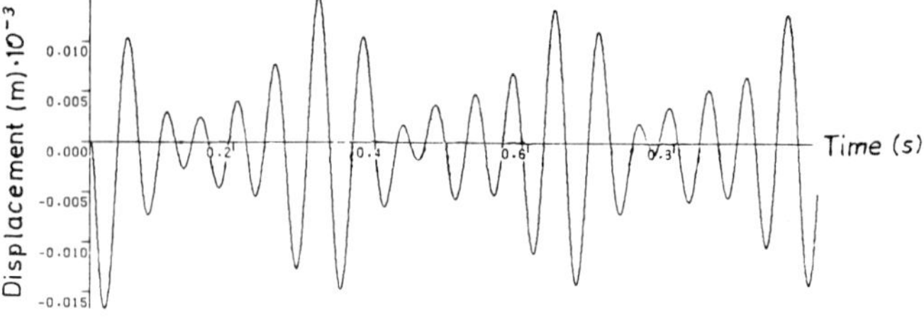

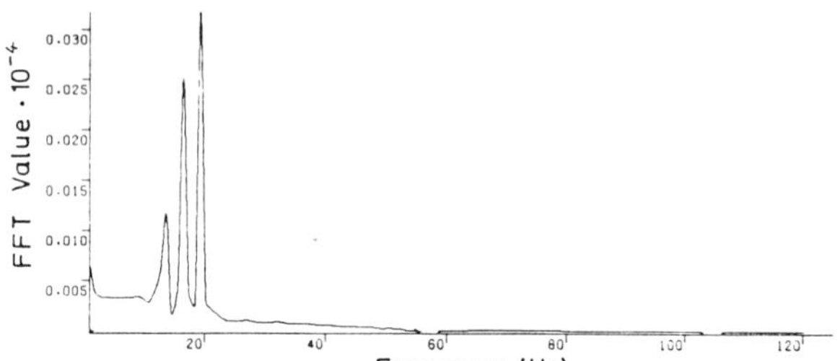

Figure 14. Dynamic vertical displacement and spectrum in the excitation point. Damping constants tuned to 12.5 Hz, 15.5 Hz, and 18.5 Hz.

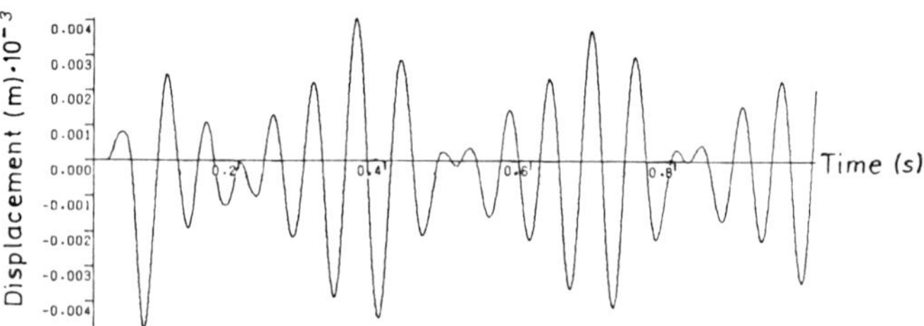

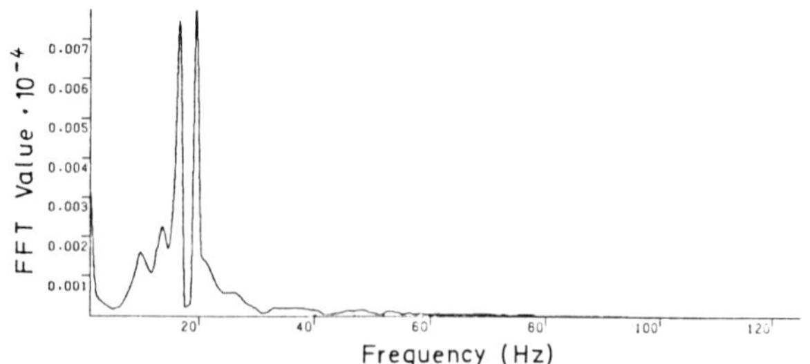

Figure 15. Dynamic vertical displacement and spectrum in point A. Damping constants tuned to 12.5 Hz, 15.5 Hz, and 18.5 Hz.

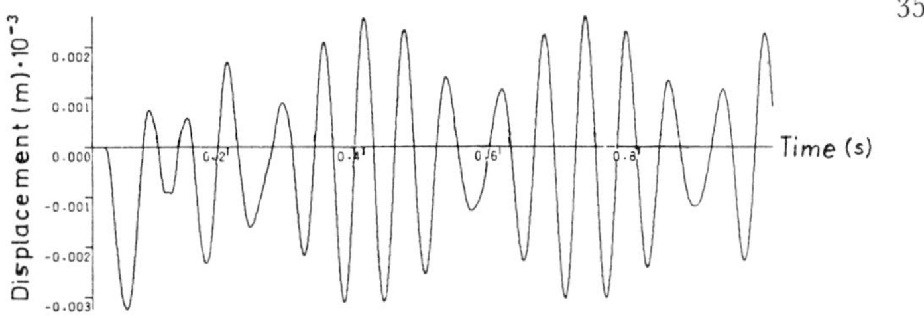

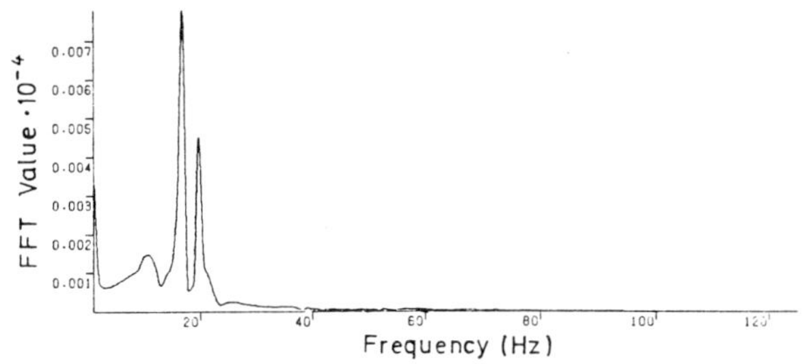

Figure 16. Dynamic vertical displacement and spectrum in point B. Damping constants tuned to 12.5 Hz, 15.5 Hz, and 18.5 Hz.

Integration of wave propagation problems will be stable if

$$\Delta t \leq \frac{1}{v\sqrt{\frac{1}{\Delta x^2} + \frac{1}{\Delta y^2} + \frac{1}{\Delta z^2}}} \tag{14}$$

in which Δx, Δy, Δz stand for the element dimensions in the x-, y-, z-directions respectively, and v is the wave velocity (in soil dynamics the compression wave velocity). Eq. (14) deals with the three-dimensional problems, and for two- and one-dimensional calculations the non-existent element dimensions vanish in the denominator. Eq. (14) is based on the considerations of Courant, Friedrichs and Lewy[4] and Forsythe and Wasow[5] (see also Bathe[2] chap. 9.4.4).

Several methods are available for the performance of the time integration. The best known are the central difference method, the Newmark method, and the Wilson-Θ method. The chosen timestep has different effects on the solutions gained with these methods. For example, for the only conditionally stable difference method it follows from the stability analysis that

$$\Delta t \leq \frac{T_n}{\pi} \tag{15}$$

has to be preserved (see Bathe[2] chap. 9.4.2 and Levy and Wilkinson[3] chap. 1.7). This condition is satisfied if the time step satisfies Eqs. (13) and (14). The Wilson-θ and the Newmark methods are unconditionally stable, i.e. the choice of Δt is subject to the demands of (13) and (14) only.

For the one-dimensional case of wave propagation, investigations have been made with the help of the FEM-programme ADINA in order to get an answer on how the time step effects the results at different distances from the excitation source. These results were received by the three integration methods named above (see Elmer and Thiede[6]). The compression wave propagation has been studied in a rod 250 m long with soil characteristics (without damping) and with a harmonic excitation force of 1 kN and 20 Hz at one end of the rod. The wave train propagating through the rod was observed every 0.5 seconds (corresponding to 10 periods) at distances from the excitation point of 0, 2.5, 5, 7.5, and 15 wave lengths (L_w). With a wave length L_w = 10 m this conforms to distances of 0, 25, 50, 75, and 150 m. It is well known that an amplitude decay occurs with the use of time integration methods for solving dynamic problems (see Bathe[2] chap. 9.4.3). For the wave propagation problem investigated, the corresponding amplitude decay with an increasing distance from the excitation source is represented in Figs. 17 and 18 for various time step-period ratios and different integration methods. The results obtained from the solution with the central difference method, and which are not plotted, show that for all the considered ratios $\Delta t/T$ the amplitude decay at a distance of 15 wave lengths is less than 5%. For all integration methods the investigation results indicate that a time step determined by Eqs. (13) and (14), i.e. $\Delta t/T \leq 0.05$, leads to acceptable solutions in soil dynamics, especially if the distance from the excitation source is not too great.

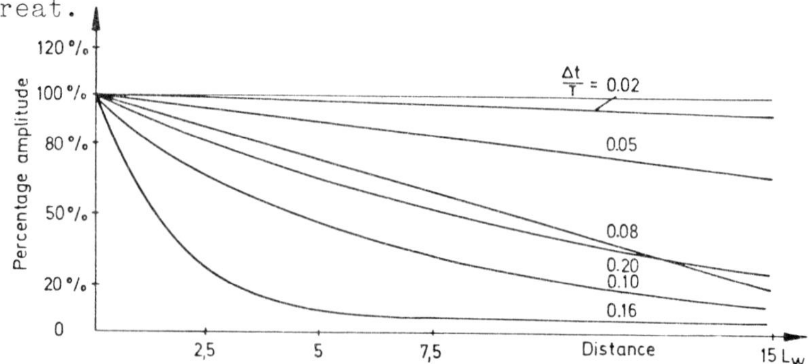

Figure 17. Average amplitude decay using the Wilson-θ method.

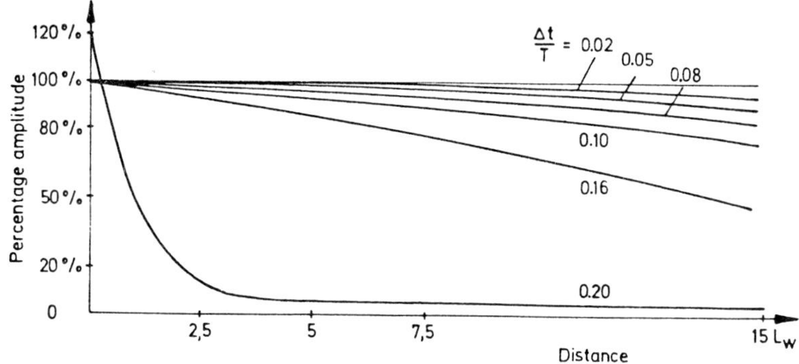

Figure 18. Average amplitude decay using the Newmark method.

THE INFLUENCE OF ELEMENT SIZE

In a similar way to the influence of the time step on the results of wave propagation calculation, the influence of the element size is investigated using the same system (see Elmer and Thiede[6]). For the three different time integration methods Figs. 19, 20, and 21 show the amplitude decay dependent on the distance from the excitation source, on the ratio b = element number/wave length, and on the ratio a = time step/ period. The curves in the figures illustrate that if the time step is chosen suitably for Eqs. (13) and (14), the choice of about 10 elements per wave length, as recommended in publications, leads to satisfactory accurate results for all the integration methods. If the excitation is transient, the element size, of course, has to match the wave length of the highest frequency of excitation taken into account. This leads to correspondingly small element dimensions, and, according to Eq. (14), to similarly small time steps.

CONCLUDING REMARKS

The calculation of wave propagation in the soil by numerical solution methods produces the problem of the satisfactory accurate modelling of the finite

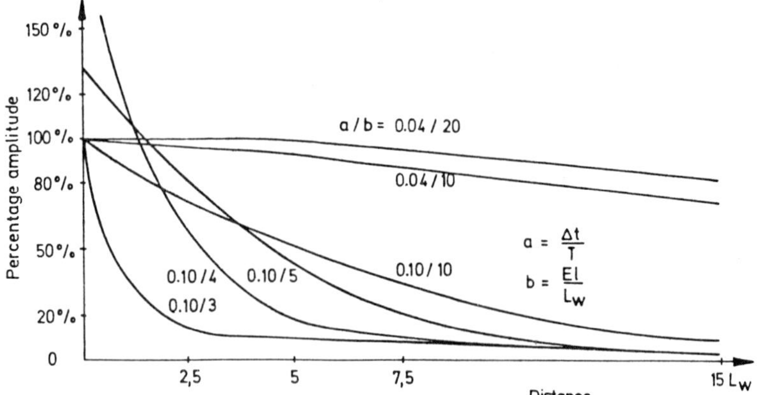

Figure 19. Average amplitude decay using the Wilson-θ method.

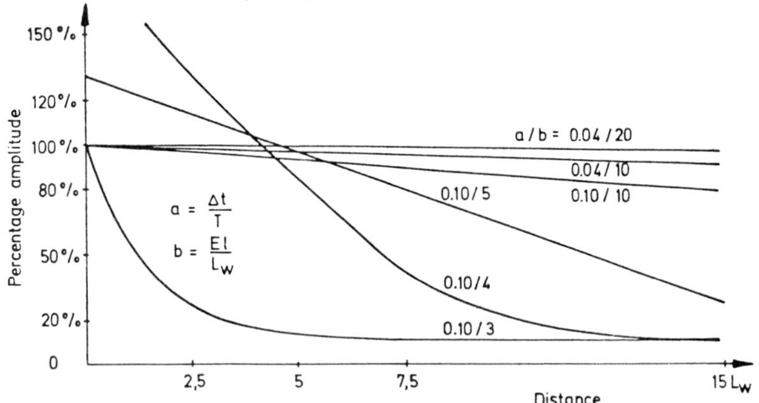

Figure 20. Average amplitude decay using the Newmark method.

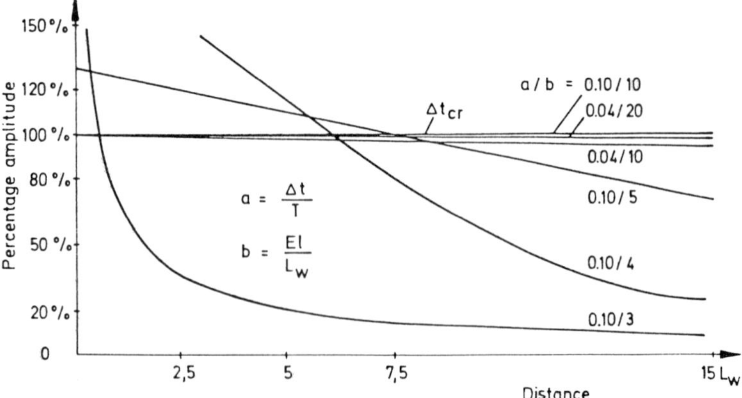

Figure 21. Average amplitude decay using the central difference method.

domain. Wave reflection at the artificial boundaries of the model must be avoided. A possible design for these boundaries is presented for the case of non harmonic excitation discretized in the frequency domain. It is characterized by placing several dashpot rows side by side at the lateral boundaries. Each of these dashpot rows is tuned to a different frequency of the excitation. The results of calculated examples demonstrate the effectiveness and usefulness of the possible method of solving the problem as suggested here. This method is very simple and it does not require great effort, either in designation or in calculation. It can be used in connection with a FEM programme (like ADINA) that is usually available in computer centres.

The results of investigations with regard to the choice of satisfactory time steps and element sizes in wave propagation calculations with finite elements and different time integration methods are also presented. Until now these investigations have only been made for spatial one-dimensional systems. The applicability of this insights to spatial multidimensional systems has not yet been found.

ACKNOWLEDGEMENTS

The authors are grateful to cand. math. U. Prells, who calculated the examples, and to Regionales Rechenzentrum Niedersachsen (RRZN), University Hannover, where all computations were performed.

REFERENCES

1. Lysmer J. and Kuhlemeyer R.L. (1969), Finite Dynamic Model for Infinite Media, Journal Eng. Mech. Div., ASCE, EM4, Vol. 95, pp. 859-877.

2. Bathe K.-J. (1986), Finite Element Methoden (German translation of "Finite Element Procedures in Engineering Analysis", Prentice Hall, 1982), Springer-Verlag, Berlin-Heidelberg-New York-Tokyo.

3. Levy S. and Wilkinson J.P.D. (1976), The Component Element Method in Dynamics, McGraw-Hill Book Comp., New York.

4. Courant R., Friedrichs K. and Lewy H. (1928), Über die partiellen Differenzengleichungen der mathematischen Physik, Mathematische Annalen, Vol. 100, pp. 32-74.

5. Forsythe G.E. and Wasow W.R. (1960), Finite-Difference Methods for Partial Differential Equations, Whiley & Sons, New York.

6. Elmer K.-H. and Thiede R. (1987), Finite Element Modelling and Analysis of Wave Propagation Mechanism Investigated with ADINA, presented at the 6th Conf. on Nonlinear Finite Element Analysis and ADINA, June 10-12, 1987, MIT, Cambridge, Massachusetts, USA.

Fast Fourier Bessel Transforms for Calculating the Green's Function for Semi-Infinite Soil Media

T. Kobori, K. Miura and T. Moroi
Kobori Research Complex, Kajima Corporation, Fl.30 Shinjuku Mitsui Bldg., Nishishinjuku, Shinjuku-ku, Tokyo 163, Japan
K. Masuda, F. Sasaki
Information Processing Center, Kajima Corporation, 2-7 Motoakasaka, 1-chome, Minato-ku, Tokyo 107, Japan

INTRODUCTION

Recently , the characteristics of the dynamic soil - structure interaction for an embedded structure which was difficult to analyse by former methods has made clear progress by virtue of BEM . BEM is the method which induces the boundary integral equation according to apply the Divergence Theorem to wave equation using Green's function . Therefore it possesses the inevitability of calculating the Green's function .

In dynamic interaction analysis , it is necessary to calculate the Green's function for semi-infinite soil media considering the free surface condition. In generally, this Green's function is given in the form of the following Fourier-Bessel transform (FBT) .

$$F_k(a_0)=\int_0^{\infty} f(\zeta)\zeta J_k(a_0\zeta)\,d\zeta$$

The above infinite integral has already been calculated by numerical integration , but large amount of computational time was required . If the calculation is accurately done in less computational time , it would become more practical to be applied for the dynamic interaction problem of BEM .

Several methods for calculating the FBT by using the FFT were reported (Siegman[1] ; Talman[2] ; Kausel and Bouckovalas[3] ; Candel[4]). The method of S. M. Candel calculates high range of degree k simultaneously , while the required degree in dynamic interaction problem is limited to the 2nd order.

This paper proposes an advanced procedure of Candel's method to lower degree FBT . Accuracy and computational time of this method are examined in comparison with the other methods .

ANALYTICAL METHOD

Green's function for semi - infinite soil media is represented by the followin Fourier - Bessel transform .

$$F_k(r)=\int_0^{\infty} f(\zeta)\zeta J_k(\zeta r)\,d\zeta \qquad k=0\ ,\ 1\ ,\ 2 \quad \cdots\cdots (1)$$

Generally , $f(\zeta)$ is a complex function

$$f(\zeta)=f^{(r)}(\zeta)+if^{(i)}(\zeta) \quad \cdots\cdots (2)$$

Let us divide eq. (1) into the real part and the imaginary part .

$$F_k(r)=G_k(r)+iH_k(r) \quad \cdots\cdots (3)$$

where

$$G_k(r)=\int_0^{\infty} f^{(r)}(\zeta)\zeta J_k(\zeta r)\,d\zeta \quad \cdots\cdots (4-1)$$

$$H_k(r)=\int_0^{\infty} f^{(i)}(\zeta)\zeta J_k(\zeta r)\,d\zeta \quad \cdots\cdots (4-2)$$

Considering the Fourier Transform of function $f^{(r)}(\zeta)\zeta$,

$$\int_{-\infty}^{\infty} f^{(r)}(\zeta)\zeta e^{i\eta\zeta}\,d\zeta \quad \cdots\cdots (5)$$

under the following condition ,

$$f^{(r)}(\zeta)=\begin{cases} f^{(r)}(\zeta) & :\zeta\geqq 0 \\ 0 & :\zeta<0 \end{cases} \quad \cdots\cdots (6)$$

Let $\phi(\eta)$ designate the one - side Fourier Tramsform of $f^{(r)}(\zeta)\,\zeta$,

$$\phi(\eta)=\int_0^{\infty} f^{(r)}(\zeta)\zeta e^{i\eta\zeta}\,d\zeta \quad \cdots\cdots (7)$$

Using the generating function of Bessel function , $G_k(r)$ can be represented in the following form .

$$G_k(r)=\int_0^{\infty} f^{(r)}(\zeta)\zeta J_k(\zeta r)d\zeta=\frac{1}{2\pi}\int_0^{2\pi}\phi(r\sin\theta)e^{-ik\theta}d\theta \quad \cdots (8)$$

Discretize $f^{(r)}(\zeta)$ into next form .

$$\hat{f}^{(r)}(n)=\begin{cases} f^{(r)}(n\Delta\zeta) & :n=0\ ,1\ ,\cdots,\ \frac{M}{2}-1 \\ 0 & :n=\frac{M}{2}\ ,\cdots,\ M-1 \end{cases} \quad \cdots\cdots (9)$$

Eq. (7) yields :

$$\phi(m\Delta\eta)=(\Delta\zeta)^2\hat{\phi}(m) \quad \cdots\cdots (10)$$

$$\hat{\phi}(m)=\sum_{n=0}^{M-1}\hat{f}^{(r)}(n)\,n e^{i\frac{2\pi nm}{M}} \quad \cdots\cdots (11)$$

where

$$\Delta\eta \cdot \Delta\zeta = \frac{2\pi}{M} \quad \cdots\cdots (12)$$

Similary , let

$$r = l\Delta r \ : \ l = 0 \ , \ 1, \cdots\cdots, L-1 \quad \cdots\cdots (13)$$

$$\theta_j = j\frac{2\pi}{S} \quad \cdots\cdots (14)$$

Eq. (8) is

$$G_k(r) = \frac{(\Delta\zeta)^2}{S}\sum_{j=0}^{S-1}\hat{\phi}(l\Delta r \sin\theta_j)e^{-i\frac{2\pi jk}{S}} \quad \cdots\cdots (15)$$

where

$$\Delta\zeta\,\Delta r = \frac{2\pi}{N}, N \leqq M \quad \cdots\cdots (16)$$

$\hat{\phi}(l\Delta r \sin\theta_j)$ is calculated from $\hat{\phi}(m)$ using linear interpolation . Let $\hat{\phi}_{l,j}$ be the value of $\hat{\phi}(l\Delta r \sin\theta_j)$.
The $\hat{\phi}_{l,j}$ have the following properties due to the Fourier coefficient and the sine function characteristics.

$$\hat{\phi}_{l,j} = \hat{\phi}_{l,\frac{S}{2}-j} \ : j = \frac{S}{4}+1, \cdots, \frac{S}{2} \quad \cdots\cdots (17-1)$$

$$\hat{\phi}_{l,j} = \hat{\phi}^{*}_{l,S-j} \ : \ j = \frac{S}{2}+1, \cdots, S-1 \quad \cdots\cdots (17-2)$$

$$\hat{\phi}_{l,j} = \hat{\phi}_{l,\frac{3}{2}S-j} \ : \ j = \frac{3}{4}S+1 \ , \cdots, S-1 \quad \cdots\cdots (17-3)$$

moreover

$$\hat{\phi}^{(r)}_{l,S} = \hat{\phi}^{(r)}_{l,0} \quad \cdots\cdots (17-4)$$

$$\hat{\phi}^{(i)}_{l,0} = \hat{\phi}^{(i)}_{l,\frac{S}{2}} = \hat{\phi}^{(i)}_{l,S} \quad \cdots\cdots (17-5)$$

where

$$\hat{\phi}_{l,j} = \hat{\phi}^{(r)}_{l,j} + i\hat{\phi}^{(i)}_{l,j} \quad \cdots\cdots (17-6)$$

$$\hat{\phi}^{*}_{l,j} = \hat{\phi}^{(r)}_{l,j} - i\hat{\phi}^{(i)}_{l,j} \quad \cdots\cdots (17-7)$$

Applying the above characteristics to eq. (15) , the next formulations are obtained .

$$G_0(l\Delta r) = \frac{(\Delta\zeta)^2}{S}\left[\, 2\,(\hat{\phi}^{(r)}_{l,0} + \hat{\phi}^{(r)}_{l,\frac{S}{4}}) + 4\sum_{j=1}^{\frac{S}{4}-1}\hat{\phi}^{(r)}_{l,j}\right] \quad \cdots\cdots (18)$$

$$G_1(l\Delta r)=\frac{(\Delta\zeta)^2}{S}\left[4\sum_{j=1}^{\frac{S}{4}-1}\widehat{\phi}^{(i)}_{l,j}\sin\theta_j+2\widehat{\phi}^{(i)}_{l,\frac{S}{4}}\right] \quad \cdots\cdots\cdots\cdots\cdots\cdots\cdots\cdots\cdots (19)$$

$$G_2(l\Delta r)=\frac{(\Delta\zeta)^2}{S}\left[4\sum_{j=1}^{\frac{S}{4}-1}\widehat{\phi}^{(r)}_{l,j}\cos2\theta_j+2(\widehat{\phi}^{(r)}_{l,0}-\widehat{\phi}^{(r)}_{l,\frac{S}{4}})\right] \quad \cdots\cdots\cdots (20)$$

$H_k(r)$ can obviously be calculated similarly from eq (15). Candel used eq (15) to simultaneous calculations to orders up to k, but in the case of dynamic soil - structure interaction, only the order 0, 1 and 2 are needed, thus the use of eq (18) to (20) saving computation time. This method is called the Fast Fourier Bessel Transform (FFBT).

INVESTIGATION OF ACCURACY AND COMPUTATIONAL TIME

To show accuracy and computational time of FFBT, the Green's function of a vertical point load applying on the surface of semi-infinite soil medium is examined. As in Fig-1, the vertical response displacement of surface in case of the point load $Pe^{i\omega t}$ applying on the surface (z=0) is given as follows.

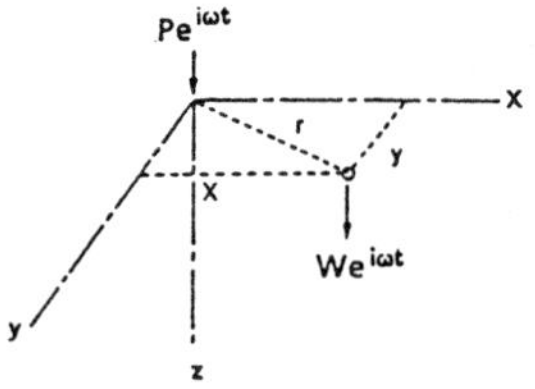

$$w=\frac{1-\nu}{2\pi G}\cdot\frac{Pe^{i\omega t}}{r}(f_1+if_2) \quad \cdots\cdot (21)$$

Fig.1 POINT LOAD ON SURFACE

where

$$f_1+if_2=-\frac{a_0}{1-\nu}g^2\int_0^{\infty}\frac{\sqrt{\zeta^2-\gamma^2 g}}{(2\zeta^2-g)^2-4\zeta^2\sqrt{\zeta^2-\gamma^2 g}\sqrt{\zeta^2-g}}\cdot\zeta\ J_0(a_0\zeta)\,d\zeta \quad \cdots (22)$$

$$a_0=\frac{\omega r}{Vs},\ \gamma=\sqrt{\frac{1-2\nu}{2(1-\nu)}},\ g=\frac{1}{1+i2h}$$

G : shear rigidity $\qquad$ Vs : shear wave velocity

ν : poisson's ratio $\qquad$ h : damping of soil

The integral part of eq (22) is FBT of the 0-th order.
Calculation flow is summarized in the following.

(1) Choose M in a power of 2 under the condition $M \geqq N$.
Choose S in a multiple of 4

(2) Choose Δa_0 and $\Delta\zeta$ under the condition $\Delta a_0 \Delta\zeta = 2\pi/N$

(3) Calculate and store $\sin\theta_j$
where

$$\theta_j=j\frac{2\pi}{S} \quad : j=0\ ,\ 1\ ,\ 2\ ,\ \cdots\cdots,\frac{S}{4}$$

(4) Calculate discrete function $f(n)$: $n = 0,1, \cdots\cdots, M\text{-}1$

$$\hat{f}(n) = \begin{cases} f(n\Delta\zeta) : n=0,1, \cdots, \dfrac{M}{2}-1 \\ 0 \quad : n=\dfrac{M}{2}, \cdots, M-1 \end{cases}$$

(5) Calculate the real part of $\hat{f}^{(r)}(n)\,n$ and the imaginary part of $\hat{f}^{(i)}(n)\,n$ of $\hat{f}(n)\,n$ using FFT

$$\hat{\phi}(m) = \sum_{n=0}^{M-1} \hat{f}^{(r)}(n)\,n\,e^{i\frac{2\pi nm}{M}}$$

$$\hat{\psi}(m) = \sum_{n=0}^{M-1} \hat{f}^{(i)}(n)\,n\,e^{i\frac{2\pi nm}{M}}$$

(6) Set $l\,(a_0 = l\Delta a_0)$

(7) Calculate $\hat{\phi}_{lj}$, $\hat{\psi}_{lj}$ using $\hat{\phi}(m)$, $\hat{\psi}(m)$

$$\hat{\phi}_{lj}, \hat{\psi}_{lj} : j = 0, 1, \cdots\cdots, \frac{S}{4}$$

(8) Calculate $G_0(l\Delta a_0)$, $H_0(l\Delta a_0)$ using eq (18)

(9) Calculate $F_0(a_0)$,

$$F_0(a_0) = \{G_0(l\Delta a_0) + iH_0(l\Delta a_0)\}$$

(10) Increment l and repeat from step (7)

The above calculation flow is also applicable to $k=1,2$ changing eq (18) in step (8) into eq (19) or eq (20). The displacement function f_1+if_2 of eq (22) is calculated along the above steps. In this case, the function is calculated with $\nu=0.4$, $\Delta a_0=0.2$, $M=2N$, and $S=4N$.

Fig.2 shows the transformed real part $f^{(r)}(\zeta)$ of function $f(\zeta)$ and Fig.3 shows the imaginary part $f^{(i)}(\zeta)$ of $f(\zeta)$. In the case of an undamped soil (h=0), the Rayleigh pole is located on the real axis, which is the path of integration, and thus the real part diverges at this point. However, when the soil has some damping, the pole is shifted from the real axis and $f^{(r)}(\zeta)$ becomes a continuous function.

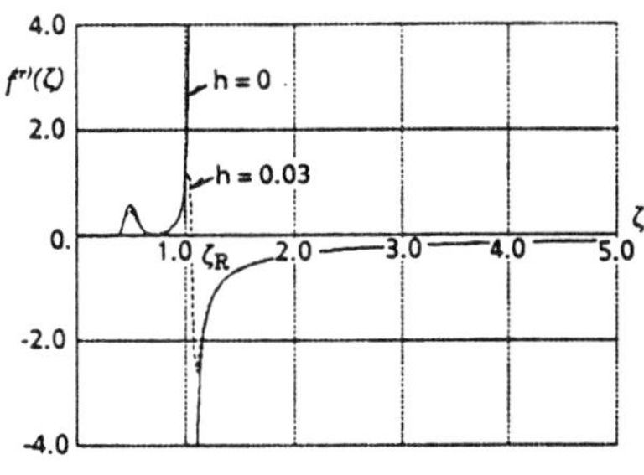

Fig.2 REAL PART OF $f(\zeta)$

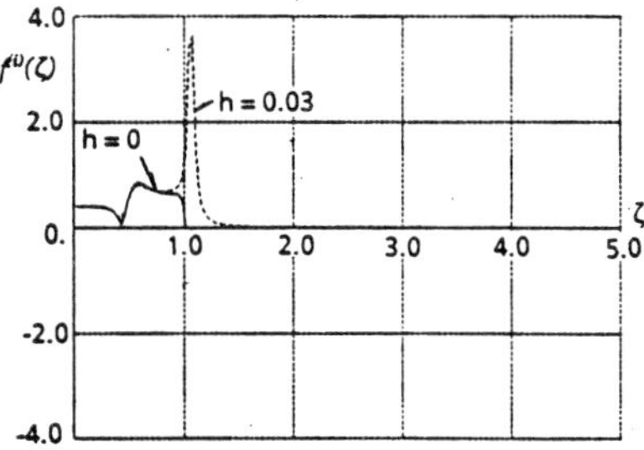

Fig.3 IMAG. PART OF $f(\zeta)$

Fig.4 and 5 show the Fourier coefficients $\hat{\phi}(m)$ at step 5 of $\hat{f}^{(r)}(n)\,n$ and $\hat{\psi}(m)$ of $\hat{f}^{(i)}(n)\,n$ up to 30 in case of $h=0.03$ soil damping with $N=1024$. Therefore $M = 2N = 2048$, $\Delta\zeta = 2\pi / N\Delta a_0 = 0.0306796$ and the domain of ζ is $\zeta_{max} = (M/2-1)\cdot\Delta\zeta = 31.3852$

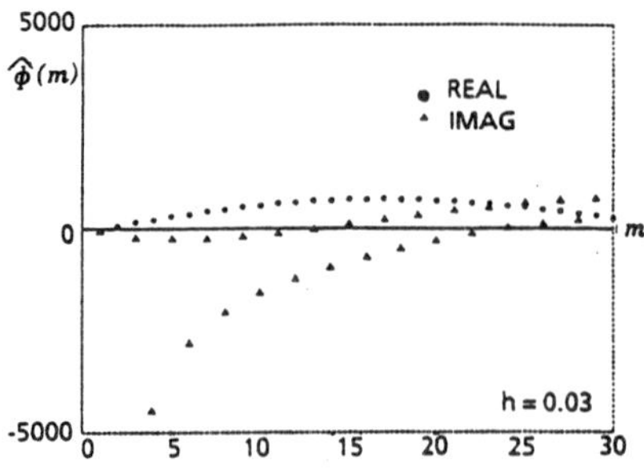

Fig.4 FOURIER COEFICIENT $\phi(m)$ OF. $f^{(r)}(n)\,n$

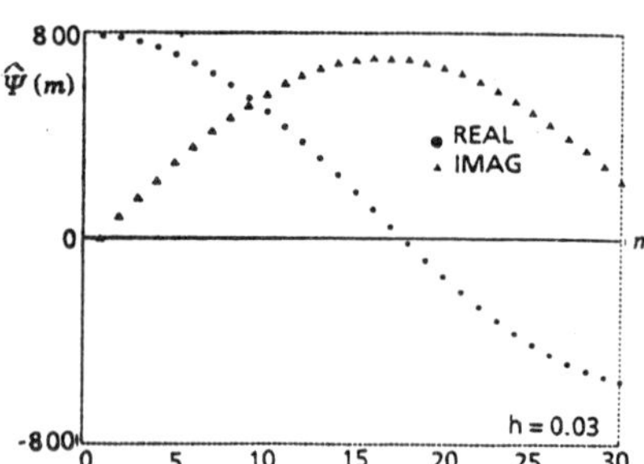

Fig.5 FOURIER COEFICIENT $\Psi(m)$ OF $f^{(i)}(n)n$

Fig.6 shows the displacement function f_1, f_2 that is calculated up to $l = 150$ ($a_0 = l\Delta a_0 = 30$). In case of $l = 0$ (i.e. $a_0 = 0$), displacement function is $f_1 = 1$, $f_2 = -2h$, however the solution is unstable with this method. So Fig.6 shows from the value $a_0 = 0.2$.

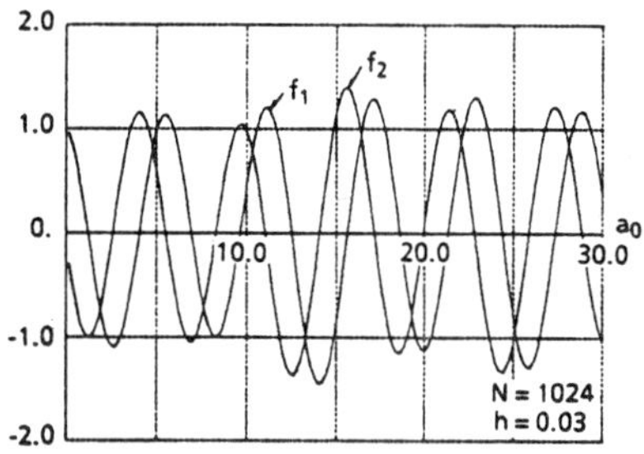

Fig.6 DISPLACEMENT FUNCTION f_1, f_2

As mentioned in the introduction, Green's function can be calculated by direct numerical integration (DIM) (Sasaki, Ohta, Masuda and Miura[5]). Fig.7 and 8 compare the displacement functions computed with DIM and FFBT.

The solutions of FFBT in case of $N= 1024$, 2048, 4096 and DIM show good agreement, but the solution of FFBT in case of $N= 512$ has some difference in the domain of large a_0. Table-1 shows the value of displacement function on typical a_0 in this case. Table-1 results imply that the stability of solution increases as the value of N increases and $N= 1024$ may be sufficient for practical use.

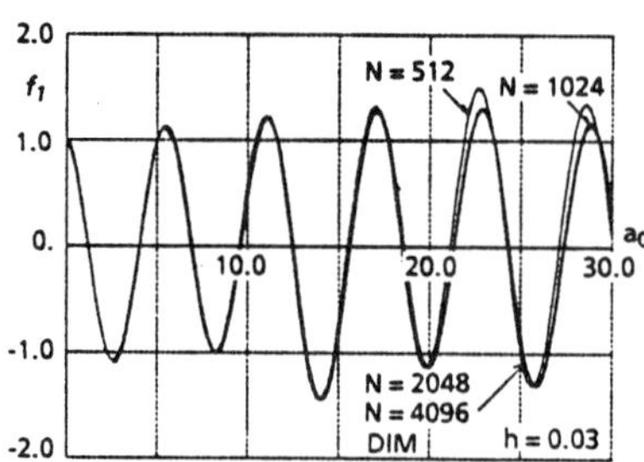

Fig.7 STABILITY OF SOLUTION f_1

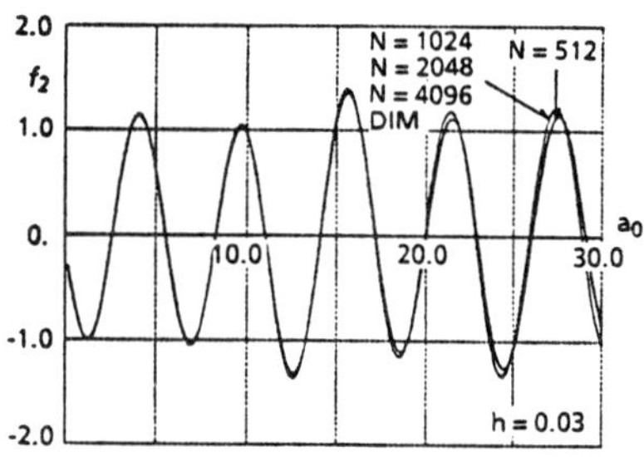

Fig.8 STABILITY OF SOLUTION f_2

TABLE-1 TYPICAL VALUE OF SOLUTION

	a_0	N 512	1024	2048	4096	DIM
f_1	5.6	1.0906	1.1303	1.1293	1.1299	1.1199
	17.0	1.3243	1.2887	1.2821	1.2808	1.2525
	28.8	1.3153	1.1806	1.1503	1.1407	1.1279
f_2	4.0	1.1446	1.1590	1.1620	1.1620	1.1534
	15.6	1.3655	1.4015	1.4085	1.4084	1.3898
	27.4	1.1429	1.2176	1.2300	1.2299	1.1989

As shown in Fig .2 the real part $f^{(r)}(\zeta)$ of $f(\zeta)$ is discontinuous at ζ_R when $h = 0$, because Rayleigh pole approaches to the real axis as soil damping decreases . To check that effect , Fig .9 -Fig .12 show the comparison of the solution of FFBT with that of DIM for the damping $h = 0$ and $h = 0.01$. In case of $h = 0$ in Fig .11 and Fig .12 , the solution of DIM are calculated by a numerical integration transforming infinite integral into finite integral (Kobayashi[6]). In the case of $h = 0.03$, the solution of FFBT was stable at $N \geqq 1024$, however in the case of $h = 0.01$ larger N ($N \geqq 2048$) is necessary to obtain stability . In the case of $h = 0$, the solution f_1 of FFBT is not stable in spite of increasing N , and the solution considerably differs from that of DIM . This is due to different integration methods , DIM requiring the use of Cauchy's principal value and the residue of Rayleigh pole .

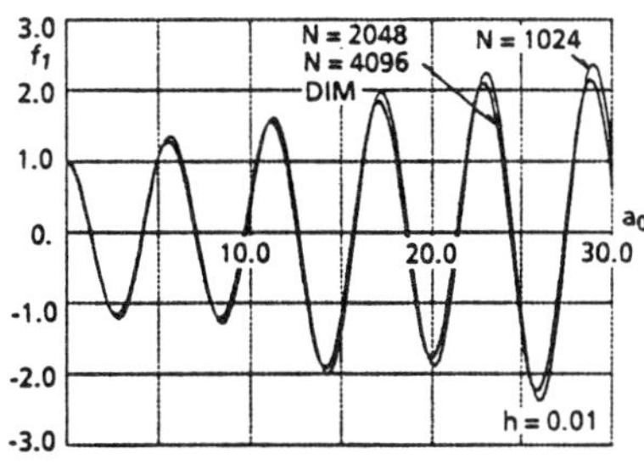

Fig.9 STABILITY OF SOLUTION f_1

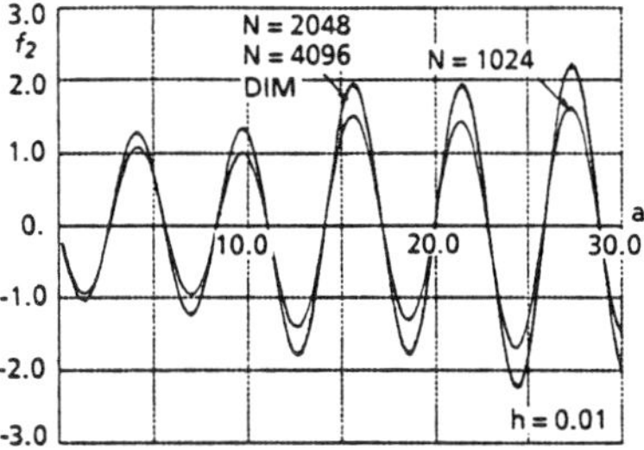

Fig.10 STABILITY OF SOLUTION f_2

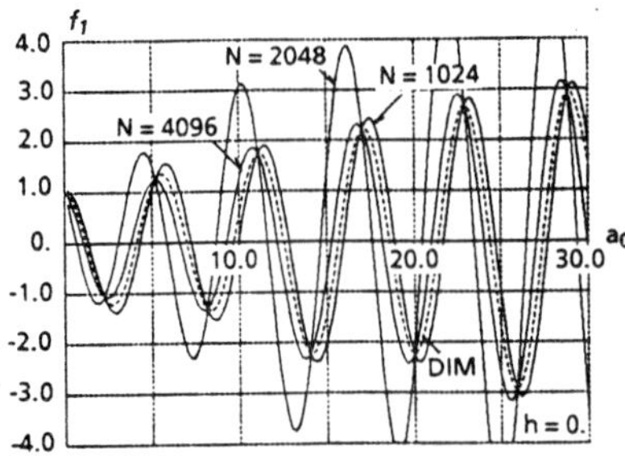

Fig.11 STABILITY OF SOLUTION f_1

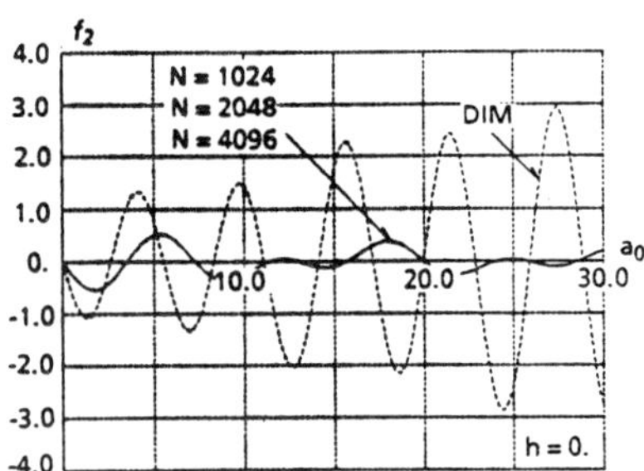

Fig.12 STABILITY OF SOLUTION f_2

Table- 2 compares the computational time for displacement function with $h = 0.03$. The computational time of FFBT increases in proportion to N , and in this example the computational time for $N = 4096$ is 1/10 of the direct numerical integration .

TABLE-2 COMPUTATIONAL TIME

METHOD			COMPUTATIONAL TIME
FFBT	N	512	0. 35 sec
		1024	0. 65 sec
		2048	1. 26 sec
		4096	2. 54 sec
DIM (h≠0)			23. 11 sec
DIM (h=0)			4. 83 sec

In case of l=150step Calculation

(a0=0 ~ 30 , Δa0=0.2)

APPLICATION FOR A POINT LOAD ON LAYERED SOIL

The above point load applied on a layered media is used to the interaction analysis of an embedded structure . The Green's function (i.e. point load) forms FBT using Haskell's matrix method (Haskell[7]) or Reflection Transmition Operater Method (Chapel and Tsakalidis[8]) , and the calculation flow applied directly .

TABLE-3 PHYSICAL CONSTANTS

STRATUM	Vs (m/sec)	ν	ρ (ton/m³)	h	H (m)
1-st	100	0. 4	2. 0	0. 03	5
2-nd	300	0. 4	2. 0	0. 03	

For example , the Green's function of a two layered soil with physical constants in Table- 3 is calculated . Fig.13 shows the transformed real part and imagenary part of function $f(\zeta)$ in 5Hz . Fig .14 , 15 show the result of FFBT . The dotted line of TLM are calculated by the Thin Layer Method

(Tajimi[9] ; Kausel and Peek[10]). L=30 is a divided number of thin layer and dividing in the depth 100 m . The results of FFBT and TLM (L= 30) agree apploximately . Computational time for calculating Fig .14 , 15 of FFBT is 2.96 sec. and TLM is 32.42 sec . Direct comparison of TLM with FFBT concerning the computational time shoudn't be discussed , because TLM can calculate all direction Green's functions at the same time for one eighvalue computation .

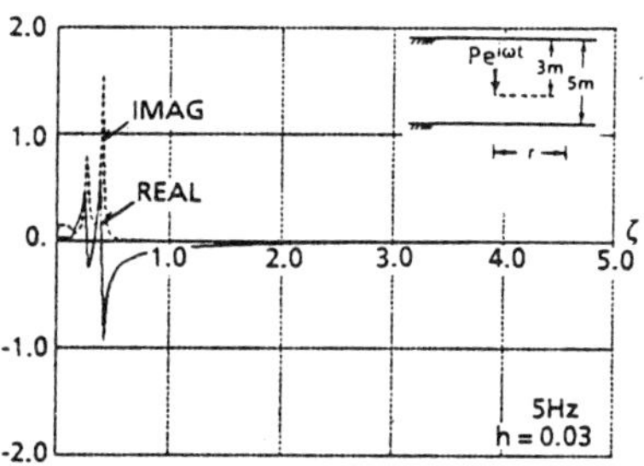

Fig.13 FUNCTION $f(\zeta)$

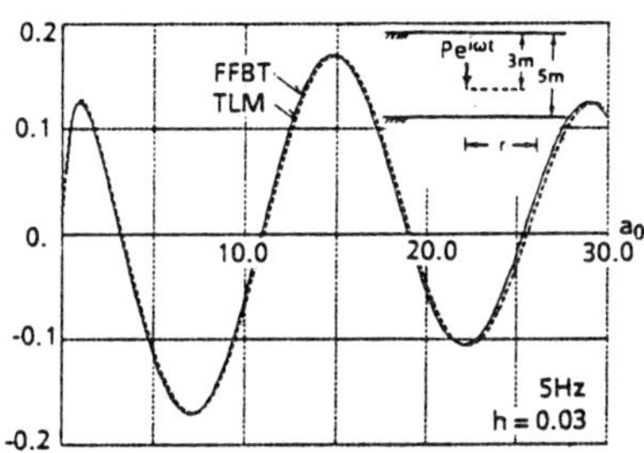

Fig.14 DISPLACEMENT FUNCTION f_1

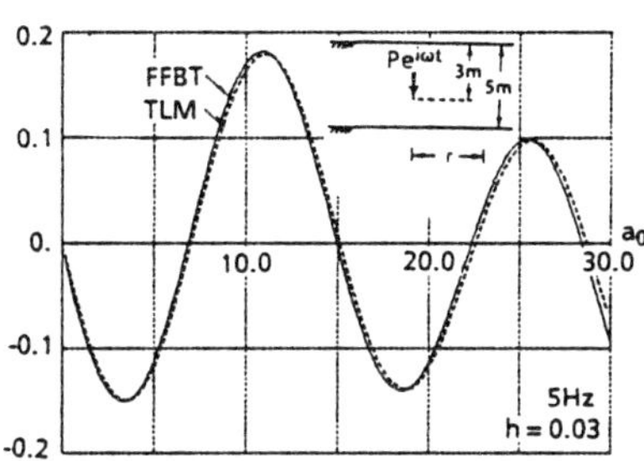

Fig.15 DISPLACEMENT FUNCTION f_2

CONCLUSION

This paper presented the FFBT method which is the advance Candel's method for calculating Green's function for semi-infinite soil media and examined the accuracy and computational time of this method .

The following remarks can be obtained .

(1) Computational time is less than the numerical integration .
(2) The method can be applied to layered soil media .
(3) It is not necessary to calculate Bessel function . Therefore it is advantageous to calculate Green's function for large value of a_0 .
(4) Only soil with damping can be analysed .

Computer which was used in this paper is HITAC M-280H .

REFERENCES

(1) Siegman . A(1977) Quasi Fast Hankel transform , Optics Letters , vol . 1, No.1 , P13 - P15
(2) Talman . J (1978) Numerical Fourier and Bessel Transforms in Logarithmic Variables , J . Computational Physics , vol . 29 , P35- P48
(3) Kausel . E and Bouckovalas . G (1979) . Computation of Hankel Transforms using the Fast - Fourier Transform Algorithm , M I T Research Report R79 - 12 , Order No. 639
(4) Candel . S . (1981) . Simultaneous Calculation of Fourier - Bessel Transforms up to order N , J . Computational Physics , vol . 44 , P 243 - P261

(5) Sasaki . F , Ohta . T , Masuda . K and Miura . K (1984) Response Analysis on an Elastic Flexible Base Mat resting on the Surface of Multi - Layered Strata , Proc . 8th WCEE vol.3 , P737 - P744
(6) Kobayashi . T (1981) Evaluation of Response to Point Load Excitation on Semi - Infinite Elastic Medium , Tran . Archi . Institute Japan , vol . 302 , P29 - P35 (in Japanese)
(7) Haskell . N (1953) The Dispersion of Surface Waves on Multilayered Media , Bull . Seis . Soc . Amer . , vol . 43 , P17 -P37
(8) Chapel . F , Tsakalidis . C (1985) Computation of the Green's Functions of Elastodynamics for a layered Half Space through a Hankel Transformation : Applications to Foundation vibration and Seismology , Fifth Int . Conf . on Numeri . Methods in Geomech ., Nagoya , P1311 -P1318
(9) Tajimi . H . (1980) A Contribution to Theoretical Prediction of Dynamic Stiffness of Surface Foundations , Proc . 7 th WCEE , vol . 5 , P105 - P112
(10) Kausel . E and Peek . R (1982) . Dynamic Loads in the Interior of a Layered Stratum : An Explicit Solution , Bull . Seis . Soc . Amer ., vol . 72 , P1459 - P1481

Diffraction and Scattering Analysis of Surface Waves by Surficial Geology

G.L. Wojcik
Weidlinger Associates, Palo Alto, California, U.S.A.

INTRODUCTION

Analytical methods for time-domain elastic wave propagation in engineering seismology can be classified as separable or non-separable, depending on the medium's degree of inhomogeneity. Separable problems include the common homogeneous or vertically stratified half-space, and are well-solved by classical methods. Weak inhomogeneities can sometimes be included via perturbation methods. However, for truly nonseparable problems, characterized here by any significant deviation from the layered half-space, no corresponding analytical methods are available. Either local geometrical solutions and/or global numerical solutions are necessary. These are provided in practice by geometrical diffraction and ray theory, as well as discrete numerical methods like finite difference and finite element wave solvers, and to a lesser extent by boundary integral methods.

This paper examines time-domain diffraction and scattering of ducted surface waves (and their body wave components) by 2-D inhomogeneities of interest in engineering seismology, i.e., prisms, basins, and other lateral variations in geology and tectonics of the upper crust. Some of these sedimentary, erosional, and tectonic features are nicely illustrated in Fig. 1, reproduced collectively from Holmes[1], showing near-surface structural variations at both local and regional scales in the Western United States. The results reported here derive from an ongoing research program on seismic wave diffraction in complex geology by Wojcik[2].

In the following, so-called generalized models are examined, where curvature is concentrated at singular points (corners or edges) on or near the free-surface. In particular, some problems of Love wave scattering by near-surface features are simulated in the time domain and illustrated by various numerical experiments. Rayleigh wave and body wave examples are also included. The principal analysis tool is a straightforward explicit finite element wave solver.

All of the results are found to be diffraction limited, which in the present context means that diffracted fields from regions of high interface curvature are as significant as the plane wave fields in general. Since corners are only modeled approximately by the finite element code, there is a question as to how well diffractions from geometric singularities are resolved. Discrete solution accuracy, presumably adequate from an engineering viewpoint, cannot be rigorously established without canonical solutions for comparison. Also, this diffraction limited behavior implies that geometrical (ray tracing) methods of interpretation and analysis are severely limited without canonical diffraction solutions for the edges of interest.

Therefore, to verify discrete solutions and also utilize geometrical methods, canonical diffraction solutions are necessary. Unfortunately, no canonical solutions presently exist due to their mathematical complexity in seismic applications, involving vector rather than scalar diffraction. Vector diffraction occurs whenever two or more wave functions are coupled across a boundary or interface with discontinuous tangent. Some progress on this problem in terms of a consistent vector wave function formulation and the resulting theory of connected boundary value problems will be discussed towards the end of the paper.

SURFACE WAVE MODELING

To explore the rudiments of surface wave diffraction and scattering by 2-D inhomogeneities, either Love waves or Rayleigh waves can be studied. Both wave types are important in practical applications, but in this research emphasis is on the simpler Love wave case. The advantage is that 2-D scalar simulations (SH or Love waves, which possess one out-of-plane unknown at each node point) require less than half the computer resources of vector simulations (P-SV or Rayleigh waves, with two in-plane unknowns at each node). In addition, scalar examples exhibit most of the relevant physics and phenomena of interest here.

One significant difference between the two surface wave types is that Rayleigh waves exhibit a fundamental mode which is not a ducted wave (a so-called inhomogeneous or evanescent mode), while all other Love and Rayleigh modes are at least partially ducted, i.e., can be decomposed into trapped body waves in the lower velocity surficial layers. Therefore, not all surface wave diffraction effects will be exhibited by studying Love waves, but from this author's viewpoint the most critical physics are included.

The canonical geophysical model is a uniform layer over a half-space, in particular, a 1 km layer of shale over a granite half-space, both homogeneous. Corresponding Love wave dispersion curves--phase and group velocity versus period--are shown in Fig. 2, with ray angle, $\sin^{-1}(\beta_1/v_p)$, replacing phase velocity v_p, in the third curve. The mode shape, S(z), is given by

$$S(z) = \begin{cases} \cos(k\nu_1 z), & z \leq h \\ \cos(k\nu_1 h)e^{-k\nu_2(z-h)}, & z > h \end{cases} \quad (1)$$

where $\nu_1=\sqrt{(v_p/\beta_1)^2-1}$ and $\nu_2=\sqrt{1-(v_p/\beta_2)^2}$. The figure also includes a cartoon illustrating the decomposition of Love waves into a beam of trapped rays reverberating within the layer at the given angle. It is readily proven that motion at any point in the layer is completely governed by the interference of upgoing and downgoing segments of the beam. This view provides a simple qualitative tool for interpreting the effect of changes in waveguide geometry (thickness) on surface wave modes, i.e., leakage or backscatter of the component ray beam.

Finite element models

In order to simulate Love waves in the 2-D model, the geology is first discretized into quadrilateral finite elements over an appropriate range and depth around the inhomogeneity of interest. The velocity mode shape from (1) is then applied on the right or left side, with an absorbing boundary condition on the opposite side and bottom, and of course a free-surface condition on the top.

The explicit finite element code used here employs bilinear shape functions over each element and is integrated forward in time using a modified leapfrog scheme (centered differences). The term, explicit (in contrast to implicit), implies that lumped masses are used, thereby eliminating the need for a mass matrix inversion, and that no assembly of a global stiffness matrix is performed. Instead, nodal forces are accumulated element by element and the nodal (lumped mass) velocity increments calculated from Newton's law in incremental form. The algorithm is stable provided the timestep is less than the wave transit time across the smallest element. This follows since all nodes are then decoupled by virtue of the fundamental hyperbolicity of the governing equations, i.e., no waves can communicate information between nodes during a single timestep.

The algorithm stores three quantities for each node--lumped mass, M(n), anti-plane velocity, V(n), and anti-plane force, F(n), where n=1 to N, with N the total number of nodes.

Velocities are defined at full timesteps while forces are defined at timestep midpoints. The basic arithmetic loop involves the following steps: (1) calculate the velocity increments ($\Delta V=M^{-1}F\Delta t$, where M is diagonal) using the nodal forces at the previous half timestep; (2) add the increments to the nodal velocity vector to obtain values at the next timestep ($V=V+\Delta V$); (3) calculate the force increment ($\Delta F=K\Delta U=K\Delta V\Delta t$); and (4) add to the nodal force vector to obtain forces at the next half timestep ($F=F+\Delta F$). This loop is repeated for the required number of timesteps.

The algorithm lends itself to efficient vectorization in "vanilla" FORTRAN (i.e., no assembly coding necessary) and approaches the peak performance levels expected on pipelined supercomputers like the CRAY machines. This is not to say that a supercomputer is required for practical calculations. Tens of thousands of elements are routinely executed on minicomputers, up to perhaps 100,000 elements, with reasonable turnaround. Models considered here have 20-40 thousand elements and generally take from two to four hours on a minicomputer.

Diffraction limits illustrated

An effective illustration of the modeling technique and diffraction phenomena is afforded by Love waves propagating over a "cliff", i.e., where the waveguide thickness abruptly increases to the bottom of the model. Results are shown in Fig. 3 for the first and second mode Love wave at periods of .5 and 1 second in a 5x10 km model (100x200 elements) of the 1 km shale waveguide over granite. The figure shows vector snapshots at intervals of 200 timesteps, with the anti-plane velocity at each node drawn vertically.

In the absence of diffraction the trapped modes would each yield a beam of SH-waves propagating into the quarter-space at the angle given in Fig. 2c. Although very short periods should approach the ideal beam, clearly this is not observed for the periods considered here. Instead the beams exhibit considerable spread due to diffractions in all directions emanating from the transition corner. The theoretical beams drawn in the last frames at 500 timesteps show how the deviation increases with period. Spreading is generally found in direct proportion to period, or equivalently, penetration of the surface wave mode shape below the waveguide given by (1). This follows since diffracted waves and the inhomogeneous part of the surface wave (below the layer) obey the same exponential decay law in space, i.e., diffraction is an inhomogeneous (evanescent) wave.

Similar results have been computed for a variety of other waveguide transitions, like finite steps and converging or diverging transitions (dipping layers). These experiments show that, although the geometrical ray beam interpretation gives a qualitative description of transition effects valid in the limit for very short periods, it fails for periods of interest in engineering seismology since diffraction, i.e., evanescent waves, are equally important.

SURFACE INHOMOGENEITIES

An interesting question in engineering seismology is to what degree waveguide inhomogeneities (surface or buried) affect propagating surface wave modes by the mechanisms of diffraction and scattering? This bears on two particular problems. The first is coupling of surface wave energy into filled basins (the interior problem), i.e., the amount and distribution of absorbed energy inside the basin. The other problem is mode coupling induced by the inhomogeneities (the exterior problem), i.e., the redistribution of surface wave energy between all modes supported by the waveguide outside the inhomogeneity.

The interior problem is of obvious interest in earthquake engineering, while the exterior problem is more relevant to engineering seismology, namely, scattering attenuation, stable mode mixing, and shielding by inhomogeneities. It appears that the role of mode coupling has not been appreciated fully in seismology, due in part to the difficulty of including diffraction and scattering effects in surface wave analysis. However, mode coupling has been studied extensively in optical fiber analysis via perturbation methods, because of its influence on long-range ducted propagation, e.g., Marcuse[3]. In the seismic environment mode coupling may have particular relevance to Lg propagation, for example.

Filled basin examples

Idealized examples of inhomogeneities in the 1 km shale layer over granite include rectangular, trapezoidal, and circular arc channels penetrating the waveguide. The cross-sections considered here are nominally .5 km deep and 2 km wide, filled with a low speed, B=0.5 km/sec, sediment. The simulations assume Love waves incident at normal incidence on an infinitely long basin with the above cross sections and fill. The actual models are 3x14 km (60x280 elements) with the .5x2 km basin in the middle.

Figure 4 presents some contour snapshots of the simulations from a 1.5x4km window around the basin. The figure shows four contour plots of the anti-plane velocity field at timestep 750--the first plot with no basin (for reference), and the other three with the example cross sections. These nicely illustrate the interior problem and the qualitative effects of basin shape. Observe that the interior voids are the antinodes, i.e., regions of high oscillatory amplitude, and the bands are nodes. Snapshots at 250 timestep intervals, not included here, show that these resonance patterns establish themselves quickly, after one or two cycles of the surface wave passage, and are stable thereafter.

Figure 5 gives vector velocity plots over the full finite element grid for the circular and rectangular cross sections. At timestep 250 the wave is just encountering the basin, at 500 it is halfway across, and at 750 it has traveled a few wavelengths past. These frames provide a quantitative comparison of velocity inside and outside the basin, and also indicate qualitative aspects of the exterior problem. In particular, it is clear that the transmitted Love wave is lower amplitude than the incident wave, and that there is a significant reflected wave as indicated by the wave train's nonuniformity to the left of the basin. Unfortunately, both of these observations are somewhat contaminated with unwanted reflections from the left and right boundaries.

Mode coupling

Figures 4 and 5 illustrate the basic methodology for simulating Love wave diffraction and scattering phenomena from surface inhomogeneities. The interior problem is readily quantified in this manner. However, comparable quantification of the exterior problem requires modification of the model as well as development of new analytical tools. First, the model's absorbing boundary conditions must either be enhanced or the left and right sides extended farther from the inhomogeneous region. The purpose is of course to eliminate virtually all nonphysical reflections (model truncation error) during the experiment. Second, a means for quantifying the mixture of reflected and transmitted modes at any station is required, and similarly, a scheme for determining the pattern of body waves scattered into the underlying half-space.

These difficulties illustrate some ubiquitous problems with numerical experiments--finite boundaries and the extraction of quantitative, global information. For the examples considered here the finite boundaries are not a serious problem since the model is small to begin with and doubling its size is not prohibitive. However, higher order boundaries that include both the prescription of incident wave fields and the absorption of scattered fields would be useful in general. Methods to accomplish this are still a subject of research, but progress has been demonstrated for such problems, e.g., Wojcik and Gustafsson[4], and will be reported elsewhere.

Modal orthogonality is utilized to quantify reflected or transmitted surface waves. Velocity-time histories are first recorded on vertical lines of nodes upstream and downstream of the inhomogeneous region. Provided the output lines are sufficiently removed from the scatterers, the recorded velocities are essentially a superposition of all surface wave modes supported by the waveguide. There are only transmitted modes downstream, and reflected modes plus one incident mode upstream. By virtue of orthogonality the participation factor, P_m, m=1 to M, for each of the modes supported is determined by multiplying its mode shape times the recorded output at a given time and integrating over depth. The integration is repeated at each timestep over a half-period for each mode and the maximum or minimum found, from which amplitude and phase of the converted modes are determined. If the waveguide is well-modeled by finite elements (sufficiently refined) then it is reasonable to use the exact mode shapes given by (1) in calculating participation factors. However, if mode shapes in the model deviate significantly due to truncation, then for consistency it is necessary to use those determined from the model itself.

The principle of superposition is utilized to calculate body wave scattering into the half-space. This is accomplished by recording velocity on an output contour surrounding the scatterer and penetrating into the half-space, e.g., a rectangle. Output is also recorded for the same model without the scatterer and the two results subtracted, yielding the scattered field. If the output is recorded well below the penetration depth of the incident surface wave then this technique gives a reasonable estimate of body wave scattering.

CORNER DIFFRACTION

The last issue to be considered in this paper is a more detailed look at diffraction phenomena from geometric singularities in a generalized cross section. Most of the effects observed in the above examples are caused by diffraction from the edges, scattering from the faces, and internal resonances associated with characteristic lengths of the inhomogeneities. Since the geometric singularity is smoothed out in discrete models, a valid question can be raised as to the solution's accuracy. Unfortunately, this question is very difficult to answer.

The only case that can be verified analytically is one of scalar diffraction, i.e., where one scalar wave function diffracts off an edge. A number of classical solutions exist for this problem, the simplest being a similarity solution due to Keller and Blank[5]. To verify finite element simulations of this case, a variety of numerical experiments have been performed for a plane SH-wave incident on reentrant wedges. An example is illustrated in Fig. 6, showing comparisons for a plane wave incident on a 270° wedge (90° void) with an offset of 22.5° from normal incidence to a face. The time history is a wavelet (exponentially damped cosine) and the wedge faces are modeled stepwise rather than using conforming elements to smooth them.

Referring to Fig. 6, agreement between theory and experiment is generally good for the main wavelet, but oscillatory errors occur at later times. It was also observed that high frequency degradation occurred, so that approximately 20 element per wavelength were required for the results in Fig. 6, rather than the usual rule-of-thumb of 10 elements/wavelength. The same example modeled with conforming elements showed that these errors are caused by stepwise approximation of the wedge faces. Thus, the discrete wave solver does a reasonable job simulating scalar diffraction, with stepwise resolution of interfaces causing the principal error.

Examples of vector diffraction

The case of 2-D vector wave diffraction occurs when either a P- or SV-wave is incident in a single wedge, or an SH-wave is incident in two coupled wedges. This would occur for a dilatational or shear wave input on the edge pictured in Fig. 6, or if the 90° void was replaced by an elastic medium and an SH-wave input. The common factor for the vector problem is two or more wave functions coupled across an interface with a geometric singularity. In no case can a finite element simulation be verified since exact solutions are mathematically intractible.

A number of vector simulations have been performed for wedge models like that in Fig. 6. Their purpose was to experimentally determine relative amplitude trends of diffracted P-waves, SV-waves, and Rayleigh waves for plane P- and SV-wave input at various incidence angles. Note that the diffracted Rayleigh wave from the edge is the inhomogeneous or evanescent fundamental mode, not a ducted mode analogous to the Love wave described earlier in the paper.

Rather than reproduce these cases here, another vector problem will be illustrated, involving diffraction of the fundamental mode Rayleigh wave by various reentrant corners (edges). A sequence of vector plot snapshots are shown in Fig. 7 for Rayleigh wave incidence from the right. The wave is generated by a vertical velocity transient on the right corner assuming symmetry boundary conditions along that side. The sequence begins at 250 timesteps, when the surface wave just reaches the edge. At this point the outgoing P-wave and its diffractions have been absorbed by the left and bottom boundaries, and the remnant of the circular shear wave from the source is seen propagating to the right at a shallow angle to the bottom absorbing boundary. The plots show that the evanescent surface wave is mainly transmitted around the corner, with very weak reflection. Most of the diffracted energy is found in the shear wave, rather than the dilatational wave. The higher amplitudes (longer arrows) observed on the surface for the smallest wedge angle (200°) are caused by the nonuniform stepwise resolution of the sloping wedge face.

On an exact formulation of vector diffraction

A major effort of the research effort summarized in this paper is a rigorous analysis of vector wave diffraction. Clearly, one use of this theory is to determine the limitations of finite element simulations for vector wave diffraction. Of course the theory has a number of other direct applications in mechanics and physics, and as a source of canonical solutions in the geometrical theory of diffraction. Since a complete development is beyond the scope of this paper, only a brief topical description will be given here.

The 2-D theory is based on self-similar wave solutions in the so-called generalized diffractor, composed of M contiguous sectors covering the plane from a common vertex. An example is drawn in Fig. 8a. Each sector supports one or more wave functions, for a total of N functions, and, by definition, each satisfies a scalar wave equation. The unknowns are represented by an N component vector governed by the diagonal form of the vector wave equation. The N functions supported in M wedges are coupled by general vector boundary conditions on the union of limit points. These include the vertex and point at infinity, as well as the wedge faces.

Because the diffractor lacks a characteristic length, self-similar solutions of the governing vector partial differential equation and boundary condition are admissible. The most effective approach to solving the self-similar form is to transform it to characteristic coordinates, yielding the normal vector form. This transforms the wedge domains into N semi-infinite strips in "parallel" complex characteristic planes illustrated in Fig. 8b. The resulting vector equations can be integrated directly giving d'Alembert's classic 1-D solution in two characteristic coordinate vectors. Application of the boundary condition gives a deceptively simple appearing vector boundary value problem coupling all wave functions on limit points of the N semi-infinite strips.

To solve the boundary value problem a so-called characteristic transformation is developed, uniformly mapping the strips to N complex half-planes shown in Fig. 8c. The problem thereby reduces to a vector Riemann-Hilbert type boundary value problem. The classical scalar formalizm for such problems is not valid for this vector case because of cross-coupling. Therefore, a more fundamental solution is derived based on Schwarz-type integral representations over the union of half-plane boundaries (real lines). The representation's density vectors are obtained by solving well-posed systems of Cauchy-type singular integral equations. Thus, the solution finally reduces to a boundary integral representation based on solutions of systems of singular integral equations.

DISCUSSION AND CONCLUSIONS

This paper has demonstrated the use of finite element wave solvers on minicomputers to simulate various nonseparable, elastic wave propagation problems in engineering seismology. These simulations have included ducted Love waves, body waves, and Raleigh waves in generalized geologic models of inhomogeneities relevant to engineering and seismic applications.

The results illustrate the strong role of diffraction and scattering at engineering frequencies, and the need for validated wave solvers to quantify so-called interior and exterior problems for isolated inhomogeneities on or near the free-surface. The exterior problem requires better absorbing boundaries (or larger models) and global processing schemes for extracting the mix of transmitted and reflected surface wave modes, as well as scattered body waves.

Accurate simulations of diffraction and scattering from generalized geologic models involve the ability to solve vector wave diffraction. Canonical solutions are necessary to validate discrete solutions, but none presently exist in the literature. For the scalar case, validation experiments show that the major error is due to interface roughness, which is removed by using conforming finite elements. At present the analyst can only assume that solvers do a reasonable job for the more common vector case.

The analytical formulation of vector wave diffraction is shown to involve known mathematical techniques, but in a vectorized form that yields multiply connected boundary value problems to be solved. These reduce to boundary integral representations after solving a system of singular integral equations for the kernels. This area of mathematical research is necessary if the analyst hopes to validate, and possibly augment, numerical wave solvers. It is absolutely essential for the practical use of geometric methods like ray tracing and geometrical diffraction theory.

The conclusion is that numerical experiments using discrete wave solvers in engineering seismology are straightforward, practical on small computers, and can yield a wealth of relevant data. Their validity for generalized models has yet to be rigorously established, and this issue needs to be pursued.

ACKNOWLEDGMENT

This research was supported by the Air Force Geophysics Laboratory under Contract F19628-84-C-0102.

REFERENCES

1. Holmes, A. (1965). Principles of Physical Geology, The Ronald Press Co., New York.

2. Wojcik, G. L. (1987). A Numerical and Theoretical Study of Seismic Wave Diffraction in Complex Media, Report to the Air Force Geophysics Laboratory, by Weidlinger Associates, Palo Alto, California.

3. Marcuse, D. (1969). Mode Conversion Caused by Surface Imperfections of a Dielectric Slab Waveguide, The Bell System Technical Journal, Dec., pp 3187, 3215.

4. Wojcik, G. L. and Gustafsson, B. (1987). Higher-order, Multi-purpose Boundaries for Wave Domain Truncation, Ongoing research under AFGL Contract F19628-84-C-0102.

5. Keller, H. B. and Blank, A.(1951). Diffracton and Reflection of Pulses by Wedges and Corners, Comm. Pure Appl. Math., 4.

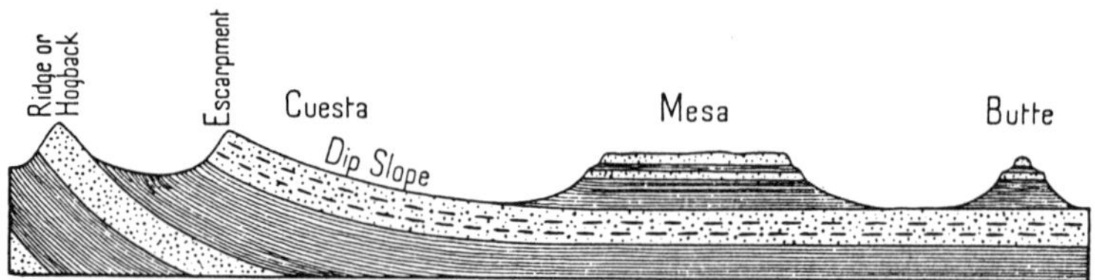

FIG. 413 Diagram to illustrate the relation of various erosional landforms to the structure and dip of the strata from which they have been carved

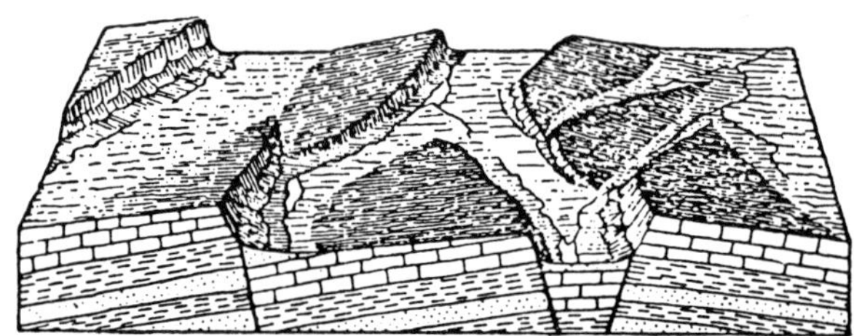

FIG. 796 Diagram to illustrate the fault-block structure of the ranges of the Great Basin, Utah (*After W. M. Davis*)

FIG. 791 Schematic section across the Cordillera of the western United States. Length of section about 1,900 miles (3,000 km.)

Figure 1. Examples of erosional, sedimentary and tectonic landforms, reproduced directly from Holmes (1965). The upper two diagrams are at local scales (1-100 km) while the bottom section is on a regional scale (1000-3000km).

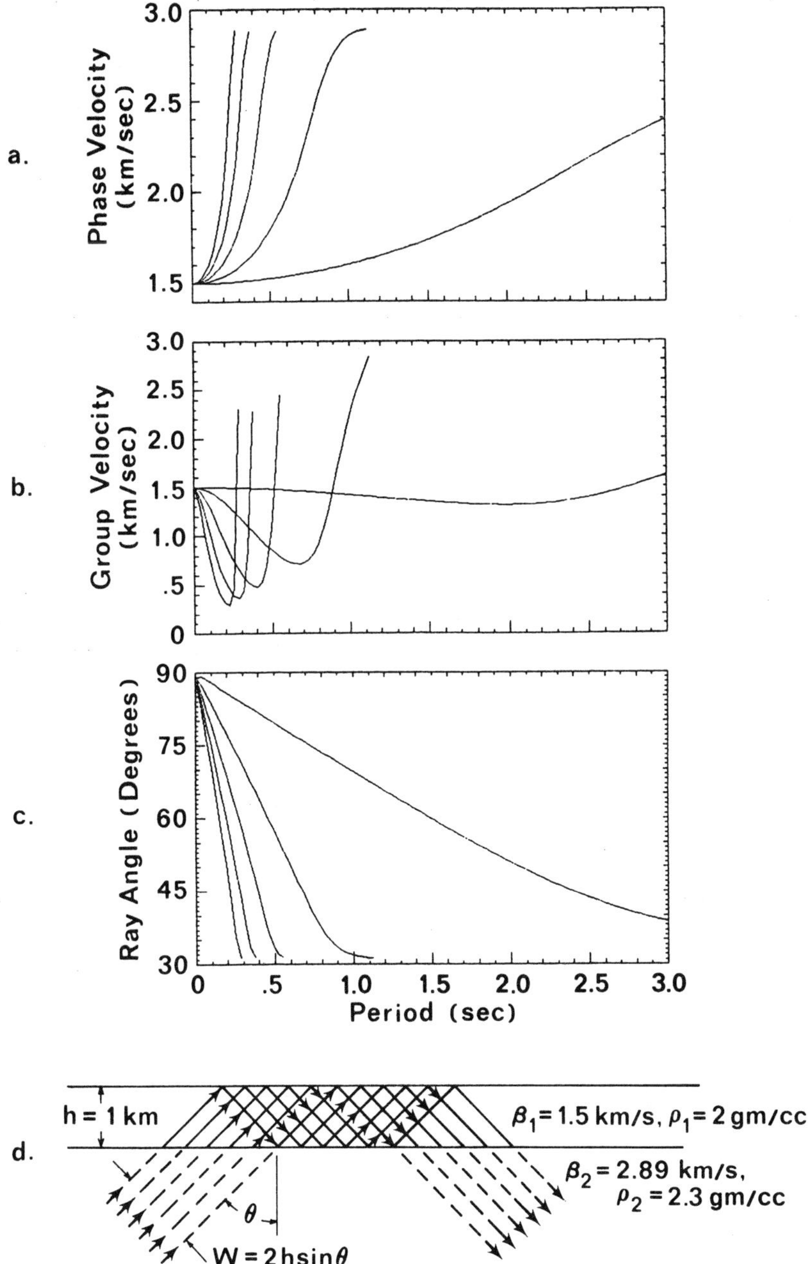

Figure 2. Dispersion curves for a 1km shale layer over a granite half-space in a. and b.. Ray angle replaces phase velocity in c. and d. shows properties and the decomposition of Love waves into a trapped beam.

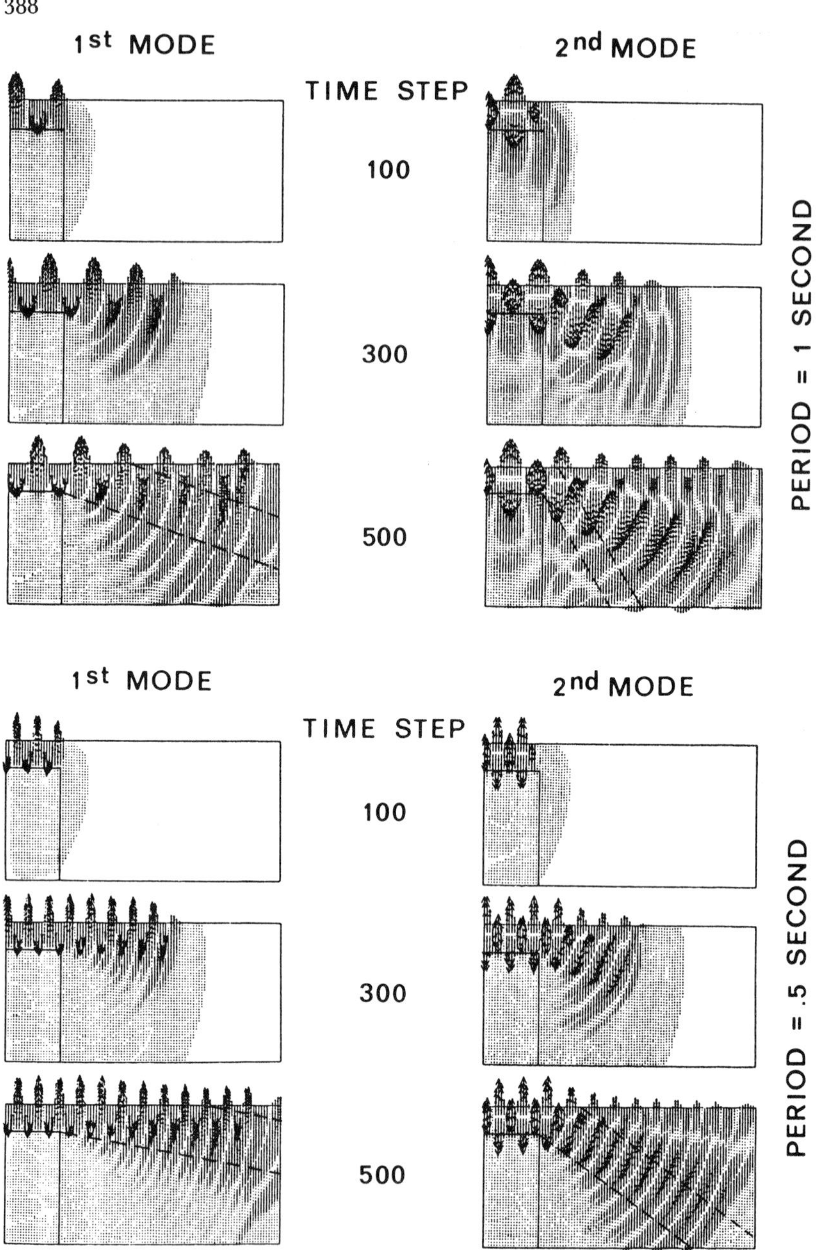

Figure 3. Simulations of Love waves propagating over a "cliff," for first and second modes at periodes of .5 and 1.0 sec. Anti-plane velocity is drawn vertically in the vector plots.

NO CHANNEL

CIRCULAR CHANNEL

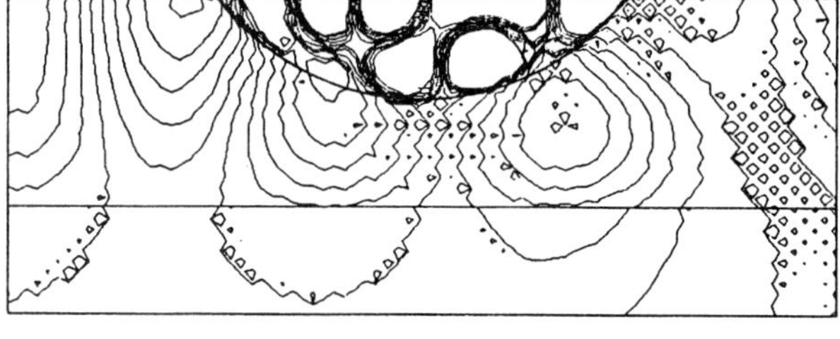

TRAPEZOIDAL CHANNEL

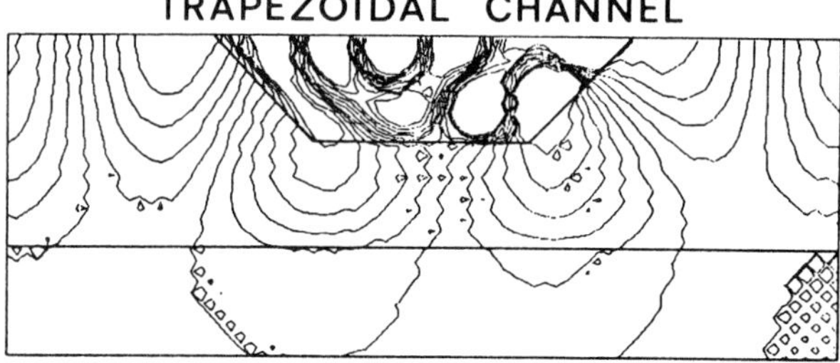

RECTANGULAR CHANNEL

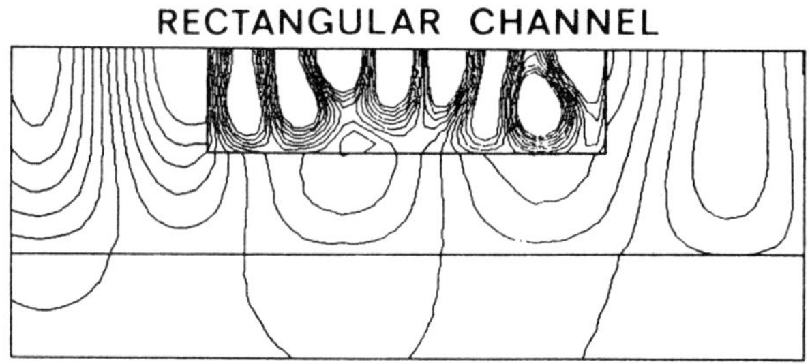

Figure 4. Contour snapshots of anti-plane particle velocity in the region around the channel, showing results with and without the three cross-sections considered.

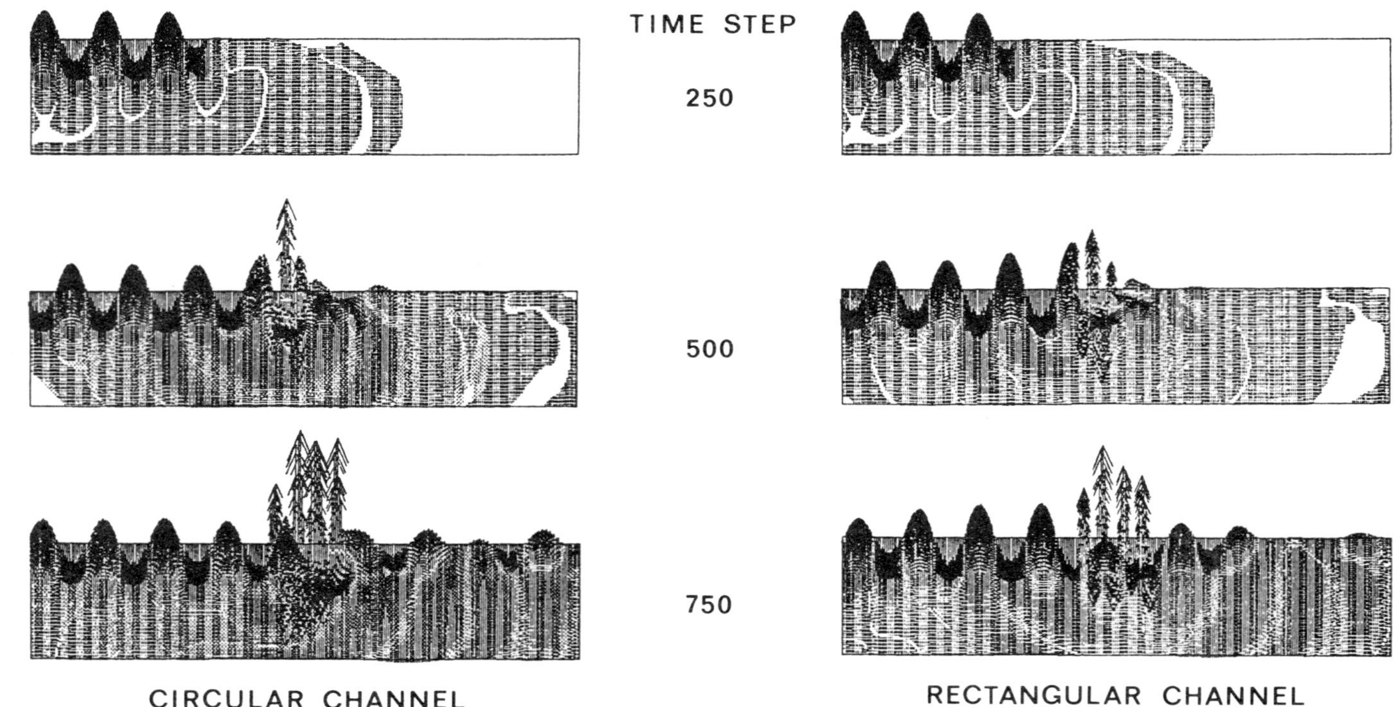

Figure 5. Vector snapshots of the fundamental mode Love wave incident on two filled channels with 1:3 contrast between fill and layer wave speed. Anti-plane velocity vector is drawn vertically.

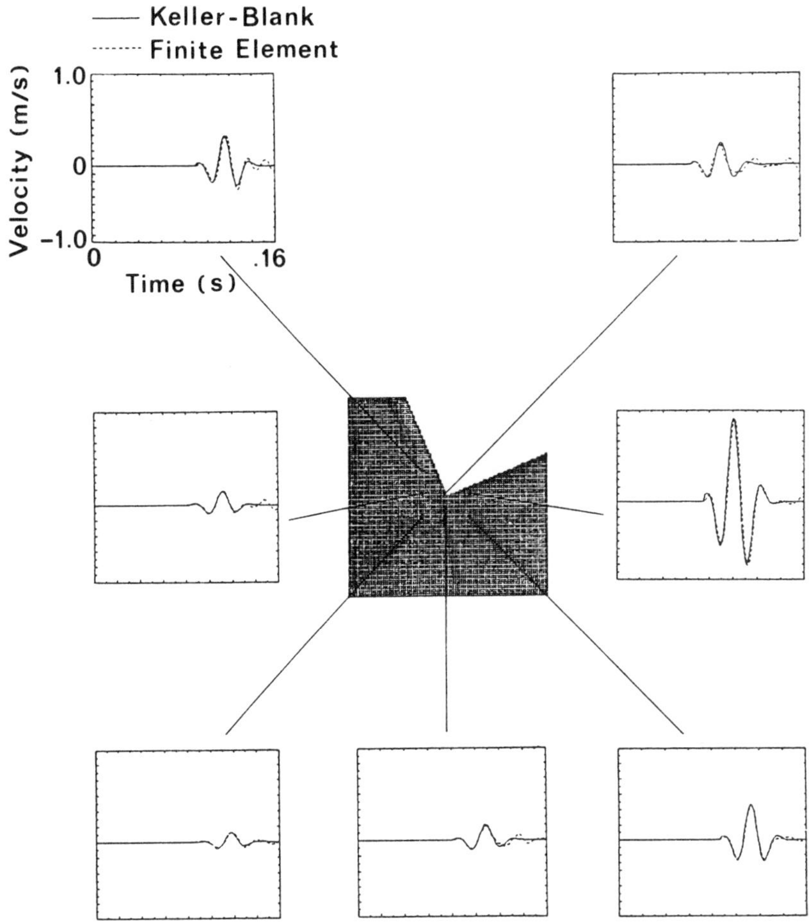

Figure 6. Comparison of exact Keller-Blank solution and finite element solution for the diffracted wave due to a plane SH-wave input from the lower boundary. Incidence angle is 22.5° from normal incidence on the right face.

10^{-7} m/sec

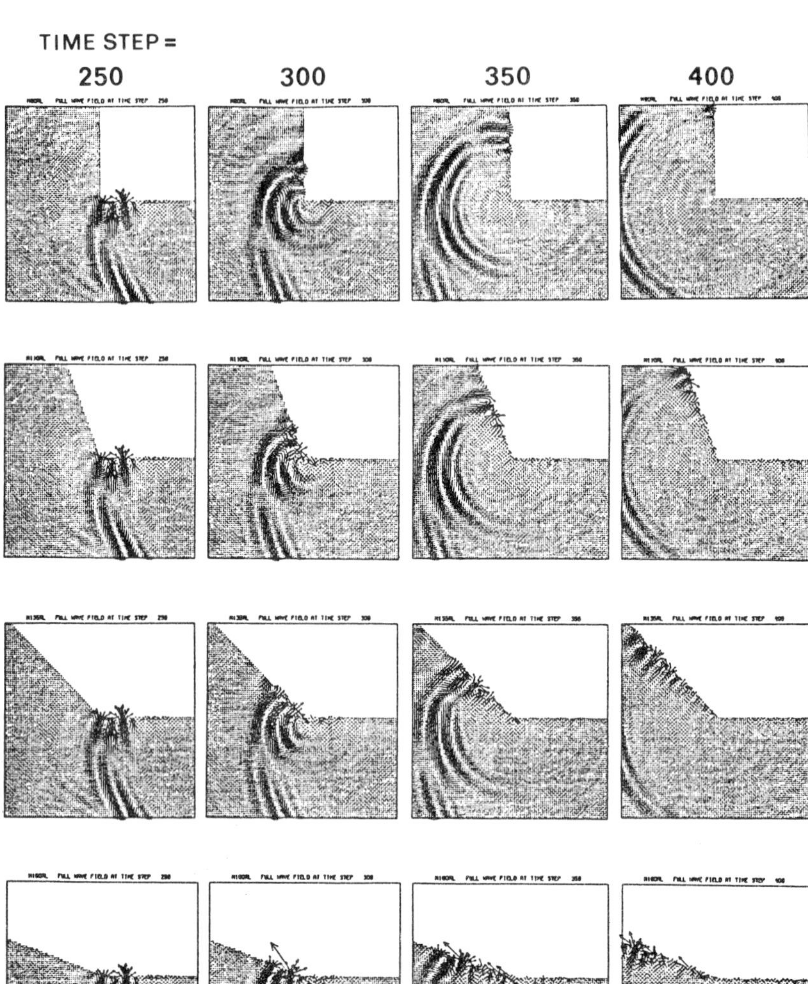

Figure 7. A sequence of snapshots from finite element simulations of Rayleigh surface wave reflection, transmission and diffraction in four reentrant corners. The Rayleigh wave is incident from the right, and each sequence shows vector plots of the velocity field.

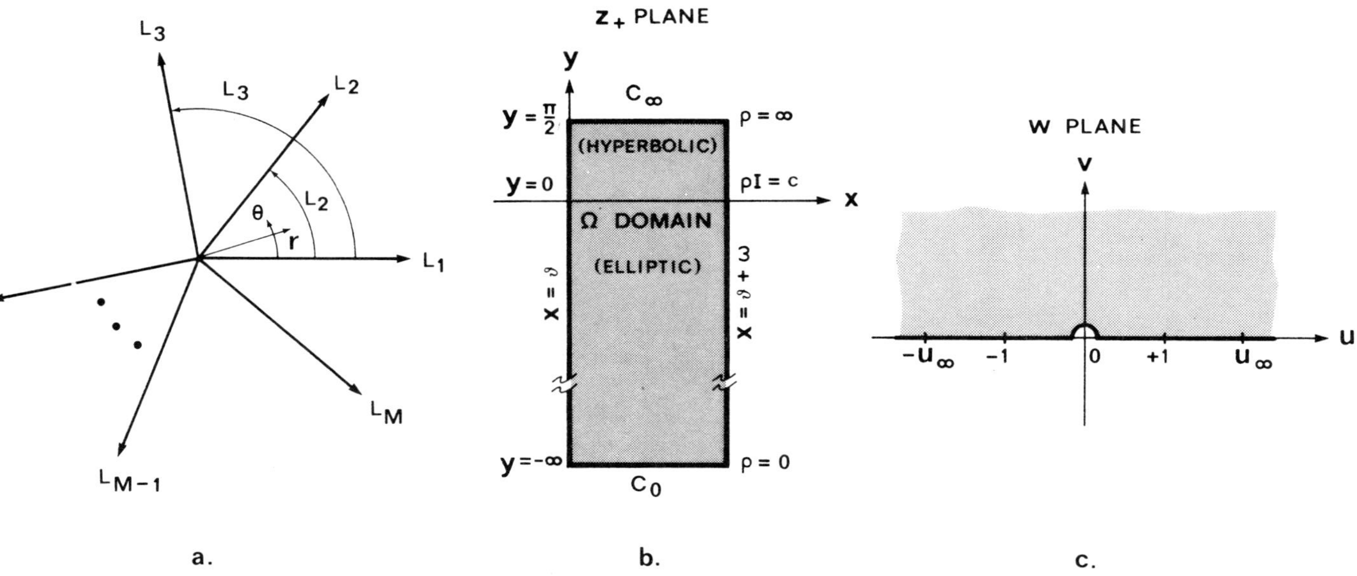

Figure 8. (a) The generalized 2-D diffractor consisting of M semi-infinite plane wedges. (b) The self-similar mapping of the wedge wave domains to N semi-infinite strips in "parallel" complex planes. (c) The characteristic mapping of the strips to N complex half-planes, yielding the final geometry of the boundary value problem.

A Simple Boundary Element Method Procedure to Solve the Diffraction of SH Waves in a Stratified Soil

F. Bettinali, R. Rangogni
ENEL, Italian National Power Agency, Italy

SUMMARY

In studying seismic waves diffraction, caused by topographical irregularities, and by stratified soils, some numerical difficulties are encountered whenever infinite extension is present.

An accurate numerical technique to solve problems defined on infinite domains is the Boundary Element Method (BEM), which gives the advantage to limit the extension of the domain to the irregularity only.

The aim of the present work is to show that accurate results can be obtained also limiting the infinite stratified domain.

In this paper an infinite stratified soil is reduced to a finite one in order to use a standard BEM technique.

INTRODUCTION

Field observations made after strong earthquakes, have shown (Esteva[1]) concentrated damage in some small areas close to other almost uninfluenced ones. This is due to various causes such as: low quality of the constructions, local geology, effects of topography, etc.

It has been suggested that the damage distribution is due to seismic wave amplification induced by the local topography and superficial soil characteristics (Wong et al[2]).

In the last few years the problem has been studied by numerous authors (Aki et al[3]; Bouchon[4]; Sanchez-Sesma and Rosenblueth[5]).

For the case of a general topography the three dimensional problem has been dealt with, by most researchers, with the help of two-dimensional models.

A formulation in terms of singular integral equations has been used to solve the scattering and diffraction of SH waves by irregularities of the surface (canyons, alluvial valley and ridges) (Bouchon[4]; Sanchez-Sesma et al[5]; Sanchez-Sesma et al[6]).

In this paper we deal with a two-dimensional problem in the case of an arbitrary surface irregularity, of finite dimension, resting on an infinite stratified soil.

Formulations in terms of transfer matrix that have the limitation to solve the problem of horizontal layers only and domain technique such as Finite Element Method (FEM), that require to limit the domain and to introduce fictitious absorbing boundaries in order to take into account radiation condition has just been used by other authors. The solution method here employed is the Boundary Element Method (Brebbia[10]) which was shown to be a rather efficient technique.

In presence of stratified soils the usual BEM technique is not applicable but if the infinite layers are reduced to a finite ones, as in (FEM), BEM becomes again an efficient tool.

The dimension of the domain, reduced from infinite to finite, that does not introduce relevant errors in the solution (named in the text optimal reduction) was just computed by (Bettinali and Rangogni[7]) for the case of an infinite surface irregularities. Here the same procedure is applied.

The authors are aware that the optimal reduction depends on many factors (shape of the irregularities, wavelength, elastic coefficients etc.), and that parametric studies, are probably necessary for every class of topographic irregularity. An estimate of the optimal reduction is also an important information in FEM mesh preparation.

In BEM technique boundary conditions give rise to a system of linear equations, which are solved in the least-squares sense. The formulation and numerical results are only presented for SH waves; extension of the method to P and SV waves is straightforward.

PROBLEM FORMULATION

In the propagation of horizontal polarized SH waves, the total displacement u(x,y,t) is governed by the wave equation (Aki and Richards[8]) (in cartesian coordinates):

$$\frac{\partial^2 u}{\partial^2 x^2} + \frac{\partial^2 u}{\partial y^2} = \frac{1}{c^2} \frac{\partial^2 u}{\partial t^2} \tag{1}$$

where $c = (\mu/\rho)^{1/2}$, μ is the shear modulus, ρ is the mass density, and t is the time. Equation (1) is defined on the domain D with boundary S, shown in Fig. 1.

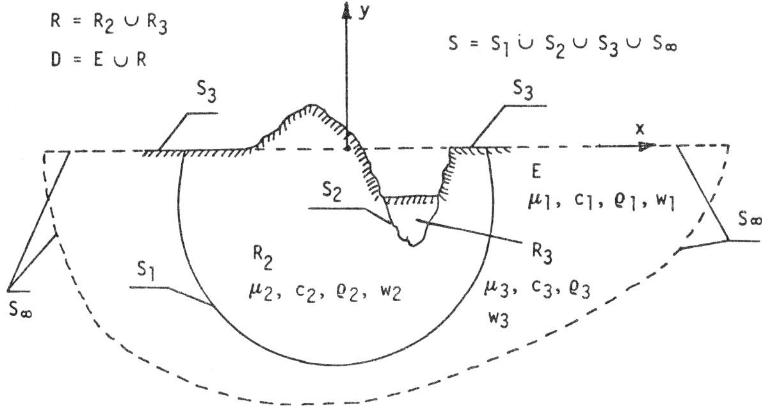

Fig. 1 - Sketch of scattering region: domain D, bounded by curves S_3 and S_∞, contains inner region R, bounded by curve S_3 and S_1, and outer region E, bounded by curve S_1, S_3 and S_∞.

If the time dependence is harmonic ($u(x,y,t) = w(x,y) * \exp(-i\omega t)$) equation (1) becomes the Helmholtz equation:

$$\frac{\partial^2 w}{\partial x^2} + \frac{\partial^2 w}{\partial y^2} + k^2 w = 0 \tag{2}$$

where $k = \omega/c$, $\omega = 2\pi/T$ being the circular

frequency and T the period of the wave.
Equation (2) governs the diffraction of SH waves caused by surface irregularities or by non-homogeneous media.

The total field w can be written in the form

$$w = w^{(o)} + w^{(d)} \tag{3}$$

where $w^{(o)}$ is the free-field displacement (without surface irregularities or non-homogeneous media) and $w^{(d)}$ is the contribution of the diffracted waves.

The diffracted component $w^{(d)}$ has to satisfy the Sommerfeld's radiation condition (Sommerfeld[9]).

The free field $w^{(o)}$ is the sum of an incident wave field $w^{(i)}$ and a reflected field $w^{(r)}$

$$w^{(o)} = w^{(i)} + w^{(r)} \tag{4}$$

For the incident wave $w^{(i)}$, we assume the expression:

$$w^{(i)} = \exp [ik (x \cos \beta + y \operatorname{sen} \beta)] \tag{5}$$

Consequently, the reflected wave $w^{(r)}$ is (see Fig. 2)

$$w^{(r)} = \exp [ik (x \cos \beta - y \operatorname{sen} \beta)] \tag{6}$$

where β is the incident angle (see Fig. 2) and $i = \sqrt{-1}$.

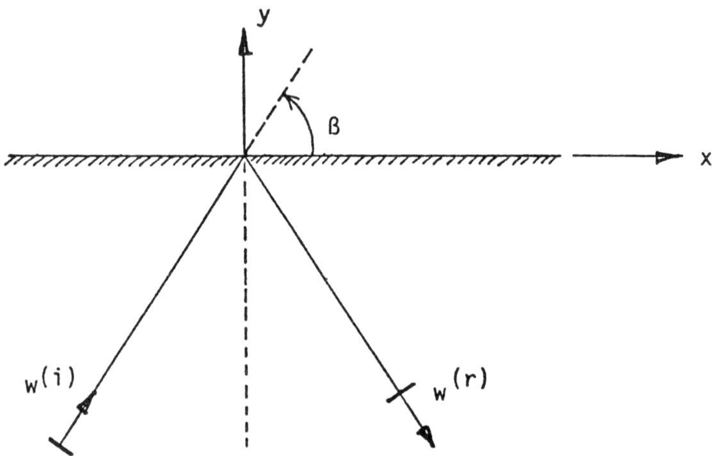

Fig. 2 - Incident and reflected plane SH waves, free-field.

To compute the total field w it is useful to subdivide the domain of interest into an interior and an exterior region, (see Fig. 1).

To define a unique solution w of eq. (2) we take into account the following boundary condition:

$$\frac{\partial w}{\partial n} = 0 \qquad \text{on } S_3 \tag{7}$$

where n is the inward normal on S_3. According to conditon (7) surface S_3 totally reflects the waves and is free to move.

On the interface between two regions, continuity conditions, for displacements and stresses have to be satisfied.

These conditions are:

$$w_1 = w_2; \quad \mu_1 \frac{\partial w_1}{\partial n_1} = \mu_2 \frac{\partial w_2}{\partial n_2} \quad \text{on } S_1$$

$$w_2 = w_3; \quad \mu_2 \frac{\partial w_2}{\partial n_2} = \mu_3 \frac{\partial w_3}{\partial n_3} \quad \text{on } S_2 \tag{8}$$

where n_j (j = 1,2,3) denotes the inward normals to region j on S_i i = 1,2 respectively.

METHOD OF SOLUTION

Suitable analytic solutions of the Helmholtz equation are usually not available for general geometries so that numerical methods have to be used. Domain methods such as finite differences or finite elements cannot be used advantageously, in this kind of problem, because of the size of the domain to be discretized, which implies solving a very large system of equations, and because of the difficulty of taking into account the radiation condition.

A more recent and alternative numerical method suitable for overcoming the aforementioned difficulties is the Boundary Element Method (BEM) (Brebbia[10]). This method will be used here. In essence, it consists in trasforming the differential equation (2) into an integral one.

A particular feature of this method is that integrals are made on the boundary S_i (i = 1,2) of the domain D.

Among the different approaches, used to write the integral equation, the one reported here makes use of the second Green identity so that eq. (2) is written in the following integral form:

$$\frac{\alpha(P)}{2\pi} w(P) = \int_S [G(P,Q) \frac{\partial w}{\partial n} - w \frac{\partial G(P,Q)}{\partial n}] dS \qquad (9)$$

$$P \in D \cup S \qquad\qquad Q \in S$$

In eq. (9) $G(P,Q)$, the Green function associated with the Helmholtz equation (2), is defined by:

$$G(P,Q) = \frac{i}{4} H_0^{(1)}(kr) = \frac{i}{4} [J_0(kr) + i\, Y_0(kr)]$$

$$r = |P - Q| \qquad (10)$$

and $\alpha(P)$ is the angle in P described by the vector Q-P when the point Q runs all over the boundary S. If $P \in D$ then $\alpha = 2\pi$; if $P \in S$, α is the interior angle between the tangents to S in P; it follows that if the boundary is smooth, $\alpha = \pi$ for $P \in S$.

In eq. (10) J_0 and Y_0 represent respectively Bessel functions of the first and second kind while $H_0^{(1)}$ is a Hankel function of first kind and order zero which represents cylindrical SH waves propagating towards infinity with speed c and satisfies Sommerfeld's radiation condition.

Because of the boundary conditions on S_3 and of the properties of the Green function G, eq. (9) allows us to study the problem only on a limited domain (i.e. only inside R).

For numerical purposes the boundary S is divided into N sub-elements S_j ($j = 1,2,..N$) on each of which functions w and $\partial w/\partial n$ are approximated by suitable shape functions. For example, in the numerical applications described below function w is approximated by a linear shape function on each S_j and $\partial w/\partial n$ is considered constant on each S_j.

Making use of these approximations the discretized form of eq. (9) is:

$$\frac{\alpha_i}{2\pi} w_i = \sum_{j=1}^{N} [q_j\, \phi_{0j} - w_j\, (\psi_{0j} - \psi_{1j}) - w_{j+1}\, \psi_{1j}]$$

$$q_j = \frac{\partial w}{\partial n}\Big|_{P=P_j} \qquad w_j = w|_{P=P_j} \qquad (11)$$

where

$$\phi_{0j} = \frac{i}{4} \int_{S_j} H_0^{(1)} (kr)\, dS \tag{12,a}$$

$$\psi_{0j} = \frac{i}{4} \int_{S_j} \frac{\partial}{\partial n} H_0^{(1)} (kr)\, dS \tag{12,b}$$

$$\psi_{1j} = \frac{i}{4} \int_{S_j} \xi \frac{\partial}{\partial n} H_0^{(1)} (kr)\, dS \tag{12,c}$$

In formula (12,c) ξ is a parameter relating to the position of generic point on S_j. Integrals (12) are computed numerically by using Gauss quadrature formulae when $P_i \notin S_j$; when $P_i \in S_j$ the integrand functions become singular; consequently the integration is performed analytically in order to avoid numerical errors.

Details of the numerical procedure can be found in Di Monaco et al[11], Rangogni and Occhi[12].

In the general algebraic equation (11) the unknowns are represented by q_j or w_j or both according to the boundary conditions and according to the position of the segment S_j on S.

By moving the fixed point P_i on the boundary S it is possible to write a sufficient number of equations of type (11) the solution of which gives the unknown values q_j and w_j.

In particular the number of equations is chosen to be greater than the number of unknowns, and the solution is found in the least squares sense.

Once the boundary values q_j and w_j are known, it is possible to compute the value of the displacement w at an interior point P by simply using the eq. (11) with $\alpha = 2\pi$ and by performing the integration only on the boundary of the subregion about P.

APPLICATIONS

It is well known that there is no problems in dealing with arbitrary surface irregularities resting on a homogenous half space (Sanchez-Sesma et al[6]).

In the present work the half space is stratified and the layers can be of arbitrary shape.

However, to reduce the computational effort

only a single layer is considered even though the mathematical formulation allows to deal with any number of layers.

So a general configuration is shown in Fig. 3.

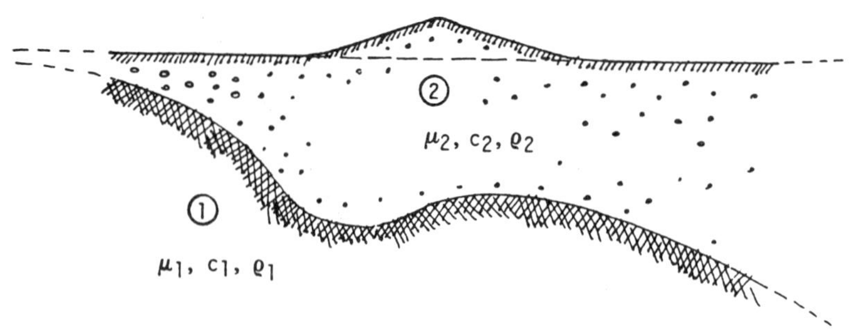

Fig. 3 - Sketch of a general stratified domain.

Without loss of generality a horizontal layer is considered.

Infinite extension of the layer is reduced to a finite one adding artificial vertical boundaries put at a distance that gives results at the control points B_i, Fig. 4, not affected by the reduction.

In a previous work (Bettinali and Rangogni [7]) an evaluation of this distance was given for a different problem.

In order to deduce the optimal reduction also in this case a well studied problem (Sanchez-Sesma et al [6]) is dealt with as a first step.

So the problem really studied is schematically shown in Fig. 4.

The computations are made for a normalized frequency η defined by:

$$\eta = k_2 \frac{b}{\pi} = \frac{2b}{\lambda_2} \tag{13}$$

The boundary discretization of the domain was performed in such a way that at least six elements for wavelength were present.

The results for homogeneous domain are summarized in Tab. 1 together with the one reported in Sanchez-Sesma et al [6].

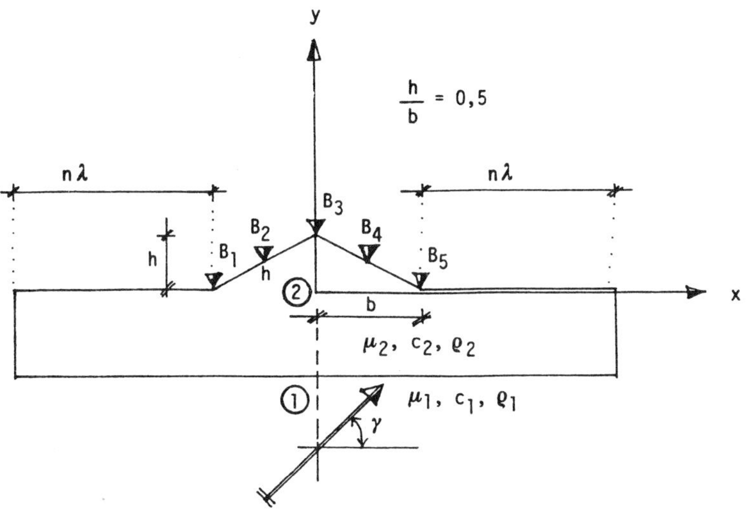

Fig. 4 - Sketch of the domain really considered. $\lambda = c\,T$ is the wave length.

Tab. 1 - Comparison between numerical results and literature ones.

γ	B_1	B_2	B_3	B_4	B_5	$\eta = 0{,}5$ $c_1 = c_2$
	1.43	2.36	2.97	2.36	1.43	Ref. 6
90°	1.43	2.38	3.00	2.38	1.43	n = 10
	1.43	2.39	3.01	2.39	1.43	n = 5
	1.23	1.96	2.82	2.52	1.90	Ref. 6
60°	1.19	1.90	2.75	2.44	1.84	n = 10
	1.19	1.90	2.75	2.44	1.84	n = 5
	1.35	1.74	2.68	2.49	2.04	Ref. 6
45°	1.26	1.64	2.52	2.33	1.90	n = 10
	1.25	1.61	2.49	2.30	1.87	n = 5
	1.60	1.46	2.44	2.39	2.10	Ref. 6
15°	1.33	1.27	2.11	2.04	1.76	n = 10
	1.43	1.34	2.25	2.18	1.90	n = 5
	1.63	1.42	2.40	2.37	2.11	Ref. 6
0°	1.48	1.35	2.24	2.20	1.93	n = 10
	1.51	1.33	2.23	2.19	1.95	n = 5

From Tab. 1 it is evident that the solution is affected by an error of few percent and an optimal length of the domain for this problem could be n = 10.

As a second step in order to evaluate the influence of the elastic characteristics of the layers on the amplification factor a parametric study was carried out changing the elastic characteristics of layer 1.

The results are shown in Tab. 2.

Tab. 2 - Amplification factor for different values of c_1/c_2, and for $\eta = 0.5$, $\gamma = 90°$, $n = 10$.

c_1/c_2	B_1	B_2	B_3	B_4	B_5
1.00	1.43	2.38	3.00	2.38	1.43
2.24	1.98	4.84	6.28	4.84	1.98
3.16	2.50	6.07	7.87	6.07	2.50
10.00	2.06	6.34	8.29	6.34	2.06
31.62	2.20	6.70	8.75	6.70	2.20

From Tab. 2 it can be seen that the amplification factor increases with the increase of c_1/c_2 and seems to have an asimptotic behaviour.

The last case, of Tab. 2, can be considered the schematization of a layer resting on a rigid one, for this case a further computations have been carried out varying the incident angle of the wave.

In Tab. 3 are shown the numerical results.

In order to verify the influence of the incident wavelength a few computations, like those reported in Tab. 1, have been carried out for a different value ($\eta = 1.0$).

The results are shown in Tab. 4.

Tab. 3 - Amplification factor for c_1/c_2 = 31.62 and for different angles of the incident wave; η = 0.5, c_1/c_2 = 31.62.

γ	B_1	B_2	B_3	B_4	B_5	
90°	2.20	6.70	8.75	6.70	2.20	n = 10
	1.64	7.44	10.44	7.44	1.64	n = 5
60°	2.87	4.21	5.54	4.71	3.48	n = 10
	2.36	5.77	8.84	6.80	2.40	n = 5
45°	3.20	2.99	3.21	2.79	3.61	n = 10
	2.84	4.47	7.32	5.94	2.87	n = 5
15°	3.11	3.01	2.65	1.00	3.20	n = 10
	3.38	2.48	4.97	4.58	3.39	n = 5
0°	3.05	3.13	2.88	1.11	3.11	n = 10
	3.45	2.20	4.65	4.40	3.45	n = 5

Tab. 4 - Amplification factor for η = 1.0 at different angles of incident wave; η = 1.0, $c_1 = c_2$, n = 5.

γ	B_1	B_2	B_3	B_4	B_5
90°	2.04	2.27	2.40	2.27	2.04
60°	1.87	2.18	2.34	2.25	2.08
45°	1.77	2.10	2.27	2.20	2.05
15°	1.71	2.06	2.26	2.20	2.07
0°	1.69	2.04	2.24	2.18	2.06

CONCLUSION

In this paper the results of a numerical study carried out in order to verify the influence of a few parameters (elastic and of the wave) on the amplification factor are shown.

The authors are aware that this study is not exhaustive and more cases should be investigated.

However it seems reasonable to take as optimal reduction the distance of 10 wavelengths.

Taking this value as optimal one, the error in the region of interest is bounded so that a standard BEM procedure can be used also for infinite stratified soils.

On the other hand the results obtained could be useful in using domain technique such as FEM and Finite Differences.

REFERENCES

1. Esteva L. (1977) Microzoning: models and reality, Proc. 6th World Conf. Earthquake Eng., New Delhi
2. Wong H.L., Trifunac M.D., Westermo B. (1977), Effects of surface and subsurface irregularities on the amplitudes of monochromatic waves, Bull. Seism. Soc. Am. 67, pp. 353-368.
3. Aki K., Larner K.L. (1970), Surface motion of a layered medium having an irregular interface due to incident plane SH waves, J. Geophys. Res., Vol. 75, pp. 933-954.
4. Bouchon M. (1973), Effect of topography on surface motion, Bull. Seism. Soc. Am., Vol. 63, pp. 715-732.
5. Sanchez-Sesma F.J., Rosenblueth E. (1979), Ground motion at Canyons of arbitrary shape under incident SH waves, Int. J. Earth. Eng. Struct. Dyn., Vol. 7, pp. 441-450.
6. Sanchez-Sesma F.J., Herrera I., Aviles J. (1982), A boundary method for elastic wave diffraction: application to scattering of SH waves by surface irregularities, Bull. Seism. Soc. Am., Vol. 72(2), pp. 473-490.
7. Bettinali F., Rangogni R. (1986) On the optimal reduction of the dimension, from infinite to finite, of a topographic irregularity under seismic SH waves, in italian, Submitted to III° Italian Congress on "Seismic engineering in Italy", Roma, 30 Sett. - 2 Ott. 1987.

8. Aki K., Richards P.G. (1980). Quantitative seismology: Theory and methods, Vol. I W.H. Freeman and Co. S.Francisco.
9. Sommerfeld A. (1949). Partial differential equation in physics, Accademic Press, Inc., New York.
10. Brebbia C.A. (1978). The boundary element method for engineers. Pentech Press, London
11. Di Monaco A., Occhi R., Rangogni R. (1984) Trattazione approssimata della rifrazione-diffrazione delle onde per mezzo del BEM. Proc. XIX Convegno di idraulica e costruzioni idrauliche. Pavia (Italy)
12. Rangogni R., Occhi R. (1984) Il boundary element method nella soluzione dell'equazione di Helmholtz per la diffrazione delle onde (in italian). Internal Rep. n. 3237 ENEL-CRIS

Circular and Rectangular Foundations on Halfspace: Numerical Values of Dynamic Stiffness Functions

Th. Triantafyllidis, B. Prange and Ch. Vrettos
Institute of Soil Mechanics and Rock Mechanics, University of Karlsruhe, Federal Republic of Germany

SUMMARY

In previous papers on the complex dynamic stiffness functions of circular and rectangular foundations on the halfspace Triantafyllidis and Prange gave the exact analytical frequency depending solutions of the respective boundary value problems in graphical form for different aspect ratios and Poissons ratios. In the paper presented the same data are published in polynomial form to allow immediate engineering application in dynamic soil-structure interaction problems. The accuracy of the polynomial approximation is of the same order (1%) as the initial presentation of the analytical solutions.

INTRODUCTION

The dynamic response of rigid foundations bonded perfectly to the surface of a linear elastic, isotropic and homogeneous halfspace can be determined by the substructure method. By this method, the soil-structure interaction problem is subdivided into the inert mass of the foundation including a possible superstructure (subsystem 1) and a rigid massless plate perfectly bonded to the surface of the halfspace (subsystem 2).

Subsystem 2 represents the solution of the dynamic mixed boundary value problem foundation-interface on halfspace which depends on the geometry of the foundation, the dynamic parameters of the halfspace and the excitation frequency.

In two papers by Triantafyllidis[1] and Triantafyllidis and Prange[2] the exact analytical solutions for the boundary value problems of rigid circular and rectangular foundations on the halfspace are given as complex dynamic stiffness functions depending on frequency. Making use of the lumped parameter concept, the real part of the stiffness function can be regarded as an equivalent elastic spring, the imaginary part as an equivalent dashpot. Hence, the halfspace can be replaced by these elastic parameters representing elastic response and wave radiation of the halfspace.

The above solutions are presented in graphical form depending on dimensionless frequency. For immediate engineering application in dynamic interaction problems a numerical presentation in the form of polynomial approximations is given in this paper. The numerical data allow an easy calculation of the solution of the coupled differential equations making use of computers.

Since the numerical values are given in the frequency domain two cases have to be distinguished:

Forced Vibrations:

Because the frequency of excitation is known the numerical values of the stiffness functions are immediately gained from the frequency input.

Free Vibrations:

Because, to start with, the eigenfrequencies of the system are unknown, an iteration procedure must be applied with a suitable frequency as a starting input. Subsequently, the calculation of the eigenfrequencies takes place with the respective values of the dynamic stiffness functions being the input of the next step of the iteration. The iteration converges quickly depending on the accuracy of the starting frequency input.

CIRCULAR FOUNDATIONS

The co-ordinate system for a rigid circular foundation is given in Fig. 1 :

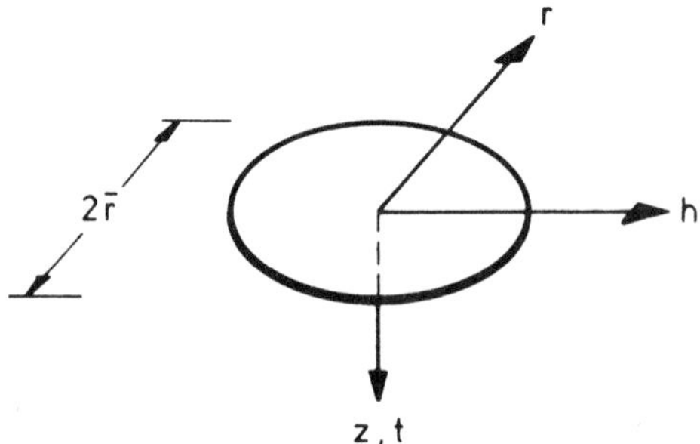

Figure 1. Co-ordinate system rigid circular plate on halfspace

Because of symmetry we find only four degrees of freedom:

horizontal translation h

vertical translation z

rotation about r (rocking)

rotation about z (torsion)

The dynamic stiffness functions depend upon dimensionless frequency a_0:

$$a_0 = \frac{\omega \bar{r}}{v_S} = 2\pi \frac{\bar{r}}{\lambda_S} \tag{1}$$

where ω - circular frequency

v_S - shear wave velocity = $\sqrt{G/\rho}$

λ_S - shear wave length

G - shear modulus

ρ - mass density

$\bar{r}$ - radius of circular plate

Real and imaginary parts of the dynamic stiffness functions are given for translation as follows:

$$Re(F_j) = u_j\, G\, \bar{r}\, K_{jj} \tag{2}$$

with $G\, \bar{r}\, K_{jj} = k_{jj}$ = spring parameter for translation

$$Im(F_j) = \omega\, u_j\, \bar{r}^2 \sqrt{G\rho}\, C_{jj} \tag{3}$$

with $\bar{r}^2 \sqrt{G\,\rho}\, C_{jj} = c_{jj}$ = damping parameter for translation

where F_j - force in the j-axis

u_j - translation in the j-axis

$\sqrt{G\,\rho}$ - wave impedance

Similarly for rotation we have:

$$Re(M_j) = \hat{u}_j\, G\, \bar{r}^3\, K_{\hat{j}\hat{j}} \tag{4}$$

with $G\, \bar{r}^3\, K_{\hat{j}\hat{j}} = k_{\hat{j}\hat{j}}$ = spring parameter for rotation

$$Im(M_j) = \omega\, \hat{u}_j\, \bar{r}^4 \sqrt{G\rho}\, C_{\hat{j}\hat{j}} \tag{5}$$

with $\bar{r}^4 \sqrt{G\,\rho}\, C_{\hat{j}\hat{j}} = c_{\hat{j}\hat{j}}$ = damping parameter for rotation

where M_j - moment about the j-axis

$\hat{u}_j$ - rotation about the j-axis

K_{jj} and C_{jj} in Eq. (2) - (4) are regarded as dimensionless spring and damping parameters, resp.

Their polynomial approximation is given by the following power expansion :

$$S_{jj} = C_0 + C_1 a_0 + C_2 a_0^2 + C_3 a_0^3 + \cdots + C_6 a_0^6 = \sum_{i=0}^{6} C_i a_0^i \tag{6}$$

where S_{jj} represents K_{jj} and C_{jj} with the respective coefficients $C_0 \ldots C_6$. In Tables 1 - 5 these coefficients ($C0 \ldots C6$ in Tables 1 - 5) are given in tabular form for the four degrees of freedom. They are valid for

$$0.05 \leq a_0 \leq 3 \tag{7}$$

or with Eq. (1) $\qquad 0.008 \leq \bar{r}/\lambda_S \leq 0.48$

according to the graphical presentations by Triantafyllidis and Prange[2] .

To achieve the same absolute accuracy of 1% as in the analytical solution different orders of polynomial approximations are needed for different stiffness functions.

Poisson's ratio enters the polynomial approximation in three values of $\nu = 0.25$, 0.33, 0.40 covering the practical range of interest ($\nu \rightarrow$ NY).

Numbers are given in decimal power form, e.g. C_2 Table 1, K_{hh}, $\nu = 0.25$:

$$C_2 = -0.50590 - 1 = -0.50590 \cdot 10^{-1} = -0.050590 \tag{8}$$

Symmetric coupling terms $K_{hr} = K_{rh}$ and $C_{hr} = C_{rh}$ arise due to the perfect bond of the plate to the surface of the halfspace, see Table 5. In consequence interaction of normal- and shear stresses takes place in horizontal translation h and rocking r, resp.

RECTANGULAR FOUNDATIONS

The co-ordinate system for a rigid rectangular plate is given in Fig. 2 :

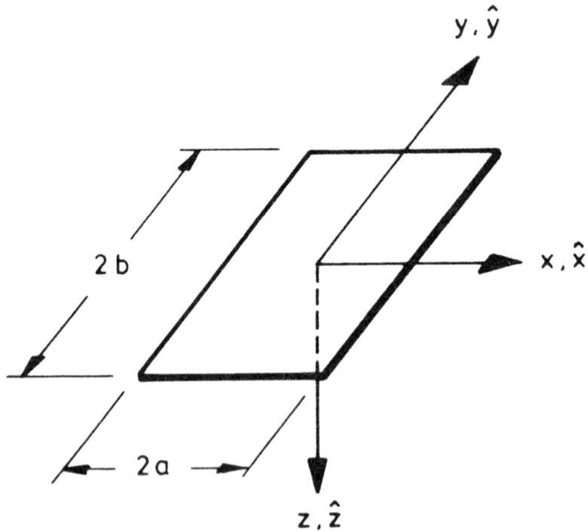

Figure 2 Co-ordinate system rigid rectangular plate on halfspace

All six degrees of freedom are relevant :

translation	x, y, z
rotation	$\hat{x}$, $\hat{y}$, $\hat{z}$

The dynamic stiffness functions depend upon dimensionless frequency r_0 :

$$r_0 = \frac{\omega\sqrt{a^2+b^2}}{v_S} = 2\pi\frac{\sqrt{a^2+b^2}}{\lambda_S} \tag{9}$$

where

a = half width of plate

b = half length of plate

Real and imaginary part of the dynamic stiffness functions are given for translation as follows:

$$Re(F_j) = u_j\, G\, \sqrt{a^2+b^2}\, K_{jj} \tag{10}$$

with $G\sqrt{a^2+b^2}\,K_{jj} = k_{jj}$ = spring parameter for translation

$$Im(F_j) = \omega\, u_j\, (a^2+b^2)\, \sqrt{G\rho}\, C_{jj} \tag{11}$$

with $(a^2+b^2)\sqrt{G\,\rho}\,C_{jj} = c_{jj}$ = damping parameter for translation

and similarly for rotation

$$Re(M_j) = \hat{u}_j\, G\, \sqrt{a^2+b^2}\, d^2\, K_{\hat{j}\hat{j}} \tag{12}$$

with $G\sqrt{a^2+b^2}\,d^2\,K_{\hat{j}\hat{j}} = k_{\hat{j}\hat{j}}$ = spring parameter for rotation

$$Im(M_j) = \omega\, \hat{u}_j\, (a^2+b^2)\, d^2\, \sqrt{G\rho}\, C_{\hat{j}\hat{j}} \tag{13}$$

with $(a^2+b^2)\,d^2\sqrt{G\,\rho}\,C_{\hat{j}\hat{j}} = c_{\hat{j}\hat{j}}$ = damping parameter for rotation

and
$$d = a \quad \text{for} \quad j = y, z \tag{14}$$
$$d = b \quad \text{for} \quad j = x$$

K_{jj} and C_{jj} in Eq. (10) - (13) are again regarded as dimensionless spring and damping parameters, resp.

The polynomial approximation by power expansions is analogous to Eq. (6) with the respective coefficients ($C0....C7$) given in Tables 6 to 11 for the 6 degrees of freedom.

To again achieve an absolute accuracy of 1% in some cases 7th order polynomials are neccessary.

The polynomial approximation is valid for

$$0.05 \leq r_0 \leq 3.5 \tag{15}$$

or with Eq. (1) $\quad 0.008 \leq \sqrt{a^2+b^2}/\lambda_S \leq 0.56$

according to the graphical presentation by Triantafyllidis[1] .

Again the three values of Poissons's ratio $\nu = 0.25;\ 0.33;\ 0.40$ enter the Tables 6 - 11.

As in the case of circular foundations, numbers are given in decimal power form, see e.g. Eq. (8).

Coupling terms $K_{\hat{y}x}$; $C_{\hat{y}x}$ and $K_{\hat{x}y}$; $C_{\hat{x}y}$ which are not symmetric if $a \neq b$ arise due to the perfect bond of the rectangular foundation to the surface of the halfspace, see Tables 12 - 13. In consequence interaction of normal- and shear stresses takes place in horizontal translation and rocking about an orthogonal axis.

In some cases, 7th order polynomials are not sufficient to achieve the absolute accuracy of 1% for the coupling terms. An asterix * denotes in such cases that the accuracy is between 1% and 6%. Since the jk-coupling-terms are much smaller than the jj-terms (diagonal terms in the complex stiffness matrix) the relative accuracy of the coupling-terms is again of the order of 1%.

EXAMPLES

Circular Foundation

The use of Tables 1 - 5 shall be illustrated by an example of a rigid circular foundation forced to vibrate in the z-translation:

frequency	f	$= 50$ Hz
geometry	$\bar{r}$	$= 1$ m
soil	v_S	$= 200$ m/s , $\rho = 1.8$ t/m^3 , $\nu = 0.33$

From Eq. (1) we find

$$a_0 = \frac{2\pi f \bar{r}}{v_S} = \frac{2\pi \cdot 50 \cdot 1}{200} = 1.57 \tag{16}$$

and $\quad G = v_s^2 \cdot \rho = 200^2 \cdot 1.8 = 72 \cdot 10^3$ kN/m^2.

Table 2 yields for $\nu = 0.33$ and $a_0 = 1.57$ according to Eq. (1) and (2) :

$$S_{zz} = \sum_{i=0}^{6} C_i \cdot a_0^i$$

Real part:

$$K_{zz} = -0.61402 \cdot 10^1 + 0.83234 \cdot 10^{-1} \cdot 1.57 + 0.30600 \cdot 10^0 \cdot 1.57^2 + 0.23483 \cdot 10^0 \cdot 1.57^3 - 0.78015 \cdot 10^{-1} \cdot 1.57^4$$

$$K_{zz} = -4.82 \tag{17}$$

Imaginary part:

$$C_{zz} = -0.48135 \cdot 10^1 - 0.36151 \cdot 10^{-1} \cdot 1.57 - 0.12768 \cdot 10^0 \cdot 1.57^2$$
$$C_{zz} = -5.18 \qquad (18)$$

Putting these dimensionless values of the dynamic stiffness function into Eq. (2) and (3) we find:

$$k_{zz} = G \cdot \bar{r} \cdot K_{zz} = -72 \cdot 10^3 \cdot 1.0 \cdot 4.82$$
$$= -347 \cdot 10^3 \quad \text{kN/m} \qquad (19)$$

lumped parameter spring for vertical translation

$$c_{zz} = \bar{r}^2 \sqrt{G\,\rho} \cdot C_{zz} = -1 \cdot \sqrt{72 \cdot 10^3 \cdot 1.8} \cdot 5.18$$
$$= -1.86 \cdot 10^3 \quad \text{kN s/m} \qquad (20)$$

lumped parameter dashpot for vertical translation

Rectangular Foundation

The use of Tables 6 - 13 shall be illustrated by an example of a rigid rectangular foundation forced to vibrate in the $\hat{y}$-rotation (rocking).

frequency	f	$= 25$ Hz
geometry	a	$= 1$ m , $b = 2$ m , $b/a = 2$
soil	v_S	$= 200$ m/s , $\rho = 1.8$ t/m^3 , $\nu = 0.33$

From Eq. (9) we find:

$$r_0 = \frac{2\pi f \sqrt{a^2 + b^2}}{v_S} = \frac{2\pi \cdot 25 \cdot \sqrt{1^2 + 2^2}}{200} = 1.76 \qquad (21)$$

Table 10 yields for $\nu = 0.33$, $b/a = 2$ and $a_0 = 1.76$ according to Eq. (12) and (13)

$$S_{\hat{j}\hat{j}} = \sum_{i=0}^{7} \cdot C_i \cdot r_0^i$$

Real part:

$$K_{\hat{y}\hat{y}} = -0.55819 \cdot 10^1 + 0.29800 \cdot 10^0 \cdot 1.76 + 0.23241 \cdot 10^0 \cdot 1.76^2$$
$$-0.47311 \cdot 10^{-1} \cdot 1.76^3$$
$$K_{\hat{y}\hat{y}} = -4.60 \qquad (22)$$

Imaginary part:

$$C_{\hat{y}\hat{y}} = -0.35858 \cdot 10^{-1} + 0.12191 \cdot 10^{0} \cdot 1.76 - 0.56612 \cdot 10^{0} \cdot 1.76^{2}$$
$$+0.30227 \cdot 10^{0} \cdot 1.76^{3} - 0.67932 \cdot 10^{-1} \cdot 1.76^{4} + 0.57385 \cdot 10^{-2} \cdot 1.76^{5}$$
$$C_{\hat{y}\hat{y}} = -0.482 \tag{23}$$

Putting these dimensionless values of the dynamic stiffness function into Eq. (12) and (13) ($d = a$ for $j = y$) we find with Eq. (14)

$$k_{\hat{y}\hat{y}} = G\sqrt{a^2 + b^2}\, d^2 K_{\hat{y}\hat{y}} = -72 \cdot 10^3 \sqrt{1^2 + 2^2} \cdot 1 \cdot 4.60$$
$$= -741 \cdot 10^3 \text{ kNm/rad} \tag{24}$$

lumped parameter spring for rocking about the $\hat{y}$-axis

$$c_{\hat{y}\hat{y}} = (a^2 + b^2)\, d^2 \sqrt{G\,\rho}\, C_{\hat{y}\hat{y}} = -(1^2 + 2^2) \cdot 1 \cdot \sqrt{72 \cdot 10^3 \cdot 1.8} \cdot 0.482$$
$$= -868 \text{ kNm s/rad} \tag{25}$$

lumped parameter dashpot for rocking about the $\hat{y}$-axis

Once the numerical data of Tables 1 - 13 are stored in a computer, the calculation of the lumped parameters k_{jk} and c_{jk} is simple and straightforward.

REFERENCES

1. Triantafyllidis Th. (1986), Dynamic Stiffnes of Rigid Rectangular Foundations on Halfspace, Earthquake Engineering and Structural Dynamics, Vol.14, pp. 391 - 411.

2. Triantafyllidis Th. and Prange B. (1987), Rigid Circular Foundation: Dynamic Effects of Coupling to the Halfspace, in print: Soil Dynamics and Earthquake Engineering.

APPENDIX : Tables of Stiffness Functions

	Khh			Chh		
NY	0.25	0.33	0.40	0.25	0.33	0.40
C0	-0.50673+1	-0.48489+1	-0.46914+1	-0.28346+1	-0.27900+1	-0.28003+1
C1	+0.27404+0	+0.16048+0	+0.20918+0	-0.94003-1	-0.89116-1	+0.24552-2
C2	-0.50950-1					-0.27672-1

Table 1. Stiffness Functions Khh (Real Part) and Chh (Imaginary Part) of Circular Foundation for Horizontal Translation h

	Kzz			Czz		
NY	0.25	0.33	0.40	0.25	0.33	0.40
C0	-0.66796+1	-0.61402+1	-0.54920+1	-0.52469+1	-0.48135+1	-0.44293+1
C1	-0.11174+1	+0.83234-1	-0.48845+0	-0.42338-1	-0.36151-1	-0.26038-1
C2	+0.42391+1	+0.30600+0	+0.10339+1	-0.15837+0	-0.12768+0	-0.98526-1
C3	-0.42496+1	+0.23483+0	-0.22400+0			
C4	+0.24668+1	-0.78015-1				
C5	-0.67817+0					
C6	+0.68804-1					

Table 2. Stiffness Functions Kzz (Real Part) and Czz (Imaginary Part) of Circular Foundation for Vertical Translation z

	Krr			Crr		
NY	0.25	0.33	0.40	0.25	0.33	0.40
C0	-0.45303+1	-0.41889+1	-0.38228+1	-0.17387-1	-0.24865-1	-0.31811-1
C1	+0.27100+0	+0.56923+0	+0.47386+0	+0.28649+0	+0.27256+0	+0.15229+0
C2	+0.79905+0	+0.17266+0	+0.17872+0	-0.20349+1	-0.18426+1	-0.12575+1
C3	-0.36608+0	-0.45346-1	-0.48259-1	+0.18237+1	+0.16423+1	+0.91185+0
C4	+0.52855-1			-0.76877+0	-0.69643+0	-0.27172+0
C5				+0.15980+0	+0.14742+0	+0.29665-1
C6				-0.13167-1	-0.12535-1	

Table 3. Stiffness Functions Krr (Real Part) and Crr (Imaginary Part) of Circular Foundation for Rotation r (Rocking)

NY	Ktt 0.25	0.33	0.40	Ctt 0.25	0.33	0.40
C0	-0.53804+1	-0.53804+1	-0.53804+1	-0.50701-2	-0.50701-2	-0.50701-2
C1	+0.55078+0	+0.55078+0	+0.55078+0	+0.12746+0	+0.12746+0	+0.12746+0
C2	+0.26645+0	+0.26645+0	+0.26645+0	-0.13252+1	-0.13252+1	-0.13252+1
C3	-0.73798-1	-0.73798-1	-0.73798-1	+0.10335+1	+0.10335+1	+0.10335+1
C4				-0.38370+0	-0.38370+0	-0.38370+0
C5				+0.71533-1	+0.71533-1	+0.71533-1
C6				-0.53733-2	-0.53733-2	-0.53733-2

Table 4. Stiffness Functions Ktt (Real Part) and Ctt (Imaginary Part) of Circular Foundation for Rotation t (Torsion)

NY	Khr 0.25	0.33	0.40	Chr 0.25	0.33	0.40
C0	+0.25143+0	+0.37137+0	+0.46774+0	+0.14933+0	+0.22189+0	+0.29202+0
C1	+0.23337+0	+0.26082+0	+0.63784+0	-0.49085+0	-0.55282+0	-0.48919+0
C2	-0.15535+0	-0.15953+0	-0.12567+1	+0.32664+0	+0.39069+0	+0.71757-1
C3	+0.91883-1	+0.62038-1	+0.14441+1	-0.11505+0	-0.14178+0	+0.21957+0
C4	-0.30939-1	-0.12877-1	-0.83947+0	+0.15250-1	+0.19795-1	-0.16061+0
C5	+0.30546-2		+0.23043+0		-0.99252-4	+0.43337-1
C6			-0.24058-1			-0.41729-2

Table 5. Stiffness Functions Khr (Real Part) and Chr (Imaginary Part) of Circular Foundation between Rotation r and Translation h

b/a = 1	Kxx			Cxx		
NY	0.25	0.33	0.40	0.25	0.33	0.40
C0	-0.38729+1	-0.40064+1	-0.42225+1	-0.20334+1	-0.20343+1	-0.19623+1
C1	+0.51467-1	+0.12610+0	+0.14719+0	+0.18304+0	+0.16255+0	-0.67402-1
C2		-0.26507-1	-0.48530-1	-0.11320+0	-0.10805+0	
C3				+0.14921-1	+0.14860-1	

b/a = 2	Kxx			Cxx		
NY	0.25	0.33	0.40	0.25	0.33	0.40
C0	-0.36577+1	-0.37543+1	-0.40218+1	-0.18268+1	-0.18434+1	-0.18177+1
C1	-0.23476-1	-0.46648-1	+0.93429-2	+0.12641+0	+0.10765+0	-0.82730-1
C2			-0.34096-1	-0.68931-1	-0.63227-1	+0.24604-1
C3				+0.11980-1	+0.11781-1	

b/a = 5	Kxx			Cxx		
NY	0.25	0.33	0.40	0.25	0.33	0.40
C0	-0.30886+1	-0.32026+1	-0.34548+1	-0.12747+1	-0.13013+1	-0.13206+1
C1	+0.13961+0	+0.14945+0	+0.18262+0	+0.48456-1	+0.42053-1	-0.57470-1
C2	-0.19874+0	-0.21171+0	-0.26010+0	-0.62873-2	-0.48516-2	+0.33992-1
C3	+0.37856-1	+0.39338-1	+0.46842-1	+0.37152-2	+0.39346-2	

b/a = 10	Kxx			Cxx		
NY	0.25	0.33	0.40	0.25	0.33	0.40
C0	-0.25033+1	-0.26001+1	-0.28110+1	-0.84200+0	-0.86274+0	-0.87111+0
C1	+0.72768-1	+0.80732-1	+0.11143+0	+0.24103-1	+0.22809-1	-0.74372-1
C2	-0.16807+0	-0.17794+0	-0.22251+0	+0.10975-1	+0.12200-1	+0.63741-1
C3	+0.30660-1	+0.31844-1	+0.39427-1			-0.75144-2

Table 6. Stiffness functions Kxx (Real Part) and Cxx (Imaginary Part) of Rectangular Foundations for Horizontal Translation x

b/a = 1	Kyy			Cyy		
NY	0.25	0.33	0.40	0.25	0.33	0.40
C0	-0.38729+1	-0.40064+1	-0.42225+1	-0.20334+1	-0.20343+1	-0.19623+1
C1	+0.51467-1	+0.12610+0	+0.14719+0	+0.18304+0	+0.16255+0	-0.67402-1
C2		-0.26507-1	-0.48530-1	-0.11320+0	-0.10805+0	
C3				+0.14921-1	+0.14860-1	

b/a = 2	Kyy			Cyy		
NY	0.25	0.33	0.40	0.25	0.33	0.40
C0	-0.35244+1	-0.36392+1	-0.37786+1	-0.16723+1	-0.16566+1	-0.15625+1
C1	+0.37065-1	+0.12124+0	+0.10201+0	+0.19600+0	+0.17016+0	-0.32193-1
C2	+0.24902-1			-0.10812+0	-0.98252-1	
C3				+0.15344-1	+0.14193-1	

b/a = 5	Kyy			Cyy		
NY	0.25	0.33	0.40	0.25	0.33	0.40
C0	-0.27739+1	-0.28210+1	-0.29163+1	-0.10545+1	-0.10350+1	-0.96416+0
C1	-0.61866-1	-0.56376-1	-0.49612-1	+0.11921+0	+0.10179+0	-0.34712-2
C2	+0.20724-1	+0.19814-1	+0.17373-1	-0.49464-1	-0.43020-1	+0.62207-2
C3				+0.79327-2	+0.71189-2	

b/a = 10	Kyy			Cyy		
NY	0.25	0.33	0.40	0.25	0.33	0.40
C0	-0.21567+1	-0.21868+1	-0.22474+1	-0.63253+0	-0.61534+0	-0.57515+0
C1	-0.59741-1	-0.55886-1	-0.51686-1	+0.43823-1	+0.32049-1	+0.45101-2
C2				-0.23795-2		+0.51967-2

Table 7. Stiffness functions Kyy (Real Part) and Cyy (Imaginary Part) of Rectangular Foundations for Horizontal Translation y

b/a = 1	Kzz			Czz		
NY	0.25	0.33	0.40	0.25	0.33	0.40
C0	-0.44018+1	-0.50132+1	-0.55920+1	-0.27973+1	-0.31480+1	-0.35498+1
C1	-0.42818+0	+0.77281-1	-0.20174+0	-0.30589+0	-0.32376+0	+0.27930+0
C2	+0.11621+1	+0.16308+0	+0.70749+0	+0.17595+0	+0.20773+0	-0.48156+0
C3	-0.83513+0		-0.20318+0	-0.54706-1	-0.63485-1	+0.20294+0
C4	+0.28932+0		+0.35301-1			-0.35038-1
C5	-0.36515-1					

b/a = 2	Kzz			Czz		
NY	0.25	0.33	0.40	0.25	0.33	0.40
C0	-0.43047+1	-0.45751+1	-0.51986+1	-0.26061+1	-0.27260+1	-0.30729+1
C1	+0.42243-1	+0.71521-1	+0.42781-1	-0.66458-1	-0.84852-1	-0.13899+0
C2	+0.77847-1	+0.94357-1	+0.20714+0			

b/a = 5	Kzz			Czz		
NY	0.25	0.33	0.40	0.25	0.33	0.40
C0	-0.35105+1	-0.37289+1	-0.42569+1	-0.17335+1	-0.18172+1	-0.20536+1
C1	+0.19073+0	+0.20027+0	+0.21882+0	-0.40958-1	-0.51377-1	-0.94541-1
C2	-0.24179+0	-0.23572+0	-0.21283+0	+0.35138-1	+0.37473-1	+0.45930-1
C3	+0.59922-1	+0.58547-1	+0.55933-1			

b/a = 10	Kzz			Czz		
NY	0.25	0.33	0.40	0.25	0.33	0.40
C0	-0.27914+1	-0.29622+1	-0.33690+1	-0.10894+1	-0.11409+1	-0.12901+1
C1	+0.75567-1	+0.90413-1	+0.13892+0	-0.12529+0	-0.12181+0	-0.81463-1
C2	-0.21228+0	-0.21899+0	-0.25685+0	+0.11516+0	+0.11223+0	+0.70899-1
C3	+0.47641-1	+0.48202-1	+0.54829-1	-0.17233-1	-0.16456-1	-0.76469-2

Table 8. Stiffness functions Kzz (Real Part) and Czz (Imaginary Part) of Rectangular Foundations for Vertical Translation z

b/a = 1	Kϑϑ			Cϑϑ		
NY	0.25	0.33	0.40	0.25	0.33	0.40
C0	-0.40211+1	-0.46414+1	-0.49731+1	-0.32500-1	-0.25025-1	-0.69543-2
C1	-0.32406-1	-0.13183+0	-0.33529+0	+0.12418+0	+0.10461+0	+0.75371-1
C2	+0.85423+0	+0.10062+1	+0.97480+0	-0.84985+0	-0.85809+0	-0.93718+0
C3	-0.36057+0	-0.38567+0	-0.32761+0	+0.51682+0	+0.52241+0	+0.57551+0
C4	+0.46307-1	+0.45162-1	+0.36900-1	-0.12889+0	-0.12989+0	-0.14294+0
C5				+0.11774-1	+0.11823-1	+0.12975-1

b/a = 2	Kϑϑ			Cϑϑ		
NY	0.25	0.33	0.40	0.25	0.33	0.40
C0	-0.33365+1	-0.35164+1	-0.39441+1	-0.21976-1	-0.20519-1	-0.64840-2
C1	+0.49158-1	+0.40060-1	+0.15263+0	+0.89644-1	+0.15416+0	+0.92680-1
C2	+0.61653+0	+0.66678+0	+0.60458+0	-0.73956+0	-0.10029+1	-0.96231+0
C3	-0.27367+0	-0.29176+0	-0.26210+0	+0.46173+0	+0.76572+0	+0.69666+0
C4	+0.35367-1	+0.37654-1	+0.34368-1	-0.11706+0	-0.27710+0	-0.23056+0
C5				+0.10853-1	+0.50242-1	+0.37362-1
C6					-0.36687-2	-0.23961-2

b/a = 5	Kϑϑ			Cϑϑ		
NY	0.25	0.33	0.40	0.25	0.33	0.40
C0	-0.21610+1	-0.22813+1	-0.25977+1	-0.90125-2	-0.80650-2	-0.42840-2
C1	-0.51704-1	-0.47956-1	+0.29054+0	+0.52019-1	+0.69213-1	+0.81810-1
C2	+0.44465+0	+0.46359+0	+0.50328-1	-0.40734+0	-0.48937+0	-0.60552+0
C3	-0.19853+0	-0.20526+0	-0.19663-1	+0.24504+0	+0.33834+0	+0.45853+0
C4	+0.25174-1	+0.25916-1		-0.59856-1	-0.10845+0	-0.16585+0
C5				+0.54395-2	+0.17323-1	+0.30483-1
C6					-0.11029-2	-0.22674-2

b/a = 10	Kϑϑ			Cϑϑ		
NY	0.25	0.33	0.40	0.25	0.33	0.40
C0	-0.15414+1	-0.16592+1	-0.19004+1	-0.40654-2	-0.37437-2	-0.21954-2
C1	-0.47532-2	+0.17354+0	+0.18804+0	+0.33051-1	+0.42781-1	+0.45011-1
C2	+0.24428+0	+0.13053-1	+0.23964-1	-0.22810+0	-0.27692+0	-0.32963+0
C3	-0.11535+0	-0.11136-1	-0.13566-1	+0.12828+0	+0.18009+0	+0.23004+0
C4	+0.14807-1			-0.29144-1	-0.54445-1	-0.77137-1
C5				+0.24875-2	+0.83339-2	+0.13408-1
C6					-0.51600-3	-0.95789-3

Table 9. Stiffness Functions Kϑϑ (Real Part) and Cϑϑ (Imaginary Part) of Rectangular Foundations for Horizontal Rotation ϑ (Rocking)

b/a = 1	$K_{\varphi\varphi}$			$C_{\varphi\varphi}$		
NY	0.25	0.33	0.40	0.25	0.33	0.40
C0	-0.40211+1	-0.46414+1	-0.49731+1	-0.32500-1	-0.25025-1	-0.69543-2
C1	-0.32406-1	-0.13183+0	-0.33529+0	+0.12418+0	+0.10461+0	+0.75371-1
C2	+0.85423+0	+0.10062+1	+0.97480+0	-0.84985+0	-0.85809+0	-0.93718+0
C3	-0.36057+0	-0.38567+0	-0.32761+0	+0.51682+0	+0.52241+0	+0.57551+0
C4	+0.46307-1	+0.45162-1	+0.36900-1	-0.12889+0	-0.12989+0	-0.14294+0
C5				+0.11774-1	+0.11823-1	+0.12975-1

b/a = 2	$K_{\varphi\varphi}$			$C_{\varphi\varphi}$		
NY	0.25	0.33	0.40	0.25	0.33	0.40
C0	-0.53032+1	-0.55819+1	-0.62629+1	-0.46009-1	-0.35858-1	-0.91154-2
C1	+0.26957+0	+0.29800+0	+0.64858+0	+0.14097+0	+0.12191+0	+0.60027-1
C2	+0.21916+0	+0.23241+0		-0.56628+0	-0.56612+0	-0.52273+0
C3	-0.44307-1	-0.47311-1		+0.30328+0	+0.30227+0	+0.23770+0
C4				-0.68839-1	-0.67932-1	-0.30299-1
C5				+0.58776-2	+0.57385-2	-0.34436-2
C6						+0.81402-3

b/a = 5	$K_{\varphi\varphi}$			$C_{\varphi\varphi}$		
NY	0.25	0.33	0.40	0.25	0.33	0.40
C0	-0.63189+1	-0.66620+1	-0.75340+1	-0.49132-1	-0.38147-1	-0.88736-2
C1	-0.10573+0	-0.55430-1	+0.38600+0	+0.68364-1	+0.54961-1	+0.25098-1
C2	+0.26349+0	+0.26039+0		-0.16172+0	-0.15480+0	-0.14539+0
C3	-0.43368-1	-0.43952-1		+0.66448-1	+0.61516-1	+0.48958-1
C4				-0.12264-1	-0.10661-1	-0.51827-2
C5				+0.87793-3	+0.70410-3	

b/a = 10	$K_{\varphi\varphi}$			$C_{\varphi\varphi}$		
NY	0.25	0.33	0.40	0.25	0.33	0.40
C0	-0.64897+1	-0.68987+1	-0.78915+1	-0.61176-1	-0.54291-1	-0.13712-1
C1	+0.14928+0	+0.16985+0	+0.19814+0	+0.59215-2	+0.11990+0	+0.69501-2
C2				-0.39571-1	-0.48173+0	-0.33007-1
C3				+0.54953-1	+0.73049+0	+0.65448-2
C4				-0.35148-1	-0.54990+0	+0.40310-3
C5				+0.96939-2	+0.21644+0	-0.17121-3
C6				-0.96574-3	-0.42813-1	
C7					+0.33616-2	

Table 10. Stiffness Functions $K_{\varphi\varphi}$ (Real Part) and $C_{\varphi\varphi}$ (Imaginary Part) of Rectangular Foundations for Horizontal Rotation φ (Rocking)

b/a = 1	Kẑẑ			Cẑẑ		
NY	0.25	0.33	0.40	0.25	0.33	0.40
C0	-0.62520+1	-0.62520+1	-0.62520+1	-0.35011-2	-0.35011-2	-0.35011-2
C1	+0.41628+0	+0.41628+0	+0.41628+0	+0.88215-1	+0.88215-1	+0.88215-1
C2	+0.19560+0	+0.19560+0	+0.19560+0	-0.89638+0	-0.89638+0	-0.89638+0
C3	-0.42496-1	-0.42496-1	-0.42496-1	+0.57776+0	+0.57776+0	+0.57776+0
C4				-0.17556+0	-0.17556+0	-0.17556+0
C5				+0.26699-1	+0.26699-1	+0.26699-1
C6				-0.16386-2	-0.16386-2	-0.16386-2

b/a = 2	Kẑẑ			Cẑẑ		
NY	0.25	0.33	0.40	0.25	0.33	0.40
C0	-0.32099+1	-0.32099+1	-0.32099+1	-0.10635-2	-0.10635-2	-0.10635-2
C1	+0.39165+0	+0.39165+0	+0.39165+0	+0.26430-1	+0.26430-1	+0.26430-1
C2	-0.35137-1	-0.35137-1	-0.35137-1	-0.32007+0	-0.32007+0	-0.32007+0
C3				+0.18995+0	+0.18995+0	+0.18995+0
C4				-0.51794-1	-0.51794-1	-0.51794-1
C5				+0.68256-2	+0.68256-2	+0.68256-2
C6				-0.33498-3	-0.33498-3	-0.33498-3

b/a = 5	Kẑẑ			Cẑẑ		
NY	0.25	0.33	0.40	0.25	0.33	0.40
C0	-0.11578+1	-0.11578+1	-0.11578+1	+0.36335-5	+0.36335-5	+0.36335-5
C1	+0.91238-1	+0.91238-1	+0.91238-1	+0.91003-3	+0.91003-3	+0.91003-3
C2	-0.76181-2	-0.76181-2	-0.76181-2	-0.44666-1	-0.44666-1	-0.44666-1
C3				+0.11908-1	+0.11908-1	+0.11908-1
C4				+0.70538-2	+0.70538-2	+0.70538-2
C5				-0.51395-2	-0.51395-2	-0.51395-2
C6				+0.12165-2	+0.12165-2	+0.12165-2
C7				-0.10290-3	-0.10290-3	-0.10290-3

b/a = 10	Kẑẑ			Cẑẑ		
NY	0.25	0.33	0.40	0.25	0.33	0.40
C0	-0.54716+0	-0.54716+0	-0.54716+0	+0.13546-3	+0.13546-3	+0.13546-3
C1	+0.16764-1	+0.16764-1	+0.16764-1	-0.31885-2	-0.31885-2	-0.31885-2
C2				+0.49252-2	+0.49252-2	+0.49252-2
C3				-0.19616-1	-0.19616-1	-0.19616-1
C4				+0.17421-1	+0.17421-1	+0.17421-1
C5				-0.71261-2	-0.71261-2	-0.71261-2
C6				+0.14241-2	+0.14241-2	+0.14241-2
C7				-0.11209-3	-0.11209-3	-0.11209-3

Table 11. Stiffness Functions Kẑẑ (Real Part) and Cẑẑ (Imaginary Part) of Rectangular Foundations for Vertical Rotation ẑ (Torsion)

b/a = 1

NY	K$\hat{y}$x 0.25	K$\hat{y}$x 0.33	K$\hat{y}$x 0.40	C$\hat{y}$x 0.25	C$\hat{y}$x 0.33	C$\hat{y}$x 0.40
		*			*	
C0	+0.49193+0	+0.44210+0	+0.21549+0	+0.27057+0	+0.24061+0	+0.10773+0
C1	+0.61584-2	-0.17491-1	+0.14182+0	-0.75297+0	-0.77275+0	-0.42455+0
C2	+0.58655+0	+0.67620+0	-0.20121-1	+0.13126+1	+0.15487+1	+0.64637+0
C3	-0.81673+0	-0.96187+0	-0.68782-2	-0.15920+1	-0.20174+1	-0.80916+0
C4	+0.45061+0	+0.54209+0		+0.10925+1	+0.14278+1	+0.57613+0
C5	-0.11254+0	-0.13743+0		-0.41047+0	-0.54499+0	-0.22273+0
C6	+0.10472-1	+0.12925-1		+0.79044-1	+0.10596+0	+0.43853-1
C7				-0.61049-2	-0.82382-2	-0.34490-2

b/a = 2

NY	K$\hat{y}$x 0.25	K$\hat{y}$x 0.33	K$\hat{y}$x 0.40	C$\hat{y}$x 0.25	C$\hat{y}$x 0.33	C$\hat{y}$x 0.40
		*			*	
C0	+0.57494+0	+0.52096+0	+0.26644+0	+0.28797+0	+0.26978+0	+0.12442+0
C1	-0.70303-1	-0.15087+0	+0.20116-1	-0.39438+0	-0.74472+0	-0.34579+0
C2	+0.69033+0	+0.97479+0	+0.15023+0	+0.35208+0	+0.17817+1	+0.57723+0
C3	-0.85367+0	-0.12307+1	-0.68250-1	-0.35199+0	-0.24756+1	-0.75671+0
C4	+0.46690+0	+0.67986+0	+0.67287-2	+0.24233+0	+0.17787+1	+0.53520+0
C5	-0.11937+0	-0.17343+0		-0.96125-1	-0.68122+0	-0.20450+0
C6	+0.11432-1	+0.16526-1		+0.19728-1	+0.13255+0	+0.39965-1
C7				-0.16136-2	-0.10299-1	-0.31260-2

b/a = 5

NY	K$\hat{y}$x 0.25	K$\hat{y}$x 0.33	K$\hat{y}$x 0.40	C$\hat{y}$x 0.25	C$\hat{y}$x 0.33	C$\hat{y}$x 0.40
				*	*	*
C0	+0.60131+0	+0.58414+0	+0.26758+0	+0.24802+0	+0.22185+0	+0.10212+0
C1	-0.73481-1	-0.10471+1	+0.20934+0	-0.12757+0	-0.12074+0	-0.13693-1
C2	+0.55501+0	+0.43580+1	-0.49234+0	+0.63068-2	-0.18781-2	-0.22312+0
C3	-0.70556+0	-0.65980+1	+0.61699+0	+0.40781-2	+0.82676-2	+0.31588+0
C4	+0.40228+0	+0.48893+1	-0.32749+0	-0.26327-3	-0.90241-3	-0.21816+0
C5	-0.10583+0	-0.18998+1	+0.77253-1			+0.79792-1
C6	+0.10383-1	+0.37168+0	-0.67578-2			-0.14693-1
C7		-0.28893-1				+0.10759-2

b/a = 10

NY	K$\hat{y}$x 0.25	K$\hat{y}$x 0.33	K$\hat{y}$x 0.40	C$\hat{y}$x 0.25	C$\hat{y}$x 0.33	C$\hat{y}$x 0.40
		*		*	*	
C0	+0.67018+0	+0.66572+0	+0.35231+0	+0.22869+0	+0.20986+0	+0.10873+0
C1	+0.15559+0	-0.99292+0	+0.39473-1	-0.46602-1	-0.79755-1	-0.99930-2
C2	+0.26318+0	+0.46656+1	+0.45086-1	+0.12941+0	+0.24287+0	-0.23071-1
C3	-0.58565+0	-0.73711+1	-0.18899-1	-0.21841+0	-0.36034+0	+0.76082-2
C4	+0.39469+0	+0.55541+1	+0.21465-2	+0.13224+0	+0.21066+0	-0.61008-3
C5	-0.11033+0	-0.21729+1		-0.34387-1	-0.54118-1	
C6	+0.11060-1	+0.42665+0		+0.32787-2	+0.51308-2	
C7		-0.33256-1				

Table 12 Stiffness Functions K$\hat{y}$x (Real Part) and C$\hat{y}$x (Imaginary Part) of Rectangular Foundations between Rotation $\hat{y}$ and Translation x

b/a = 1

NY	$K_{\hat{x}y}$ 0.25	0.33	0.40	$C_{\hat{x}y}$ 0.25	0.33	0.40
				*	*	
C0	-0.49121+0	-0.47205+0	-0.21606+0	-0.27013+0	-0.24037+0	-0.10783+0
C1	-0.63044-2	+0.75123+0	-0.13934+0	+0.75045+0	+0.77131+0	+0.42387+0
C2	-0.58608+0	-0.35602+1	+0.17563-1	-0.13057+1	-0.15452+1	-0.64407+0
C3	+0.81569+0	+0.54883+1	+0.74586-2	+0.15826+1	+0.20127+1	+0.80562+0
C4	-0.45004+0	-0.40576+1		-0.10860+1	-0.14245+1	-0.57354+0
C5	+0.11236+0	+0.15715+1		+0.40802+0	+0.54374+0	+0.22173+0
C6	-0.10450-1	-0.30704+0		-0.78567-1	-0.10572+0	-0.43651-1
C7		+0.23893-1		+0.60673-2	+0.82180-2	+0.34323-2

b/a = 2

NY	$K_{\hat{x}y}$ 0.25	0.33	0.40	$C_{\hat{x}y}$ 0.25	0.33	0.40
		*		*	*	
C0	-0.40761+0	-0.36841+0	-0.18443+0	-0.19787+0	-0.17323+0	-0.78964-1
C1	-0.27244-1	+0.26319-1	-0.10166+0	+0.48368+0	+0.45251+0	+0.29164+0
C2	-0.39859+0	-0.57457+0	+0.23005-1	-0.45876+0	-0.46699+0	-0.22028+0
C3	+0.62631+0	+0.85364+0	+0.15479-1	+0.29790+0	+0.34670+0	+0.12868+0
C4	-0.35588+0	-0.48412+0	-0.37318-2	-0.12242+0	-0.15679+0	-0.53162-1
C5	+0.89831-1	+0.12257+0		+0.26543-1	+0.36081-1	+0.11950-1
C6	-0.85000-2	-0.11595-1		-0.22808-2	-0.32273-2	-0.10652-2

b/a = 5

NY	$K_{\hat{x}y}$ 0.25	0.33	0.40	$C_{\hat{x}y}$ 0.25	0.33	0.40
		*		*	*	
C0	-0.25237+0	-0.24300+0	-0.11130+0	-0.97426-1	-0.87919-1	-0.37660-1
C1	-0.37992-1	+0.38901+0	-0.69589-1	+0.26271+0	+0.32895+0	+0.16762+0
C2	-0.18783+0	-0.18271+1	+0.22844-1	-0.19878+0	-0.53327+0	-0.12228+0
C3	+0.33516+0	+0.28524+1	+0.22212-1	+0.99069-1	+0.63672+0	+0.73229-1
C4	-0.18875+0	-0.20973+1	-0.83725-2	-0.37207-1	-0.44829+0	-0.34554-1
C5	+0.46125-1	+0.80701+0	+0.60014-3	+0.80320-2	+0.17197+0	+0.85372-2
C6	-0.42448-2	-0.15711+0		-0.69167-3	-0.33582-1	-0.79817-3
C7		+0.12198-1			+0.26198-2	

b/a = 10

NY	$K_{\hat{x}y}$ 0.25	0.33	0.40	$C_{\hat{x}y}$ 0.25	0.33	0.40
		*				
C0	-0.16616+0	-0.14281+0	-0.70724-1	-0.47627-1	-0.41145-1	-0.18484-1
C1	+0.15109+0	+0.12062-1	-0.31261-1	+0.14406+0	+0.13792+0	+0.99624-1
C2	-0.79981+0	-0.22756+0	-0.10361-1	-0.92194-1	-0.10552+0	-0.69433-1
C3	+0.12662+1	+0.34359+0	+0.38437-1	+0.38298-1	+0.62814-1	+0.42513-1
C4	-0.92438+0	-0.18811+0	-0.13918-1	-0.15094-1	-0.29412-1	-0.21966-1
C5	+0.35335+0	+0.45461-1	+0.13738-2	+0.37042-2	+0.73095-2	+0.57593-2
C6	-0.68646-1	-0.41280-2		-0.34973-3	-0.68538-3	-0.55368-3
C7	+0.53372-2					

Table 13 Stiffness Functions $K_{\hat{x}y}$ (Real Part) and $C_{\hat{x}y}$ (Imaginary Part) of Rectangular Foundations between Rotation $\hat{x}$ and Translation y

Boundary Element Applications in Soil Dynamics

C.A. Brebbia
Computational Mechanics Institute, Southampton, U.K.
W. Mansur
Universidad Federal de Rio de Janeiro, Brazil

INTRODUCTION

Cruse and Rizzo [1,2] appeared to have been the first researchers to solve elastodynamics problems using boundary integral equations and Laplace transform. As an extension of their work, Manolis and Beskos[3] much more recently used the Fourier's transform rather than Laplace's and concluded that this formulation gives better numerical results than that employed by Cruse and Rizzo.

Time stepping techniques have also been used to solve transient two dimensional elastodynamics problems. The approach employed by Niwa et al.[4] consists of solving two dimensional problems using the simpler three dimensional fundamental solution. The approach in this case consists of using the third spatial coordinate playing the role of a time related variable.

Steady state solution of elastodynamics problems was studied in Japan by Kobayashi, Niwa, Nishimura and others. The transient response could then be obtained based on the frequency domain solution. This work concentrated on geomechanical applications and the study of stress states around cavities[5,6,7]. In reference [8] Kobayashi and Nishimura discussed the formal derivation of Green's tensor, using the Laplace transform. Mindlin's solution for steady state elastodynamics was also obtained. In 1983, the same authors[9] investigated the case of soil-structure interaction problems in two dimensions applying a semi-infinite element.

Dominguez[10] studied the use of frequency dependent solutions for soil-structure interaction problems and presented some interesting numerical results for different types of foundations.

Mansur and Brebbia[11,12] have applied the boundary element method to analyse transient problems governed by the two dimensional equations of elastodynamics. The present paper is an extension of that work.

Finally it is interesting to mention that a new procedure for the boundary element solution of dynamic problems has been presented by Brebbia and Nardini[13]. The technique permits the formulation of a mass matrix in function of the boundary nodes only. This allows for a range of elastodynamics problems to be treated in a similar way as in finite elements or finite differences, i.e. the problem is reduced to a set of time-dependent differential equations expressed in matrix form. The free vibrations case for instance can then be reduced to the solution of an algebraic eigenvalue problem. The main advantage of the new approach is that the boundary integrals need to be computed only once as they are frequency independent, hence the procedure is very economic for free vibrations when compared to the one previously presented. The technique also allows for the general elastodynamics case to be solved in time rather than in the transform domain employing the simple fundamental solution of elastostatics. The method, which is now called the Dual Reciprocity Method is explained in detail in reference [14].

2. GOVERNING EQUATIONS

The equation governing the dynamic behaviour of a homogeneous isotropic, linearly elastic material under small displacements can be written as follows,

$$(c_1^2 - c_2^2)u_{i,ij} + c_2^2\, u_{j,ii} + \frac{b_j}{\rho} = \ddot{u}_j \qquad \text{in } \Omega \tag{1}$$

where u_i indicates the displacement component in the x_i direction; b_j are the body force components. Commas indicate space derivatives and the dots time derivatives. Ω is the domain and Γ its boundary, which has a normal called n. The propagation velocities of the dilatational - c_1 - and distortional - c_2 - waves are given as,

$$c_1 = \sqrt{\frac{\lambda+2\mu}{\rho}} \quad ; \quad c_2 = \sqrt{\frac{\mu}{\rho}} \tag{2}$$

where λ and μ are the Lamé's constant and ρ is the mass density of the material.

Strain components are defined as,

$$\varepsilon_{ij} = \tfrac{1}{2}(u_{i,j} + u_{j,i}) \tag{3}$$

The material is assumed to obey Hooke's law, i.e.

$$\sigma_{ij} = 2\mu\varepsilon_{ij} + \lambda\varepsilon_{kk}\delta_{ij} \tag{4}$$

where δ_{ij} is the Kronecker delta function.

In addition to the above equations the problem has to satisfy initial and boundary conditions. The initial conditions can be written as follows,

$$u_i(x,t) = u_i^\circ(x)$$
$$\dot{u}_i(x,t) = v_i^\circ(x) \qquad \text{for } t = t_o \text{ in } \Omega+\Gamma \tag{5}$$

where $u_i^\circ(x)$ and $v_i^\circ(x)$ are prescribed displacement and velocity at time $t = 0$.

The solution also has to satisfy boundary conditions on Γ of the type called essential and natural. The essential - or displacement - boundary conditions in this case are,

$$u_i(x,t) = \bar{u}_i(x,t) \qquad \text{for } t > t_o \text{ on } \Gamma_1 \tag{6}$$

and the natural - or traction - conditions can be written as,

$$p_i(x,t) = \sigma_{ij}\, n_j = \bar{t}_i(x,t) \qquad \text{for } t > t_o \text{ on } \Gamma_2 \tag{7}$$

where $\Gamma = \Gamma_1 + \Gamma_2$.

The bar indicates that these values are known. n_j is the direction cosine of the normal to the boundary with respect to the x_j coordinate axis.

The propagation of the c_1 and c_2 waves is a rather complex time dependent phenomenon even when the medium is homogeneous. In addition, when propagation waves meet a discontinuity, then reflection, refraction and diffraction are produced and the resultant wave motion is the superposition of all these components. Furthermore for free surface or surface of material discontinuities new types of waves are produced. These are the Rayleigh waves at the free surface or the Lamé's waves in the case of layered half-spaces.

The solution of elastodynamic problems using integral equation techniques can be carried out using different approache i.e. i) Laplace transform; ii) Fourier transforms for steady state elastodynamics; iii) Time dependent integral formulations; iv) Combination of boundary elements in space with finite difference on time and v) Approximate functions within the integral equations to simplify the computational solution.

The first approach involves the removal of the time dependence in the governing equations by using the Laplace's transform In this way the initial conditions of the problem become part of the governing differential equations which are function of the transform parameter. The equations can then be put into integral form and solved as for the standard case of elliptical equations. Once the solution has been found for a range of values of the transform variable one needs to backtransform to return to the time domain. Different types of inverse transform methods have been proposed but the main drawback of the technique is that in many engineering applications the use of time dependent boundary conditions or loads can make the backtransformation difficult in practice.

Steady state elastodynamics solutions using Fourier transform are specially convenient when one needs to predict the dynamic response of the system under harmonic excitation. In these cases the response is function of the exciting frequency and the initial conditions can be neglected assuming that a sufficiently long time has elapsed so that the steady state is reached. Notice that in this case it is not necessary to carry out a backtransformation as required in the Laplace's approach.

If the fundamental solution can be expressed as the response of an infinite medium to a unit impulse at a certain time one can write an integral statement which is integrated in time as well as space. This formulation is mathematically more complex but can give very accurate results. Its numerical solution can be attempted in two ways, i.e. as a series of integrations from t to $t+\Delta t$ or as an integral statement to be computed always from time 0 to the time under consideraiton. Because of the way in which time is split one can refer to the first as the 'tangent' approach and the second as the 'secant' method.

Combination of finite difference with boundary elements can be applied for closed domains if one uses the fundamental solution corresponding to the static case and form the mass matrix by integrating over the domain. The resulting matrix system of equations with time derivatives can then be integrated on time by applying a finite difference type scheme.

An interesting alternative procedure for closed domain has been proposed by Brebbia and Nardini[13] under the name of Dual Reciprocity Method. The technique also consists of applying the static fundamental solution but instead of integrating over the domain to obtain the mass matrices one reduces those integrals to the boundary. This reduction is achieved by preparing approximate functions over the domain of a type that not only can be easily transformed to the boundary but also is based on the same type of influence coefficients as the static case. Once this is done the matrices can be used as in the previous approach, one can solve the problem using a step by step scheme, or find the eigenvalues and eigenvectors solution or use modal superposition to integrate the response. The dual reciprocity method has found important applications not only in elastodynamics but also in other types of time dependent problems, such as diffusion [19].

The approach used in this paper is the time and space dependent integral formulation, where the integrals are always considered to start at the original t_o and at the time under consideration. This can be called the 'secant' approach as different from the step by step technique - 'tangent' approach - which requires the updating of the initial conditions from step by step.

3. BOUNDARY INTEGRAL FORMULATION

This paper discusses the solution of elastodynamics problems using the time and space dependent integral approach.

The fundamental solution of the elastodynamics equation represents a unit load concentrated at a point ξ in the x_i direction and acting in an infinitely short time,

$$(c_1^2 - c_2^2)u^*_{ik,kj} + c_2^2 u^*_{ij,kk} - u^*_{i,j} = -\frac{\delta_{ij}}{\ell}\Delta(\xi,x)\Delta(\tau,t) \tag{8}$$

The two and three dimensional fundamental solutions are well known and documented[1] . They are written in function of Heaviside and Dirac delta functions.

By using weighted residual - or reciprocity - methods in time and space, one can obtain the boundary integral equation for the time-dependent formulation, i.e.

$$c_{ij}u_j + \int_{t_o}^{\tau}\int_{\Gamma} p^*_{ij}u_j \, d\Gamma \, dt = \int_{t_o}^{\tau}\int_{\Gamma} u^*_{ij} \, p_j \, d\Gamma \, dt$$

$$+ \int_{o}^{\tau}\int_{\Omega} u^*_{ij}b_j \, d\Omega \, dt + \rho\int_{\Omega} [u^*_{ij}v_j]_{t_o} \, d\Omega \tag{9}$$

$$+ \rho\int_{\Omega} [v^*_{ij}u_j]_{t_o} \, d\Omega$$

where for smooth surface $c_{ij} = \frac{1}{2}$ and for interior points $c_{ij} = 1$. This equation is the starting point for the boundary element formulation.

In many practical cases one can start with an initial state with $\dot{v}_j^o$ and u_j^o components equal to zero and without body forces. For these cases formula (9) reduces to,

$$c_{ij}u_j + \int_{t_o}^{\tau}\int_{\Gamma} p_{ij}^* \, u_j \, d\Gamma \, dt = \int_{t_o}^{\tau}\int_{\Gamma} u_{ij}^* \, p_j \, d\Gamma \, dt \tag{10}$$

The interpolation functions used for u_j and p_j components are in terms of space and time. Let us consider that the time from t_o to τ can be divided into 'n' constant Δt time steps and then the time integrals in (10) can be written as,

$$c_{ij}u_j + \sum_{k=o}^{n} \int_{t_k}^{t_{k+1}}\int_{\Gamma} p_{ij}^* \, u_j \, d\Gamma \, dt = \sum_{k=o}^{n} \int_{t_k}^{t_{k+1}}\int_{\Gamma} u_{ij}^* \, p_j \, d\Gamma \, dt \tag{11}$$

Analytical time integration can be carried out resulting in expressions which are too long to be included here. The integrations over the Γ boundary are carried out numerically using Gauss quadrature formula at all nodes but those with the singularity in the space variable r, for which an analytical integration is performed.

The final system of equations can be cast in the usual form, i.e.

$$\tilde{H}\,\tilde{U} = \tilde{G}\,\tilde{P} \tag{12}$$

where $\tilde{H}$ and $\tilde{G}$ are square matrices, $\tilde{U}$, $\tilde{P}$ are vectors. When boundary conditions at the time t_n are considered, equation (12) can be reordered as,

$$\tilde{A}\,\tilde{X} = \tilde{B} \tag{13}$$

where the vector $\tilde{X}$ is formed by unknowns u_j and p_j at the nodes.

Displacements at internal points can be calculated using the boundary integral equations at those points. To compute internal stresses one can use a triangular cell over which linear interpolation functions are used to approximate the displacement components (similar to what is done over a finite element). The displacements are then computed at the vertices of the triangle and one can then find their derivatives and stresses. Those stresses are not as accurate as those which would be obtained using the correct integral expressions for the derivatives of the displacements but are much easier to calculate.

4. APPLICATIONS

The technique described in this paper has been applied to solve two representative soil dynamics problems, and results compared with those obtained using other numerical methods. In the examples reference is made to the parameter β, which is defined as,

$$\beta = \frac{c_d \, \Delta t}{\ell_j} \tag{14}$$

where c_d is the dilatational celerity, Δt the time step and ℓ_j the element length.

The examples studied here are

i. Half-plane under Discontinuous Prescribed Boundary Velocity.

ii. Half-plane and Continuous Prescribed Traction Distribution.

EXAMPLE i) Half Plane Under Discontinuous Prescribed Boundary Velocity

This example has been previously studied[1,2,17] using the Laplacė's transform which requires the solution for different values of the Laplace parameter and then a numerical inversion to find the results in the time domain.

The problem under study is the half plane of figure 1, initially at rest and suddenly undergoing a constant velocity at t = 0. The values of the constants were

$$\lambda = G = 10^6 \text{ psi}$$

$$c_d = 3.27 \times 10^4 \text{ ips} \quad ; \quad c_s = 1.86 \times 10^4 \text{ ips} \tag{15}$$

$$b = 3000 \text{ inches}$$

The boundary discretization and cell first used in the analysis is shown in figure 2. It consisted of 20 equal boundary elements, each haing a length of 6000 inches. Four values of β (0.13, 0.25, 0.5 and 1) were considered. Notice that if β is too large, errors due to contradicting the causality property and those as a result of bad time interpolation will contribute to reduce the accuracy of the results. Because of this a value of $\beta = 0.25$ was chosen to be the best for this analysis. When evaluating the stresses it is important to recognize that the bigger the cell the less representative they will be. Conversely, very small cells must also be avoided because when the differences between node displacements are too small their contribution to stresses due to numerical errors can become excessively large.

The vertical displacements and the stress σ_{22} at the internal point D are plotted in figures 3 and 4 respectively. Inspection of these two figures shows that the numerical results are in agreement with the simple straight line prediction. At $t = b/c_d$ the stress σ_{22} jumps from zero to $-\rho c_d \dot{u}_o$ as expected. Notice that the agreement with Cruse's solution is better for displacements than for stresses.

In figures 5 and 6 tractions at the boundary points E(-6b,0), A(-4b,0) and C(0,0) are plotted. Tractions at points E and A obtained with both boundary element techniques were not as close to each other as the displacements in figure 3. At points C and D both numerical techniques gave the predicted results. The time-stepping scheme results oscillated slightly around the analytical solution. This fact had already been noticed by the authors, when investigating problems governed by the scalar wave equation[18]. Apparently oscillation can occur whenever boundary displacements are prescribed and β is too small.

Another analysis in which β was regarded as being equal to 0.75 and the size of the elements taken to be 2000 inches was also undertaken. Displacements and stresses at D were similar to the ones obtained with the first discretization, however tractions varied. A comparison of figures 5 and 7 demonstrates that the boundary discretization depicted in figure 2 is too coarse, resulting in bad numerical results for tractions at the boundary points A and E.

In figure 8 tractions at point C are plotted. This figure shows that by using $\beta = 0.75$ the oscillation of the numerical results was practically eliminated.

Finally as far as this problem is concerned it can be concluded that i) the displacements obtained using the time-stepping technique agreed with the results obtained by using the Laplace's transform; ii) excessively small values of β should be avoided in problems in which displacements are prescribed over portions of the boundary; iii) both, the time-stepping and the Laplace transform techniques, yielded results which very closely followed the predicted physical behaviour of the problem analysed.

EXAMPLE ii) Half-plane under Continuous Prescribed Stress Distribution

This example also studies the wave propagation in a half-plane as shown in figure 9. The half plane is initially at rest and its surface is disturbed by a vertical traction which is continuous in both time and space.

The following numerical values were adopted for the constants of the problem.

$$E = 200 \text{ ksi} \quad ; \quad \nu = 0.15$$
$$c_d = 3.288 \times 10^4 \text{ ips} \quad ; \quad c_s = 2.112 \times 10^4 \text{ ips} \qquad (16)$$

The boundary element technique reported in this paper is compared with results obtained using the finite different model implemented by Tseng et al. [19]. In that report a transmitting boundary was developed and used together with the generalized lumped parameter model.

The criterion given by Tseng[19] to choose the finite difference mesh shown in figure 10 requires that

$$t_r > 2 \frac{\Delta x}{c_d} \qquad (17)$$

where t_r is rise or decay time of the applied pressure and x gives the mesh refinement. When t_r = 20 m sec, $\Delta x \leq 27.4$ ft is obtained. Tseng chose Δx = 10 ft and the discretization as depicted in figure 10 where the position selected for the gradient wave transmitting body can also be seen. The boundary element discretization and cells used in the analysis are shown in figure 11.

According to [19] the time increment t used in this finite difference analysis must obey the following formula,

$$t \leq 0.433 \frac{\Delta x}{c_d} \qquad (18)$$

Consequently a value Δt = 1 m sec was adopted.

The discrete mass approach was also used by Lima[20] who solved the same problem using finite elements. A similar mesh as shown in figure 10 was used for the discretization but without employing transmitting boundaries. Instead of this the results reported were for cases for which the reflected waves did not arrive to the region of interest.

For the boundary element analysis, β was taken to be equal to ½, which gives

$$\Delta t = 3.65 \text{ m sec} \tag{19}$$

The time history of the displacements shown in figures 12 to 13 shows a similar agreement for the time interval considered, the major difference occurring for the point on the transmitting boundaries (figure 15).

In his research Tseng carried out another analysis using a pair of transmitting boundaries which enclosed a small rectangular region whose side lengths were equal to 90 ft and 150 ft. The two finite-different analyses showed that the larger the region enclosed by the transmitting boundaries, the closer finite difference and boundary element results were. Therefore it is justified to suppose that the major proportion of the difference between the displacements obtained with these two numerical methods is caused by errors generated at the transmitting boundaries.

The time history for the vertical displacement for the point G (150', 10') obtained with the 90' × 150' rectangular region is shown in figure 14. As G is located exactly on the transmitting boundary the finite difference results are expected to be less accurate. The point G is also a critical one in the boundary element analysis because it is too close to the boundary of the half plane. As it was expected the agreement is not as close as recorded previously.

Figures 15 to 17 describe the time history of stresses at points A(45',75'); B(75',75') and C(5',75'). Results for the three techniques agree reasonably well and tend to converge to the static solution.

From this example it can be concluded i) that the solutions used with the three different methods are in good agreement; and ii) the time increment required by boundary elements was bigger than that necessary for finite differences or finite elements.

5. CONCLUSIONS

The boundary element method can be applied to solve two dimensional elastodynamics problems using a space and time dependent fundamental solution. Such an approach in combination with the time integration of the equations starting always from the initial time produces a technique which can give accurate results to study soil dynamics problems.

REFERENCES

[1] CRUSE, T.A. and RIZZO, F.J. "A Direct Formulation and Numerical Solution of the General Transient Elasto-Dynamic Problem, Part I" Journal of Mathematical Analysis Applic., 22, 1968.

[2] CRUSE, T.A. "A Direct Formulation and Numerical Solution of the General Transient Elastodynamic Problem, Part II" Journal of Mathematical Analysis and Applications, 22, 1968.

[3] MANOLIS, G.D. and BESKOS, D.E. "Dynamics Stress Concentration Studies by Boundary Integrals and Laplace Transform". Int. J. Num. Meth. Engng. 17, 573-599, 1981.

[4] NIWA, Y., FUKUI, T., KATO, S. and FUJIKI, K. "An Application of the Integral Equation Method to Two-dimensional Elastodynamics". Theoret. and Appl. Mech. 28, 281-290, Univ. of Tokyo Press 1980.

[5] NIWA, Y., KOBAYASHI, S. and AZUMA, N. "An Analysis of Transient Stresses Produced around Cavities of an Arbitrary Shape during the Passage of Travelling Wave ". Memo. Fac. Eng., Kyoto Univ., 37, 28-46, 1975.

[6] NIWA, Y., KOBAYASHI, S. and FUKUI, T. "Application of Integral Equation Method to Solve Some Geomechanical Problems". Proc. 2nd Int. Conf. Numerical Meth. Geomech. 120,131, ASCE, 1976.

[7] KOBAYASHI, S. and NISHIMURA, N. "Dynamics Analysis of Underground Structures by the Integral Equation Method". Proc. 4th Int. Conf. Num. Method. Geomech, 1, 401-409, balkema, 1982.

[8] KOBAYASHI, S. and NISHIMURA, N. "Green's Tensor for Elastic Half-space - An Application of Boundary Integral Equation Method". Memo Faculty of Eng., Kyoto Univ., Vol. 42, Pt. 2, pp.228-241, 1980.

[9] KOBAYASHI, S. and NISHIMURA, N. "Analysis of Dynamic Soil-Structure Interaction by Boundary Integral Equation Method" Vol.1, (Ed. P. Lascaux), pp.353-362, Pluralis, Paris, 1983

[10] DOMINGUEZ, J. and ALARCON, E. "Elastodynamics". Chapter 7 of Progress in Boundary Element Methods - 1. (Ed. C.A. Brebbia), Pentech Press, 1981.

[11] MANSUR, W.J. and BREBBIA, C.A. "Transient Elastodynamics using a Time-stepping Technique". Proc. of the Fifth Int. Seminar on Boundary Element Methods in Engg. (C.A.Brebbia (Ed.), Springer-Verlag, Heidelberg, new York, 1983.

[12] MANSUR, W.J. and BREBBIA, C.A. "Transient Elastodynamics". Chapter 5 in Topics in Boundary Element Research, Vol.2, Time Dependent and Vibration Problems (Ed. C.A. Brebbia), Springer Verlag, Berlin & NY, 1985.

[13] BREBBIA, C.A. and NARDINI, D. "Dynamic Analysis in Solid Mechanics by an Alternative Boundary Element Procedure". Soil Dynamics & Earthquake Engg. Journal 2(4), 1983.

[14] NARDINI, D. and BREBBIA, C.A. "Boundary Integral Formulation of Mass Matrices for Dynamic Analysis". Chapter 7 in Topics in Boundary Element Research, Vol.2, Time Dependent and Vibrations Problems, Springer-Verlag, Berlin & NY, 1985.

[15] WROBEL, L. and BREBBIA, C.A. "Dual Reciprocity Boundary Element Formulation for Nonlinear Diffusion Problems" To be published in Computer Methods in Applied Mechanics and Engineering.

[16] BREBBIA, C.A., TELLES, J. and WROBEL, L. "Boundary Element Techniques - Theory and Applications in Engineering" Springer-Verlag, Berlin, NY, 1984.

[17] CRUSE, T.A. "The Transient Problem in Classical Elastodynamics solved by Integral Equations". Ph.D. Thesis, University of Washington, 1967.

[18] MANSUR, W. and BREBBIA, C.A. "Further Developments on the Solution of the Transient Scalar Wave Equation". Chapter 4 in Topics in Boundary Element Research, Vol.2, Time Dependent and Vibrations Problems (C.A. Brebbia, Ed.), Springer Verlag, Berlin & NY, 1985.

[19] TSENG, M.N. and ROBINSON, A.R. "A Transmitting Boundary for Finite-Difference Analysis of Wave Propagation in Solids." Project No. NR 064-183, University of Illinois, Urbana, Illinois, 1975.

[20] LIMA, E.C.P. "LORANE/DINA - Uma Linguagem Orientada para Análise Dinâmica de Estruturas". Report PTS 14-77, COPPE/UFRJ, Brazil, 1977.

[21] MANSUR, W.J. and BREBBIA, C.A. "Foundation of the Boundary Element Method for Transient Problems Governed by the Scalar Wave Equation". Appl. Math. Modelling, Volume 6, pp.307-311, 1982.

[22] MANSUR, W.J. and BREBBIA, C.A. "Numerical Implementation of the Boundary Element Method for Two-Dimensional Transient Scalar Wave Propagation Problems". Appl. Math. Modelling, Volume 6, pp.299-306, 1982.

[23] MANSUR, W.J. "A Time-Stepping Technique to Solve Wave Propagation Problems using the Boundary Element Method", Ph.D. Thesis, Southampton University, 1983.

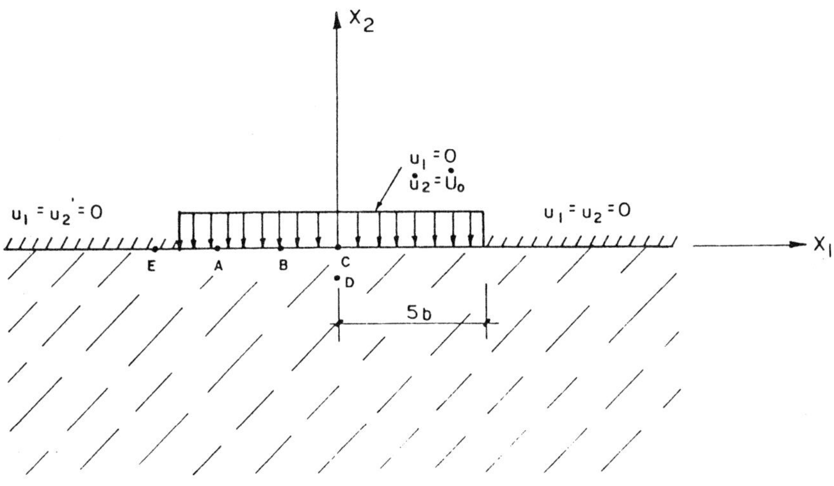

Figure 1 Boundary conditions for the half-plane under imposed boundary velocity

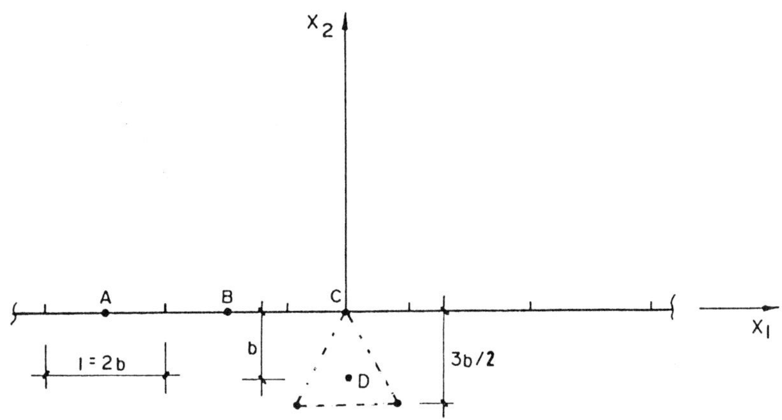

Figure 2 Boundary discretization and internal cell for the half-plane under discontinuous boundary stress distribution

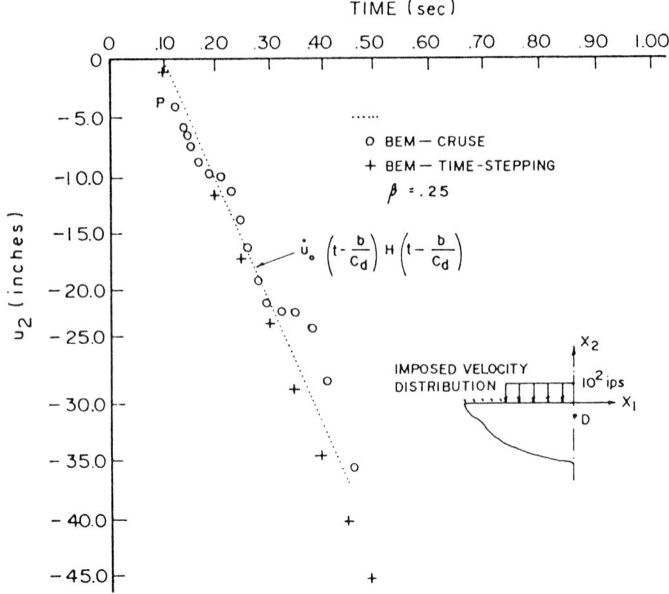

Figure 3 Half-plane under imposed boundary velocity. Vertical displacement at the internal point D(0,-b)

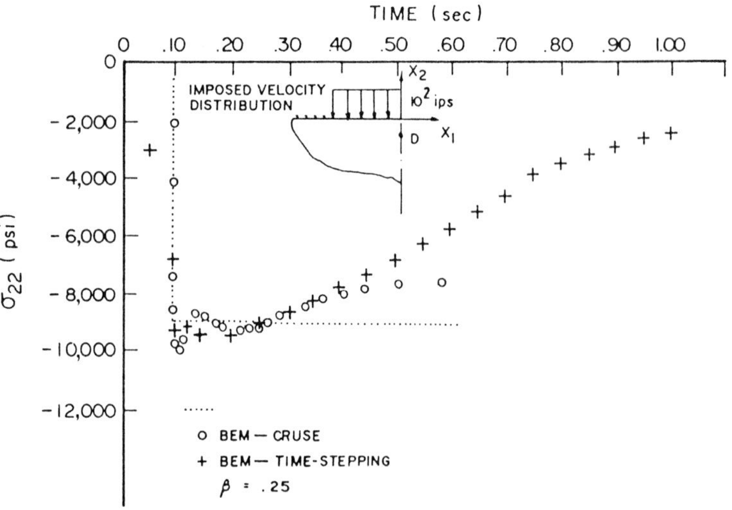

Figure 4 Half-plane under imposed boundary velocity. Stress σ_{22} at the internal point D(0,-b).

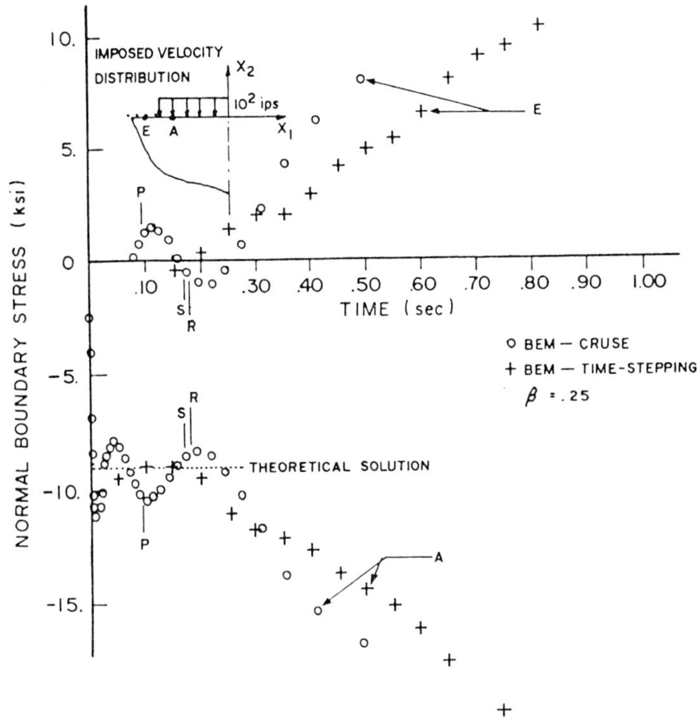

Figure 5 Half-plane under imposed boundary velocity. Normal boundary stress at the points E(-6b,0) and A(-4b,0)

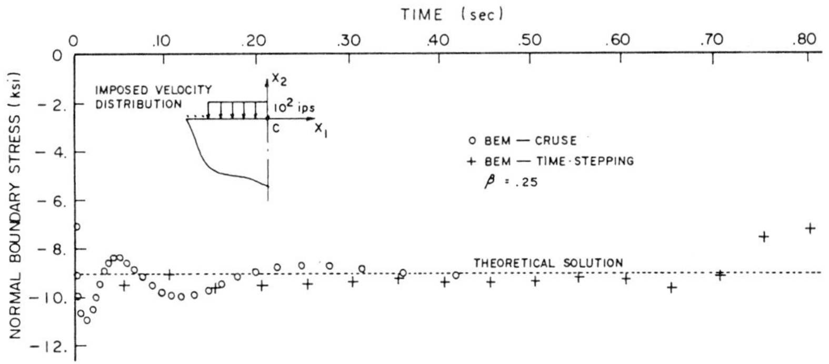

Figure 6 Half-plane under imposed boundary velocity. Normal stress at the point C(0,0).

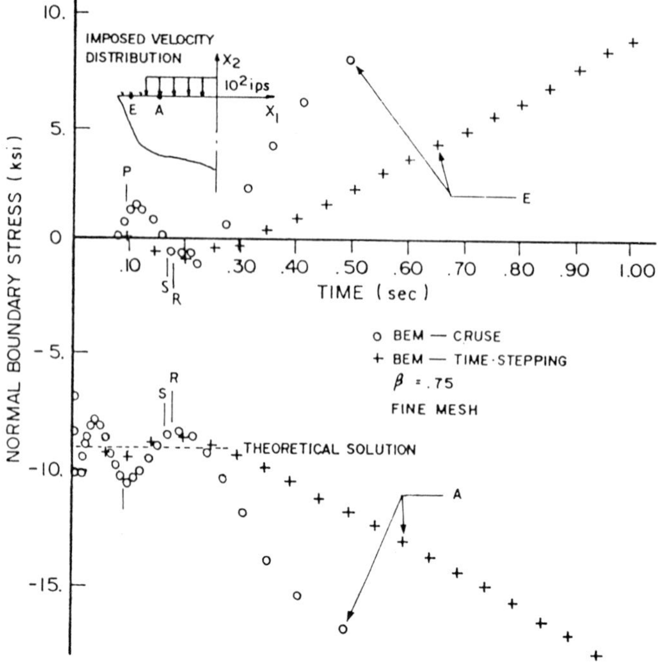

Figure 7 Half-plane under imposed boundary velocity. Normal boundary stress at the points E(-6b,0) and A(-4b,0).

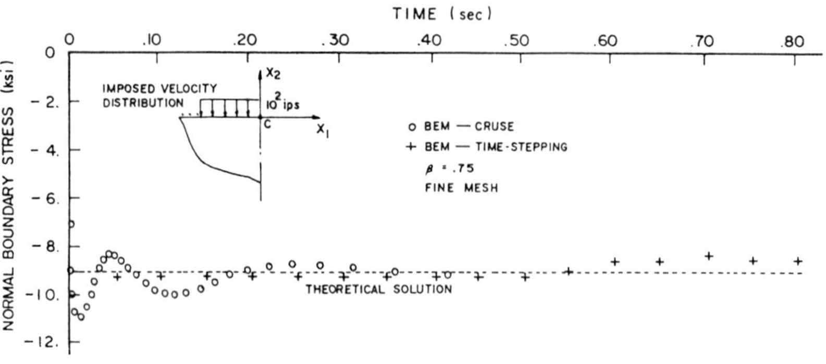

Figure 8 Half-plane under imposed boundary velocity. Normal boundary stress at the point C(0,0).

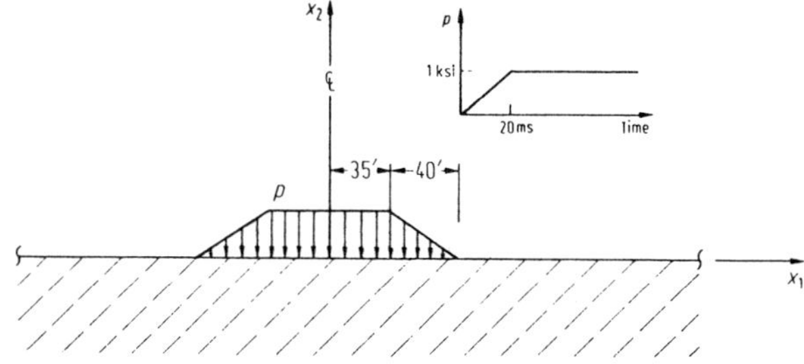

Figure 9 Load for the half-plane under continuous prescribed stress distribution

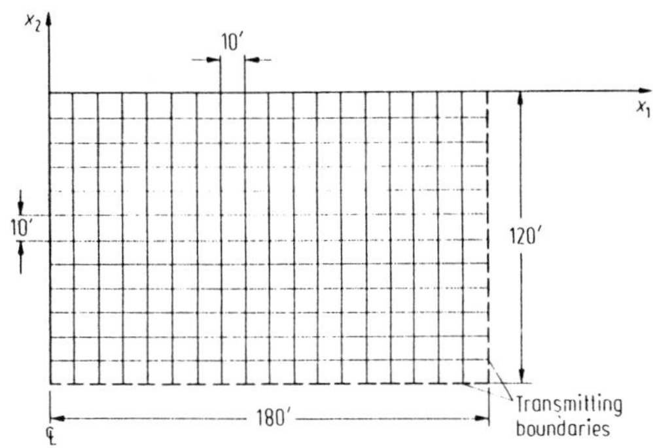

Figure 10 Finite-difference mesh for the half-plane under continuous prescribed stress distribution

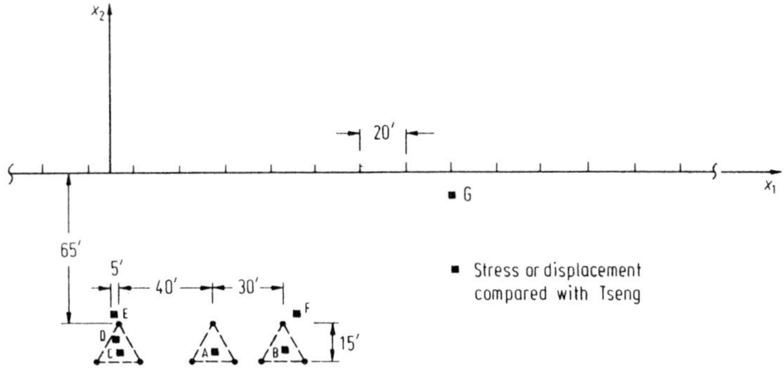

Figure 11 Boundary element discretization for the half-plane under continuous prescribed stress distribution

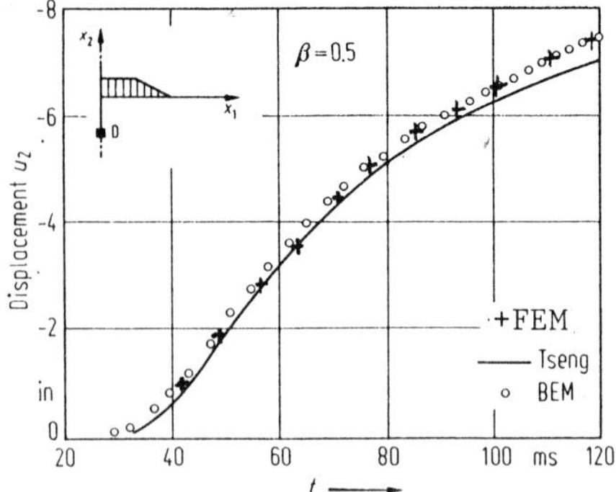

Figure 12 Half-plane under continuous prescribed stress distribution. Displacement u_2 at the internal point D(0',70')

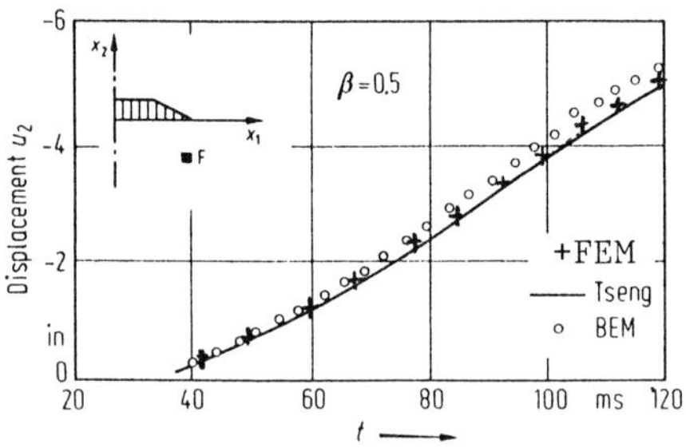

Figure 13 Half-plane under continuous prescribed stress distribution. Displacement u_2 at the internal point F(80',60')

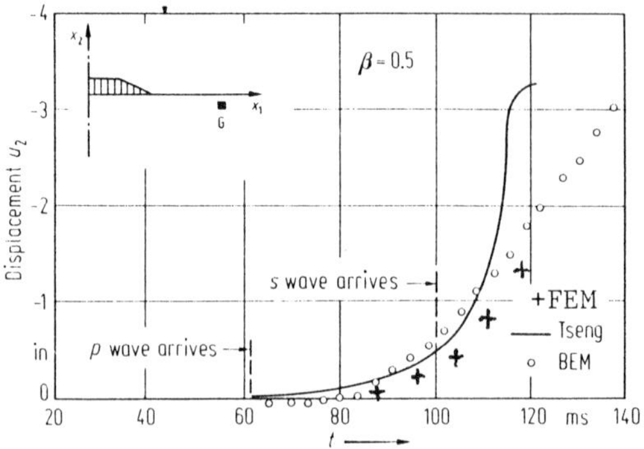

Figure 14 Half-plane under continuous prescribed stress distribution. Displacement u_2 at the internal point G(150',10')

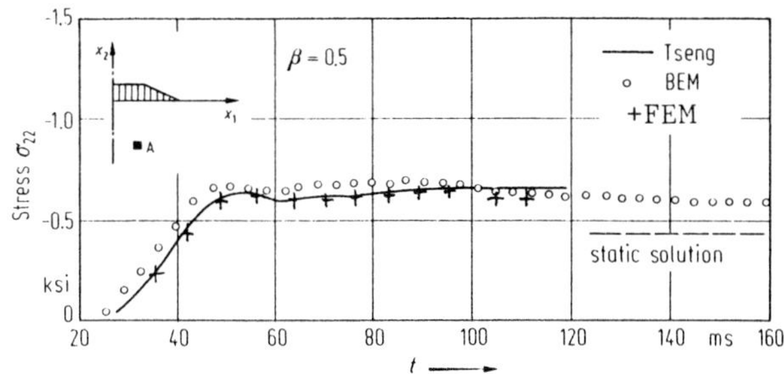

Figure 15 Half-plane under continuous prescribed stress distribution. Stress σ_{22} at the internal point A(45',75')

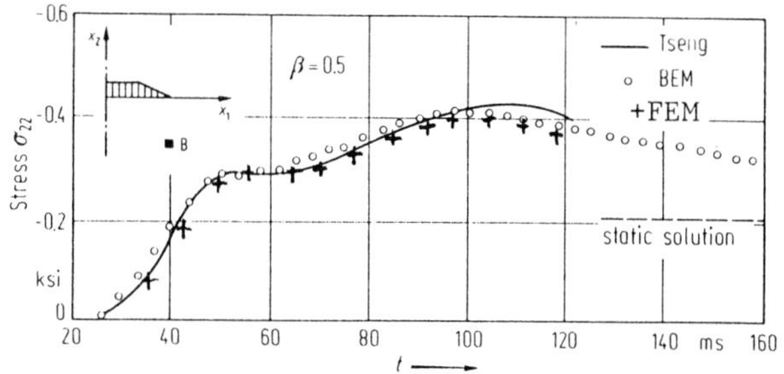

Figure 16 Half-plane under continuous prescribed stress distribution. Stress σ_{22} at the internal point B(75',75')

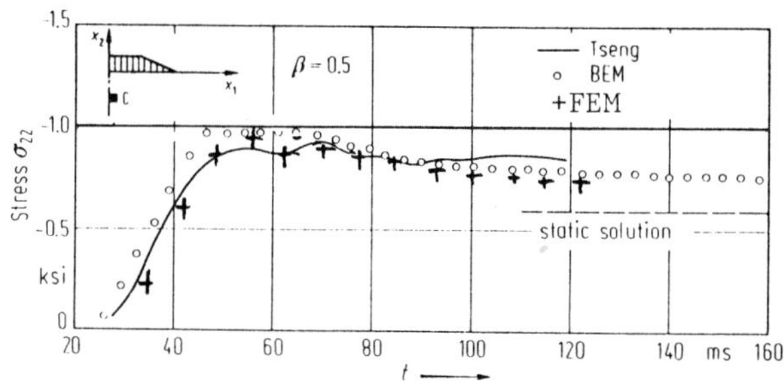

Figure 17 Half-plane under continuous prescribed stress distribution. Stress σ_{22} at the internal point C(5',75')

SECTION 5: GROUND MOTION

PGA, RMSA, PSDF, Duration, and MMI

F.K. Chang, A.G. Franklin
U.S. Army Engineer Waterways Experiment Station, Corps of Engineers, Vicksburg, Mississippi 39180, U.S.A.

PURPOSE AND SCOPE

There are unique relationships connecting peak ground acceleration (PGA), root-mean-square acceleration (RMSA), or average acceleration, power spectral density function (PSDF), and duration (record length) of strong-motion earthquake records. The purpose of this study is to determine the site dependence of the PSDF and its scaling factor, which could be correlated with the damage intensity or Modified Mercalli Intensity (MMI), from the above relationships.

Factors affecting the ground motion at a particular site include the source mechanism (nature of fault movement and magnitude of energy release), propagation path characteristics, the direction of the site relative to the fault rupture, and local geological and soil conditions. This study, however, deals only with the influence of local geological and soil conditions on ground motion.

PREVIOUS WORK

Seed, Ugas, and Lysmer[1] and Kiremidjian and Shah[2] presented similar results of a statistical analysis of the site-dependent response spectral shapes from ground motion accelerograms obtained mostly in the western United States. The analysis shows clear differences in spectral shapes for different soil and geological conditions. Chang and Krinitzsky[3] studied the duration and spectral content of strong-motion records from the western United States according to site conditions and found that the predominant frequencies are in the range of 1.0 to 6.67 Hz and the spectral shape depends on the source spectrum function (magnitude), distance, and geological conditions.

Considering the earthquake ground motion to be random in nature, Arnold[4] and Arnold and Vanmarcke[5] studied the influence of site azimuth relative to source fault orientation and local soil conditions on earthquake ground motion spectra for the San Fernando, California, earthquake of 9 February 1971 using PSD functions. The result showed that local soil conditions and site azimuth, as well as epicentral distance, can have a significant effect on both the intensity and the frequency content of ground motions. This study demonstrated the potential value of PSD functions as a tool for comparing and studying variations in ground motion characteristics and also showed the great utility of PSD functions as input to random vibration analyses of structural response.

POWER SPECTRAL DENSITY FUNCTION (PSDF)

In the application of the random vibration theory of linear systems for evaluation of the effects of variations in ground motion characteristics on structural response to earthquake excitation, the ground motion may be defined in the form of a PSDF. The PSDF, $G(\omega)$, is defined as a measure of the ground motion power or energy per unit time as a function of frequency ω (Figure 1). Usually, estimates of the PSD are obtained from the squared amplitudes of the Fourier transform, or the squared Fourier amplitude spectrum. In Figure 1, A_j is the amplitude of the Fourier transform at frequency ω_j.

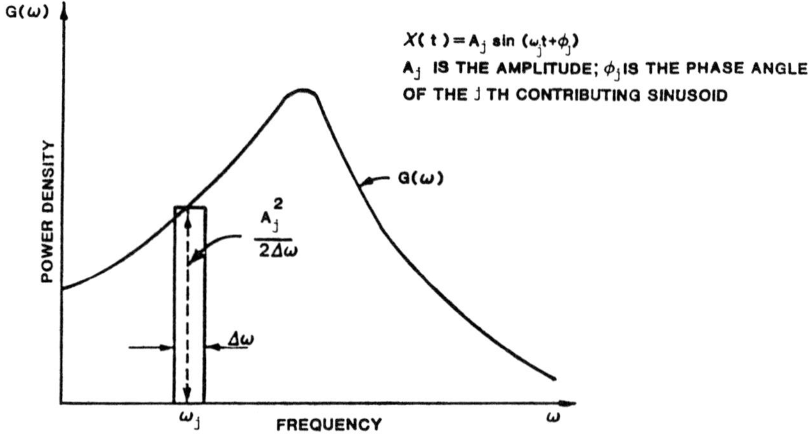

Figure 1. The PSDF $G(\omega)$

FOURIER TRANSFORM

Generally, the given earthquake ground motion function, such as an acceleration versus time plot, can be represented in the time domain as $a(t)$ and in the frequency domain as $F(\omega)$:

$$F(\omega) = \int_0^{t_o} a(t)e^{-i\omega t}\,dt\ ,\quad i = \sqrt{-1} \tag{1}$$

and

$$a(t) = \frac{1}{\pi}\int_{\omega_1}^{\omega_o} F(\omega)e^{i\omega t}\,d\omega\ ,\quad i = \sqrt{-1} \tag{2}$$

Equations 1 and 2 are called the Fourier transform pair; i.e., $F(\omega)$ is the Fourier integral or Fourier transform of $a(t)$, and $a(t)$ is the inverse Fourier transform of $F(\omega)$. The symbols t and t_o denote the instant in time and total duration, and ω, ω_1, and ω_o represent the frequency, lower frequency bound $\omega_1 = 2\pi/t_o$, and maximum frequency (in radians per second), respectively. For practical purposes, ω_1 can usually be taken as zero.

TOTAL INTENSITY, AVERAGE POWER, AND PSDF

By Parseval's theorem, the relation between the energy of the motion as expressed in the time domain and in the frequency domain can be represented in the following equations. For a nonperiodic function, the total intensity (total energy) or Arias intensity I_o delivered by the source is given by

$$I_o = \int_0^{t_o} |a(t)|^2 dt = \frac{1}{\pi}\int_0^{\omega_o} |F(\omega)|^2\, d\omega \tag{3}$$

The mean-square average value is expressed

$$\frac{1}{t_o}\int_0^{t_o} |a(t)|^2 dt = \frac{1}{\pi}\frac{1}{t_o}\int_0^{\omega_o} |F(\omega)|^2\, d\omega = \int_0^{\omega_o} G(\omega)\, d\omega \tag{4}$$

in which $G(\omega)$ is the energy per unit time (power), or the power spectral density of the function $a(t)$. The integrand on the right side of Equation 4 can be written as

$$G(\omega) = \frac{1}{\pi}\frac{1}{S_o}\,|F(\omega)|^2 \tag{5}$$

if S_o (duration of strong motion) is substituted for t_o, and

$$\frac{1}{\pi}\int_0^{\omega_o} |F(\omega)|^2\, d\omega = S_o\int_0^{\omega_o} G(\omega)\, d\omega$$

The quantity $|F(\omega)|^2$ is called the power (energy) spectrum or power spectral density function of $a(t)$ (Hsu[6]). The left side of Equation 4 gives the statistical (PSDF) average power λ_o^2 of the function $a(t)$ over the total duration of the motion t_o. From Equation 4, also note that λ_o^2 is equal to the area under the curve of $G(\omega)$ in Figure 1. Therefore, the average power can be written as

$$\lambda_o^2 = \int_0^{\omega_o} G(\omega)\, d\omega \tag{6}$$

A more effective way for dealing with the frequency content of ground motion is through the normalized spectral density function $G^*(\omega)$:

$$G^*(\omega) = \frac{1}{\lambda_o^2}\, G(\omega) \tag{7}$$

If $G_i^*(\omega)$ is an individual normalized PSD function, the statistical mean NPSD curve will be

$$\bar{G}^*(\omega) = \frac{1}{n}\sum_{i=1}^{n} G_i^*(\omega)\ , \quad i = 1,2,\ldots n \tag{8}$$

In practice, curves of $\bar{G}^*(\omega)$ computed from suites of actual earthquake records show large and irregular fluctuations. To isolate the systematic component from the random variations, frequency smoothing with a "Hanning" window (Blackman and Tukey[7]) was used.

DATA SELECTION AND SITE CLASSIFICATIONS

A total of 421 horizontal ground accelerograms from 89 earthquakes, mostly in the western United States and Japan (with a few records from Russia, Rumania, and India), were selected for this analysis. Based on the site classifications of Seed, Ugas, and Lysmer[1], these records have been divided into four groups: (a) 56 records for rock sites, (b) 131 records for stiff soil sites (depth <150 ft), (c) 120 records for deep cohesionless soil sites (depth >250 ft), and (d) 114 records for soft to medium clays with associated strata of sands or gravels. One hundred seventy-three of the 421 records were obtained from the California Institute of Technology[8] Volume II-Corrected Accelerograms (1971-75), and 220 uncorrected records were provided by the Port and Harbour Research Institute, Japan. The digitized Gazli (USSR) and Bucharest (Rumania) records were provided by Dr. A. G. Brady, U. S. Geological Survey. All 421 corrected and uncorrected records were adjusted to zero mean before processing the PSD.

DEFINITION OF AVERAGE POWER AND AVERAGE ACCELERATION

The main approach used in this study was to determine the normalized mean and the mean plus one standard deviation PSD shape (NPSD) for each group, and the average acceleration λ_o which is the root-mean-square of the average power λ_o^2 for each raw record. Figure 1 shows $G(\omega)$, whose value at ω is equal to $A^2/2\Delta\omega$, so that λ_o^2 is actually the power or energy density in an accelerogram for a finite frequency band ($0 \leq f \leq 10$ Hz in this study), and A_j is the amplitude of the jth frequency component in centimetres per second squared. The total or average power will become equal to the area under the continuous curve $G(\omega)$.

RECORD LENGTH, INCREMENT FREQUENCY, AND NPSD FUNCTION

The record length and spectral content (spectral amplitude and frequency range) are two basic elements for controlling the spectral intensity. The incremental frequency Δf, used in the PSD computer program, depends on the total record length. Since it is necessary to use the same value for Δf throughout, all accelerograms have been processed to give them a duration of 163.82 sec or 8192 (2^{13}) digital points with an equal time interval Δt of 0.02 sec, which gives $\Delta f = 0.006104$ Hz. Outside the time of the actual record, the amplitudes at extended times were set to zero. In this study, the PSD function $G(f)$ has been defined to include only the frequency range of 0 to 10 Hz, since this is also the natural frequency-range of many engineering structures, so that

$$\lambda^2 = \int_0^{10\,\text{Hz}} G(f)\, df \tag{9}$$

Since Δf is 0.006104 Hz, there are 1640 points in the raw PSD function for each accelerogram. The NPSD function is defined as the PSD amplitude divided by the area λ^2 under the power density versus frequency curve. This area may also be called the spectral density or spectral intensity.

ANALYSIS OF SITE-DEPENDENT PSD SPECTRAL SHAPE

The NPSD functions for the horizontal accelerograms in each group were first determined and then were analyzed statistically to obtain the average NPSD spectra and the average plus one standard deviation (σ) NPSD spectra (about 84 percentile). Figure 2 present these mean and mean plus one standard deviation NPSD spectra for the four different site conditions. The mean NPSD spectra for the different site conditions are compared in Figure 2a, and the mean plus one standard deviation NPSD spectra are compared in Figure 2b. Both NPSD spectra in Figures 2a and 2b are smoothed 500 times.

LEGEND

GROUP 1 ROCK SITES (56 RECORDS; 11 EARTHQUAKES)
GROUP 2 STIFF SOIL SITES (131 RECORDS; 39 EARTHQUAKES)
GROUP 3 DEEP COHESIONLESS SOIL SITES (120 RECORDS; 47 EARTHQUAKES)
GROUP 4 SOFT TO MEDIUM CLAY AND SAND SITES (114 RECORDS; 28 EARTHQUAKES)

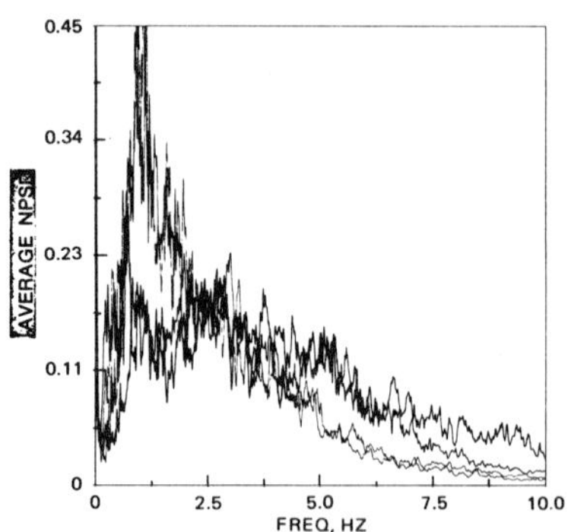

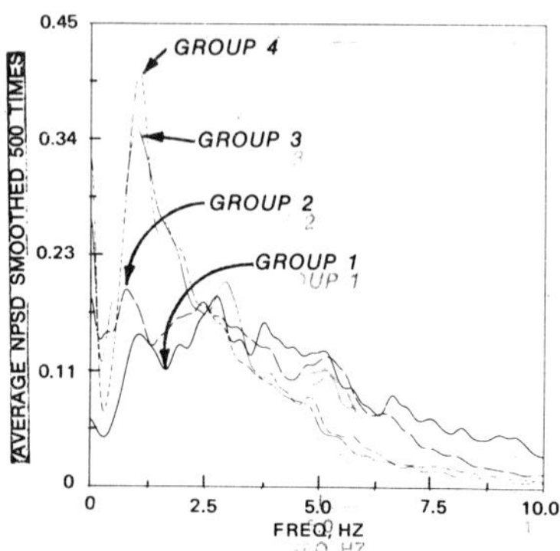

Figure 2a. Comparison of mean NPSD curves of four soil groups, raw (left) and smoothed 500 times (right)

LEGEND

GROUP 1 ROCK SITES (56 RECORDS; 11 EARTHQUAKES)
GROUP 2 STIFF SOIL SITES (131 RECORDS; 39 EARTHQUAKES)
GROUP 3 DEEP COHESIONLESS SOIL SITES (120 RECORDS; 47 EARTHQUAKES)
GROUP 4 SOFT TO MEDIUM CLAY AND SAND SITES (114 RECORDS; 28 EARTHQUAKES)

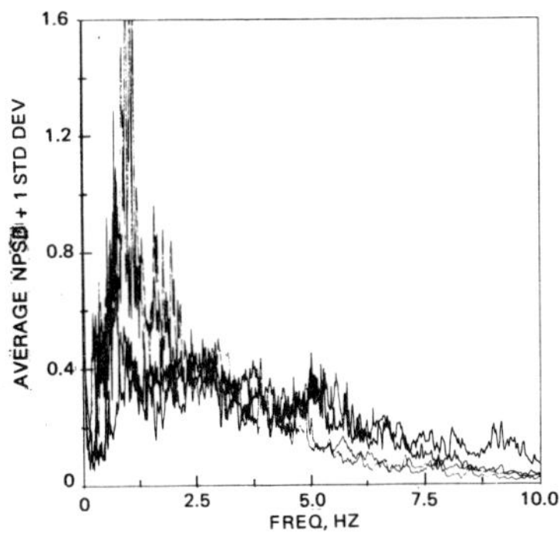

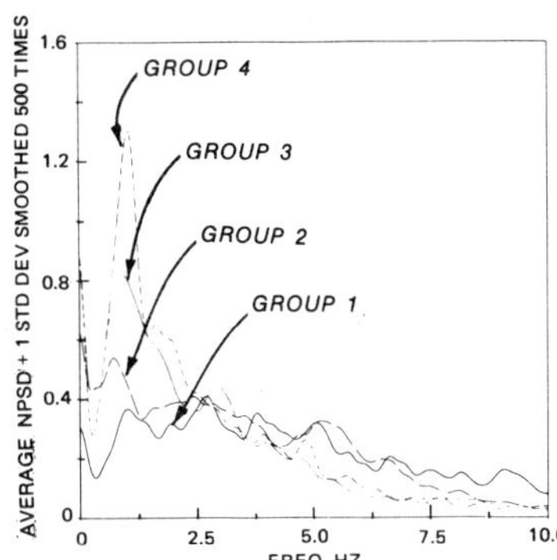

Figure 2b. Comparison of mean + σ NPSD curves of four soil groups, raw (left) and smoothed 500 times (right)

It is clear that the differences in PSD spectral shapes depend on the site conditions. In particular, two categories can be distinguished: sites having soft to medium clays and sands or deep cohesionless soils (>250 ft) are similar and form one category, the soft group; stiff soil and rock sites form another category, the hard group. A dividing line on the frequency axis appears at 2.5 Hz (0.4-sec period). In the frequency range below 2.5 Hz, spectral amplifications for the soft group are much higher than those for the hard group; in the frequency range above 2.5 Hz, spectral simplifications for the hard group are higher than those for the soft group. In the soft group, the energy peaks for the soft to medium clays and sands and the deep cohesionless soils both occur at about the same frequency, 1 Hz, but the amplifications are slightly different (0.4 and 0.35, respectively).

The average NPSD spectrum of the deep cohesionless soil sites has a large hump at 2.8 Hz (0.36-sec period), but the spectrum of the soft to medium clay and sand sites does not. The large amplitude at 0 Hz is believed to be caused by a digitization error, particularly that due to the uncorrected Japanese strong-motion data in the soil site group. For the hard site group, in the frequency range below 2.5 Hz, the spectral amplitude for the stiff soil is higher than that for the rock; but in the frequency range above 2.5 Hz, the spectral amplitude for the rock is slightly higher than that for the stiff soil. The largest energy peaks for the rock sites and the stiff soil sites are at 2.75 Hz (0.36 sec) and 0.8 Hz (1.25 sec), respectively.

STATISTICAL CHARACTERISTICS OF THE EARTHQUAKE GROUND MOTIONS

Table 1 lists the statistical characteristics of the earthquake ground motions.

Table 1

Statistical Characteristics of Earthquake Ground Motion

Condition	Maximum Ground Acceleration, g			Frequencies of Selected Peaks, PSD	Nor. Max. PSD		Average Acceleration, cm/sec^2		
	Mean	S.D.	Var.	Hz	Mean	Mean + σ	Mean	S.D.	Var.
Rock	0.196	0.234	0.054				30.84	38.35	1444.8
				1.06	0.146	0.369			
				2.75	0.184	0.412			
				3.80	0.160	0.353			
				5.17	0.133	0.321			
Stiff soil	0.050	0.369	0.137				17.5	12.8	162.4
				0.80	0.192	0.547			
				2.45	0.180	0.412			
				5.20	0.127	0.327			
Deep cohesionless soil	0.063	0.272	0.074				11.2	8.3	68.5
				1.00	0.350	0.827			
				2.80	0.190	0.445			
Soft to medium clay and sand	0.054	0.046	0.002				9.3	8.4	70.5
				1.05	0.405	1.315			
				4.58	0.089	0.266			

It is apparent from Table 1 that statistical characteristics of the ground motions are strongly site-dependent. The rock sites produce an average maximum ground acceleration of about 0.20 g for the entire suite of 56 records. The rock site group shows the highest maximum acceleration and average acceleration of the four site groups. The PSD function estimates are almost uniform over the peak frequencies of 1.06, 2.75, 3.80, and 5.17 Hz.

The average maximum ground accelerations and PSD spectral intensities (or squared average accelerations) for the other three site groups--stiff soils, deep cohesionless soils, and soft to medium clays and sands--are relatively close together. However, the spreads of the standard deviation of maximum accelerations for the stiff soil and deep cohesionless soil groups are wider than for the soft to medium clays and sand group. This large spread is believed to be caused by the different magnitudes of earthquakes and the different epicentral distances. The group of accelerograms for soft to medium clays and sands shows relatively low ground acceleration but the highest PSD function estimates at the frequency of 1 Hz among the four groups. One second (1 Hz) is probably near the predominant natural period of sites on soft to medium clays and sands. It seems that the acceleration and the PSD spectral intensity at 1 Hz are roughly in proportion and inverse proportion, respectively, to the degree of stiffness of the site material. In conclusion, the average acceleration or the mean spectral intensity is site-dependent.

SITE-DEPENDENT NPSD FUNCTIONS AND SCALING FACTORS

In the previous sections, the four statistical site-dependent NPSD functions of rock, stiff soil, deep cohesionless soil, and soft to medium clay and sand based on the analysis of 421 horizontal ground accelerograms have been established and are shown in Figures 2a and 2b. The scaling factor, λ^2 is defined as the area under each PSD, G(f) versus f; its physical meaning in an engineering sense is the average power of spectral intensity. It is logical to use λ^2 as the scaling factor of the four NPSD spectra. It is also closely related to the degree of damage to engineering structures. Although earthquake strong ground motions are of primary concern in engineering problems, we used an extended ground motion duration, t_o = 163.82 sec, throughout the analysis. The extended duration was made by adding zeros to the trailing part of the accelerogram.

From Equations (3), (4), and (9), we have

$$\frac{I_o}{t_o} = \int_o^{\omega_o} G(\omega)\, d\omega = 2\pi \int_o^{10\ \mathrm{Hz}} G(f)\, df = 2\pi\cdot\lambda^2 \tag{10}$$

For a given earthquake ground motion

$$I_o = 2\pi\cdot\lambda^2\cdot t_o = 2\pi\cdot\lambda_o^2\cdot S_o \tag{11}$$

i.e.

$$\lambda_o^2 = \lambda^2\cdot(t_o/S_o) \tag{12}$$

where λ_o^2 is defined as the scaling factor for the earthquake strong-motion and S_o, the strong-motion duration for a design earthquake. The total duration t_o is fixed as 163.82 sec. If the PGA of a design earthquake is known, λ can be determined from the linear relation in the curves of PGA versus λ in Chang's study[9], or Figure 3 in this paper. Therefore, λ_o^2 can be calculated. Figure 3 presents the mean values obtained from Chang's[9] PGA versus λ plots for the four site groups. They reflect the site-dependency of the average acceleration, λ, or RMSA.

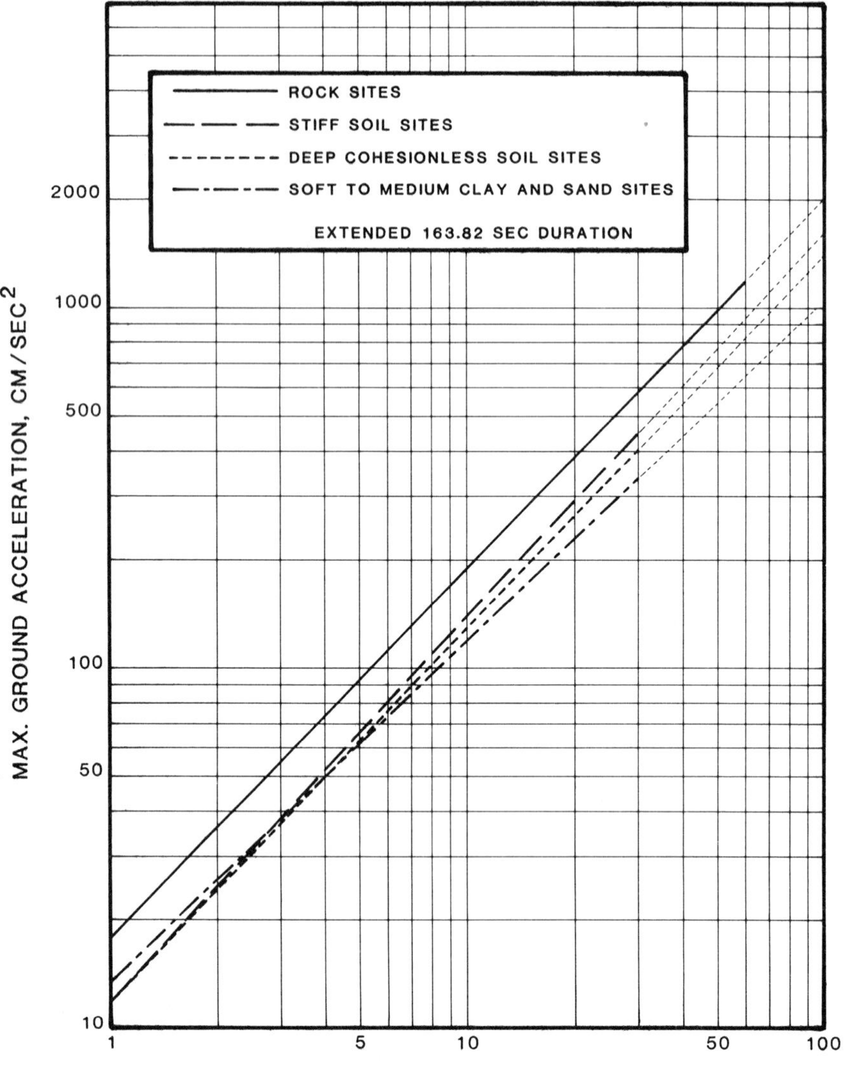

Figure 3. Correlation of maximum ground acceleration (g) versus "base" average acceleration based on extended 163.82-sec duration. Dashed lines show extrapolation beyond existing data

GENERATION OF EARTHQUAKE ACCELERATION TIME-HISTORY

An important characteristic was found in the study by Chang[9]. The value of λ , or root mean square acceleration (RMSA) for t_o = 163.82 sec, which was defined as the base average acceleration to distinguish from the RMSA in the strong-motion duration S_o , was in linear relation with the PGA or A_{max}. This is shown in Figure 3. From Equation (1), we have

$$F(\omega) = \int_0^{t_o} a(t)e^{-i\omega t}\, dt = \int_0^{163.82\ \text{sec}} a(t)e^{-i\omega t}\, dt \qquad (13)$$

Based on Equation (13), if a satisfying statistical results in terms of the Fourier spectrum can be achieved, then using inverse Fourier transform, a(t) can be derived. This has been demonstrated by Chang, et al.[10], using the Chang's statistical model of the NPSD, $\bar{G}^*(f)$ (Figures 2a and 2b) and the linear relation of PGA versus the average acceleration, λ (Figure 3), with the computer program EQGEN to generate a synthetic earthquake acceleration time-history.

CORRELATION OF RMSA AND MMI

It will be of much benefit to the engineering community if a quantitative earthquake intensity scale, such as root-mean-square acceleration (RMSA) of the strong-motion duration, can be correlated with the Modified Mercalli Intensity (MMI) (Figure 4 for hard sites and Figure 5 for soft sites).

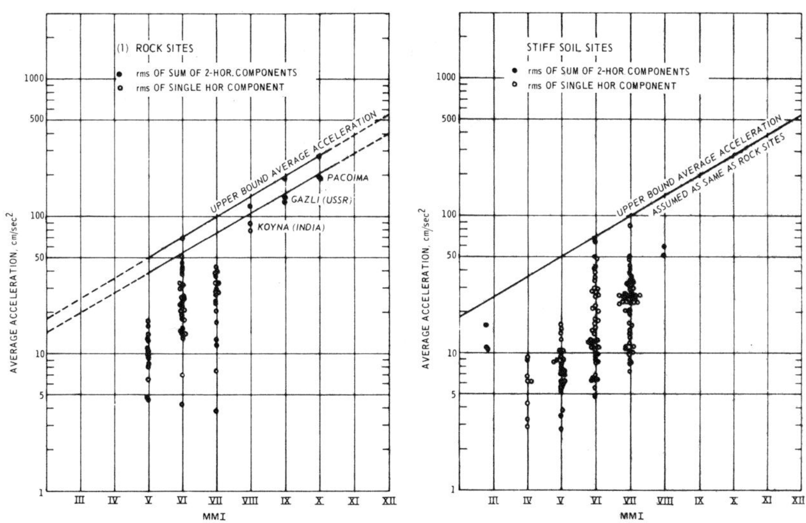

Figure 4. Correlation of upper-bound average acceleration versus MMI - hard sites (rock and stiff soil)

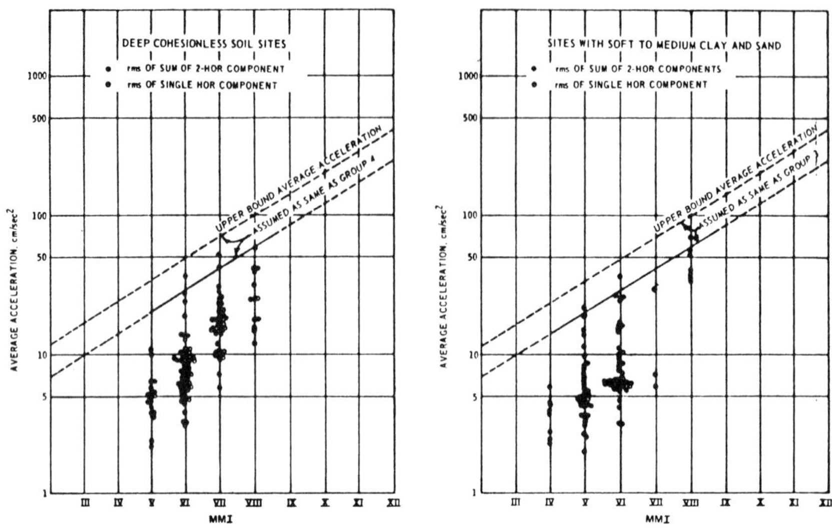

Figure 5. Correlation of upper-bound average acceleration versus MMI - soft sites (deep cohesionless soil and soft to medium clay and sand sites)

Table 2 shows the upper bound of site-dependent RMSA intensity (square root of the sum of two horizontal average powers, λ_o^2) versus MMI.

Table 2. Correlation of Upper-Bound RMSA versus MMI

Upper bound of the site-dependent RMSA intensity, cm/sec^2 (square root of the sum of 2-horizontal variances, λ_o^2)

MMI	Hard Sites	Soft Sites
XII	400-500	250-400
XI	290-400	200-250
X	200-290	150-200
IX	150-200	100-150
VIII	100-150	70-100
VII	70-100	50-70
VI	50-70	30-50
V	40-50	20-30

To obtain the bounds shown in the above table, the rock and stiff soil sites were combined as hard sites and the deep cohesionless soil and soft soil sites as soft sites. The Pacoima Dam, Karakyr Point (Gazli earthquake), Koyna Dam, and Lake Hughes Array No. 12 sites were located in the epicentral regions and near the faults (3 km $\le$ R $\le$ 20 km). They possessed the maximum average accelerations seen and these might be called epicentral average accelerations. It seems from these limited data that the maximum average acceleration at the epicentral region might not be over 550.0 cm/sec^2. However, the power λ_o^2 or the average acceleration λ_o^2 is inversely proportional to the duration, i.e., the smaller the duration of strong shaking, the higher the average acceleration or RMSA.

CORRELATION OF I_o AND MMI

The correlations of total intensity I_o with the MMI based on the data of hard sites (Group 1 and 2) and soft sites (Group 3 and 4), are plotted in Figures 6 and 7, respectively. The upper bound line of Figure 6 is established by five earthquakes (San Fernando, Gazli, Parkfield, Koyna, and Tokachi Oki). There are four sites located in the epicentral region, so the values are named as epicentral intensities. The extrapolation from these values, the upper-bound epicentral seismic intensities versus the MMI, are listed as in Table 3 below:

Table 3. Upper-Bound Seismic Epicentral Intensity

MMI	Hard Sites 10^4(cm^2/sec^3)	Soft Sites 10^4(cm^2/sec^3)
XII	135-160	120
XI	82-92	70
X	52-54	40
IX	30-34	24
VIII	17-20	14
VII	10-13	8
VI	6-8	5
V	3-5	3
IV	2-3	2

The upper-bound line of Figure 7 indicates that the greatest seismic intensity at soft sites is lower than at hard sites. Also, the rate of attenuation is lower for soft sites than for hard sites. Of course, the upper-bound seismic intensities for both the hard sites (Figure 6) and the soft sites (Figure 7) are in the near field. The data under the upper-bound line (both Figures 6 and 7) spread widely because of various earthquake magnitudes and distances.

Damage to structures in the epicentral area is generally more severe on soft sites than on hard sites, based on past experience and observations. However, this study showed the seismic total intensity (seismic energy) at soft sites to be lower than at hard sites. Thus, the degree of damage to structures does not correlate with the seismic total intensity in the epicentral region. Furthermore, the predominant frequencies at soft sites are in the range of 0 to 2.5 Hz; the seismic energy in this low-frequency range deserves further investigation as it relates to structural damage.

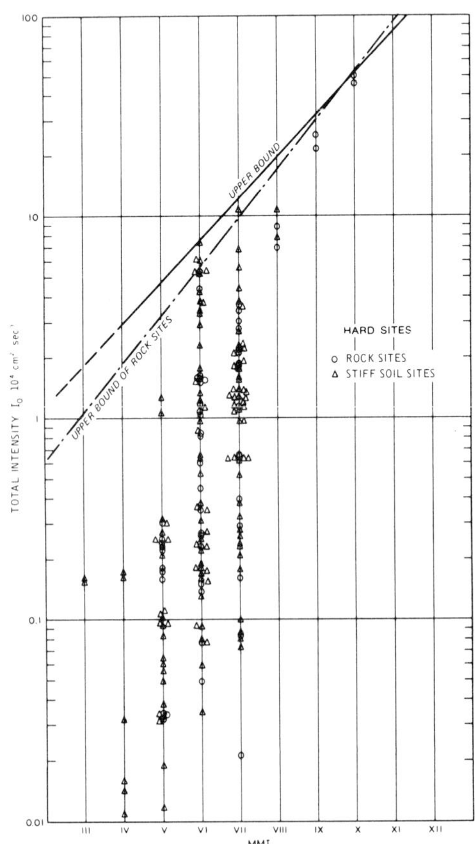

Figure 6. Total seismic intensity I_o at epicentral region - hard sites

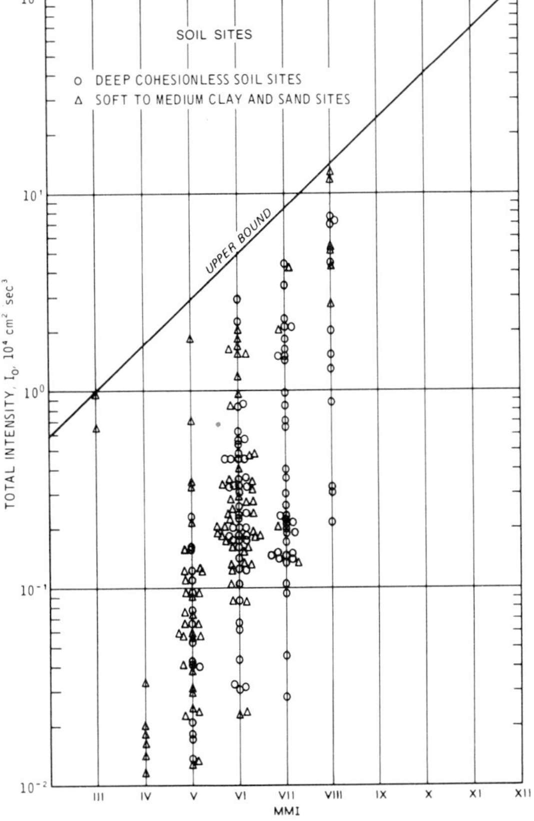

Figure 7. Total seismic intensity I_o at epicentral region - soft sites

CONCLUSIONS

The statistical analysis of 421 accelerograms shows clear differences in spectral shapes for different soil and geological conditions. Within the high-frequency range of 2.5 to 10 Hz, the spectrum for the rock sites contains the highest energy or intensity; the spectrum of the stiff soil sites is slightly lower than for the rock sites; and the spectra of the soft clay and sand sites and the deep cohesionless soil sites are lower still and almost the same. However, in the low-frequency range of 0 to 2.5 Hz, the reverse relation exists: the soft sites indicate the highest energy, the deep cohesionless soil sites are next, the stiff soil sites are third, and finally, the rock sites. Generally, the spectra of rock sites and stiff soil sites of similar characteristics can be classified together as hard sites; the other two site types can be classified together as soft sites.

The site dependence of NPSD spectra have been established by statistical analysis as expected. The most significant finding of the study is the approximate linear correlation of the PGA (a_{max}) and base average acceleration for an arbitrary record length of 163.82 sec (λ). However, the area, λ_o^2 under the PSDF curve of a strong-motion duration S_o, or any record length (S_o < 163.82 sec) can be determined by the following relation:

$$\lambda_o^2 = \lambda^2 \cdot \frac{163.82}{S_o}$$

λ can be found from the established charts. If the PGA and S_o for the design earthquake are given, the scaling factor, λ_o^2 for the NPSDF curve can also be obtained. Therefore, the standard NPSD spectrum can be amplified by λ_o^2 to become a design PSD spectrum. Furthermore, the design PSD spectrum can be used to generate a design accelerogram with the computer program-EQGEN by Chang, et al.

ACKNOWLEDGMENT

This study was conducted as part of the work at the U. S. Army Engineer Waterways Experiment Station under the Military Works Program, Project 4A161102AT22, Work Unit 00296, which was monitored for the Office, Chief of Engineers (OCE) by Mr. A. F. Muller.

REFERENCES

1. Seed, H. B., Ugas, C., and Lysmer, J. (1976), Site-Dependent Spectra for Earthquake-Resistant Design, Bulletin, Seismological Society of America, Vol. 66, pp. 221-244.

2. Kiremidjian, A. S., and Shah, C. H. (1978), Probabilistic Site-Dependent Response Spectra, Report No. 29, John A. Blume Earthquake Engineering Center, Department of Civil Engineering, Stanford University, Stanford, California.

3. Chang, F. K., and Krinitzsky, E. L. (1977), State-of-the-Art for Assessing Earthquake Hazards in the United States; Duration, Spectral Content, and Predominant Period of Strong-Motion Earthquake Records from Western United States, Miscellaneous Paper S-73-1, Report 8, U. S. Army Engineer Waterways Experiment Station, CE, Vicksburg, Mississippi.

4. Arnold, P. (1975), The Influence of Site Azimuth and Local Soil Conditions on Earthquake Ground Motion Spectra, M.S. thesis, Massachusetts Institute of Technology, Cambridge, Massachusetts.

5. Arnold, P., and Vanmarcke, E. H. (1977), Ground Motion Spectral Content: The Influence of Local Soil Conditions and Site Azimuth, Proceedings, Sixth World Conference on Earthquake Engineering, New Delhi, India, Vol 2, pp. 113-118.

6. Hsu, H. P. (1967), Outline of Fourier Analysis, Simon and Schuster, Inc., New York.

7. Blackman, R. B., and Tukey, J. W. (1958), The Measurement of Power Spectral, Dover Publications, Inc., New York.

8. California Institute of Technology, (1971-1975), Strong-Motion Earthquake Accelerograms; Corrected Accelerograms and Integrated Ground Velocities and Displacements, Vol. 2, Parts A-N, Earthquake Engineering Research Laboratory, Pasadena, California.

9. Chang, F. K. (1981), Site Effects on Power Spectral Densities and Scaling Factors, Miscellaneous Paper GL-81-2, U. S. Army Engineer Waterways Experiment Station, CE, Vicksburg, Mississippi.

10. Chang, N. Y., Huang, M. J., Lien, B. H., and Chang, F. K. (1986), EQGEN-A User-Friendly Artificial Earthquake Simulation Program, Vol. I, pp. 439-450, Proceedings of the Third U. S. National Conference on Earthquake Engineering, Charleston, South Carolina, U.S.A., August 24-28, 1986. Earthquake Engineering Research Institute, El Cerrito, CA.

11. Vanmarcke, E. H., and Lai, S. S. (1977), Strong-Motion Duration and R.M.S. Amplitude of Earthquake Records, Bulletin, Seismological Society of America, Vol. 70, No. 4, pp. 1293-1307.

Irregular Ground Analysis to Interpret Time-Characteristics of Strong Motion Recorded in Mexico City during 1985 Mexico Earthquake

H. Kawase
Ohsaki Research Institute, Shimizu Construction Co., Ltd., Tokyo 100, Japan

INTRODUCTION

A destructive earthquake of Ms 8.1 occurred near the coast of the Michoacan State of Mexico on September 19, 1985. This eathquake caused severe damage in Mexico City, which is located about 370 km far away from the ruptured area. The strong ground motion records were obtained by the research group of Universidad National Autonoma de Mexico (Prince et al.[1]) at several points in and around the city. As summerized in their reports, the characteristics of these records are very remarkable, especially the long duration times recorded on the soft soil deposit. Because of the relatively thick and soft lacustrine clay layer, it seems reasonable that frequency-characteristics of observed records, that is, the large amplification with the long predominant period, may be attributed to the existence of this clay layer. However its time-characteristics including long duration time and large amplitude of the later part are very unusual even if the effect of the soft surface layer is taken into account. The reason why such time-characteristics were generated is very important because they may be strongly related to the severe damage, especially collapse of tall buildings (Hall et al.[2]).

The author intends to interpret these time-characteristics of the strong ground motion recorded on the soft soil deposit in Mexico City. Since the information about the structure of ground in and around the city is limited, the analysis is to be a kind of a sample study rather than a simulation. However it may be important to distinguish possible and impossible ways of interpretation. The author presents two different soil models both of which have two different media contacting with a simple irregular boundary. The SH-type wave field is assumed in this paper.

OBSERVED RECORDS

Figure 1 shows acceleration time histories of the Mexico Earthquake of September 19, 1985 observed at several points in Mexico City shown in Figure 2. The stations CDAO, CDAF and SCT1 are located on an alluvial lacstrine layer, while TACY on a hard rock. As already reported by other researchers, the amplification factor of the stations on the soft layers to that on hard rock is remarkable as are the long predominant periods. These features reflect the properties of the lacustrine layer under these observation points, that is, softness with shear wave velocities ranging from 40 m/s to 80 m/s and thickness ranging from 30m to 50m. However, it should be noted that the duration time is also longer on the soft layer than that on hard rock and an easily distingushable later phase exists, especially at CDAO. As is shown later, it is difficult to generate such a later arrival phase by using a horizontally layered model assuming the record on hard rock as an input motion at the bottom of the layer.

To see the frequency-characteristics of various observed records, Fourier amplitude spectra of 8 observation points are shown in Figure 3. Figure 3a corresponds to the NS-component, while Figures 3b and 3c correspond to EW- and UD-component, respectively. The vertical axis is shifted by multiplying by 10 for one different site but it is the absolute value for CDAF. The effect of the alluvial layer can clearly be seen in this figure for NS- and EW-components. The peak frequencies of second- and third-order appeared in the spectra of CDAF, CDAO and SCT keep the ratio 3 times or 5 times to the first-order peak frequency. The first-order peak frequency of CDAO is longest which reflects the largest thickness of the layer at that site. Relative amplification factor to the hard rock site ranges from 15 to 40 at the first resonant peak, as reported by Singh et al.[3]. However, spectra of UD-component for different sites are almost identical and no clear effect of layering can be seen. This fact suggests that only the shear wave played a deterministic role during the earthquake.

1-DIMENSIONAL ANALYSIS

First the one-dimensional wave propagation analysis is carried out since the amplitude and frequency characteristics seem to be strongly affected by the surface layer. The information of the clay layer of Mexico City is limited so the model assumed here is very simple one, a single layer overlying a half-space. Referring published papers such as Marsal et al.[4], Herrera et al.[5], Muris[6], etc. , shear wave velocity of 50m/s is assumed for the layer and 250m/s for the half-space. The thickness of the layer differs from site to site and is chosen as 30m, 40m, and 50m for SCT1, CDAF AND CDAO, respectively. Here only the case of vertical incidence is presented.

Figure 4 shows the resultant acceleration time histories obtained by the calculated transfer functions of the layer by solid lines. As an input motion at the bottom of the surface layer, the author adopted the record at TACY. The observed records are also plotted by dotted lines for comparison. It is apparent that at SCT1 and CDAO the former part of time histories up to 30 seconds is similar to those recorded, but that the later part is quite different from each other. This simple model could not generate at all the time-characteristics of the obsreved records at those sites. On the other hand, for CDAF, it seem to be simulated well by this model. However, the accelerograms at this site aren't considered as fully recorded ones (Prince et al.[7]) so that it is dangerous to conclude that the later phase did not exist at CDAF.

2-DIMENSIONAL ANALYSIS FOR SURFACE LAYER

Since the simple layer model of the alluvium could not generate the later phase observed at SCT1 and CDAO, more complicated model should be considered. First an irregularity of the boundary between the alluvium and the half-space is considered. From boring data (UNAM[8]) the thickness of the alluvial layer seems to be changing so slowly that it might be impossible to attribute the cause of later phase to the irregularity of the alluvium. Also it is not likely that the surface wave generated by such an irregularity propagates a long distance. To confirm this, the irregularity stronger than actual ground may possess is considered. Figure 5 shows a model of the irregular ground for CDAO. The thickness of the layer changes from 50m to 0m within a distance of 100m with a elliptic shape. The boundary forms a kind of open-ended valley. The method used here is a direct boundary element method that has already been applied widely to irregular ground analyses (e.g.

Wong et al.[9], Kobori et al.[10], Kawase et al.[11]).

Figure 6 shows the response of the ground along the surface for the case of the vertical incidence of a Ricker wavelet with a characteristic period of 4 sec. Hereafter maximum calculation frequency is set to 1 Hz. The uppermost time history corresponds to that of a one dimensional case with a single horizontal layer. The distance shown in the figure is measured from the edge of the irregular boundary. By comparing the response of the irregular ground with that of a horizontal layer, it is concluded that the surface wave generated by an irregularity of the alluvial layer is not sufficiently large to interpret the time-characteristics of the observed records. Also it is worthwhile to note that the time difference between direct- and reflected-wave arrival and surface-wave arrival is too small to distinguish because the distance from the edge of the irregularity is assumed to be relatively small. If much more distance is assumed, however, the surface wave contribution will become negligible. Even assuming inclined incidence up to $\pm 10^\circ$ the result is affected very little.

2-DIMENSIONAL ANALYSIS FOR DEEPER STRUCTURE

Since it seems difficult to attribute the cause of the strong later phase to the irregularity of the surface layer, the effect of a deeper irregularity of the ground is examined to determine whether it can generate this phase or not. Again a simple structure, shown in Figure 7, is considered; a closed elliptical valley with 20 km width and 1km depth. The shear wave velocity is assumed to be 1000m/s and 2500m/s for the valley and the half-space, respectively. Although there doesn't exist sufficient information on the deeper structure of the ground underneath Mexico City, the profile assumed here may be reasonable one. Figure 8 shows the response due to the incident Ricker wavelet at several points along the surface of the valley. The angle of incidence θ is set to 30°. The uppermost time history corresponds to the result of a horizontally layered model with the depth of the valley by a vertically incident wave. The surface wave generated at the edge of the valley is clearly seen propagating along the surface of the valley. Its amplitude is relatively large compared with the case of a surface layer irregularity.

Although the model studied here is rather simple, it is worthwhile to calculate the response of the valley due to one of the observed records on the hard rock site. Figure 9 shows an example of such calculation for points P08, P10 and P12 in Figure 7 using TACY as an input motion after filtering up to 1 Hz. Two different angles of incidence are chosen, 30°, 60°. The absolute level of calculated acceleration is smaller than that for observed records and it should be because here the amplification of the soft surface layer is neglected. On the other hand, time-characteristics of the calculated waves are rather similar to those recorded. The duration time is prolonged by the arrival of the subsequent surface wave generated at the edge of the valley. However, the observed records at CDAO have longer later phases than calculated by this simple model. The effects of the valley become larger as the angle of incidence become larger, which reflects the fact that the efficiency of surface-wave generation at the edge of the valley strongly depends on the form of the incident wave field. Therefore, it is crucial to know what kind of an incident wave or waves is dominant at the distance of 370km from the ruptured area for the frequency range from 0.2Hz to 0.5Hz . Also more detailed information on the ground structure in and around Mexico City is essential.

CONCLUSIONS

After studying the effects of irregularities by using simple models to determine the possibile cause of the time-characteristcs observed in Mexico City, it is concluded that the irregularity of the surface layer may not be considered as a principal cause of the time-characteristcs because it generates rather small surface-wave with small arrival time differences, but that the deeper structure is a possibility because it generates large amplitude surface-waves with relatively large differences of arrival times. The mode convergence rate from a incident body-wave to a surface-wave is highest in case of horizontal incidence, which suggests that the incident surface-wave will generate the time-characteristics as much as or more than a horizontally incident body-wave. Since both the structure of the ground in and around Mexico City and the form of an incident wave have a great uncertainties, simulation analyses with more realistic models could not be carried out at present. Further investigation is necessary to clarify what caused the observed characteristics of strong motion in Mexico City from the view point of seismic microzonation or hazard analysis for possible earthquakes in future.

ACKNOWLEDGMENTS

I wish to acknowledge the offer of the strong ground motion records from Prof. J. Prince and his coliegue (MT No. MX0056, AEDP, Japan). Also I would like to express my gratitude to Prof. K. Aki of University of Southern California and Dr. P. Bard of Grenoble University for their helpful suggestions. This paper is prepared on leave from Ohsaki Research Inst. of Shimizu Construction Co., Ltd. to Department of Geological Sciences, University of Southern California as a visiting scholar.

Reference

1. Prince J. et al. (1985) Acelerogramas en Ciudad Universitaria del Sismo del 19 de Septiembre de 1985, Instituto de Ingenieria, UNAM INFORME IPS-10A~D

2. Hall J.F. and J.L. Beck (1986) Structural Damage in Mexico City, Geophys. Res. Lett., Vol.13, No.6, pp. 589-592

3. Singh S.K. et al. (1987) Some Aspects of Source Characteristics of the 19 September, 1985 Michoacan Earthquake and Ground Motion Amplification in and near Mexico City from Strong Motion Data, submitted to Bull. Seism. Soc. Am.

4. Marsal J.R. et al. (1969) El Hundimento de la Ciudad de Mexico y Proyecto Texcoco (Nabor Carillo), SHCP, Mexico

5. Herrera I et al. (1965) Earthquake Spectrum Prediction for the Valley of Mexico, Proc. of 3rd World Conf. Earthq. Eng., Vol.1, pp. 61-74

6. Muris R.C. (1978) Ciudad de Mexico, El Subsuelo y la Ingenieria de Cimentaciones en el Area Urbana del Valle de Mexico, SMMS Simposio

7. the same as 1

8. UNAM research group (1985) El Temblor del 19 de Septembre de 1985 y sus Efectos en las Construcciones de la Ciudad de Mexico, Instituto de Ingenieria, UNAM

9. Wong H.L. and P.C. Jennings (1975) Effect of Canyon Topography on Strong Ground Motion, Bull. Seism. Soc. Am., Vol.65, pp. 1239-1257

10. Kobori T. and Y. Shinozaki (1978) Danamic Soil-Structure Interaction under Topographical Site Condition, Proc. of 5th Japan Earthq. Eng. Symp., pp. 489-496

11. Kawase H. et al. (1985) Analytical and Observational Study on Seismic Scattering Property of a Site with Geological Irregularity, Proc. of 8th Int. Conf. Struct. Mech. in React. Tech. Vol.K(a), pp. 1-6

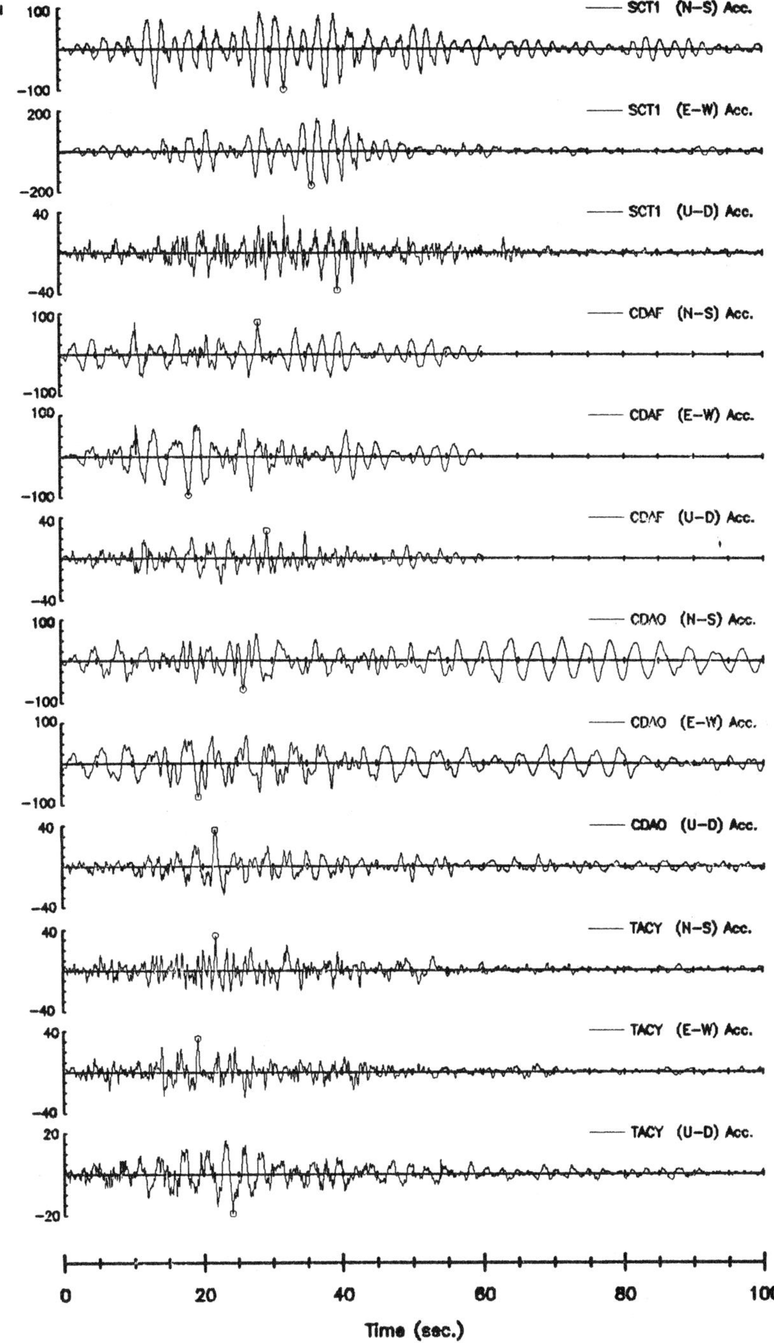

Figure 1: Acceleration time histories observed in Mexico City

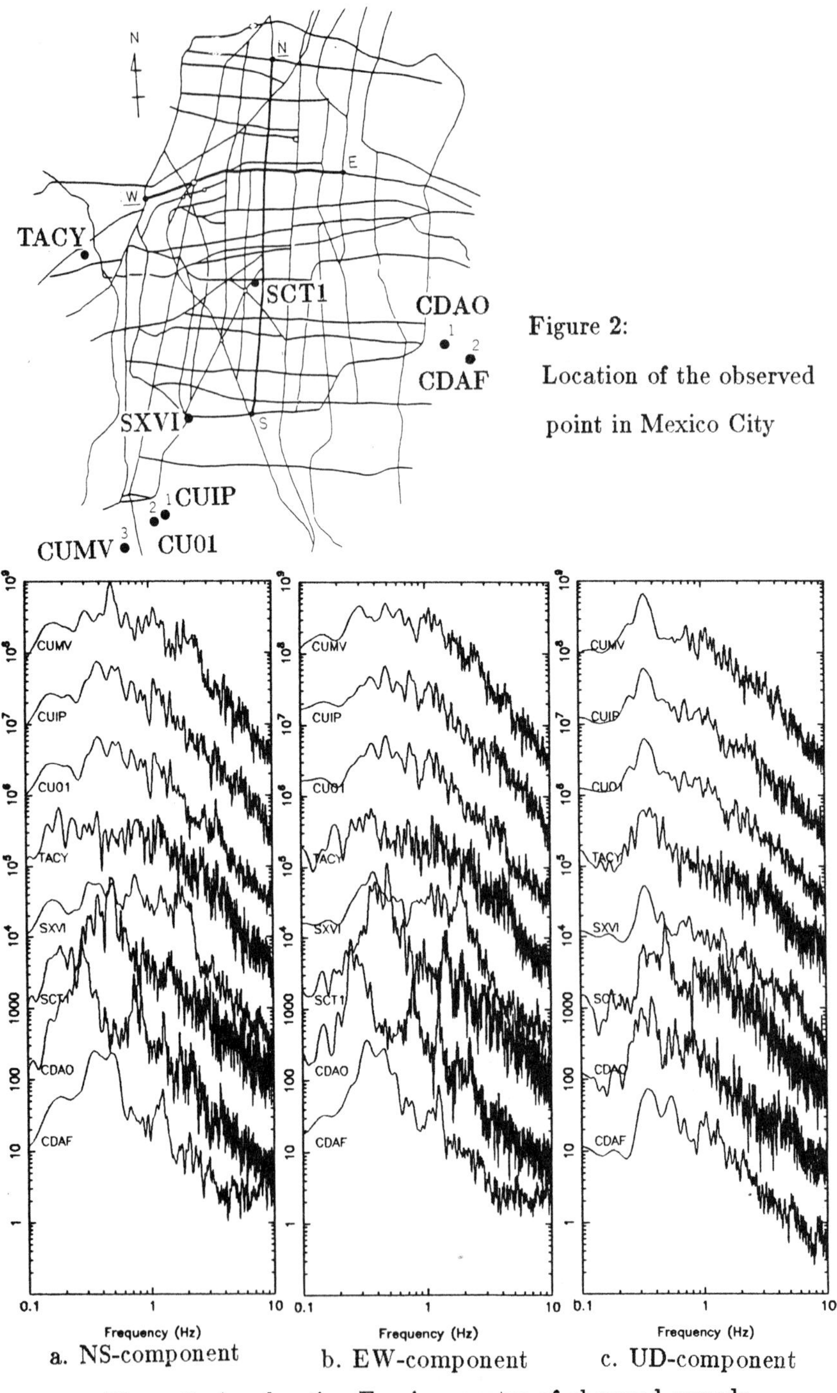

Figure 2: Location of the observed point in Mexico City

a. NS-component b. EW-component c. UD-component

Figure 3: Acceleration Fourier spectra of observed records

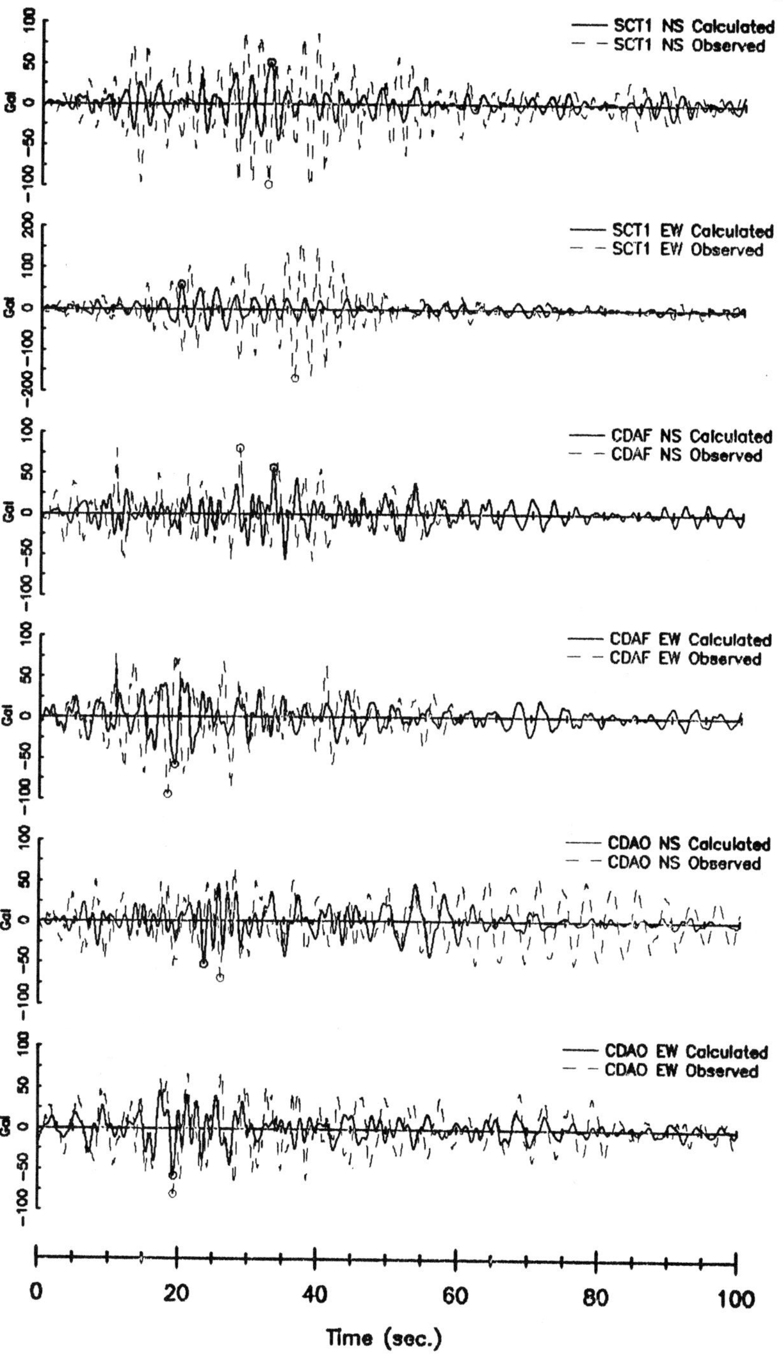

Figure 4: Acceleration time histories calculated by 1-D model

Vs=50m/s ρ=1.25gr/m³ Q=25
−200
−100
0
100
200
50m
Q=50
Vs=250m/s ρ=2.50gr/m³

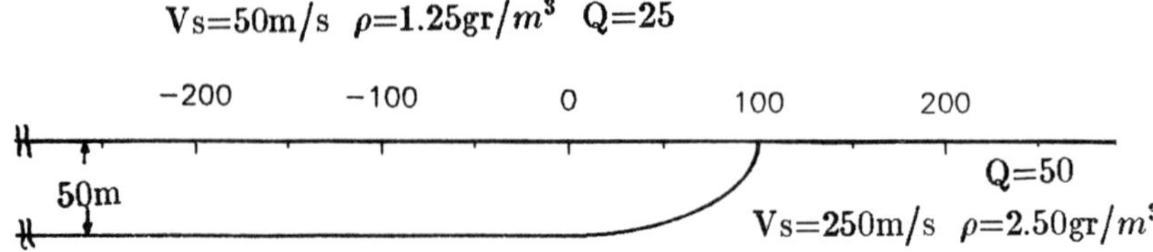

Figure 5: 2-D model for surface layer irregularity

Horizontal Layer
Point P01 D=−240.0m
Point P02 D=−220.0m
Point P03 D=−200.0m
Point P04 D=−180.0m
Point P05 D=−160.0m
Point P06 D=−140.0m
Point P07 D=−120.0m
Point P08 D=−100.0m
Point P09 D=− 80.0m
Point P10 D=− 60.0m
Point P11 D=− 40.0m
Point P12 D=− 20.0m
Point P13 D= 0.0m
0
3
6
9
12
15
18
21
Time (sec.)

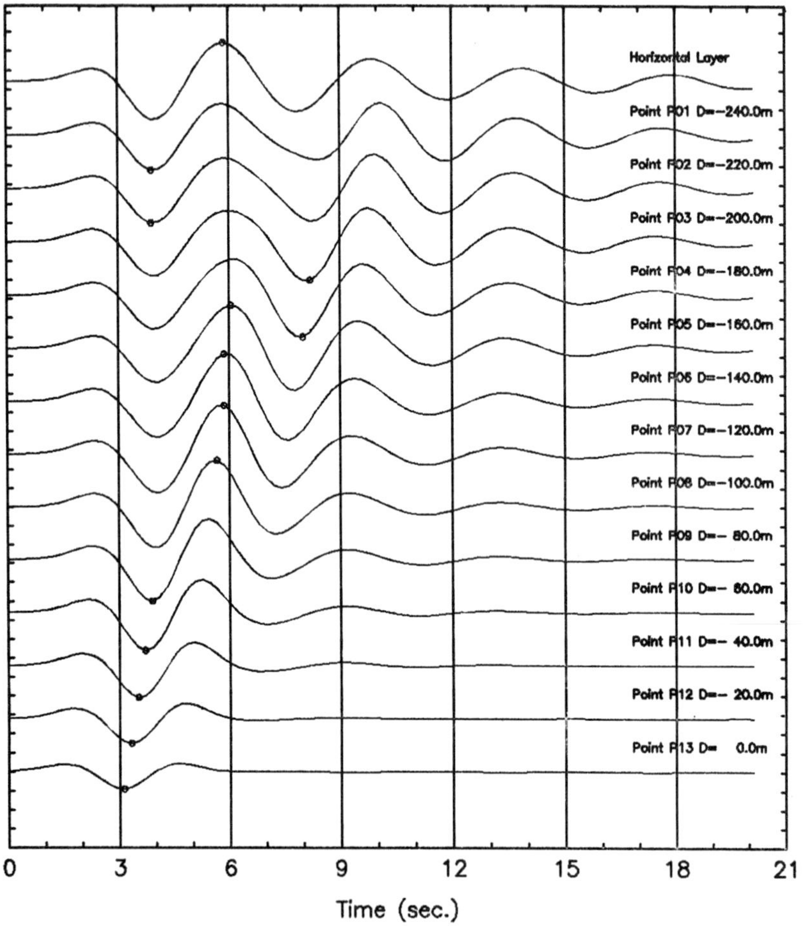

Figure 6: Response of 2-D model for surface layer irregularity due to incident Ricker wavelet

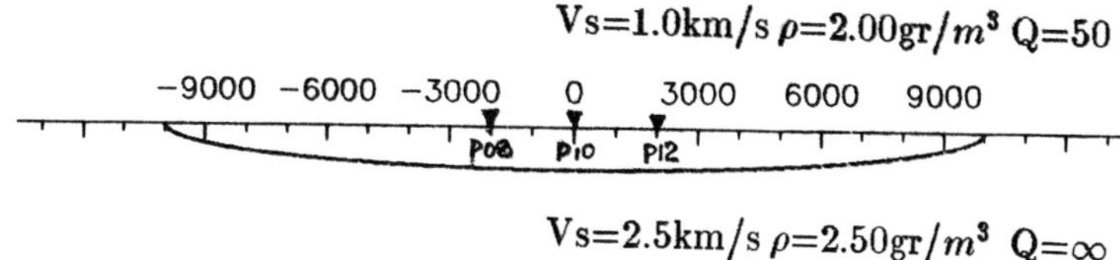

Figure 7: 2-D model for deeper structure irregularity

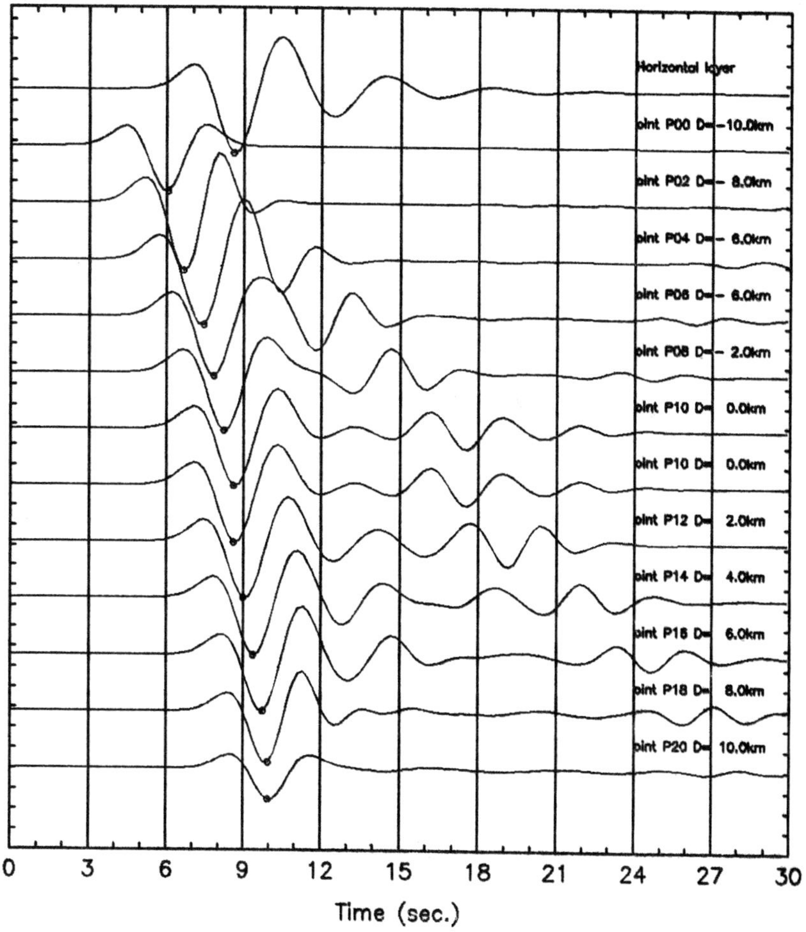

Figure 8: Response of 2-D model for deeper structure irregularity due to incident Ricker wavelet $\theta = 30^\circ$

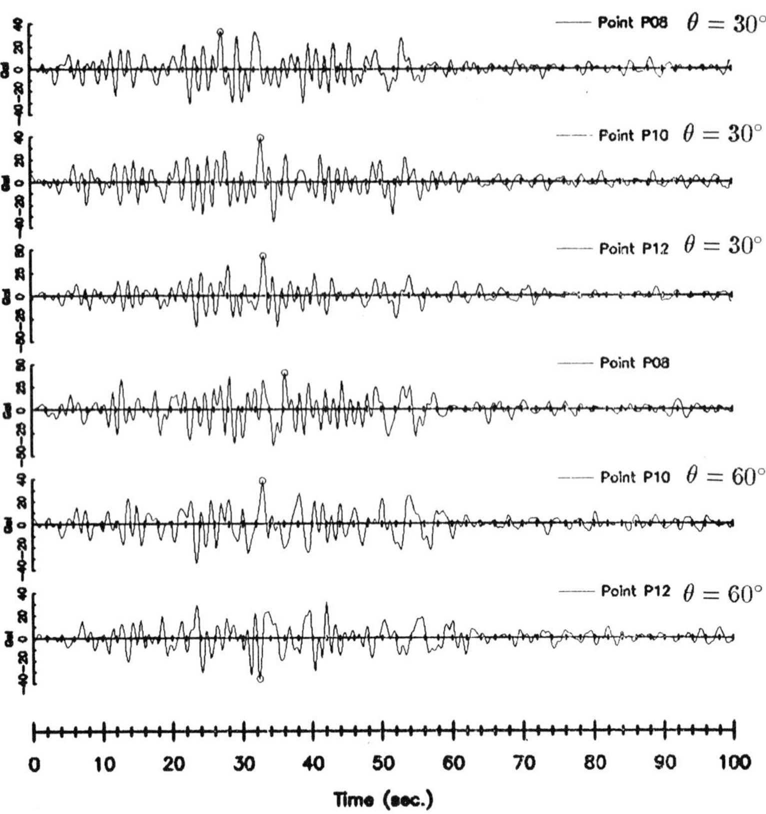

Figure 9: Response of 2-D model for deeper structure irregularity due to TACY NS-component

Numerical Determination of Critical Hardening for Shear Band Localization in Geological Materials

J-G Beliveau
Civil Engineering and Mechanical Engineering Department, University of Vermont, Burlington, Vermont, 05405, U.S.A.
H. Ameziane-Hassani
Departement de genie civil, Universite de Sherbrooke, Sherbrooke, Quebec, J1K 2R1, Canada
J. Desrues
Institut de Mecanique de Grenoble, BP68, Saint Martin d'Hères, 38402, Franc

INTRODUCTION

The localization of shear bands in materials subjected to large stresses and having correspondingly large or finite strains has been observed in both man-made and natural materials. Often these localized instabilities are sudden, occur with very little warning, and can have catastrophic results, i.e. earthquakes.

In a recent thesis Ameziane-Hassani[1] pursued the concept of using stationary acceleration waves(Hill[2])in quantifying instability and confirmed many results in the literature for plane strain conditions(Beliveau and Ameziane-Hassani[3].) The criteria for instability based on quasi-static considerations(Rudnicki and Rice[4])was also obtained by a wave propagation analysis.

Rather than assuming imperfections as is commonly done for global geometric instability, material instability is investigated by considering waves of small amplitude superimposed on an existing displacement field. An important result is that, for a homogeneous stress field, the stationary wave speed corresponding to instability when it becomes zero is a gradual and not a rapidly varying function of the hardening parameter of a general three dimensional formulation, whereas the other two wave speeds were relatively insensitive to this parameter. This suggests that the measurement of wave speeds may be a valid experimental means of predicting and perhaps controlling material instability.

In this paper we discuss three numerical results. First, we show that a three-dimensional wave propagation study of a hypoelastic material subjected to biaxial loading yields a critical load for local instability lower than that obtained by the one dimensional analysis of the Considere global instability for uniaxial tension. This suggests that studies of local material instability should be performed within a three-dimensional framework and not be limited to one or two dimensions.

For more realistic material behavior at large stresses and strains, we show a substantial difference for critical hardening than that obtained by an approximate formula given in (Rudnicki and Rice[4]) for certain combination of the material parameters of a Drucker-Prager material and for uniaxial strain, typical of geologic materials subjected to an overburden. This is because the large stresses become important relative to the incremental moduli.

The Mohr-Coulomb envelop for yield of geotechnical material is often approximated quite successfully by a Drucker-Prager model with parameters scaled to represent yield correctly(Chen[5].) We show that this is not adequate for material instability because plastic deformation is so sensitive to the normal of the plastic potential and not so dependent on the yield function.

The paper is divided into four basic sections. The equations for determining wave speeds and instability are developed in both a one-dimensional and a three-dimensional framework. Thirdly, we present instability criteria for constitutive relationships commonly used for geologic material, which include the possibility of non normal plastic deformation. Lastly, we present the three examples cited earlier.

ONE DIMENSIONAL WAVE PROPAGATION

One of the basic premises of this work is that local material instability in the form of shear bands is characterized by an incremental constitutive law in three dimensions. Special loading conditions such as plane stress or plane strain, do not control instability per se though they do affect the level of stress and strain in the material. In this light, wave propagation analyses for material instability should be based on three dimensions. Nevertheless, it is usefultolookfirst at wave propagation in one dimension, both for small motion of a material satisfying Hooke's law and for large strain global instability.

Equation of Motion

Newton's equation of motion for the element of length, dx , area A, mass density, ρ , shown in Figure 1, is

$$\rho A dx\, \partial^2 u/\partial t^2 = \partial P/\partial x\, dx \quad (1)$$

u , is the displacement along the length of the member, the x axis, and the axial force P is proportional to the normal stress, σ , and area A, which are both assumed to depend on x

$$P = A\sigma \quad (2)$$

$$\partial P/\partial x = \partial A/\partial x\, \sigma + A\, \partial\sigma/\partial x \quad (3)$$

Wave Propagation

For Hooke's law in which A is assumed independent of x and for which stress is linear in strain assumed to be small

$$\sigma = F\varepsilon = F\partial u/\partial x \quad (4)$$

Eqn. 1 reduces to the classical one dimensional wave equation

$$\rho\, \partial^2 u/\partial t^2 = F\, \partial^2 u/\partial x^2 \quad (5)$$

For a planar wave propagating along the axis of the member, the x-axis, the displacement is assumed to depend on one parameter, $u(\psi)$

$$\psi = x - ct \quad (6)$$

and Eqn. 5 becomes

$$(F-\rho c^2)\, \partial^2 u/\partial\psi^2 = 0 \quad (7)$$

which yields the classical speed, c , for longitudinal waves in a bar (F=E in Eqn. 4), for E , Young's modulus of elasticity,

$$c = \sqrt{E/\rho} \quad (8)$$

Considère Instability

For large strain motion of an imcompressible material satisfying an exponential stress-strain law

$$\sigma = K\varepsilon^n \quad (9)$$

the critical strain for global Considère instability is Semiatin and Jonas[6]

$$\varepsilon_{cr} = n \tag{10}$$

which for linear elastic imcompressible material satisfying Hooke's law with $K = E$, $n = 1$ and (Poisson's ratio) $\nu = 1/2$ yields a critical stress in uniaxial tension.

$$\sigma_{cr} = E = 2(1+\nu)G = 3G \tag{11}$$

From a wave propagation analysis this corresponds to a zero wave speed since Eqn. 5 is still satisfied but with

$$F = \sigma(n/\varepsilon-1) \tag{12}$$

which is zero at critical Considère strain $\varepsilon = n$.

THREE DIMENSIONAL WAVE PROPAGATION

The constitutive equations used for a thre-dimensional formulation are incremental rather than based on total stress and strain as in the one-dimensional formulation. The resultant acceleration waves correspond to superimposed small motion on an existing stress/strain field. The total strain rate, $d\varepsilon_{ij}$, is assumed to be a linear combination of elastic, $d\varepsilon^e_{ij}$ and plastic components, $d\varepsilon^p_{ij}$.

$$d\varepsilon_{ij} = d\varepsilon^e_{ij} + d\varepsilon^p_{ij} = 1/2\ (\partial v_i/\partial x_j+\partial v_j/\partial x_i) \tag{13}$$

The equation of motion for incremental motion in a homogeneous stress field may be expressed in tensor notation as (Beliveau and Ameziane-Hassani[3])

$$\rho\partial^2 v_i/\partial t^2 = M_{ijk\ell}\partial^2 v_\ell/\partial x_j \partial x_k \tag{14}$$

in which the tensor M includes terms proportional to stress level when objective stress rates are used

$$M_{ijk\ell} = L_{ijk\ell} + 1/2(\delta_{j\ell}\sigma_{ik}+\delta_{i\ell}\sigma_{jk}-\delta_{jk}\sigma_{i\ell}-\delta_{ik}\sigma_{j\ell}) \tag{15}$$

The second portion is negligible when stresses are small compared to elastic moduli. $L_{ijk\ell}$ is the material tensor for an elasto-plastic material with stress hardening, h , and a plastic potential, g , which may be different than the yield function, f . For isotropic elastic behavior it has the following form when plastic flow is assumed Chen[5]

$$L_{ijk\ell} = E_{ijk\ell} - \frac{E_{ijtu}\ \partial g/\partial\sigma_{tu}\ \partial f/\partial\sigma_{rs}\ E_{rsk\ell}}{h+\partial f/\partial\sigma_{mn}\ E_{mnpq}\ \partial g/\partial\sigma_{pq}} \tag{16}$$

E is the well known elasticity tensor which may be written as a function of the two Lamé Constants Λ and G

$$E_{ijk\ell} = \Lambda\ \delta_{ij}\delta_{k\ell}+G\delta_{ik}\delta_{j\ell}+G\delta_{i\ell}\delta_{jk} \tag{17}$$

δ is the Kronecker delta and the yield function f, plastic potential g, and hardening, h, satisfy the following equations

$$f = 0 \tag{18}$$

$$d\varepsilon^p_{ij} = \lambda\ \partial g/\partial\sigma_{ij} \tag{19}$$

$$h = -\ (\partial f/\partial\varepsilon^p_{ij}+\partial f/\partial k\ \partial k/\partial\varepsilon^p_{ij})\ \partial g/\partial\sigma_{ij} \tag{20}$$

A p superscript represents the plastic component. λ is a scale factor and the hardening parameter, h, is related to the slope of pure shear stress/strain with constant axial stress as shown in Figure 2 (Rudnicki and Rice[4].) The hardening, h, is very large for small strain and becomes smaller as a function of shear strain γ. k is a parameter of the yield function as will be seen later. L is symmetric in ij and in $k\ell$, but is not in general in ik, $i\ell$, jk and $j\ell$.

Eigenvalue Formulation

Analogous to the development for the one dimensional problem, a plane wave is assumed propagating along a direction with direction cosines n_1 n_2 and n_3 such that the increment of displacement, $v(\psi)$, is a function of one variable ψ (See Fig. 3).

$$\psi = n_i x_i\ -\ ct \tag{21}$$

$$n_1^2 + n_2^2 + n_3^2 = 1 \tag{22}$$

Substituting this into Eqn. 13 there results a matrix eigenvalue problem with the three eigenvalues proportional to the square of the three wave speeds for a three dimensional solid

$$[B\ -\lambda I]\ \{y\} = 0 \tag{23}$$

$$\lambda_i = \rho c_i^2 \tag{24}$$

$$B_{i\ell} = M_{ijk\ell}\ n_j n_k \tag{25}$$

$$y_i = \partial^2 v_i/\partial\psi^2 \tag{26}$$

These speeds are independent of boundary or loading conditions and depend solely on the local material parameters, the direction of propagation, and the level of stress.

INSTABILITY

Much as in the case of one dimensional instability, instability in three dimensions is obtained when a combination of parameters of the material, M , loading S , and direction, N , causes at least one of the wave speeds to go to zero yielding a stationary wave. In other words, the matrix [B] is singular or it s determinant becomes zero. The shear band shown in Fig. 3a is the plane wave with direction cosines n_1 , n_2 , n_3.

$$|B| = D(M,N,S) = 0 \tag{27}$$

This is the same criterion as has been developed based on quasi-static consideration Rudnicki and Rice[4]. For given material parameters and stress levels Eqn. 27 reduces to an algebraic expression in the three direction cosines. Because of Eqn. 22, this reduces to one polynomial expression in one variable when the normal to the shear band is assumed in one plane i.e. $n_3 = 0$. In general this may not be assumed and a critical material parameter, say the hardening, may be obtained by setting the partial derivatives of Eqn. 27 with respect to two variables associated with two of the direction cosines equal to zero Ameziane-Hassani[1].

Critical Hardening

A much simpler criterion for critical hardening which neglects terms proportion to stress i.e. Eqn. 15, has been obtained as a function of the intermediate principal eigenvalue of the matrix P, associated with the plastic potential. For a Drucker-Prager material this may be expressed as a non dimensional hardening parameter H

$$H \equiv h/G \approx (\beta-\alpha)^2(1+\nu)/(1-\nu) - (1+\nu)(2P_{II}+\alpha-\beta)^2/2 \tag{28}$$

in which P_{II} is the intermediate eigenvalue of P whose elements are

$$P_{ij} = \partial g/\partial \sigma_{ij} \tag{29}$$

and

$$f = \sqrt{J_2} + \alpha I_1 - k \tag{30}$$

$$g = \sqrt{J_2} + \beta I_1 - k \tag{31}$$

Expressions for P are given elsewhere Beliveau & Ameziane-Hassani[3], J_2 is the second invariant of the deviatoric stress tensor and I_1 is the first invariant of the stress.

$\beta = \alpha$ is the case of normality with $\alpha \neq \beta$ corresponding to non-associative plastic flow. α is three times the internal friction coefficient and β is three times the dilatancy factor used in (Rudnicki and Rice[4];) (Desrues[7].)

Mohr-Coulomb

The Drucker-Prager yield curve of Eqn. 28 is shown in Fig. 4. Normality imposes the plastic deformation to be perpendicular to this curve. The Mohr-Coulomb criterion also depends on the normal stresses. Whereas Drucker-Prager is a generalization of the Prandtl-Reuss J_2 flow theory which uses von Mises yield function, Mohr Coulomb is a generalization of the Tresca yield criterion, and as such has discontinuities in slope. It is a criterion which is independent of the intermediate principal stresses σ_{II} and may be written in the following form representing the maximum shear stress τ as a function of the average of the largest σ_I and smallest, σ_{III} principal stresses, $\bar{\sigma}$ (See Fig. 5).

$$\tau = c - \bar{\sigma} \tan \phi \tag{32}$$

in which c is the cohesion and ϕ is the friction angle of the material $(\sigma_I \geq \sigma_{II} \geq \sigma_{III})$.

$$\tau = (\sigma_I - \sigma_{III})/2\cos\phi = R/\cos\phi \tag{33}$$

$$\bar{\sigma} = (\sigma_I + \sigma_{III})/2 \tag{34}$$

R is the radius of the Mohr's circle corresponding to the two extreme principal stresses and $\bar{\sigma}$ is the average of these two principal stresses.

The corresponding yield function and plastic potential are

$$f = R + \bar{\sigma} \sin \phi - 2c \cos\phi \tag{35}$$

$$g = R + \bar{\sigma} \sin \phi' - 2c' \cos \phi' \tag{36}$$

ϕ and ϕ' are the mobilized friction angle and the dilatancy angle respectively (Vermeer[8])

The partial derivatives required for Eqn. 16 are readily expressed as a function of the eigenvalue sensitivity which depend on the particular eigenvector, $\{v\}$

$$\partial\sigma_I/\partial\sigma_{ij} = \{v_I\}'[\partial\sigma/\partial\sigma_{ij}]\{v_I\} \tag{37}$$

in which the eigenvectors are orthonormal.

$$[[\sigma] - \sigma_I[I]]\ \{v_I\} = \{0\} \tag{38}$$

$$\{v_J\}^T\{v_I\} = \begin{cases} 1 & J=I \\ 0 & J \neq I \end{cases} \tag{39}$$

EXAMPLES

Material instability as determined here is based on a homogeneous stress field with no geometric imperfections. Small motion in the form of a plane wave is assumed to propagate along a particular direction. These motions are small and are superimposed on an existing displacement field possibly having large stresses and finite strains. Using an objective stress rate and incremental laws of elasto-plasticity incorporating hardening and non-normal plastic deformation, material instability is characterized by a stationary wave, i.e. one with zero wave speed for a particular combination of material parameters, existing stresses, and direction of motion. By setting the determinant of the [B] matrix equal to zero, this yields a numerical procedure for determining critical hardening and the corresponding orientation of the shear band for a given stress state and material parameters under the assumption of continued plastic flow. The examples in this section demonstrate three aspects of this approach; three dimensionality, finite strains, and proper constitutive modeling. The loading is assumed monotonic in all three examples, and, the stress level used for instability in the last two examples corresponds to first yield.

Biaxial Loading of a Hypoelastic Material

We will first consider an element loaded in the 1, 2 plane as shown in Fig. 3b with the following stress ratio (σ_3 = 0)

$$\sigma_1 = \sigma \tag{40}$$

$$\sigma_2 = m\sigma \tag{41}$$

$$-1 \leq m \leq 1 \tag{42}$$

and satisfying a linear incremental relationship with $L_{ijk\ell} = E_{ijk\ell}$ in Eqns. 16 and 17. The matrix [B] for this case reduces to the following

$$[B] = \begin{bmatrix} (\Lambda+G)n_1^2+G & (\Lambda+G)n_1n_2 & (\Lambda+G)n_1n_3 \\ +\sigma/2(n_1^2+mn_2^2-1) & +\sigma/2(1-m)n_1n_2 & +\sigma/2n_1n_3 \\ & & \\ (\Lambda+G)n_1n_2 & (\Lambda+G)n_2^2+G & (\Lambda+G)n_2n_3 \\ +\sigma/2n_1n_2(m-1) & +\sigma/2[n_1^2+m(n_2^2-1)] & +\sigma/2mn_2n_3 \\ & & \\ (\Lambda+G)n_1n_3 & (\Lambda+G)n_2n_3 & (\Lambda+G)n_3^2+G \\ -\sigma/2n_1n_3 & -\sigma/2mn_2n_3 & +\sigma/2(n_1^2+mn_2^2) \end{bmatrix} \tag{43}$$

The determinant of this matrix depends on the material parameters Λ nd G, on the stress level σ and ratio m, and lastly on the three direction cosines, n_1, n_2, and n_3. For the three specific directions along the three axis $n_1 = 1$, $n_2 = 1$ and $n_3 = 1$ it reduces to the following three special cases

$$|B_1| = \begin{vmatrix} \Lambda+2G & 0 & 0 \\ 0 & G+\sigma(1-m)/2 & 0 \\ 0 & 0 & G+\sigma/2 \end{vmatrix} \quad \text{for } n_1=1 \tag{44}$$

$$|B_2| = \begin{vmatrix} G+\sigma(m-1)/2 & 0 & 0 \\ 0 & \Lambda+2G & 0 \\ 0 & 0 & G+m\sigma/2 \end{vmatrix} \quad \text{for } n_2=1 \tag{45}$$

$$|B_3| = \begin{vmatrix} G-\sigma/2 & 0 & 0 \\ 0 & G-m\sigma/2 & 0 \\ 0 & 0 & \Lambda+2G \end{vmatrix} \quad \text{for } n_3=1 \tag{46}$$

These are zero for the following stress levels

$$\sigma = \begin{bmatrix} -2G/(1-m) & \text{for } n_1 = 1 \\ -2G & \end{bmatrix} \tag{47}$$

$$\sigma = \begin{bmatrix} -2G/(m-1) & \text{for } n_2 = 1 \\ -2G/m & \end{bmatrix} \tag{48}$$

$$\sigma = \begin{bmatrix} 2G & \text{for } n_3 = 1 \\ 2G/m & \end{bmatrix} \tag{49}$$

These are plotted in Figure 6 along with the corresponding Mohr's circle ($\sigma_3 = 0$ in all cases). For biaxial compression in the fourth quadrant, the stability boundary is given by $n_1 = 1$ i.e. the shear band is perpendicular to the direction of loading and is independent of the ratio of second stress, m. For biaxial tension i.e. the first quadrant $n_3 = 0$ is critical suggesting that the shear band is in the plane of loading with unit normal out of the plane,(Nemat-Nasser[9].) In uniaxial tension, the critical stress is at 2G and not at 3G as determined by the one dimensional approach (see Eqn. 11). The lowest load ±G is for pure shear ($\sigma_2 = -\sigma_1$). In all instances the plane of the shear band does not have a normal perpendicular to the direction of the intermediate principal stress. Inspection of Eqns. 44-46 for the case of small stress ($\sigma \simeq 0$) leads to the well known velocities for three dimensional solids

$$c_1 = \sqrt{(\Lambda+2G)/\rho} \tag{50}$$

$$c_2 = c_3 = \sqrt{G/\rho} \tag{51}$$

all of which are different than the velocity based on a one dimensional propagation along a rod, Eqn. 8, $c = \sqrt{E/\rho}$.

Uniaxial Strain of Drucker-Prager Material

As stated in (Rudnicki and Rice[4],) the approximate expression for critical hardening for Drucker-Prager material, Eqn. 28, is valid only when terms in Eqn. 15 due to stress are small relative to those in the constitutive tensor L. Earlier results based on stationary wave analysis confirmed this result(Beliveau & Ameziane-Hassani[3].)

Geologic materials in earth's gravity are loaded in a state of uniaxial strain ($\varepsilon_2 = \varepsilon_3 = 0$) there being no strain in both horizontal directions. Based on stationary waves, critical hardening was determined for this loading case and compared to that predicted by Eqn. 28. These results are plotted in Fig. 7 for $\beta = +.1$ and $\nu = .2$ with a good comparison over a large range of α, but with large discrepancies as α gets closer to the value $\alpha = .29$.

This difference is explained by the fact that at this value of the friction angle, stresses required to bring the material to yield are large since for a uniaxial compression yield ($\sigma_2 = \sigma_3 = 0$) equal to P_y, the required uniaxial compressive load in uniaxial strain is given by

$$p = p_y \frac{(1-\nu)(1-\alpha\sqrt{3})}{(1-2\nu)-\alpha\sqrt{3}(1+\nu)} \tag{52}$$

which for $\nu = .2$ requires infinite stresses at α given by

$$\alpha = (1-2\nu)/\sqrt{3}/(1+\nu) = .29 \tag{53}$$

Thus terms proportional to stress in Eqn. 14 are not small compared to elastic moduli and Eqn. 28 is no longer valid.

Drucker-Prager Approximation of Mohr-Coulomb

Under various loading situations, parameters α and k of the Drucker-Prager model may be determined to approximate a Mohr-Coulomb type of material with parameters c and ϕ. Two special cases are when the Drucker-Prager cone circumscribes the Mohr-Coulomb hexahedron along the compressive meridians of the π plane, and when it matches the tension meridians,(Chen[5].)

$$\alpha = \begin{cases} 2\sin\phi/\sqrt{3}/(3-\sin\phi) & \text{compression} \quad (54) \\ 2\sin\phi/\sqrt{3}/(3+\sin\phi) & \text{tension} \quad (55) \end{cases}$$

$$k = \begin{cases} 6c\cos\phi/\sqrt{3}/(3-\sin\phi) & \text{compression} \quad (56) \\ 6c\cos\phi/\sqrt{3}/(3+\sin\phi) & \text{tension} \quad (57) \end{cases}$$

Equivalent Drucker-Prager materials are based on Eqn. 54-57. With $c = 40$, $\nu = .2$, and the normality assumption ($g = f$ of Eqns. 35 and 36), results for critical hardening assuming the corresponding inclination of the shear plane, $n_3 = 0$ and $N = n_1/n_2$ are given in Fig. 8. These clearly show that although parameters of Drucker-Prager model may be determined such that yield for a Mohr-Coulomb material is bounded by pure tension and pure compression, this is not so for critical hardening or the corresponding orientiation of the shear band. This is due to the very different direction of plastic flow of the two criteria.

CONCLUSION

Stationary wave propagation has proved to be a useful and accurate criteria to investigate local material instability in the form of shear bands, using general three dimensional incremental formulation of the constitutive equations, i.e. the flow equations of elasto-plasticity. The results obtained confirm earlier results based on quasi-static analysis. In this study it has been shown that in this approach (1) a three dimensional formulation must be used, (2) the effect of large stresses may be important under certain loading conditions thus necessitating objective rates of stresses, (3) a proper model of the constitutive equations is required to determine instability and not an approximate model based on matching only a yield criterion. Local material instability including the orientation of the shear band is a function of the material parameters and stress level. Conditions of loading such as plane strain affect only the level of stress and strain which then affect the stability in an implicit manner.

Measurement of wave speeds could be a convenient means of determining material parameters and more importantly in predicting and possibly preventing or enhancing material instability. Of course, a more complete analysis would include the effect of strain rate and temperature as well as hardening. The form of the equations used here is nevertheless general and includes phenomena typical of geologic materials such as compressibility, non associative plastic flow, and hardening.

ACKNOWLEDGEMENTS

The authors wish to thank the National Science and Engineering Research Council of Canada for supporting this work, under Grant No. A-9306.

REFERENCES

1. Ameziane-Hassani, H. (1985) Etude de la localization de la deformation dans les matériaux elastoplastiques par ondes stationaires, Masters Thesis Université de Sherbrooke.

2. Hill, R. (1962) Acceleration Waves in Solids in Journal of the Mechanics and Physics of Solids, Vol. 10, pp. 1-16.

3. Beliveau, J-G and Ameziane-Hassani, H., (1985) Shear-Band Instability of Non-Associative Elasto-Plastic Materials by Stationary Wave Analysis, in Material Nonlinearity in Vibrations Problems, AMD Vol. No. 71, ASME, pp. 1-8.

4. Rudnicki, J.W. and Rice, J.R., (1975) Conditions for the Localization of Deformation in Pressure-Sensitive Dilatant Materials, in Journal of the Mechanics and Physics of Solids, Vol. 23, pp. 371-394.

5. Chen, W.F., (1982) Plasticity in Reinforced Concrete, McGraw-Hill, New York.

6. Semiatin, S.L. and Jonas, J-J. (1984) Formability and Workability of Metals: Plastic Instability and Flow Localization, A.S.M., Metals Park, Ohio.

7. Desrues, J., (1984) La localization de la deformation dans les materiaux granulaires, Thesis for Doctor of Science, Universite Scientifique et medicale de Grenoble.

8. Vermeer, P.A., (1982) A Simple Shear-Band Analysis Using Compliances, in Conference on Deformation and Failure of Granular Materials, IUTAM, Delft, pp. 493-499.

9. Nemat-Nasser, S. (1979), Finite Deformation Plasticity and Plastic Instability, 25th Conference of Army Mathematicians, Baltimore.

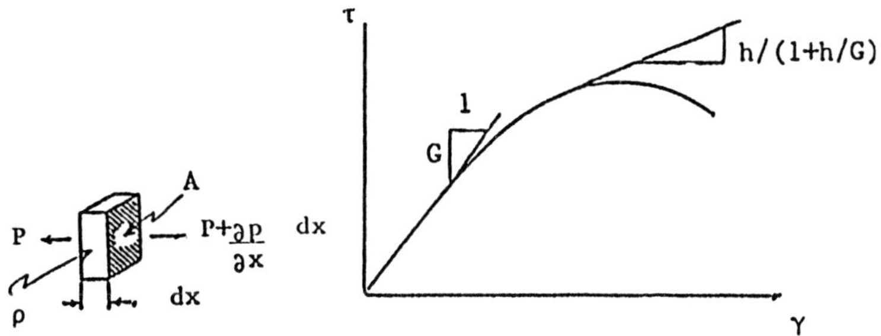

Figure 1. One Dimensional Wave Propagation

Figure 2. Strain Hardening for Constant Normal Stress (σ = constant)

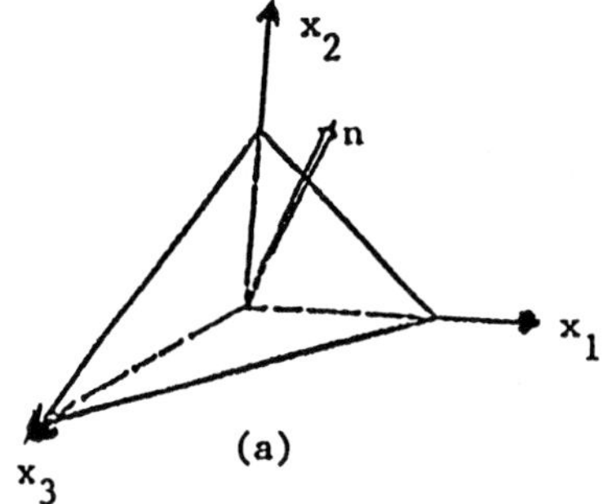

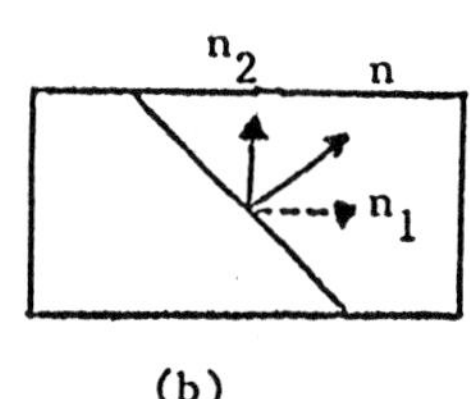

Figure 3. Geometry of Shear Band (a) Three Dimension
(b) Plane (n_3 = 0)

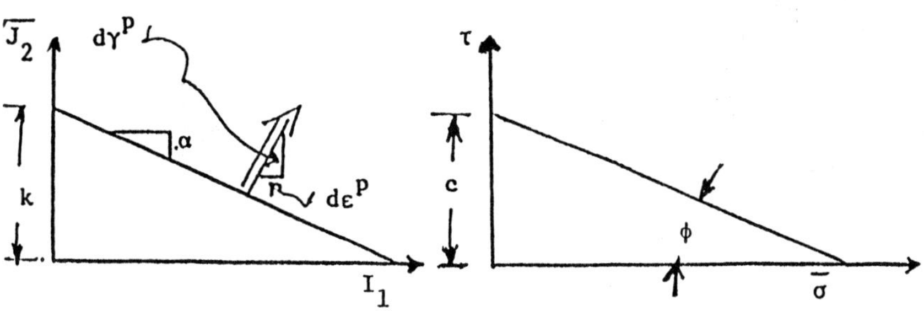

Figure 4. Drucker-Prager Model

Figure 5. Mohn-Coulomb Yield

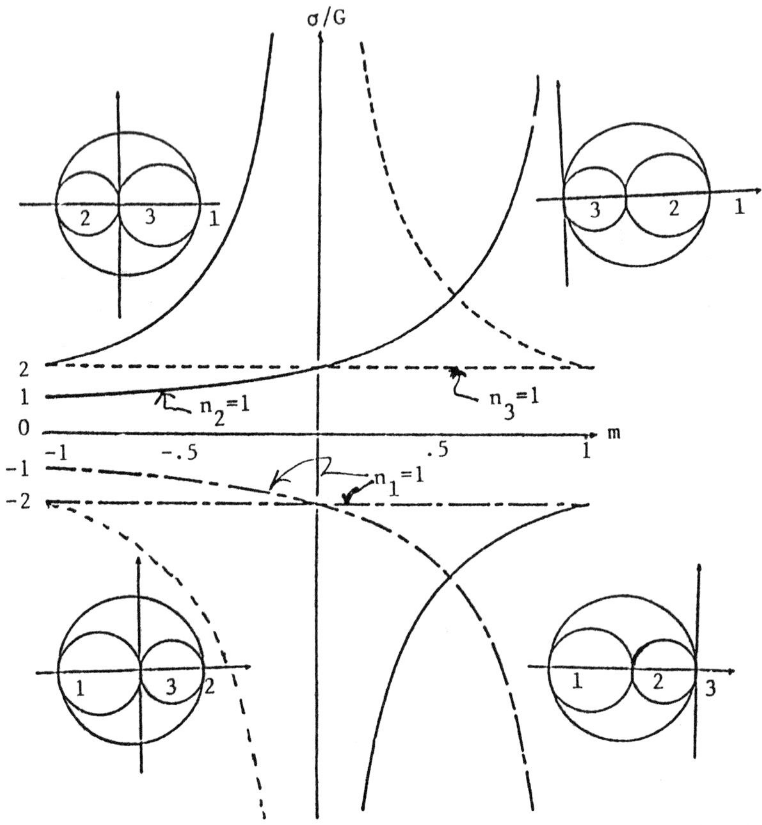

Figure 6. Stability Boundaries for Hypoelastic Material in Biaxial Stress ($\sigma_2 = m\sigma$, $\sigma_1=\sigma$, $\sigma_3=0$) ($n_1=1$, ←-- ⟶; $n_2=1$, ⟶; $n_3=1$, ----)

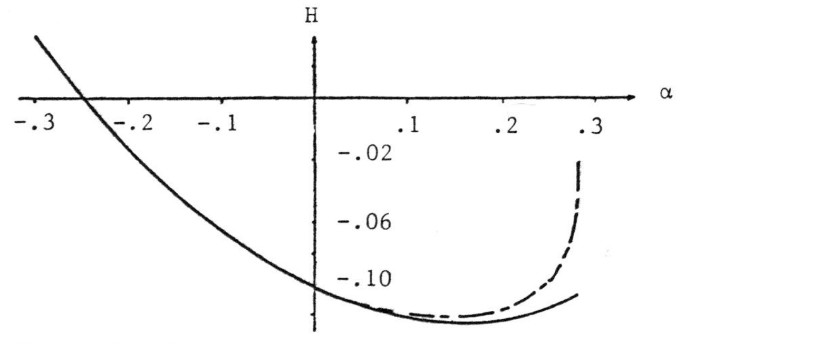

Figure 7. Critical Hardening for Uniaxial Strain of Drucker-Prager Material (Eqn. 27, — —; Eqn. 28 —)

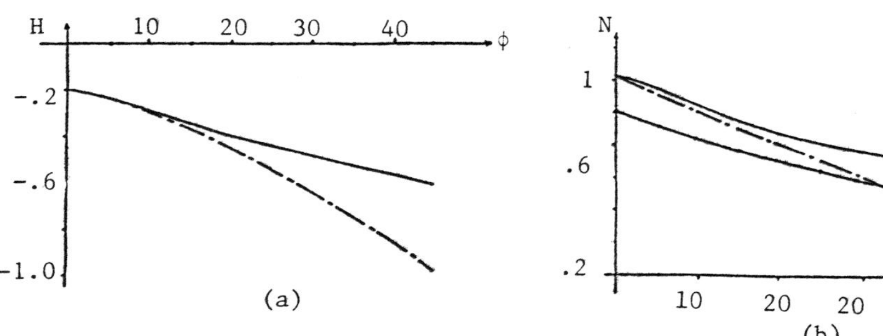

Figure 8. Comparison of Mohr-Coulomb with Equivalent Drucker-Prager
(a) Critical Hardening, (b) Direction Cosine
(D-P Tension, —; D-P Compression, — - —; Mohr-Coulomb, -•-)

A Comparison of Published Ground Motion Parameters

A. Kropp
Alan Kropp & Associates, Inc., U.S.A.

INTRODUCTION

Extensive research has been performed in the United States in the last 15 to 20 years to evaluate strong ground motion characteristics. As summarized by Campbell (Campbell[1]), at least 30 strong motion attenuation relationships have been published in the last 10 years. Other researchers have developed their own attenuation relationships and then applied them to the ground response for particular earthquakes (USGS[2]). Often there is considerable variation in the assumptions used to produce the published relationships or ground response maps because of the complex interaction of the many factors which influence the ground motion characteristics at a site. These factors include: (1) source conditions (type, stress conditions, rupture dimensions and fault slip, complexity of source, depth, seismic wave radiation patterns, and seismological phases), (2) travel path characteristics (variations in geology, incoherency, phase conversion, scattering and geometrical spreading), and (3) local conditions (topography and site soils) (EERI[3]). The purpose of this paper is to compare the possible peak strong ground parameters predicted by a number of published relationships and ground response maps for two sites in the San Francisco Bay Area.

BAY AREA FAULTS

The San Francisco Bay Area is one of the most active seismic regions in the United States. The significant earthquakes which occur in the Bay Area are generally associated with crustal movements along well defined faults which trend in a northwesterly direction. A fault map of the Bay Area is presented on Figure 1. The estimated maximum magnitude and recurrence interval for the maximum event (based on estimated slip rate) for the labelled faults shown on Figure 1 are summarized on Table 1. It should be noted that there is considerable variation in the estimated recurrence intervals for Bay Area faults (Anderson[4]).

TABLE 1 - MAXIMUM MAGNITUDE AND RECURRENCE INTERVAL OF BAY AREA FAULTS (From ABAG[5])

Fault Name	Maximum Magnitude	Recurrence Interval (Years)
Antioch	6.4	*
Calaveras	7.0	100
Concord	7.3	200
Green Valley	7.0	200
Hayward	6.9	200
Healdsburg Rogers Creek	6.8	200
San Andreas	8.4	100

* Slip rate assumed very low (equal to or less than 0.1 cm/yr); yields recurrence interval for even moderate earthquakes in the thousands of years.

Preliminary seismic exposure studies were performed for two sites in the Bay Area. These two sites are shown on Figure 1. Site 1 is located in Hayward, near the center of the Bay Area fault activity. Site 2 is located on Bethel Island, east of general Bay Area, and on the fringe of the major fault activity. Thus, a comparison of results of published peak ground motion relationships applied to a central Bay Area and another comparison of applying these relationships to a site on the fringe seems to be an appropriate test of the consistency of the published relationships and maps over a broad range of seismicity.

SITE 1

Site 1 is located on the eastern margin of the San Francisco Bay. Soft marshland deposits and Bay Mud underlie the site; these materials are up to about 15 feet thick. Stiff sandy clay layers with occasional sand layers are present below these soft soils and extend down to Franciscan Formation bedrock. The bedrock is believed to be present about 1000 to 1500 feet below the site. As shown on Figure 1, Site 1 is located between two of the Bay Area's most active faults, the San Andreas fault and the Hayward fault. The San Andreas fault passes about 14 miles southwest of Site 1 while the Hayward fault passes about 4 miles to the northeast. Two of the largest historic earthquakes in the Bay Area have impacted this site according to historic records. The Hayward Earthquake of 1868 had a magnitude of 6.7 and it has been reported that "At Hayward, almost every building was severely damaged, many (were) completely wrecked" (Coffman, et al[6]). During the 1906 San Francisco Earthquake, numerous instances of damage were reported just south or southeast of the site (Youd and Hoose[7]; Nason[8]).

In order to provide a preliminary estimate of the level of shaking that might occur at the ground surface or at the bedrock below Site 1, various estimates of shaking from the 1868 and 1906 earthquakes were obtained and compared with estimates from published attenuation relationships and ground response maps. These estimates of peak strong motion parameters are presented on Table 2. It appears that the maximum earthquake on the Hayward fault will produce larger peak motions than the maximum event on the San Andreas fault. During the maximum earthquake on the Hayward fault, Modified Mercalli Intensity levels of VIII to IX might occur, with peak ground surface accelerations of 0.42g to 0.50g.

SITE 2

Site 2 is located in the Sacramento-San Joaquin Delta region, on the eastern edge of the greater Bay Area. This site is underlain by about 50 feet of eolian (wind blown) sand deposits and then about 500 feet of heterogeneous layers of silt, sand and clay. These soils are underlain in turn by about 15,000 to 30,000 feet of sedimentary materials deposited in marine environments about 25 to 175 million years ago. The site is generally located a considerable distance from the most active Bay Area faults (see Figure 1). Therefore, to assess the relative level of shaking that might be produced at the site by many of the Bay Area faults, the peak bedrock acceleration below the site caused by the maximum earthquake on several faults was estimated using the attenuation relationships presented by Seed and Idriss[13]. The results of this comparison are presented on Table 3. Although the largest bedrock acceleration was caused by the Antioch fault, this fault has a very long recurrence interval (see Table 1) and it was assumed the Concord fault posed the greatest threat to the site during a reasonable recurrence interval.

Preliminary estimates of peak strong motion parameters for Site 2 were developed in a manner similar to those computed for Site 1. An assumption of a 500 year return period was used for Thenhaus, et al[17] and Kiremidjian and Shah[18] because that was assumed to be a reasonable recurrence interval for the maximum design level earthquake. The estimated peak values are presented on Table 4. From this table, it appears that during the maximum design earthquake, Modified Mercalli Intensity levels of V to VI might occur, with peak ground surface accelerations of 0.17g to 0.20g.

TABLE 2 - PEAK STRONG MOTION PARAMETERS - SITE 1

Parameter	Author (s)	Value	Fault Causing Earthquake	Conditions Assumed
Modified Mercalli Intensity	Lawson et al.[9]	VIII-IX	San Andreas	From Intensity Map of 1906 Earthquake; Converted from Rossi-Forel Intensity 9
	Borcherdt et al.[10]	IX	San Andreas or Hayward	San Francisco Intensity B Converted to Modified Mercalli Intensity
	ABAG[5]	VIII	Any Bay Area Fault	San Francisco Intensity C Converted to Modified Mercalli Intensity
	USGS[2]	VIII	San Andreas	From Intensity Map of Earthquake Equivalent to 1906 Event; Converted from Rossi-Forel Intensity 8
	USGS[2]	IX	Hayward	From Intensity Map of Earthquake Equivalent to 1868 Event; Converted from Rossi-Forel Intensity 9
	Davis et al.[11]	VIII-IX	San Andreas	Rossi-Forel Intensity 9 Converted to Modified Mercalli Intensity
	Toppozada and Park[12]	IX	Hayward	From Intensity Map of 1868 Earthquake

TABLE 2 - (CONTINUED)

Parameter	Author (s)	Value	Fault Causing Earthquake	Conditions Assumed
Peak Bedrock Acceleration	Seed & Idriss[13]	0.41g	San Andreas	Magnitude 8.4 Earthquake 14 Miles Away
	Seed & Idriss[13]	0.55g	Hayward	Magnitude 6.9 Earthquake 4 Miles Away
Peak Surface Acceleration	Campbell[14]	0.42g	Hayward	Mean Value - Magnitude 6.9 Earthquake 4 Miles Away
	Joyner[15]	0.48g	Hayward	Mean Value - Magnitude 6.9 Earthquake 4 Miles Away
	Krinitzsky and Marcuson[16]	0.47g	- -	Near-Field, Large Magnitude Earthquake Producing Modified Mercalli Intensity IX on a Hard Site
	Seed & Idriss[13]	0.50g	Hayward	Bedrock Acceleration Converted to Stiff Soil Acceleration

TABLE 3 - PEAK BEDROCK ACCELERATIONS AT SITE 2 (From Seed and Idriss[13])

Fault Name	Distance Fault (Miles)	Maximum Magnitude	Peak Bedrock Acceleration (g's)
Antioch	9	6.4	0.31
Calaveras	29	7.3	0.19
Concord	21	7.0	0.22
Green Valley	26	7.0	0.17
Hayward	33	6.9	0.12
Rogers Creek	46	6.8	0.07
San Andreas	50	8.4	0.15

CONCLUSIONS

Sites 1 and 2 possess very different geologic settings and the level of seismicity varies substantially. Nonetheless, both Site 1 and Site 2 each possess a surprising agreement of the estimated peak strong motions even though a wide range of published ground motion relationships were used at each site. This is particularly unusual because many of the estimates used depended on conversion factors from one parameter to another (i.e., Rossi-Forel Intensity to Modified Mercalli Intensity), and compounded errors can be introduced if more than one conversion is made (as with the value for Krinitzsky and Marcuson[16]).

TABLE 4
PEAK STRONG MOTION PARAMETERS - SITE 2

Parameter	Author (s)	Value	Conditions Assumed
Modified Mercalli Intensity	ABAG[5]	VI	San Francisco Intensity E Converted to Modified Mercalli Intensity.
	USGS[2]	V-VI	Rossi-Forel Intensity 6 to 7 Converted to Modified Mercalli Intensity for Maximum Event on Hayward Fault.
Peak Bedrock Acceleration	Seed & Idriss[13]	0.22g	Magnitude 7.0 Earthquake on Concord Fault 21 Miles Away
	Thenhaus, et al[17]	0.25g	For Return Period of 500 Years.
Peak Surface Acceleration	Seed & Idriss[13]	0.19g	Bedrock Acceleration Converted to Deep Cohesionless Soil Deposit Acceleration.
	Joyner & Boore[15]	0.20g	Mean Value - Concord Earthquake
	Campbell[14]	0.18g	Mean Value - Concord Earthquake
	Kiremidjian & Shah[18]	0.18g	For Return Period of 500 Years.
	Krinitzsky & Marcuson[16]	0.17g	Mean Value for Near-Field, Soft Site with Modified Mercalli Intensity of VI.

REFERENCES

1. Campbell, Kenneth, 1985, "Strong Motion Attenuation Relationships; A Ten-Year Perspective," Earthquake Spectra, Vol. 1, No. 4.

2. United States Geological Survey, 1981," Scenarios of Possible Earthquakes Affecting Major California Population Centers, with Estimates of Intensity and Ground Shaking", Open File Report 81-115.

3. Earthquake Engineering Research Institute, 1986, Reducing Earthquake Hazards: Lessons Learned From Earthquakes, Publication No. 86-02.

4. Anderson,, John G., 1984, "Synthesis of Seismicity and Geological Data in California," U.S. Geological Survey, Open File Report 84-424.

5. Association of Bay Area Governments, 1982 (Revised 1983), "Using Earthquake Intensity and Related Damage to Estimate Earthquake Intensity and Cumulative Damage Potential from Earthquake Ground Shaking", Working Paper #17

6. Coffman, Jerry L., Von Hake, Carl A., and Stover, Carl W., 1982, "Earthquake History of the United States," U.S. Department of Commerce and U.S. Geological Survey, Publication 41-1.

7. Youd, T.L., and Hoose, S.N., 1978, "Historic Ground Failures in Northern California Triggered by Earthquakes," U.S. Geological Survey, Professional Paper 993.

8. Nason, Robert, 1982, "Damage in Alameda and Contra Costa Counties, California, in the Earthquake of 18 April 1906", U.S. Geological Survey, Open File Report 82-63

9. Lawson, Andrew C., 1908, "The California Earthquake of April 18, 1906 - Report of the State Earthquake Investigation Commission," published by the Carnegie Institution of Washington.

10. Borcherdt, R.D., Gibbs, J.F., and Lojoie, K.R., 1975, "Maps Showing Maximum Earthquake Intensity Predicted in the Southern San Francisco Bay Region, California, for Large Earthquakes on the San Andreas and Hayward Faults," U.S. Geological Survey, Map MF-709.

11. Davis, James F., Bennett, John H., Borcherdt, Glenn A., Kahle, James E., Rice, Salem J., Silva, Michael A., 1982, "Earthquake Planning Scenario for a Magnitude 8.3 Earthquake on the San Andreas Fault in the San Francisco Bay Area," California Division of Mines and Geology, Special Report 61.

12. Toppozada, T.R., Real, C.R., Bazore, S.P., and Park, D.L., 1979, "Compilation of Pre-1900 California Earthquake History," California Division of Mines and Geology, Open File Report 70-6.

13. Seed, H. Bolton, and Idriss, I.M., 1983, Ground Motions and Soil Liquefaction During Earthquakes, Earthquake Engineering Research Institute.

14. Campbell, Kenneth, 1981, "Near-Source Attenuation of Peak Horizontal Acceleration", Bulletin of the Seismological Society of America, Volume 71, Number 6.

15. Joyner, William B., and Boore, David M., 1981, "Peak Horizontal Acceleration and Velocity from Strong - Motion Records Including Records from the 1979 Imperial Valley, California, Earthquake", Bulletin of the Seismological Society of America, Volume 71, Number 6.

16. Krinitzsky, E.L., and Marcuson, W.F., III, 1983, "Considerations in Selecting Earthquake Motions for the Engineering Design of Large Dams", U.S. Geological Survey, Open File Report 83-636.

17. Thenhaus, Paul C., Perkins, David M., Ziony, Joseph I., and Algermissen, S.T., 1980, "Probabilistic Estimates of Maximum Seismic Horizontal Ground Motion on Rock in Coastal California and the Adjacent Outer Continental Shelf," U.S. Geological Survey, Open File Report 80-924.

18. Kiremidjian, A.S. and Shah, H.C., 1978 "Seismic Risk Analysis for California State Water Project," Department of Water Resources, State of California, J.A. Blume Earthquake Center, Report No. 33.

19. Brown, Robert D., Jr., and Kockelman, William J., 1983, "Geologic Principles for Prudent Land Use," U.S. Geological Survey, Professional Paper 946.

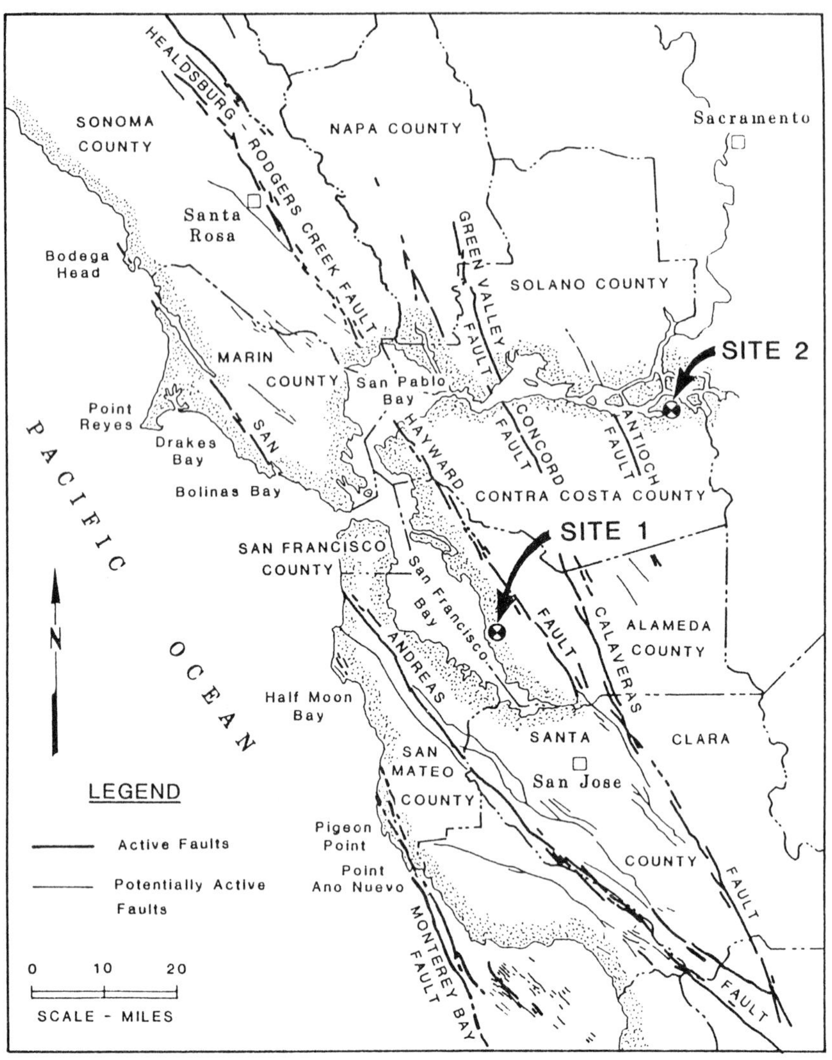

Figure 1 – Bay Area Fault Map (after Brown and Kockleman[19])

Shear Wave Induced Motions in Recorded Accelerograms

S. Sharma
Department of Civil Engineering, University of Idaho, Moscow, U.S.A.
J.L. Chameau
School of Civil Engineering, Purdue University, West Lafayette, U.S.A.

INTRODUCTION

In current earthquake engineering practice, soil structure interaction (SSI) analyses are performed using input excitation that simulates vertically propagating seismic waves. This input consists of accelerograms depicting horizontal or vertical motions to simulate the effects of shear (SV, SH) or compressional (P) seismic waves, respectively. Often, time histories recorded during previous earthquakes are be used for the SSI analysis. However, these recorded accelerograms represent the combined contribution of vertical and inclined P, SH and SV-waves, as well as surface Rayleigh and Love waves arriving at the recording station.

The three component accelerogram which is recorded at a strong motion accelerograph (SMA) station will thus implicitly include the contribution of all seismic waves arriving from many directions. Subsequent examination of such records has provided details regarding the relative contribution of shear waves. Evidence for the existence and importance of body and surface waves has been reported by many researchers. One of the first reports was based on a study by Housner[1] which examined response of structures during the 1952 Kern County earthquake, and presented strong evidence showing the presence of different wave types. Supporting evidence has been provided by Anderson[2], Bolt[3], Boore and Zoback[4], Bouchon and Aki[5], Crouse[6], Duke, et al.[7], Hanks[8], Swanger and Boore[9], Trifunac[10], Yamahara[11]. Thus, it is not possible to ignore the existence of motions in SMA records due to non-vertically propagating waves. The significance of these wave types for the design of structures has been discussed in several papers such as Gazetas and Yegian[12], Gomez-Masso, et al.[13] and Scanlan[14].

In this paper a technique for the decomposition of SMA records into their contributing wave types is discussed in order to assess the current design philosophies. The decomposition may be made into body and surface waves using the theory of wave propagation in an elastic half space. Due to the limited scope of this paper, only features relevant to SV shear waves will be discussed but interested readers may refer to Sharma and Chameau[15] for a general presentation applicable to other wave types.

THEORY

The proposed filtering technique is based on the solution of differential equations formed for a harmonic vibratory source in an elastic half space. The formulation is simplified by rotating the horizontal components into radial (R) and transverse (T) directions based on the epicenter and recording station azimuth. For such a coordinate transformation, the P, SV and R-wave motions will be confined to the Z-R plane while the SH (and Love) wave motions occur in the transverse plane.

The differential equations of motion can be formulated for harmonic waves in an elastic half space and their solution expressed as a function of the surface amplitudes of incident

SV-waves. The main features relevant to SV waves is the ratio of the radial and vertical components at the free surface. For incident SV-waves, the ratio is given as a function of the incident angle:

$$\left[\frac{z_o}{r_o}\right]_{SV} = \frac{-2k_p}{1 - k_s^2} \tag{1}$$

where :

r_o = radial displacement at the surface

z_o = vertical displacement at the surface

$$k_p = -i\left[1 - \frac{V_a^2}{V_p^2}\right]^{0.5}$$

$$k_s = -i\left[1 - \frac{V_a^2}{V_s^2}\right]^{0.5}$$

V_p = P-wave velocity

V_s = SV-wave velocity

V_a = apparent wave velocity, $V_a = V \sec(\psi)$

i = the complex variable, $\sqrt{-1}$

ψ = angle of incidence (from horizontal)

Figure 1 illustrates this ratio graphically, showing the expected range of the vertical to radial components for different angles of incidence and Poisson's ratio values of 0.25 and 0.45. For SV-wave motion there are three distinct regions with contrasting z_o/r_o ratios according to the angle of incidence.

The amplitude ratio for $\psi_{SV} < 45°$ is a positive, imaginary value which ranges between about 1.65 and ∞ for angles of incidence of 0° and 45°, respectively. The remaining two regions are delineated according to the critical angle, ψ_c, defined as the smallest angle at which incident SV-waves will generate both reflected P and SV-waves at the boundary of the half space. The critical angle is a function of Poisson's ratio, ranging between 54° and 72° for most geological media. The z_o/r_o ratio for the region $45° \leq \psi_{SV} \leq \psi_c$ is an imaginary value ranging between minus infinity and zero for angles of incidence of 45° and ψ_c, respectively. For angles $\psi_{SV} \geq \psi_c$, the amplitude ratio is a real value in the range 0 - 0.65.

In summary, if the z_o/r_o ratio can be computed for the surface motions caused by incident waves, it is possible to determine the nature of the seismic wave type. For example, if the z_o/r_o ratio is a negative, imaginary number, it is possible to conclude that the motions were induced by SV-waves incident at an angle in the $45° \leq \psi_{SV} \leq \psi_c$ range. However, although an exact ψ_{SV} value can be computed for a "pure" harmonic SV-wave, in general the exact angle cannot be determined for non-harmonic motions of the form recorded during earthquakes. Thus, it is possible to assess whether the motions have been induced by one of the following three waves types according to the angle of incidence :

1. Incident SV-waves, $\psi_{SV} \leq 45°$ (SV1-waves)
2. Incident SV-waves, $45° \leq \psi_{SV} \leq \psi_c$ (SV2-waves)
3. Incident SV-waves, $\psi_{SV} \geq \psi_c$ (SV3-waves)

These wave types can be determined from recorded motions using a filter which either enhances or attenuates motions according to how closely their amplitude ratios compare with typical values. The filter operators are designed to operate in the frequency domain according to an approach suggested by Kanasewich[16] and further extended by Sharma and Chameau[15]. The transformation from the recorded time domain to the frequency domain may be achieved using the Fast Fourier Transform (Cooley, et al.[17]). Operations in the frequency domain allow the use of solutions developed for harmonic loads to be applied to recorded accelerograms.

The filter operates on a small time segment, $N\delta t$, where δt is the time interval between two discrete points. For each such segment, the amplitude and phase of each harmonic are computed for each component of the transformed ground motion (i.e. Z, R, T). Figure 2 displays the vector representation, in three dimensional space of the three Fourier amplitudes, A_i (where i corresponds to either Z, R or T), determined from the transform. The two angles, β and ξ define the vector in space and may be calculated for each discrete frequency, nf. The angle ξ, gives the horizontal azimuth, while β describes the eccentricity between the vertical and horizontal components of surface motion. Another important angle in characterizing the wave content is α, the phase difference between the vertical and radial components. This angle is readily computed from the phase angles of the respective harmonics of the transformed R and Z components.

Table 1, Summary of Theoretical Filter Operators

WAVE TYPE	α^*	β^*	ξ^*
SV1-waves	$\pi/2$	0 - 1.23	0.0
SV2-waves	$-\pi/2$	0 - 0.52	0.0
SV3-waves	0.0	0 - $\pi/2$	0.0

$^*\alpha$, β and ξ values are in radians

The values of α, β and ξ expected for the three SV-wave types under consideration are presented in Table 1. These values are based on the phase differences and amplitude ratios, such as given in Figure 1, for typical values of Poisson's ratio. From this table, the typical values of the three angles for each wave type can be compared directly. The filtering technique makes use of these values to discriminate between ground motion components by "weighting" the computed Fourier amplitudes of the original time histories.

The filter operators (or transfer functions) have been designed to modify the original amplitudes A_Z, A_R and A_T, according to the general form :

$$A_i^* = A_i \cdot [F_{ij}^{(1)} \cdot F_{ij}^{(2)} \cdot F_{ij}^{(3)}] \tag{2}$$

where :

i = Z, R, T

j = P, SV1 SV2, SV3, R and SH

A_i = Original Fourier Amplitude

A_i^* = Modified Fourier Amplitude

$F_{ij}^{(1)}$ = Filter operator for phase control

$F_{ij}^{(2)}$ = Filter operator for eccentricity control

$F_{ij}^{(3)}$ = Filter operator for horizontal azimuth control

The filtering operator values range between 1.0 and 0.0 according to how closely the recorded amplitudes resemble the anticipated values (Table 1) based on theory. Thus if the recorded motions display values of α, β and ξ close to the values in the first row of Table 1, the SV1 motion will be preserved. However, if values are drastically different, the SV1 motion will be considerably attenuated. These operators are selected to provide a smooth function over the anticipated range of angles, with a pronounced bias towards the theoretical value.

The phase operators, based on phase angles, are shown in Figure 3 as a function of the computed phase difference between the radial and vertical components. The operator has a maximum value of 1.0 at 0°, 90° and 270° which correspond to in-phase, retrograde, and prograde motions, respectively. Thus during the first step of the filtering scheme, the original amplitudes will be modified according to the type of phase difference being examined with respect to the wave type. The filter operators used in the second step are shown in Figure 4 as a function of the eccentricity angle, β. These values are used to modify the recorded amplitudes in the same manner as the phase operator. Such shapes allow complete preservation in the theoretical range of β-values with steep exponential decay beyond the preferred zones. The final modifying operator provides the horizontal control and is applied to all wave types.

In summary, the proposed approach compares computed and theoretical values of α, β and ξ in the frequency domain and the amplitudes are subsequently modified by the filter operators discussed above. Only the amplitudes are modified, the original phase angles being retained and used for the inverse transform to the time domain. The shapes of the proposed filter operators do not have a theoretical basis and thus cannot be considered to be unique. However, the selected shapes were found to be the most appropriate when the potential effects of heterogeneity are considered. Also, since there is an amplitude reduction for all cases, except for pure wave types, a correction is performed on the filtered amplitudes to ensure that the root mean square values of the combined motions remain constant. This scheme has been implemented into a digital Seismic Wave Filter (SWF) program for examining the wave contents of recorded accelerograms.

PRACTICAL APPLICATIONS

Since seismic waves are continuously arriving at the recording station, the frequency contents will vary with time. This feature is addressed by performing the SWF analysis on small "windows", with lengths ranging between four and six times the longest wave period of interest (Landisman, et al.[18]). Such a window is then moved down the record, repeating the entire procedure, with the final amplitudes being computed as the arithmetic average of all overlapping windows. If the initial and final 10 percent of the points in the window are smoothed by the use of a "cosine-bell" function, the accuracy of the Fourier transform may be improved considerably (Newland[19]).

The recording made at Slack Canyon SMA station, California, during the May 2, 1983, Coalinga earthquake is used to illustrate the application of the proposed techniques for the identification of inclined SV-waves. The location of the recording station and epicenter is shown in Figure 5 (McJunkin and Shakal[20]). The Richter Magnitude 6.5 earthquake was associated with a low angle thrust fault that had a general NW - SE trend, with the southwest side up (Eaton[21]). The Slack Canyon station is 37 km from the epicenter and the focal depth was estimated to be 10.5 km.

The recorded accelerograms were transformed to comply with the radial and transverse directions based on the azimuthal angle of 229° between the epicenter and recording station. In this configuration, the vertical component is positive downwards and the radial and transverse components are positive in azimuthal directions of 229° and 319°, respectively. The time histories of the transformed motions are shown in Figure 6 and have peak accelerations of 52, 167 and 126 cm/s/s in the Z, R, T directions, respectively.

The results of the filtering process performed on the transformed accelerograms are presented in Figure 7, where the original scale of the time histories has been retained for a direct comparison. The time histories displaying SV1-wave motion indicate that negligible energy was recorded by the SMA at Slack Canyon. This may be attributed to the relatively shallow focus of the earthquake. The dominant contributions are made for the SV2 and SV3-wave types, representing incidence angles in the range 45°- 72° and 72° - 90°, respectively. The energy contained in the SV2 and SV3-wave motions is approximately equal. This result shows that the use of the original record to represent vertically propagating shear waves for a SSI analysis would be erroneous as the motions suggest a strong contribution by non-vertical waves, with angles of inclination between 45° and 72°.

CONCLUSIONS

A method which can quantitatively determine the relative contribution of different wave types has been presented. Although this concept is based on the relatively simple half space model, there are plans to extend the procedure to account for layered materials. The results obtained using the SWF technique can be adopted for use in earthquake engineering analyses where the structure is designed to resist a seismic environment consisting of SV1, SV2, SV3 waves in addition to P and R-waves (Sharma and Chameau[15]). Techniques for performing analyses where the seismic environment is represented by a combination of wave types are already available (e.g. Chen[22]).

The SWF technique and the results obtained from its use are some of the first such results published in the literature. However, the "robustness" of this method has not been verified so far, but there are plans to validate the SWF technique by examining data from the SMART 1 array in Taiwan. Once the technique has been reliably validated, a more realistic definition of the seismic environment may be used in earthquake engineering analysis and design.

ACKNOWLEDGEMENTS

The authors wish to acknowledge the financial support provided by the University of Idaho and the NSF's Presidential Young Investigator's Award (Grant # ECE-8451026) to Prof. J.L. Chameau, Purdue University.

REFERENCES

1. Housner, G.N. (1957), Interaction of a Building and Ground during an Earthquake, Bull. Seism. Soc. Am., Vol. 47, No. 3, pp. 179-186.

2. Anderson, J. (1974), A dislocation Model for the Parkfield Earthquake, Bull. Seism. Soc. Am., Vol. 64, No. 3, pp. 671-686.

3. Bolt, B.A. (1981), Interpretation of Strong Ground Motion Records, Misc. Paper S-73-1, Report 17, USAE-WES, Vicksburg, Mississippi, 215 pages.

4. Boore, D.M. and Zoback M.D. (1974), Two Dimensional Kinematic Fault Modeling of the Pacoima Dam Strong Motion Recordings of February 9, 1971, San Fernando Earthquake, Bull. Seism. Soc. Am., Vol. 67, No. 3, pp. 555-570.

5. Bouchon, M. and Aki, K. (1977), Discrete Wave Number Representation of Seismic-source Wave Fields, Bull. Seism. Soc. Am., Vol. 67, No. 2, pp. 259-277.

6. Crouse, C.B. (1973), Engineering Studies of the San Fernando Earthquake, EERL 73-04, Calif. Inst. of Tech., Pasadena.

7. Duke, C.M., Luco, J.E., Carriveau, A.R., Hradilek, P.J., Lastrico, R. and Ostrom, D. (1970), Strong Earthquake Motion and Site Conditions : Hollywood, Bull. Seism. Soc. Am., Vol. 60, pp. 1271-1289.

8. Hanks, T.C. (1975), Strong Ground Motion of the San Fernando, California, Earthquake : Ground Displacements, Bull. Seism. Soc. Am., Vol. 65, No. 1, pp. 193-225.

9. Swanger, H.J. and Boore, D.M. (1978), Simulation of Strong Motion Displacements using Surface Wave Model Superposition, Bull. Seism. Soc. Am., Vol. 68, pp. 247-263.

10. Trifunac, M.D. (1971), Response Envelope Spectrum and Interpretation of Strong Earthquake Ground Motion, Bull. Seism. Soc. Am., Vol. 61, No. 2, pp. 343-356.

11. Yamahara, H. (1970), Ground Motions During Earthquakes and the Input Loss of Earthquake Power, Japan Soc. of Civil Eng., Vol. 10, No. 2, pp. 605-650.

12. Gazetas, G. and Yegian, M.K. (1979), Shear and Rayleigh Waves in Soil Dynamics, ASCE, 106, GT12, pp. 1455-1470.

13. Gomez-Masso, A., Chen, J.C., Pecker, A. and Lysmer, J. (1982), Seismic Pressures on Embedded Structures in Different Seismic Environments, Proc., Conf. on Soil Dynamics and Earthquake Engineering, Southampton, England, pp. 179-191.

14. Scanlan, R.H. (1976), Seismic Wave Effects on Soil Structure Interaction, Earthquake Eng. and Structural Dynamics, 4, pp. 379-388.

15. Sharma, S. and Chameau, J.L. (1987), Relative Wave Contents from Recorded Accelerograms, submitted for publication in the Bull. Seism. Soc. Am.

16. Kanasewich, E.R. (1973), Time Sequence Analysis in Geophysics, The University of Alberta Press, Edmonton, Alberta, Canada.

17. Cooley, J.W., Lewis, P.A.W. and Welch, P.D. (1969), The Fast Fourier Transform and its Applications, IEEE, Trans. on Education, Vol. 12, No. 1, pp. 27-34.

18. Landisman, M., Dziewonski, A. and Sato, Y. (1969), Recent Improvements in the Analysis of Surface Wave Observations, Geoph. Journal, R. Astron. Soc., Vol. 17, pp. 369-403.

19. Newland, D.E. (1984), An Introduction to Random Vibrations and Spectral Analysis, pp. 146-149, Longman Group Ltd. (Pub), London, England, Second Edition.

20. McJunkin, R.D. and Shakal, A.F. (1984), Strong Motion Data, Coalinga, California, Earthquake, in Reconnaissance Report No. 84-03, EERI, pp. 71-88.

21. Eaton, J.P. (1984), Seismic Setting, Location, and Focal Mechanism of the May 2, 1983, Coalinga Earthquake, in Reconnaissance Report No. 84-03, EERI, pp. 18-21.

22. Chen, J.C, Lysmer, J. and SEED, H.B. (1981), Analysis of Local Variations in Free Field Seismic Ground Motion, UCB/EERC-81/03, University of California, Berkeley, 247p.

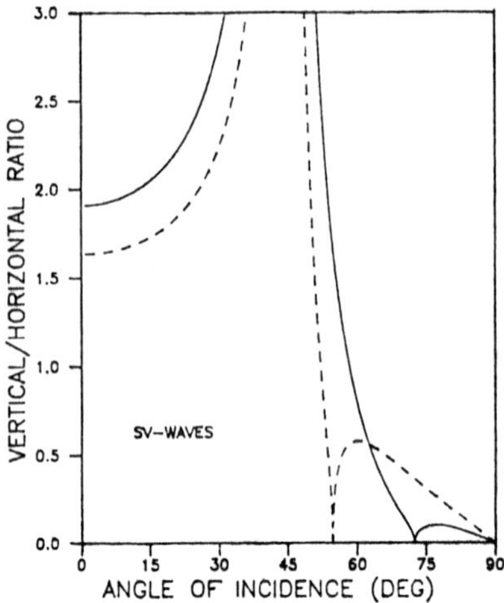

Figure 1, SV-wave Amplitude Ratios

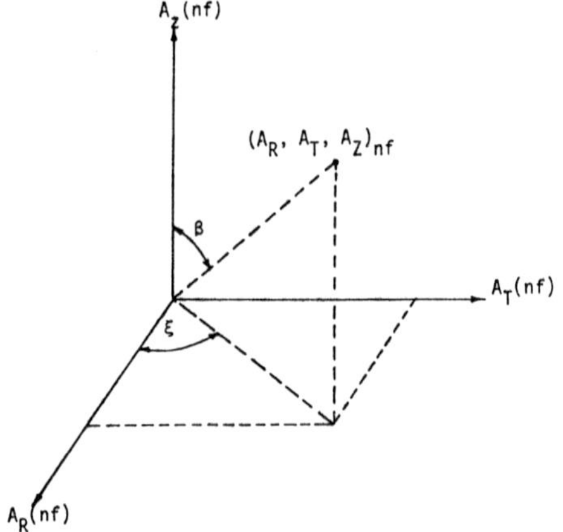

Figure 2, 3-D Component Amplitudes in Frequency Domain

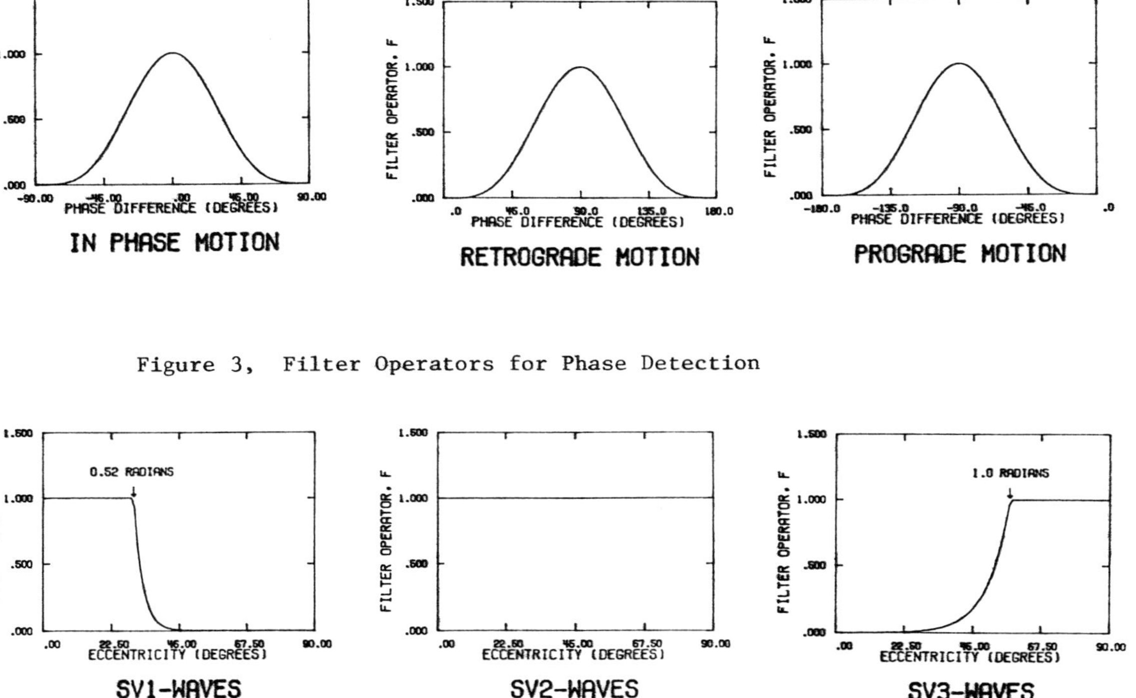

Figure 3, Filter Operators for Phase Detection

Figure 4, Filter Operators for Eccentricity Detection

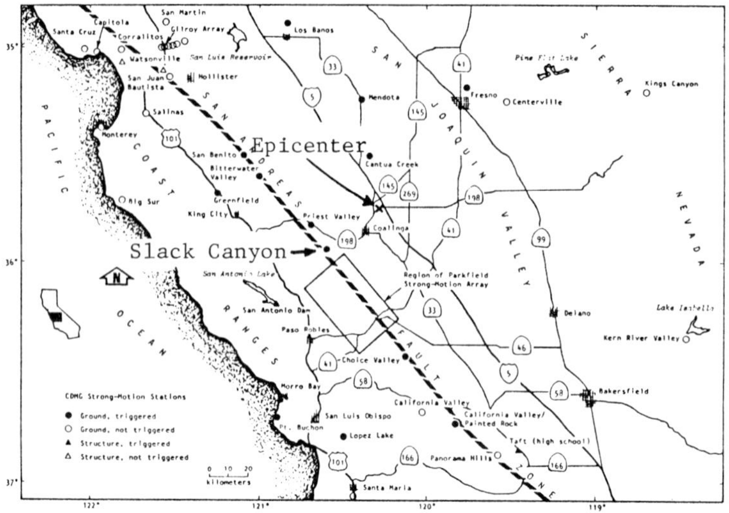

Figure 5, Location Plan for Slack Canyon Station

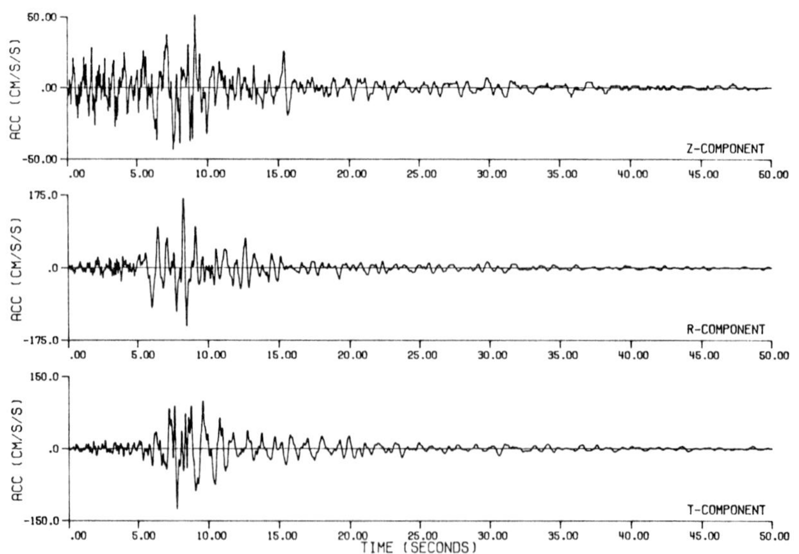

Figure 6, Transformed Accelerograms, Slack Canyon

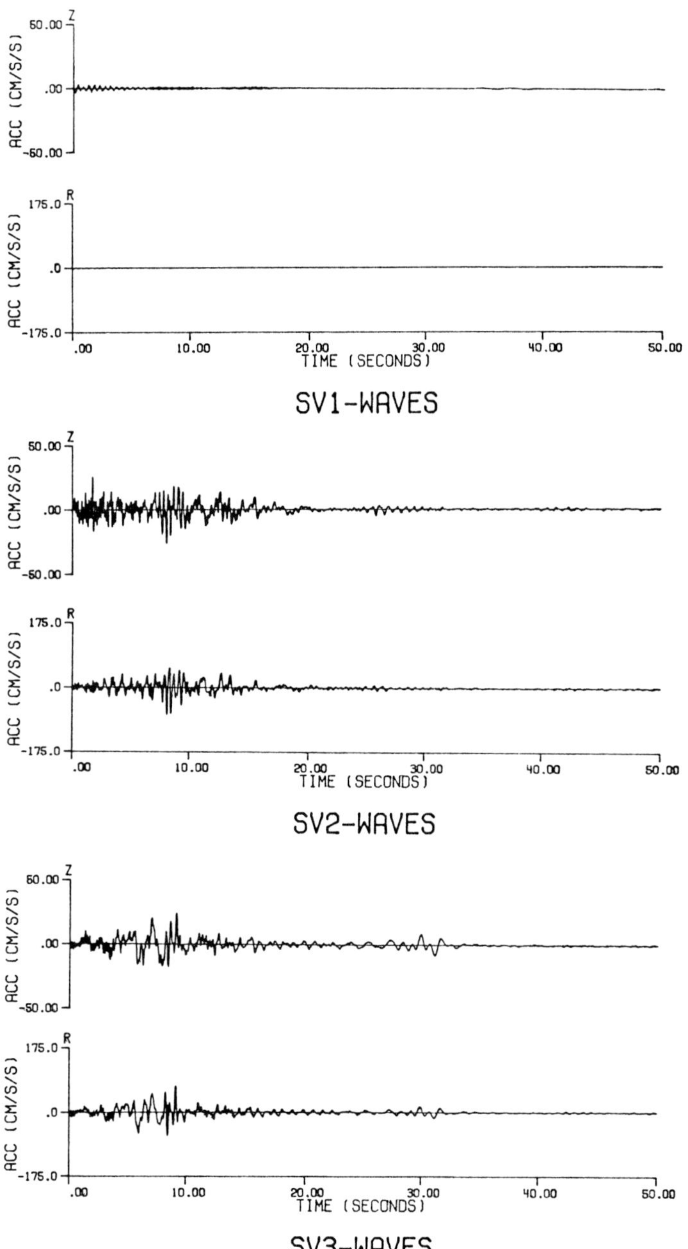

Figure 7, Accelerograms processed using the SWF

A Band-Limited, Windowed, White-Noise Process for Modeling Earthquake Motions

E. Şafak
U.S. Geological Survey, Menlo Park, California 94025, U.S.A.

INTRODUCTION

Simulation of earthquake ground motions in the time domain has been one of the popular research topics in earthquake engineering. A large number of models have been suggested, varying from a white-noise sequence (Bycroft[1]) to random pulse train model (Lin and Yong[2]). The most popular one has been the filtered white-noise modulated by an envelope function, generally referred as the Kanai-Tajimi model (Kanai[3]; Tajimi[4]). It has been shown that the Kanai-Tajimi model possesses built-in low-frequency errors (Safak and Boore[5]). There are also physical models based on the seismological description of the earthquake mechanism (i.e., Boore[6], Hanks and McGuire[7]).

In this paper another model for ground accelerations is presented. The model is a band-limited, windowed, white-noise process, developed based on the accumulation of ground motion energy in the frequency and time domains, and uses only four parameters of the motion. The paper presents first the analytical derivations for the model and the corresponding response spectra, and then an application of the results to a set of earthquake records. The main conclusion of the paper is that response spectra can be estimated with sufficient accuracy by using only four parameters of the motion.

MODEL FOR GROUND ACCELERATION

The model that will be presented for the ground acceleration is based on two functions, one in time domain and the other in frequency domain, given by the following equations:

$$U(t) = \frac{\int_0^t a^2(s)\,ds}{\int_0^\infty a^2(s)\,ds} \qquad \text{and} \qquad V(f) = \frac{\int_0^f A^2(g)\,dg}{\int_0^\infty A^2(g)\,dg} \tag{1}$$

$U(t)$ and $V(f)$ will be termed as the cumulative squared acceleration (CSA), and the cumulative squared amplitude spectrum (CSAS), respectively. $a(s)$ and $A(g)$ are the acceleration time series and corresponding Fourier amplitude spectrum, and f is the cyclic frequency. $U(t)$ has previously been used by other researchers to study the intensity and effective duration of earthquakes (e.g., Trifunac and Brady[8]). $U(t)$ gives the accumulation of the RMS (root mean square) acceleration in time, whereas $V(f)$ gives the same (because of the Parseval's equality) in frequency. A typical earthquake acceleration time history from the 1971 San Fernando earthquake and the corresponding CSA and CSAS curves are given in Figure 1.

To investigate the characteristics of $U(t)$ and $V(f)$, several data sets obtained from earthquakes in the Western United States are studied. This study shows that both $U(t)$ and $V(f)$ have the shape of a growth curve, the so-called S curve, characterized by a sharp build-up at the beginning, then a long steady (straight) region, and a sharp decay at the end. This observation suggests that, in the time domain, there is an effective time window for ground motions, represented by the straight part of $U(t)$, in which mean-square acceleration is nearly constant, and the part of the motion specified by this window provides almost all the contribution to the total energy. Similarly, in the frequency domain, there is an effective frequency band, represented by the straight part of $V(f)$, in which the amplitude spectrum is nearly constant, and the components of the motion in this frequency band provide almost all the contribution to the total energy. Therefore, it will be assumed that the CSA and CSAS curves can approximately be represented by straight lines. Figure 2 gives a schematic representation of this approximation. Different critera for the line fit can be used. The criterion used here is that the line will be fit to the part between 5- and 95-percentage points of each curve. The use of values 5 and 95 percent is rather arbitrary. They are chosen because same values have been used previously to define the effective duration of earthquakes (Trifunac and Brady[8]). As will be shown later they are not always the best values to use. The straight-line assumption for CSA and CSAS means that ground motions can be approximated by two rectangles, one in the time domain representing the region of constant mean-square acceleration, and one in the frequency domain representing the region of constant amplitude spectrum. This is also shown schematically in Figure 2. In the time domain, the height, σ_a, and the width, $T = t_2 - t_1$, of the rectangle represent the RMS value and the effective duration of the ground acceleration. In the frequency domain, they represent the power spectral density function (PSDF), $S_a(f)$, and the effective frequency band, $f_2 - f_1$. In terms of random process theory, the model corresponds to a band-limited, windowed, white-noise process. The shape of the model is similar to that used by Hanks and McGuire[7], but the parameters of the model are determined differently. The studies on recorded ground motions showed that using a rectangle in the frequency domain gives satisfactory estimates of the response spectra for frequencies between f_1 and f_2, but underestimates the response for frequencies less than f_1. Thus, for frequencies less then f_1, the model in the frequency domain was modified by adding a fourth order polynomial of the form

$$S_a(f) = \left(\frac{f}{f_1}\right)^4 S_o \qquad \text{for} \quad f \leq f_1 \tag{2}$$

The reason for using a fourth order polynomial is to be consistent with the seismological source model, so-called ω^2-model (Brune[9]). The spectral level S_o is determined from the relation that the area under the PSDF is equal to the mean square acceleration. This gives for S_o

$$S_o = \frac{5\sigma_a^2}{5f_2 - 4f_1} \tag{3}$$

RESPONSE SPECTRA

From the ground-motion model described above, the expected response of a single degree of freedom oscillator can easily be calculated by using the methods of random vibration theory (Lin[10]). The mean square value, σ_y^2, of the relative oscillator displacement with respect to ground is

$$\sigma_y^2 = \int_0^\infty S_a(f)\, |H(f)|^2\, df \tag{4}$$

where $H(f)$ is the frequency response function of the oscillator. By using the expressions for $S_a(f)$ and $H(f)$, and then integrating, Eq. 4 becomes (Safak, et al[11].)

$$\sigma_y^2 = \frac{S_0}{32\pi^2\xi_0 f_0^3} \; I(r_1, r_2, \xi_0) \tag{5}$$

where $r_1 = f_1/f_0$ and $r_2 = f_2/f_0$, and f_0 and ξ_0 denote the natural frequency and the damping ratio of the oscillator. The ratio $S_0/32\pi^2\xi_0 f_0^3$ on the right hand side of Eq. 5 is the response to white-noise excitation. The second term, $I(r_1, r_2, \xi_0)$, gives the effect of the spectral corners, f_1 and f_2, on the response. Note that $I(r_1, r_2, \xi_0)$ is not a function of the corner frequencies, but of their ratios to the oscillator frequency. $I(r_1, r_2, \xi_0)$ can be separated into two parts, as

$$I(r_1, r_2, \xi_0) = I_1(r_1, \xi_0) + I_2(r_2, \xi_0) \tag{6}$$

where $I_1(r_1, \xi_0)$ and $I_2(r_2, \xi_0)$ denote partial effects due to f_1 and f_2, respectively. They are given by the equations (Safak et al[11])

$$I_1(r_1, \xi_0) = \frac{4\xi_0}{\pi r_1^3} - \frac{\xi_0(3 - 4\xi_0^2 + r_1^4)}{2\pi\xi_d r_1^4} \ln\left(\frac{r_1^2 + 2\xi_d r_1 + 1}{r_1^2 - 2\xi_d r_1 + 1}\right) + \frac{1 - 4\xi_0^2 - \pi r_1^4}{\pi r_1^4} \tan^{-1}\frac{2\xi_0 r_1}{1 - r_1^2} \tag{7}$$

and

$$I_2(r_2, \xi_0) = \frac{\xi_0}{2\pi\xi_d} \ln\left(\frac{r_2^2 + 2\xi_d r_2 + 1}{r_2^2 - 2\xi_d r_2 + 1}\right) + \frac{1}{\pi} \tan^{-1}\frac{2\xi_0 r_2}{1 - r_2^2} \tag{8}$$

where $\xi_d = \sqrt{1 - \xi_0^2}$.

$I(r_1, r_2, \xi_0)$ is plotted in Figure 3 as a function of r_1 and r_2 ($r_1 \leq r_2$), for $\xi_0 = 0.05$. First, note that its value varies between zero and one. Also note that $I(r_1, r_2, \xi_0)$ is almost unity for oscillators whose frequencies are between the corner frequencies of the ground motion (i.e. $r_1 < 1.0$ and $r_2 > 1.0$). Therefore, the effect of the corners on the RMS response is negligible, and the RMS response is approximately equal to that from a white-noise excitation. For $r_1 > 1.0$ or $r_2 < 1.0$, $I(r_1, r_2, \xi_0)$ sharply decreases to a much smaller value than one.

The spectral displacement, y_p, can approximately be calculated from the RMS displacement, σ_y, by using the equation (Cartwright and Longuet-Higgins[12])

$$y_p = \sqrt{2\ln(2f_0T)} \cdot \sigma_y \tag{9}$$

where T is the effective duration, defined earlier as $T = t_2 - t_1$. This expression is based on the assumptions that the response time history is a narrow-band, stationary random variable with independently arriving peaks. Since our ground motion model is basically a windowed band-limited white-noise process (except for the region $f < f_1$) and the damping is low, the first two of these assumptions are valid. The assumption that peaks arrive independently represents an approximation.

It should be emphasized here that the spectral response calculated with the suggested model uses only four parameters of the ground motion, f_1, f_2, T, and σ_a. For prediction purposes, the model parameters should be correlated with source and site parameters, such as magnitude, distance, and site conditions. As will be shown in the next section, there are strong indications that such correlations exist.

EXAMPLES

The analytical approach presented above has been applied to several data sets taken from major earthquakes in the Western United States. Because of space limitations, the results of only one set, those from an aftershock of the May 2, 1983, Coalinga, California earthquake, will be presented here. Other results including those from simulated motions can be found in Safak et al[11].

The Coalinga records were collected by the USGS with portable digital instruments during an aftershock following the magnitude 6.5 main shock (Mueller et al[13]). For each record, first the cumulative squared acceleration and the amplitude spectrum curves, CSA and CSAS, were calculated; and parameters t_1, t_2, and f_1, f_2 were determined by fitting a straight line to the segment between 5- and 95-percentage points. The RMS acceleration, σ_a, was calculated using the effective duration $T = t_2 - t_1$. Then, the response spectrum for each record was calculated by using Eqs. 5 and 9. The calculated response spectrum was compared to the exact one, obtained by the time history analysis.

The results, for the best matched record and the worst matched record (from visual inspection), are summarized in Figures 4 and 5, respectively. For the best matched record, Figure 4.a gives the acceleration time history and the corresponding CSA curve, whereas Figure 4.b gives the Fourier amplitude spectrum and the corresponding CSAS curve. The Fourier amplitude spectrum is presented in an unconventional form, where the product of frequency times the squared amplitude is plotted against the logarithm of frequency. This way of plotting has two advantages: (a) the area under the curve still gives the measure of the contribution to RMS acceleration from each frequency, the same as a linear-linear plot, because of the identity that $A_a^2(f)df \equiv fA_a^2(f)d(\log f)$; and (b) there is a much better resolution for low frequencies in comparison to a linear-linear plot, so that both corners of the effective frequency band can be seen in the plot. Figure 4.c gives the comparison of the calculated response spectra with the exact one. The match is satisfactory although the motion is modeled by only four parameters.

Similarly, for the worst matched record, Figure 5.a gives the acceleration time history and the corresponding CSA curve, and Figure 5.b gives the Fourier amplitude spectrum and the corresponding CSAS curve. The comparison of the response spectra is given in Figure 5.c (solid lines). As the figure shows the match is not good for frequencies below the low-frequency corner f_1. The reason for this is that the straight line in Figure 5.b overestimates the accumulation of energy in low frequencies (i.e., the ordinates of the straight line are much larger than the ordinates of CSAS curve at low frequencies). In other words, for this record, the energy is not concentrated between 5- and 95-percent points, therefore a different criterion should be used to draw the straight line. A better alternative is to determine the turning points of the CSAS curve (i.e., the points where the second derivative of $V(f)$ has peaks), and then fit the straight line to the segment between these points. This was done and given by the dashed line in Figure 5.b. Corresponding response spectra are given, also by the dashed line, in Figure 5.c. As seen from the figure the match is now much better.

As is well known, the local site conditions alter the frequency characteristics of ground motions significantly. In order to see this, the scattering of f_1 and f_2 are plotted in Figure 6. Note that all the records are from the same event. As the figure shows, the scattering in f_2 is much wider than that in f_1. This observation supports the suggestion that the high frequency corner is a site effect, whereas the low frequency corner is a source effect (Hanks[14]).

SUMMARY AND CONCLUSIONS

This paper presents a simple model for earthquake ground accelerations that can estimate the response spectra with sufficient accuracy. The model is a band-limited, windowed, white-noise, developed based on the variations of cumulative squared acceleration and cumulative squared amplitude spectrum. Only four parameters of the ground motion are used in the model: RMS acceleration, effective duration, and the two corner frequencies that describe the beginning and the end of the effective frequency band. Response spectra are calculated by using the methods of random vibration theory. The methodology is applied to a data set from an aftershock of May 2, 1983 Coalinga, California earthquake. The results show that a satisfactory estimate of response spectra can be obtained by using only four parameters of the ground motion.

REFERENCES

1. Bycroft, G.N. (1960). White-noise representation of earthquakes, ASCE, Jour. of Eng. Mec. Div., EM2, April, 1-16.

2. Lin, Y.K. and Y. Yong (1986). Evolutionary Kanai-Tajimi type earthquake models, Report CAS-86-7, Center for Applied Stochastic Research, Florida Atlantic University, Boca Raton, Florida.

3. Kanai, K. (1957). Semi-empirical formula for the seismic characteristics of the ground, Bull. Earthq. Res. Inst., Univ. of Tokyo, 35, 308-325.

4. Tajimi, H. (1960). A statistical method of determining the maximum response of a building structure during an earthquake, Proc. 2nd World Conf. on Earthq. Eng., Tokyo and Kyoto, Japan, 781-797.

5. Safak, E. and D.M. Boore (1986). On nonstationary stochastic models for earthquakes, proc. Third U.S. National Conference on Earthquake Engineering, Vol.1, pp.137-148.

6. Boore, D.M. (1983). Stochastic simulation of high-frequency ground motions based on seismological models of the radiated spectra, Bull. Seism. Soc. Am. 73, 1865-1894.

7. Hanks, T.C. and R.K. McGuire (1981). The character of high frequency strong ground motion, Bull. Seism. Soc. Am. 71, 2071-2095.

8. Trifunac, M.D. and A.G. Brady (1975). A study on the duration of strong earthquake ground motion, Bull. Seism. Soc. Am. 65, 581-626.

9. Brune, J.N. (1970). Tectonic stress and the spectra of seismic shear waves from earthquakes, J. Geophys. Res. 75, 4997-5009.

10. Lin, Y.K. (1976). Probabilistic theory of structural dynamics, Robert E. Krieger Pub. Comp., Huntington, N.Y.

11. Safak, E., C. Mueller, and J. Boatwright (1987). A simple model for strong ground motions and response spectra, submitted for publication, Bull. Seism. Soc. Am.

12. Cartwright, D.E. and M.S. Longuet-Higgins (1956). The statistical distribution of the maxima of a random function, Proc. Roy. Soc. London, Ser. A237, 212-223.

13. Mueller, C.S., E. Sembera, and L. Wennerberg (1984). Digital recordings of aftershocks of the May 2, 1983 Coalinga, California earthquake, USGS Open-File Report, 84-697, 57 pp.

14. Hanks, T.C. (1982). f_{max}, Bull. Seism. Soc. Am. 72, 1867-1879.

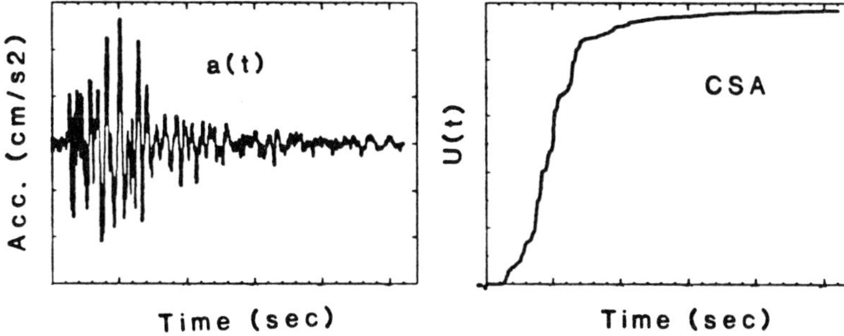

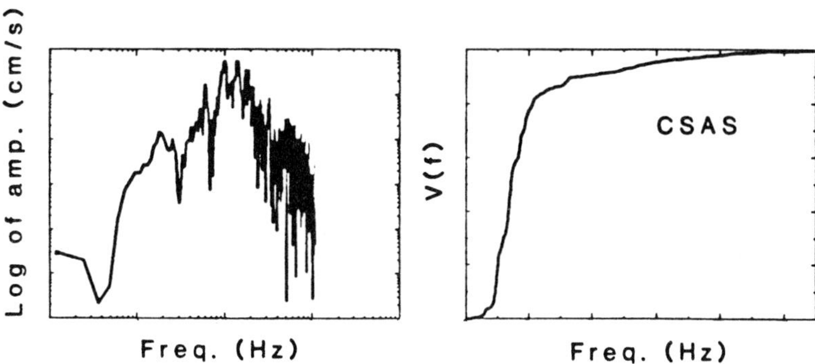

FIGURE 1. A typical acceleration time history, and corresponding cumulative squared acceleration (CSA) and cumulative squared amplitude spectrum (CSAS) curves (the record is from 1971 San Fernando earthquake).

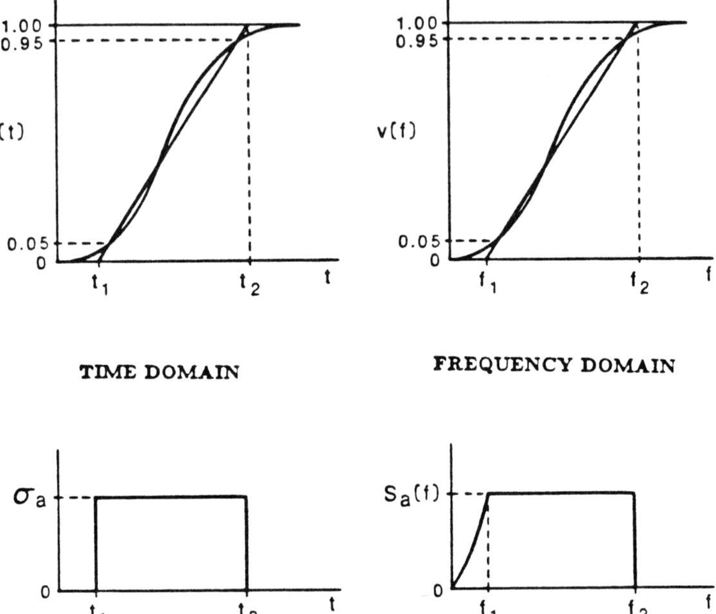

FIGURE 2. Schematic of the development of the model in time and frequency domains for ground acceleration.

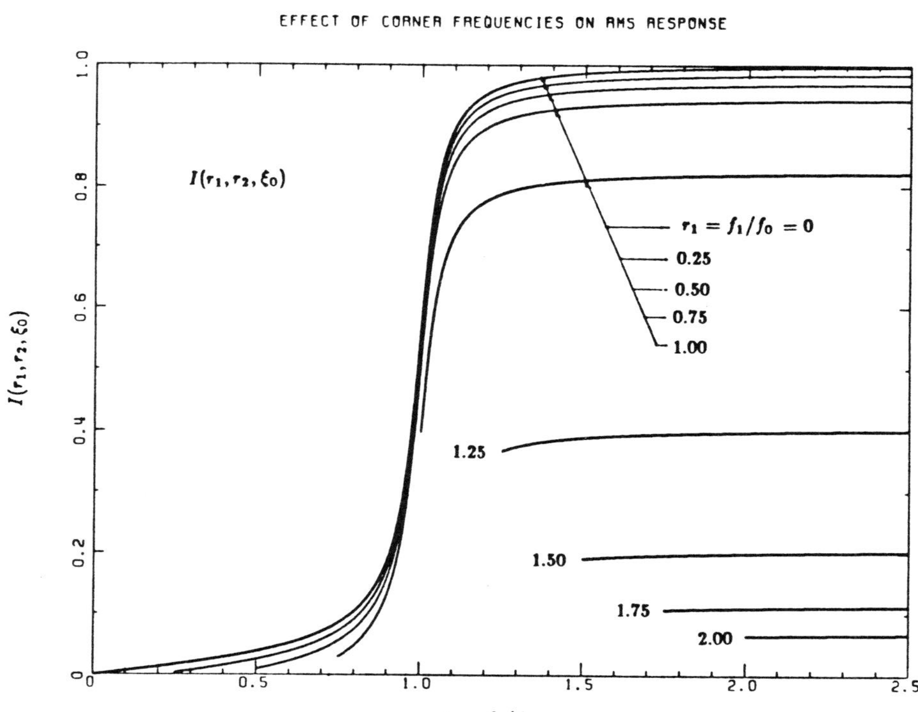

FIGURE 3. $I(r_1, r_2, \xi_0)$, the effect of corner frequencies on the RMS displacement response for 5-percent damping.

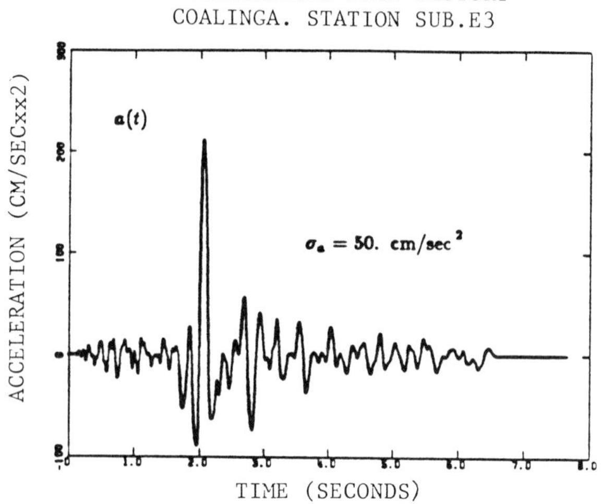

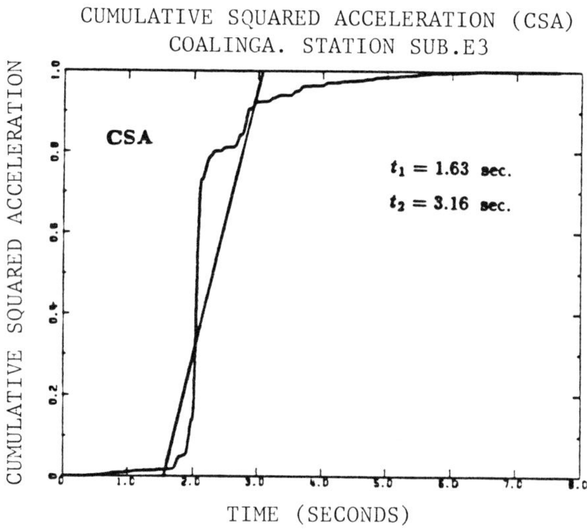

FIGURE 4.a Acceleration time history and CSA for the station that gives the best match for the response spectra (station SUB).

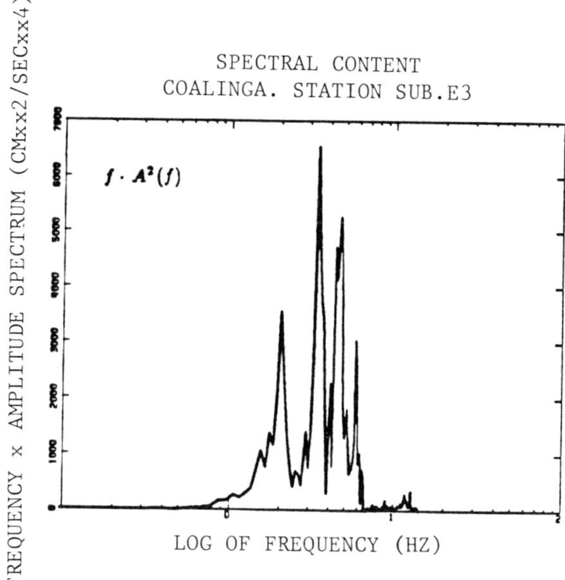

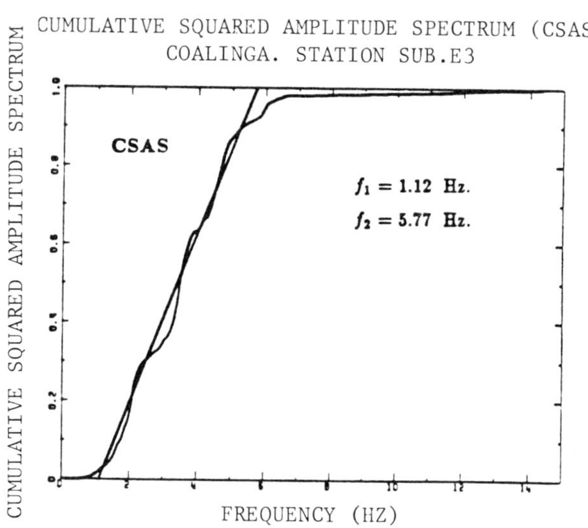

FIGURE 4.b Squared amplitude spectrum and CSAS for the station that gives the best match for the response spectra (station SUB).

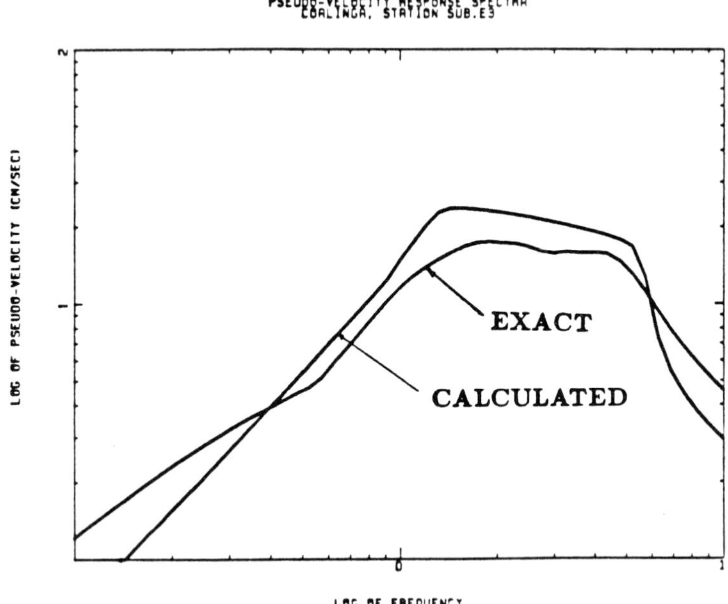

FIGURE 4.c Comparison of calculated response spectra to exact response spectra for 5-percent damping for the station that gives the best match (station SUB).

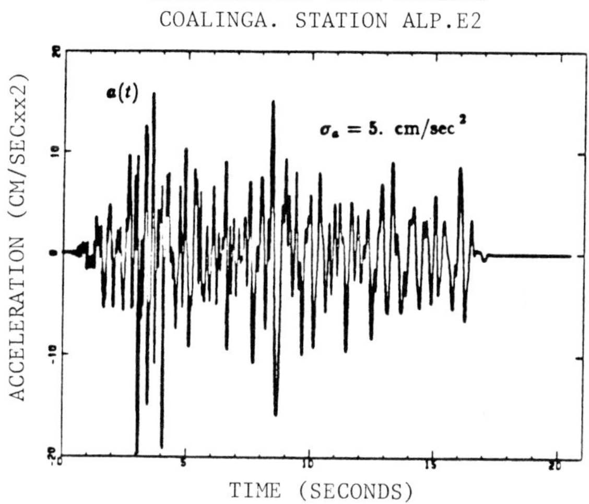

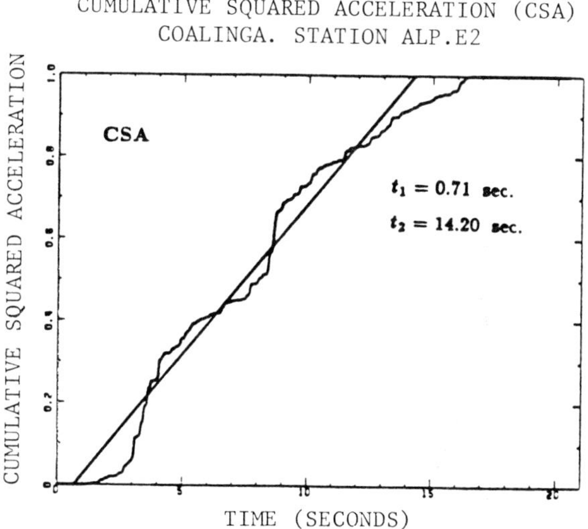

FIGURE 5.a Acceleration time history and CSA for the station that gives the worst match for the response spectra (station ALP).

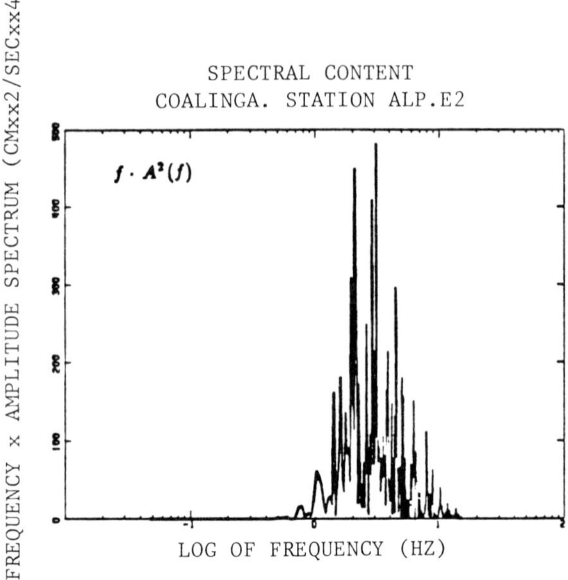

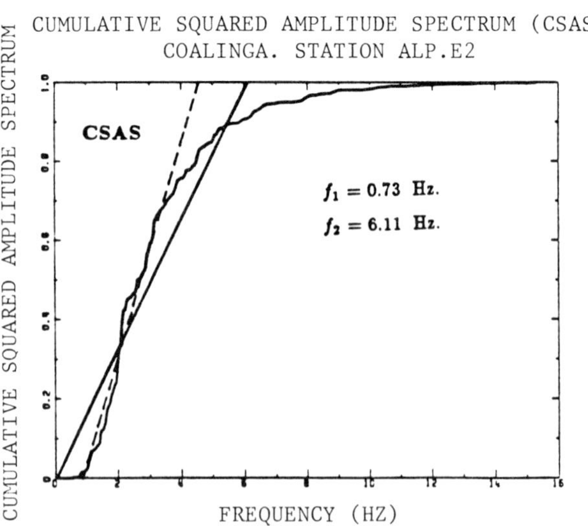

FIGURE 5.b Squared amplitude spectrum and CSAS for the station that gives the worst match for the response spectra (station ALP). Dashed line shows the improved straight-line approximation to CSAS.

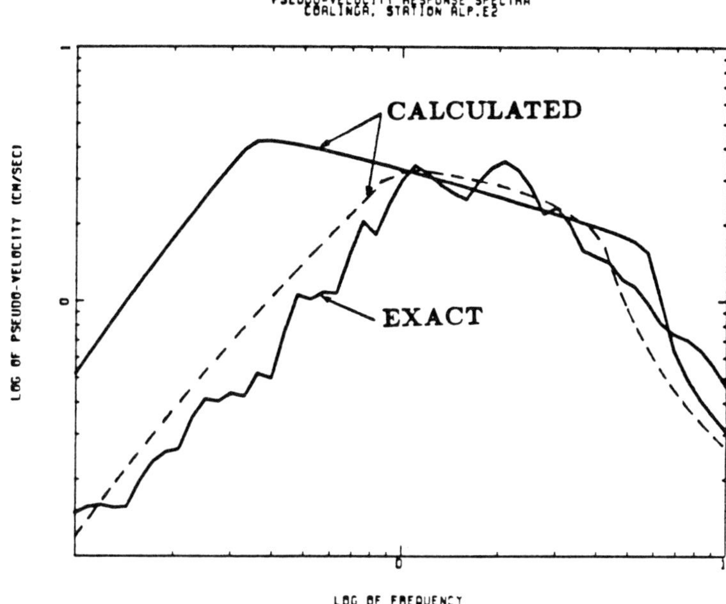

FIGURE 5.c Comparison of calculated response spectra to exact response spectra for 5-percent damping for the station that gives the worst match (station ALP). Dashed line shows the calculated response spectra for the improved straight-line approximation.

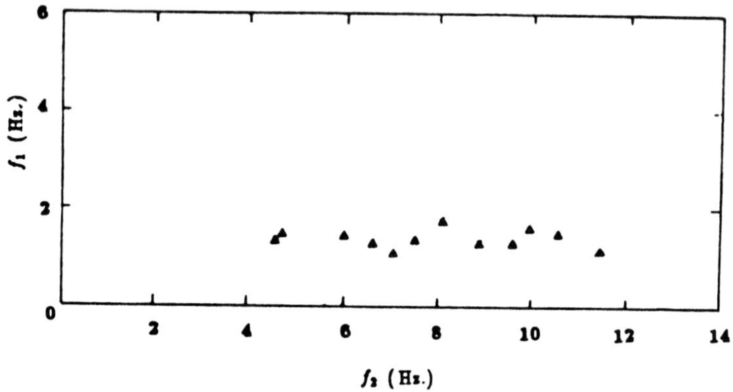

FIGURE 6. Scattering of corner frequencies f_1 and f_2 for Coalinga data set.

Ground Motion Input Through Equivalent Stationary Motion

J-S. Lin
Civil Engineering Department, University of Hartford, West Hartford, CT 06117, U.S.A.

INTRODUCTION

One of the major issues that confront engineers in the seismic design and analysis of structures is the determination of input ground motions. To accomodate the uncertainties about the characteristics of the possible future motions, generally a group of input motions are employed in the analysis, and based upon the outcome engineering judgements rendered, the analysis and design completed. This very often is done by picking up several historical strong motions and scaling them to a certain amplitudes according to the designated return period. Another frequently adopted approach is to use numerically simulated motions which are compatible to a certain response spectra as dictated by codes. Similarly, several simulated motions are generated, but the latter approach also involves additional decisions as to what duration and what shape of the acceleration amplitude envelope are to be used for the simulated motions.

An alternate approach to input a large suite of ground motions is presented in this paper. This approach utilizes the equivalent stationary motion proposed by Vanmarcke and Lai[6]. In that model each motion is characterized by four parameters. This provides a basis for describing a large number of ground excitations by specifying the joint distribution of these parameters. The current study does not intend to provide a complete distribution which will necessarily be very complicated and require a very comprehensive data base. Instead this study attempts to come up with an approximate distribution which is compatible to the present engineering practice, and at the same time is fairly easy to apply.

This study starts by examining the correlations between the ground motion parameters and based upon the result separate the independent parameters. Following the current practice that ground motions are grouped according to site soil conditions, parameters obtained from motions recorded on different site soils conditions are studied. The effects of magnitude on the parameters are derived by transforming a frequently employed magnitude-dependent liquefaction potential relationship proposed by Seed et. al.[5] Finally the advantages of the current approach are addressed.

EQUIVALENT STATIONARY MOTION

Equivalent stationary motion belongs to the category of the probabilistic ground motion modelling. A ground motion as recorded is viewed as only one of the possible realizations of a whole families of motions that possess the similar properties. Separable considerations are utilized in that the evolution of the ground motion characteristics with time are decoupled and defined separately in the time domain and in the frequency domain.

In the frequency domain, the equivalent stationary motion uses Kanai-Tajimi spectral density function in fitting the frequency characteristics of a motion. This is equivalent to assume that ground acts predominantly as a single degree of freedom structure that modifies the highly erratic white noise motion arrived at the bed rock into the shape as recorded. Three parameters therefore are needed in the frequency domain characterization, two for the single degree of freedom system idealization and the third one for describing the energy arrived at the site. One choice for these parameters may be the ground frequency, ω_g, the ground damping, ζ_g, and the root-mean-square acceleration, σ .

In the time domain, instead of presenting a modulating function which may require many parameters depending on how well detail are to be modelled, the equivalent stationary motion uses only a single parameter--the equivalent stationary duration. A motion is transformed into a time-limited stationary duration and the envelope function is therefore a "boxcar" type function with a unit amplitude all over and is cut off at the equivalent stationary duration, S.

The equivalent stationary motion preserves several essential properties of a motion: 1)the energy as defined by the Arias itensity, 2) the relative frequency composition and 3) the maximum acceleration , A_{max}, is preserved in the mean sense that it is expected to be exceeded once in the equivalent motion.

To process a record, first the Arias intensity of a motion is calculated. The spectral density function is obtained and normalized and a normalized Kanai-Tajimi spectral density function is fitted which gives the ground frequency, ω_g, and the ground damping, ζ_g. The root-mean-square acceleration, σ, and the equivalent stationary duration S are obtained by solving two simultaneous equations derived from the preservation criteria 1) and 3). Since S is a function of A_{max} , ω_g , ζ_g and σ , A_{max} can also be used instead of S as a basic parameter in the model.

It is important to note that σ and S so obtained are dependent to each other. A common misuse of the equivalent stationary motion is to treat these two parameters independently and compare them with parameters of seemingly similar nature from other models.

Because that stationarity is assumed for the equivalent motion, the model is not an appropriate choice if the time-dependent structural response statistics are a major concern. But in general engineering application, the interest is often about the peak response, in this respect the use of equivalent motion has been conclude to provide compatible average response spectra (Lin and Tyan[2]). As for the statistics in the ground motion itself, the model has also been concluded to give compatible liquefaction potentials (Lin et. al.[3]).

A JOINT DISTRIBUTION OF GROUND MOTION PARAMETERS

Within the framework of the equivalent motions, it is a natural approach in trying to define a group of motions by specifying the joint distribution of the four ground motion parameters. At present time it would be unrealistic to develope a comprehensive and rigorous distribution. Even if this is possible its practicality in applications may be questionable. On the other hand, although an approximation will necessarily simplify or neglect some inter-relationship between parameters, it still may serve the purpose of engineering analysis without the loss of generality since not all the parameters nor their inter-relationship have the same impact on the structural response. It is along this line that this study first looks into the possibility of breaking the distribution into multiplication of independent functions. This is done by examining the correlations between the ground motion parameters.

A proper data base is important in finding out the correlations. For this purpose a data base that was used by McGuire[4] in developing attenuation laws was adopted. There are 140 records in this data set and is considered to cover motions over a wide spectrum relatively homogeneously. The ground motion parameters for these motions has been obtained by Lai[1]. The correlations of the four ground motion parameters based upon the data set are calculated in this study and are listed as a correlation matrix below,

$$\begin{array}{c} \\ A_{max} \\ \sigma \\ \omega_g \\ \zeta_g \end{array} \begin{array}{c} \begin{array}{cccc} A_{max} & \sigma & \omega_g & \zeta_g \end{array} \\ \left[\begin{array}{cccc} 1.0 & 0.967 & 0.067 & 0.039 \\ 0.967 & 1.0 & 0.067 & -0.029 \\ 0.067 & 0.067 & 1.0 & -0.223 \\ 0.039 & -0.029 & -0.223 & 1.0 \end{array} \right] \end{array} \tag{1}$$

Based upon this correlations, A_{max} and σ are almost perfectly correlated, and the correlation between ω_g and ζ_g are much larger than their correlation with A_{max} and σ. It is therefore proposed as a first approximation to simplify the joint distribution $f(A_{max}, \sigma, \omega_g, \zeta_g)$ into two independent ones as followings,

$$f(A_{max}, \sigma, \omega_g, \zeta_g) = f_{A_{max},\sigma}(A_{max}, \sigma)\, f_{\omega_g,\zeta_g}(\omega_g, \zeta_g) \tag{2}$$

SITE SOIL EFFECTS ON THE DISTRIBUTION

It has long been recognized that site soil conditions as reflected in the frequency contents of ground motions affect significantly structural responses. As ω_g and ζ_g are two parameters that reflect the frequency composition of equivalent motions, this study looks into the effects of different site soil conditions on these two parameters first. To exclude the influence of magnitude, 98 motions with magnitudes around 6.6 mostly recorded during the San Fernando 1971 earthquake are used.

Fig. 1 summarizes the ω_g and ζ_g data pair from these 98 records. As reflected in the preceding negative correlation between ω_g and ζ_g, Fig. 1 does show that ground damping decreases as ground frequency increases. Data from various site soil conditions show similar trend. Some of this phenomenum may due primarily to the single degree of freedom idealization of ground. Three points more or less on the mean curve of Fig. 1 are used to construct three Kanai-Tajimi spectral density functions, namely, ω_g =10 rad/sec, ζ_g =0.6; ω_g =20 rad/sec, ζ_g =0.4 ; and ω_g=40 rad/sec, ζ_g=0.3. These functions are depicted in Fig. 2 all normalized to the same power. This plot illustrates that when ground frequency is small it requires larger damping to reflect realistic band-width.

From Fig. 1 an average relationship between ω_g and ζ_g can be found. But considering the large scatter in the ground damping with respect to ground frequency, the improvement by adding into the picture this relationship does not seem to be necessary. More importantly, it has been concluded that the possible scatter of ζ_g has negligible effects comparing with other parameters on the variability of structural responses. Even by assigning a constant to ground damping, satisfactory results were obtained (Lin and Tyan[2]).

This allows the further simplification of the joint distribution of ω_g and ζ_g into the followings,

$$f_{\omega_g,\zeta_g}(\omega_g,\zeta_g) = f_3(\omega_g)\, f_4(\zeta_g) \qquad (3)$$

Although one may use some histograms of ζ_g as the distribution f (ζ_g), a simple constant or a deterministic function depending upon ω_g will be sufficient.

The sensitivity of structural responses to ground frequency is mostly dominated by the parameter ω_g . Table 1 summarizes the site-related means and standard deviations of these parameters. As expected ground frequencies for rock sites are much larger, while for deep-cohesionless-soil sites and stiff soil sites the differences are less significant. It is proposed that ω_g be considered as site-dependent. Figs. 3 to 5 are the histograms of ground frequencies studied here, they may be used as the distribution function f (ω_g). As for sites where soil conditions are not known, larger uncertainties or scatter in ω_g should be accomodated and the ω_g histogram from the above mentioned 140 records are suggested to be used which is designated as mixed sites and is depicted in Fig. 6.

TABLE 1 Site-Related Statics for Ground Motion Parameters (Earthquake Magnitude= 6.6)

Parameters	Rock Sites		Stiff Soil Sites		Deep Cohesionless Soil Sites	
	Mean	S.D.	Mean	S.D.	Mean	S.D.
σ/A_{max}	0.359	0.023	0.346	0.028	0.348	0.034
ω_g	23.96	11.58	20.00	7.53	19.02	10.19
ζ_g	0.45	0.22	0.42	0.18	0.46	0.22

Knowing that A and σ are almost perfectly correlated, the following question would be what kind of linear relationship can be established. A plot of σ/A_{max} versus A_{max} based upon 96 records (excluding A_{max} larger than 1 g for clarity) is presented as Fig. 7. The σ/A_{max} ratio appears to be a constant independent of the magnitude of A_{max} .

This makes it possible to break the joint distribution of A_{max} and σ as a multiplication of a

σ/A_{max} distribution with an A_{max} distribution as follows,

$$f_{A_{max},\sigma}(A_{max},\sigma) = f_1(A_{max})f_2(\sigma/A_{max}) \quad (4)$$

This σ/A_{max} ratio appears to be independent of site soil conditions as shown in the statistics in Table 1. From all 98 records the histogram of σ/A_{max} is obtained and is presented in Fig. 8 as a probability density function. The distribution of A_{max} can readily be found through seismic hazard analysis, a proper attenuation law may be chosen to reflect the site soil conditions.

EFFECTS OF EARTHQUAKE MAGNITUDES AND EPICENTRAL DISTANCES

No attempt is made to relate the changes of frequency contents with respect to earthquake magnitude and epicentral distance or a distance to energy center. It is considered here that by adopting the frequency composition for records obtained on similar site soil conditions as a possible distribution, the possible variation in the ground frequency is already exaggerated. If imprecise information as to the effects of magnitudes and distance to energy center is in addition to be imposed, which although add up to a sense of completeness, will unecessarily complicate the evaluation of structural responses and is difficult to be justified from the practical point of view.

For maximum ground acceleration, A_{max}, as mentioned previously that by using proper attenuation laws and a seismic hazard study, its distribution can properly reflect the effects of distance and magnitude.

As for σ/A_{max} ,physically one would expect the far-field motion to be more uniform in the excitation amplitude and therefore will have a larger σ/A_{max} than near field records. From the San Fernado records which have epicentral distances mostly between 30 to 80 km, the data scatter and this trend can not be distinctively observed. Considering the small standard deviation of this ratio, no effort is put to reduce the uncertainty in this ratio by incoporating the effects of epicentral distance.

Seed et. al.[5] have presented a magnitude dependent relationship for liquefaction potentials. These potentials are obtained by counting the number of peaks in the records at different intervals and then transformed them according to a failure criteria into uniform cycles. Lin and Tyan[2] have shown that such earthquake-magnitude dependent relationship can be converted into σ/A_{max} vs. magnitude relationship. Seed et al's relationship is shown to give

Earthquake Magnitude	σ/A_{max}
5	0.434
5.5	0.414
6	0.391
6.5	0.371
7	0.349
7.5	0.334
8	0.320

Notice again that σ and S are inter-related, if frequency content remain similar, the smaller σ/A_{max}, the longer S becomes. The above results simply imply that if two motions from different magnitudes with same frequency composition are both scaled to the same A_{max}, the smaller magnitude motion will be more uniform in amplitude but will be much shorter in duration.

By using a different data base, different such relationship will be emerged. Since Seed's curve is widely used and accepted, one may choose to use these derived magnitude dependent mean values and fix the standard deviation of σ/A_{max} to be a constant. For all 98 magnitude 6.6 records the mean and standard deviation of σ/A_{max} are 0.351 and 0.029 respectively as shown in Fig. 8. One may choose to use the magitude dependent σ/A_{max} as the means and by keeping standard deviation as a constant 0.029 to scale Fig. 8 in deriving magnitude dependent $f(\sigma/A_{max})$ distribution.

TO INPUT A LARGE SUITE OF MOTIONS

The approximate joint distribution of the ground motion parameters can be written as,

$$f(A_{max},\sigma,\omega_g,\zeta_g) = f_1(A_{max})f_2(\sigma/A_{max})f_3(\omega_g)f_4(\zeta_g) \quad (5)$$

To input a large suite of motions, a seismic hazard analysis is carried out to find out the distribution of A_{max} for each interested earthquake magnitude. Depending upon the site soil conditions a proper ω_g distribution is selected, and a scaled σ/A_{max} function is found for the earthquake magnitude. As all these functions are discretized and by mutilpying them together, the joint distribution is found. In other words, all the possible combination of A_{max}, σ, ω_g, and ζ_g are found with a proper weight.

CONCLUSIONS

The present approach has been used to find out site-dependent response spectra and compared well with results obtain directly by integration a large suite of real motions (Lin and Tyan[2]). This approach was also used to construct magnitude dependent response spectra. In using this approach, the weightings of each combination of ground parameters and where it stands in the whole family of possible motions are very clear. It especially suits for overall seismic risk evaluation of a structure. From the analysis point of view the present approach does provide an reasonable workable alternative and is compatible to the present practice.

REFERENCES

1. Lai, S. P. (1982) Statistical Charaterization of Strong Ground Motions Using Power Spectral Density Function, 72, 1:259-274, BSSA.
2. Lin, J-S and Tyan J.Y. (1986) Equivalent Stationary Motion and Average Response Spectra, 14,267-279, Earthquake Engrg. & Str. Dyn.
3. Lin, J-S, Whitman, R.V. and Vanmarcke, E. H. (1983) Equivalent Stationary Motion for Liquefaction Study, J. Geotech. Eng. Div.,ASCE 109, 1117-1121.
4. McGuire, R. K. (1978) Seismic Ground Motion parameters, J. Geotech. Eng., ASCE, 104, 481-490.
5. Seed, H.B., Idriss, I.M.,Makdisi F. and Banerjee (1975) Representation of Irregular Stress Time History by Equivalent Uniform Stress Series in Liquefaction Study, Report EERC75-29, University of California, Berkeley, CA.

6. Vanmarcke, E.H. and Lai S.P.(1980) Strong Motion Duration and rms amplitude of Earthquake Records, BSSA, 70, 1293-1307.

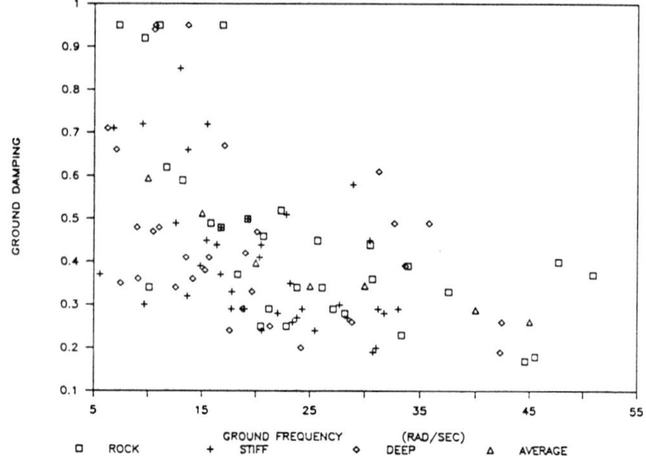

Figure 1. Ground Frequency vs. Ground damping for Records over Various Site Soil Conditions

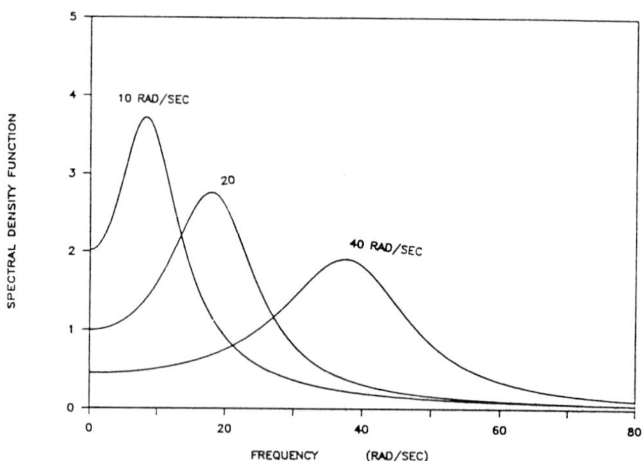

Figure 2. Kanai-Tajimi Spectral Density Functions for Different Damping and Frequency Combination.

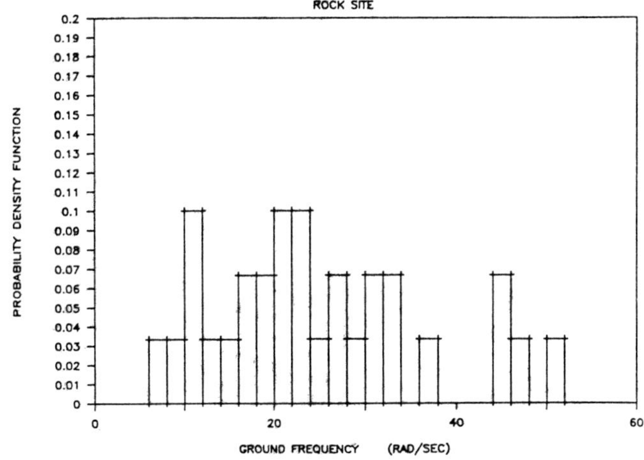

Figure 3. Histogram of the Ground Frequency from Rock Site Records

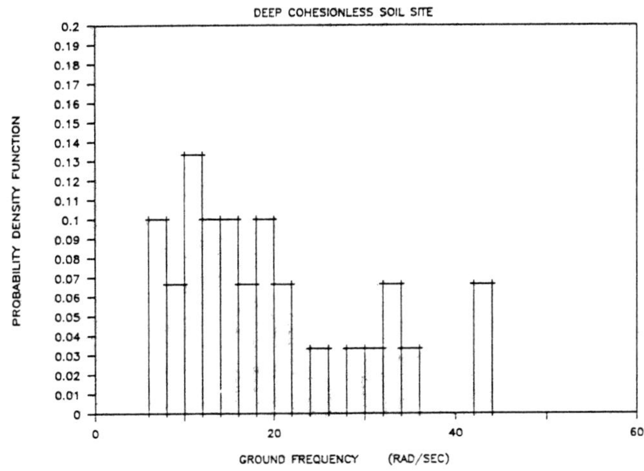

Figure 4. Histogram of the Ground Frequency from Deep Cohesionless Soil Site Records

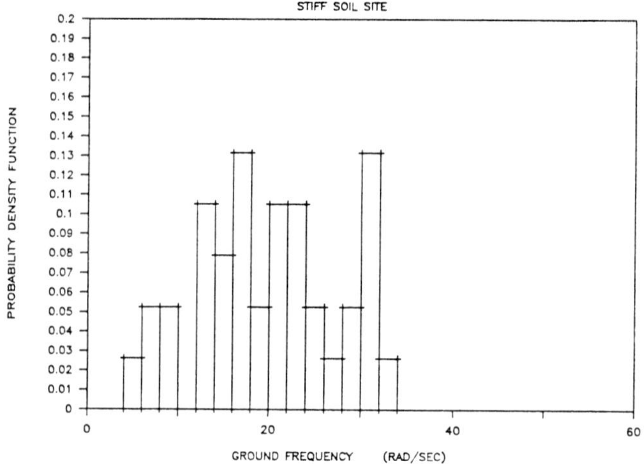

Figure 5. Histogram of the Ground Frequency from Stiff Soil Site Records.

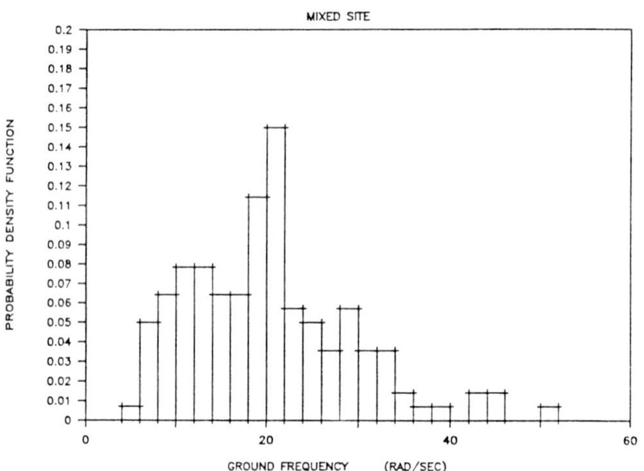

Figure 6. Histogram of the Ground Frequency from Mixed Site Records.

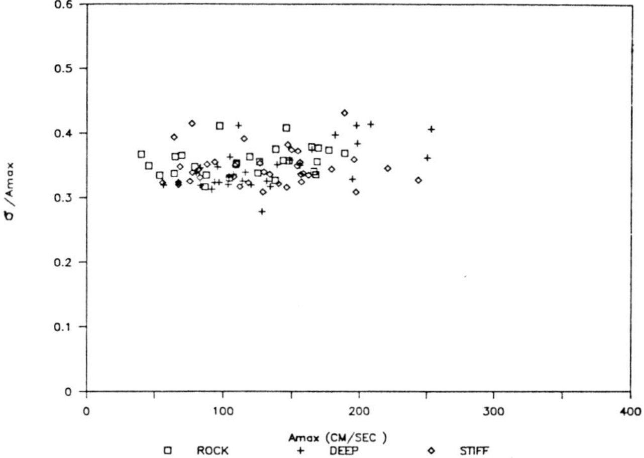

Figure 7. The Relationship between σ/A_{max} and A_{max} from Magnitude 6.6 Earthquake Records.

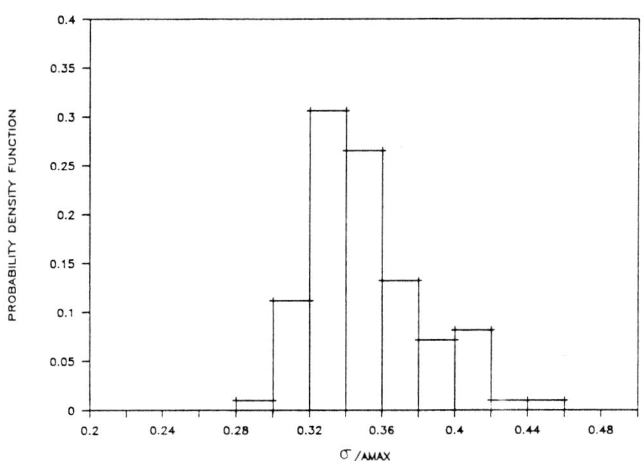

Figure 8. Histogram of σ/A_{max} from magnitude 6.6 Earthquake Records.

Modelling Earthquake Ground Motions in Seismically Active Regions Using Parametric Time Series Methods

G.W. Ellis, A.S. Çakmak and J. Ledolter[1]
Department of Civil Engineering, Princeton University, Princeton, NJ 08544, U.S.A.

ABSTRACT

Strong motion accelerograms can be efficiently modelled using time series analysis. Because of the nonstationarity of accelerograms, transformations must first be applied to stabilize the variance and frequency content. An ARMA model can then be fit to the stabilized series. By analyzing a large number of accelerograms in California, these ARMA and transform parameters have been related to physical variables. Thus it is possible to generate simulated ground motions in the area studied. The validity of the modelling process is measured by comparing the simulations and their frequency spectra to actual records.

INTRODUCTION

The damage incurred by structures during an earthquake depends upon both the nature of the structure and the properties of the ground motion. However, relatively few sites have recordings of strong ground motion. To provide input motions to structural models for sites for which no strong ground motion data exist, it is necessary to simulate accelerograms. These simulations must have realistic duration, frequency content, and intensity to match the physical conditions of the site.

Because seismic waves are initiated by irregular faulting and then travel through complex ground formations with random properties, resulting in many reflections, refractions, and attenuations before reaching the recording station, a stochastic approach has been taken to model the accelerograms. The application of Autoregressive Moving Average (ARMA) processes to model earthquake accelerograms has proven effective in

---

[1] On leave from Department of Statistics and Actuarial Science, University of Iowa, Iowa City, IA 52242.

numerous studies. The difficulty in modelling accelerograms using ARMA models is the nonstationarity of the variance and frequency content of the records. In past studies, several approaches have been taken: (1) fitting time-varying ARMA parameters to the original records (Kozin, 1977; Jurkevics and Ulrych, 1979; Gersch and Kitagawa, 1985), (2) fitting time-invariant ARMA parameters to short segments of the time series (Chang, et al., 1979), and (3) fitting time-invariant ARMA parameters to a series stabilized by transformations (Cakmak, et al., 1985). In this study the last approach has been applied. After fitting an ARMA model, it is then possible to reverse the procedure and generate many simulations of the original record.

Although for some purposes repeated realizations of a given record are useful, the ground motion expected from different earthquakes at any given site is needed. By relating the modelling parameters to physical variables such as earthquake magnitude, epicentral distance, and geological site conditions it is possible to generate simulations for any size earthquake at any given site.

AUTOREGRESSIVE MOVING AVERAGE MODELS

In the processing of accelerograms the records are digitized at uniformly spaced time intervals, normally 0.02 seconds in length. Such a sequence of n observations (Z_t, $t = 1,2,\ldots,n$) is called a discrete time series. If the time series is regarded as a realization from a stochastic process, then it is possible to generate many realizations from the same process having the same probabilistic structure as the original record.

Stochastic processes can be approximated by autoregressive-moving average (ARMA) models. In the autoregressive model the current deviation of the process $(Z_t - \mu)$, from its mean value μ, is expressed as a function of previous deviations $(Z_{t-1} - \mu)$, $(Z_{t-2} - \mu)$, ..., $(Z_{t-p} - \mu)$, and the past-value of the shocks a_t by

$$Z_t - \mu = \phi_1(Z_{t-1} - \mu) + \phi_1(Z_{t-2} - \mu) + \ldots + \phi_p(Z_{t-p} - \mu) + a_t - \theta_1 a_{t-1} - \theta_2 a_{t-2} - \ldots - \theta_q a_{t-q} \quad (1)$$

where
- p = order of AR parameters
- q = order of MA parameters
- ϕ_k = autoregressive parameter at lag k, k=1,2,...,p
- θ_k = moving average parameter at lag k, k=1,2,...,q
- μ = mean level
- a_t = white noise sequence with variance, σ_a^2

The procedures for identifying, fitting, and validating ARMA models are discussed in detail in Box and Jenkins (1976).

ANALYSIS AND SIMULATION PROCEDURE

Because one of the primary goals of this study is to relate modelling parameters to physical variables affecting the ground motion, a modelling procedure using single-valued parameters based on the methodology in Polhemus and Cakmak, 1981 and Cakmak et al., 1985 was used. Several modifications, however, were made to these previous studies, including: a frequency envelope allowing the frequency content to change during the duration of a simulation; a constrained ARMA model allowing no response at zero frequency; and physical principles relating the modelling parameters to physical variables. The first step is to determine the duration of the earthquake, T, by eliminating the first 1% and final 2% of the cumulative energy. Fig. 1 illustrates the stabilization procedure for a horizontal component of the Kern County earthquake (7/21/52) recorded at the Taft Lincoln School (USGS No. 95). The shortened accelerogram is then divided by a standard deviation envelope, thus stabilizing the variance of the series to 1.0. However, most accelerograms still exhibit nonstationary frequency content. The frequency content is stabilized by multiplying the time increment by a frequency envelope. This results in a time series that is stationary in both variance and frequency content. An ARMA model can now be fitted to the stabilized series.

From the transformation and modelling procedure, parameters are generated to describe the standard deviation envelope, frequency envelope, and the ARMA model for each record analyzed. These parameters were related to physical variables such as earthquake magnitude, epicentral distance, geographic location, and soil type. Also, correlations were found among the parameters making it possible to reduce the complexity of the model.

Using the physical relationships developed, it is possible to reverse the procedure to generate simulations for any set of physical variables. First the modelling parameters are calculated from the physical variables. From these statistical parameters a time series, standard deviation envelope, and frequency envelope are generated. By transforming the time scale of the time series and multiplying by the variance envelope, a realistic simulation can be generated possessing similar characteristics to the original accelerogram.

The validity of the results was then assessed by comparing the acceleration time histories, response spectra, and Fourier spectra of the original earthquake and several simulations.

The procedure can be summarized by the following steps.

1. Shorten the accelerogram
2. Calculate a standard deviation envelope
3. Stabilize the variance
4. Calculate the frequency envelope

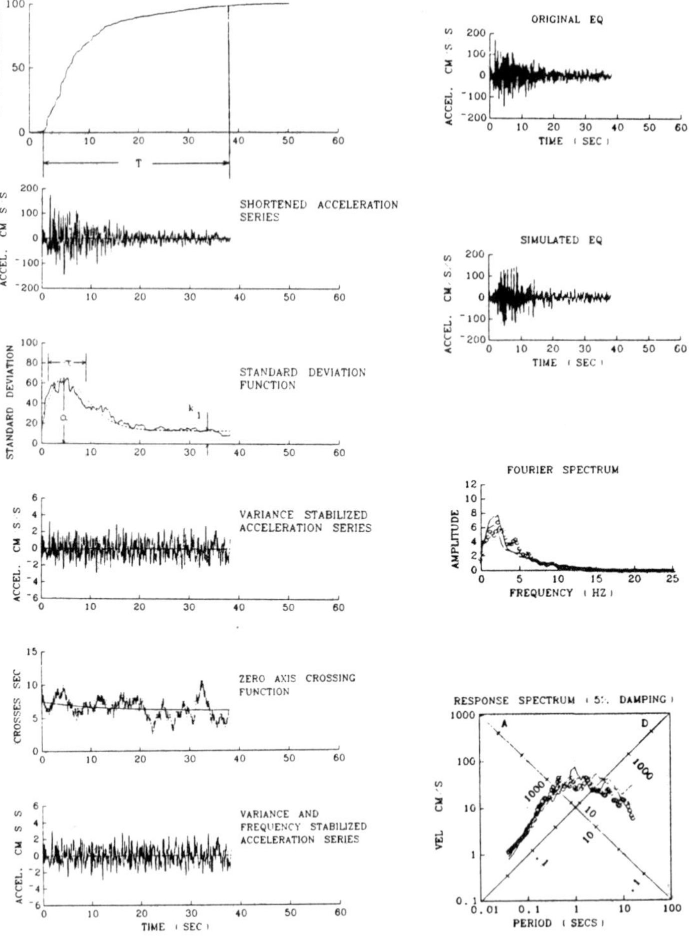

Fig. 1 Stabilizing the accelerogram

Fig. 2 A simulated record

5. Stabilize the frequency content
6. Fit an ARMA model to the stabilized series
7. Relate the ARMA and transform parameters to physical variables
8. Using the relationships estimated in (7), generate simulations
9. Compare the time histories and frequency spectra of the original and simulated series to validate the procedure.

ESTIMATION OF THE STANDARD DEVIATION ENVELOPE

By squaring the acceleration and calculating its running average using a two-second time window, an estimate of the variance of the series at any given time may be found. The square root of the variance provides an estimate of the standard deviation envelope $\sigma_z(t)$. By dividing the shortened accelerogram by the standard deviation envelope, a time history with a stationary variance of approximately 1.0 is obtained.

To relate the standard deviation envelope to physical variables, a smooth function is fitted as

$$\hat{\sigma}_z(t) = c_1(\alpha - k_1)\left(\frac{t}{\tau}\right)^3 e^{-(c_2/\tau)t} + k_1$$

where

$$c_1 = \frac{8e^3}{3\sqrt{3}}$$

$$c_2 = 2\sqrt{3} \qquad (2)$$

The maximum intensity of the strong shaking is measured by α. The value of the weak shaking, k_1, is estimated as the average of the standard deviation envelope during the final 1/3 of the record. Finally, the length of strong shaking, τ, is estimated such that the energy of the accelerogram is conserved.

ESTIMATION OF THE FREQUENCY FUNCTION

Because the arrival of high frequency P-waves preceeds the arrival of S-waves and surface waves, the variance stabilized series has a nonstationary frequency content. Most records have initially high predominant frequencies which quickly decrease with time. One measure of this phenomenon is the number of zero axis crossings per second, $F_c(t)$. This envelope was calculated using an equally weighted two second window.

A smooth function F_c', is fitted to the zero axis crossings.

$$F_c'(t) = (c_0 - k_2)\exp(-bt) + k_2 \qquad (3)$$

where c_0 = The initial value of the zero axis crossings

b = The rate of decay

and k_2 = The zero axis crossing of the weak shaking.

$F_c'(t)$ is fitted by assigning k_2 as the average value of the zero axis crossings during the final 1/3 of the record. A least squares fit may then be calculated by

$$\ln(F_c - k_2) = a - bt \tag{4}$$

c_0 is then

$$c_0 = k_2 + \exp(a) \tag{5}$$

The frequency function, $F_c'(t)$, may be used to change the time increments by

$$\Delta t' = (\Delta t)F_c'(t) \tag{6}$$

where Δt = The original time increment of 0.02 seconds
$\Delta t'$ = The new time increment.

Then the transformed record is reduced to the same length as the initial record by

$$\Delta t'' = \Delta t' \times \frac{\text{length of the original record}}{\text{length of the transformed record}} \tag{7}$$

and digitized to 0.02 second intervals. Physically, Eqs. (6) and (7) expand the time increment in the beginning of the record, where the higher frequencies occur, and decrease the time increment at the end of the record, which is dominated by the lower frequencies. This results in a time history with a constant predominant frequency.

CHOOSING AN ARMA MODEL

With the variance and frequency content stabilized, it is now possible to fit an ARMA model to the series. Because the original earthquakes were filtered to allow no response at zero frequency, a model was sought which had no response at zero frequency and fit the data well.

A constrained ARMA (2,2) model was found to meet these requirements. By setting the equation for the Fourier spectrum of an ARMA (2,2) model

$$F(f)=\left[2\sigma_a^2 \frac{1 + \theta_1^2 + \theta_2^2 - 2\theta_1(1-\theta_2)\cos 2\pi f - 2\theta_2 \cos 4\pi f}{1 + \phi_1^2 + \phi_2^2 - 2\phi_1(1-\phi_2)\cos 2\pi f - 2\phi_2 \cos 4\pi f}\right]^{1/2} \tag{8}$$

equal to zero at zero frequency $(f = 0)$, the constraint $\theta_1 + \theta_2 = 1$ was found. This constraint violated the invertibility requirement

$$\theta_1 + \theta_2 < 1 \tag{9}$$

for the ARMA (2,2) model. It was found that the constraint

$$\theta_1 + \theta_2 = 0.99 \tag{10}$$

allowed negligible response at zero frequency, while also satisfying Eq. (9).

To fit the constrained ARMA model $Z_t = \phi_1 Z_{t-1} + \phi_2 Z_{t-2} + a_t - \theta_1 a_{t-1} - \theta_2 a_{t-2}$ to the data, a nonlinear constrained optimization program written by Quandt and Goldfeld (1985) was utilized. The parameters, ϕ_1, ϕ_2, θ_1, and θ_2 were estimated to minimize the variance of the residuals, σ_a^2, while also satisfying the imposed constraint and the invertibility and stability requirements of an ARMA (2,2) model.

Of the five parameters estimated, only three can be chosen independently: ϕ_1, ϕ_2, and θ_1. Equation (10) relates θ_1 and θ_2. Also, a relationship exists between σ_a^2 and the variance σ_z^2 of the time series Z_t,

$$\sigma_a^2 = \frac{(1-\phi_2)(1-\phi_2^2) - \phi_1^2(1+\phi_2)}{(1-\phi_2)c_1 + (1+\phi_2)\phi_1 c_2 + (1-\phi_2)\phi_2 c_3}\,\sigma_z^2 \tag{11}$$

where
$$c_1 = 1 - \theta_1(\phi_1-\theta_1) - \theta_2[\phi_1(\phi_1-\theta_1) + \phi_2 - \theta_2]$$
$$c_2 = -\theta_1 - \theta_2(\phi_1-\theta_1)$$
$$c_3 = -\theta_2$$

Because the variance of the time series was stabilized to be 1.0, Eq. (11) expresses σ_a^2 as a function of ϕ_1, ϕ_2, and θ_1.

RELATING THE MODEL PARAMETERS TO PHYSICAL VARIABLES

The modeling procedure described above was applied to nine different seismic regions in California as shown in Fig. 3. For more information on most of these records see Hudson (1976). From these regions 99 horizontal and 49 vertical records were examined. For each record nine parameters were calculated: α, τ, k_1, c_0, b, k_2, ϕ_1, ϕ_2 and θ_1. The results were first examined for correlations among the model parameters. The parameters which were found to be unrelated to other parameters were then related to physical variables. The functional relationships estimated are shown in Table 1.

To understand the physical meaning of the ARMA parameters, the theoretical Fourier spectrum of the ARMA model was plotted using Eq. (8). The relative amount of high and low frequencies was controlled mainly by ϕ_1 for the values calculated in this study. As ϕ_1 approached a theoretical maximum of 2, the relative amount of lower frequencies increased. Because high

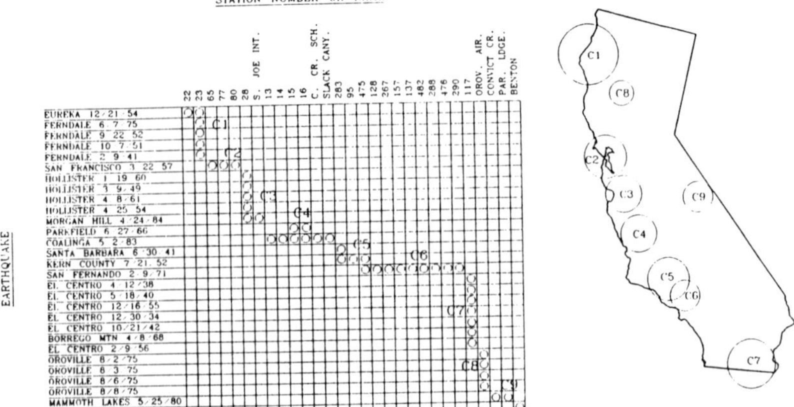

Fig. 3 Records analyzed

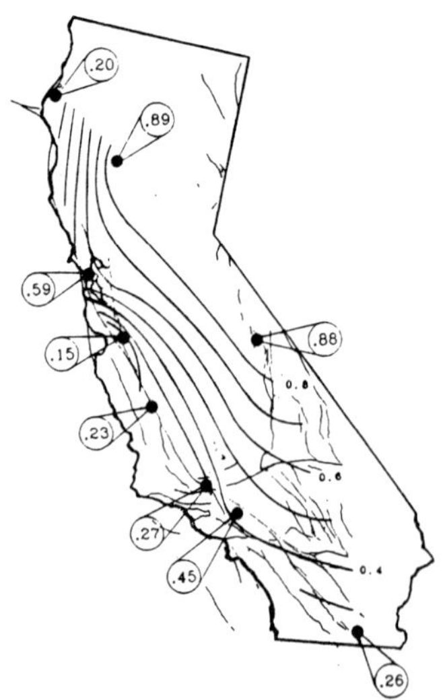

Fig. 4 Contour map of k_ϕ values

Table 1 Parametric Relations

H = Horizontal Components
V = Vertical Components

M = magnitude
d = distance
k_ϕ = location parameter

Parameter	Relation	Component	r^2
$\log \alpha$	$= 0.60 + 0.23 \log \frac{10^M}{k_\phi d^2}$	H	$r^2 = 0.38$
	$= 0.52 + 0.19 \log \frac{10^M}{k_\phi d^2}$	V	$r^2 = 0.27$
	$\alpha^V = 0.41\, \alpha^H$		$r^2 = 0.60$
τ	$= 0.55 + 0.73 \sqrt{d}$	H	$r^2 = 0.39$
	$= 1.05 + 0.121 \sqrt{d}$	V	$r^2 = 0.62$
	$\tau^V = 1.6 + 0.90\, \tau^H$		$r^2 = 0.58$
$\log k_1$	$= -0.60 + 0.85 \log \alpha$	H	$r^2 = 0.59$
	$= -0.54 + 0.80 \log \alpha$	V	$r^2 = 0.56$
$1/c_0$	$= -0.15 + 0.014 \frac{M}{k_\phi^{0.33}}$	H	$r^2 = 0.36$
	$= -0.01 + 0.010 \frac{M}{k_\phi^{0.33}}$	V	$r^2 = 0.18$
	$c_0^V = 3.1 + 1.2\, c_0^H$		$r^2 = 0.42$
b	$= 0.18$	H	
	$= 0.19$	V	
	$b^V = b^H$		
k_2	$= 0.89 + 11.6\, k_\phi$	H	$r^2 = 0.55$
	$= 0.50 + 14.4\, k_\phi$	V	$r^2 = 0.56$
	$k_2^V = 1.12\, k_2^H$		$r^2 = 0.74$
ϕ_1	$= 2 - \frac{k_\phi}{d^{0.32}}$	H	$r^2 = 0.72$
	$= 2 - 0.91 \frac{k_\phi^{1.1}}{d^{0.13}}$	V	$r^2 = 0.55$
	$\phi_1^V = -0.99 + 1.5\phi_1^H$		$r^2 = 0.73$
ϕ_2	$= 0.80 - 0.91\, \phi_1$	H	$r^2 = 0.97$
	$= 0.99 - 0.97\, \phi_1$	V	$r^2 = 0.99$
θ_1	$= 0.22$	H	
	$= 0.33$	V	
	$\theta_1^V = 1.5\, \theta_1^H$		

frequencies damp out more quickly with distance than low frequencies (Trifunac and Brady, 1975), ϕ_1 was modelled as a function of distance. However, due to the physical properties of the soil at the recording station and the geology of the path through which the seismic waves travel, it is expected that local conditions of each area studied will influence the value of ϕ_1 as well. Thus, ϕ_1 was modelled as a function of epicentral distance and also of a site parameter estimated for each group of sites as

$$\text{Log}(2 - \phi_1) = a_s - b \text{ Log } d \tag{12}$$

where a_s = constant estimated for each group of sites

b = constant estimated for all California.

By setting $k_\phi = 10^{a_s}$, Eq. (12) can be simplified as

$$\phi_1 = 2 - \frac{k_\phi}{d^b} \tag{13}$$

The value of b was estimated to be 0.32 for the horizontal components, and the value of k_ϕ estimated for each site is shown in Fig. 4. For the vertical case, ϕ_1 was regressed on distance and the value of k_ϕ calculated from the horizontal component. The result is shown in Table 1.

In Fig. 4, contour lines are drawn for k_ϕ values in order to estimate k_ϕ for areas not included in this study. High k_ϕ values estimated at Oroville and Mammoth Lakes areas indicate more high frequency content in these records compared to accelerograms recorded along the San Andreas fault. An explanation of this tendency is that the medium in which the waves propagate is fractured at the San Andreas fault due to the crushing effect of the relative plate movements, while away from the plate boundary the earth remains intact.

The site conditions at many recording stations in California have been classified by Trifunac and Brady (1975) into three types:

Classification	Site Geology
0	soft alluvial deposits
1	hard sedimentary rock or an intermediate site between 0 and 2
2	basement or crystalline rock

A regression relationship was estimated between k_ϕ and the site classification.

$$k_\phi = 0.287 + 0.230(\text{site classification}) \qquad r^2 = 0.26 \quad (14)$$

Equation (14) indicates higher frequency content at stiffer sites.

In examining the other ARMA parameters, it was found that the second auto-regressive parameter, ϕ_2, varied linearly with ϕ_1. The value of θ_1 used in Eq. (8) was found to have virtually no effect upon lower frequencies and only a small effect upon the frequencies between about 15 Hz and 25 Hz. Because of this insensitivity, θ_1 was estimated to be a constant.

Because α and the maximum acceleration are closely related, the functional relationship for α was based upon the definition of local magnitude developed by Richter (1935) as

$$M = \log(a) + 3\log(d) - 2.92 \qquad (15)$$

where a = maximum acceleration
d = epicentral distance

Thus α was expected to be a function of the ratio of 10^M divided by a power of d. It was found that d^2 best described the data, suggesting body-wave attenuation. Also, the site parameter, k_ϕ, was found to affect α. The amplitude of the weak shaking, k_1, was found to be highly correlated to the amplitude of the strong shaking. Because the velocity of wave propagation for P, S, and Surface waves differ, the time between their arrival and the length of shaking depends upon the distance the waves travel. Therefore, the length of strong shaking, τ, was related to the epicentral distance. Taking the square root of the distance made the relation linear.

Finally, the parameters of the frequency envelope were examined. The initial frequency of zero axis crossings, c_0, was found to be a function of k_ϕ and magnitude. A study by Terashima (1968) relating the spectral peak of P-waves to magnitude suggested the functional form. Because the dynamic receptance of the site affects the response, k_ϕ was also included. The effect of P-waves is significant only during the early shaking; the frequency of zero axis crossings during the weak part of the shaking, k_2, was found to be a function of the site only. The rate of decay, b, was found to be a constant.

RELATIONSHIP BETWEEN VERTICAL AND HORIZONTAL PARAMETERS

Linear relationships between the independent vertical and horizontal parameters are also shown in Table 1. The amplitude of strong shaking for the vertical case is slightly less than half that of the horizontal case. This is in agreement with a study by Street (1978) who found the maximum vertical acceleration in California to be approximately one half the maximum horizontal acceleration. The length of strong shaking, τ, was found to be approximately equal for the vertical and horizontal case. Thus the standard deviation envelope for the vertical case varies only in amplitude from the horizontal case.

In examining the frequency envelope, it was found that both the initial zero crossing frequency, c_0, and weak shaking frequency, k_2, were greater for the vertical case. The decay rate, b, was found to be approximately the same. This indicates that the frequency envelope for the vertical case has the same shape as the horizontal case and varies only in magnitude.

Finally, the value of ϕ_1 was found to be lower for the vertical component. This again indicates the higher frequency content in the vertical direction. The greater frequency content in the 15 Hz to 25 Hz range of the vertical component is indicated by a higher value of θ_1.

SIMULATIONS OF THE ORIGINAL RECORDS

ARMA modelling has a distinct advantage over other schemes in the ease in which simulations can be generated by reversing the modelling process. The modelling parameters needed to generate the simulations may be found by one of two methods. First the parameters estimated by the modelling procedure may be used. Simulations generated from these parameters will be multiple realizations of the original accelerogram. Second, for a given site k_ϕ may be found from Fig. 4 or Eq. (14). Using this value along with the earthquake magnitude and epicentral distance for which the simulation is to be generated, the modelling parameters may be found from the parametric relations summarized in Table 1.

Given all the model parameters it is possible to reverse the modelling procedure to create simulations. First, a series with stable variance and frequency content is generated at 0.02 second increments by

$$Z_t = \phi_1 Z_{t-1} + \phi_2 Z_{t-2} - \theta_1 a_{t-1} - \theta_2 a_{t-2} + a_t \tag{16}$$

A zero crossing frequency envelope, $F_c'(t)$, is then calculated from Eq. (3) to rescale the time axis of Z_t as

$$\Delta t'_s = \frac{0.02}{F'_c(t)} \tag{17}$$

After changing the time scale, it is necessary to reduce the record to the original length by

$$\Delta t''_s = \Delta t'_s \times \frac{\text{length of the original simulation}}{\text{length of the transformed simulation}} \tag{18}$$

The series is then digitized into equal increments of 0.02 seconds. From Eq. (2) a standard deviation envelope, $\hat{\sigma}(t)$, is calculated. By multiplying the time series by this envelope, a simulation nonstationary in both variance and frequency content is created.

VALIDATION OF THE MODELLING PROCESS

To assess the validity of the model, simulations were generated from the parameters fitted to the original series (Fig. 2) and from parameters calculated as a function of physical variables (Figs. 5 and 6). Because the white noise sequence a_t is random, each generated simulation will be different. Thus several simulations were generated to compare with the original record.

By examining Figs. 2, 5, and 6, it can be seen that the simulations closely match the original records. This is especially evident in the Fourier and response spectra. As expected, the simulations that were generated from parameters fitted to the original data matched the original records more closely than the simulations generated from the parametric relations in Table 1.

CONCLUSIONS

Application of the Autoregressive-Moving Average (ARMA) process to strong motion accelerograms, after processing by variance and frequency stabilizing transformations, is an efficient method for describing the ground motions through a small number of parameters. By relating these parameters to physical variables, it is possible to reasonably predict the ground motion of a site in California for which no strong motion data has been recorded.

A natural extension of this analysis for California is the study of accelerograms recorded in areas that possess different soil conditions and types of faulting, such as Japan, Mexico, Taiwan, and Turkey. Also, methods of modelling the three accelerogram components simultaneously, using multivariate modelling procedures to compute the correlation structure among the components, are being investigated.

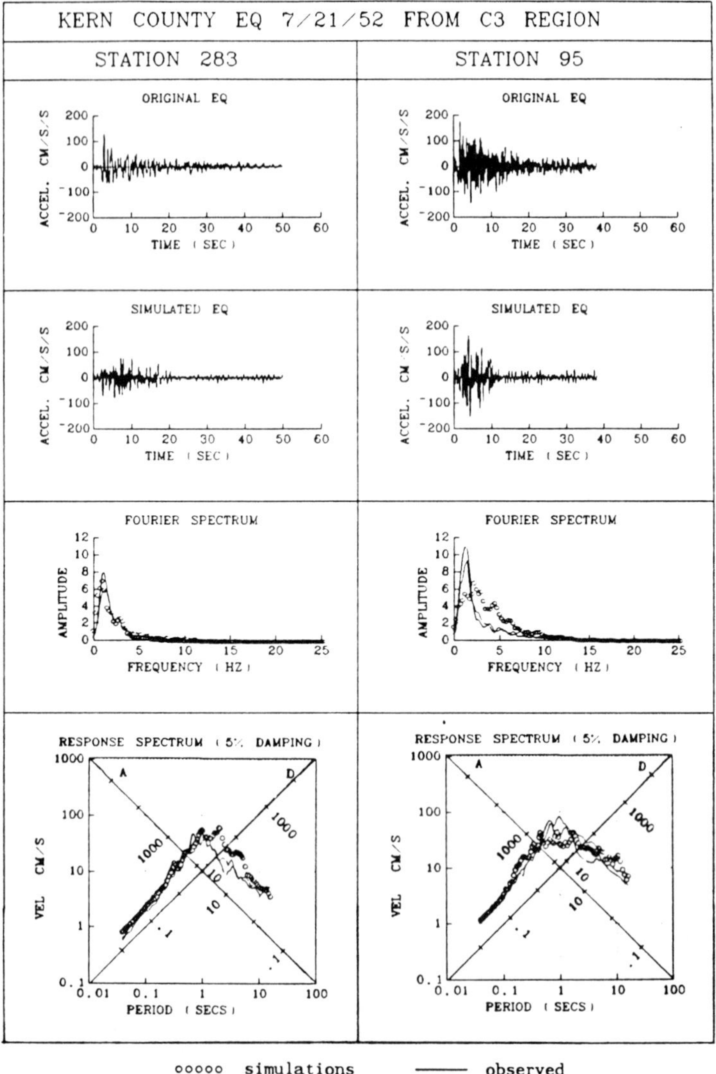

Fig. 5 Simulations of one earthquake recorded at two stations.

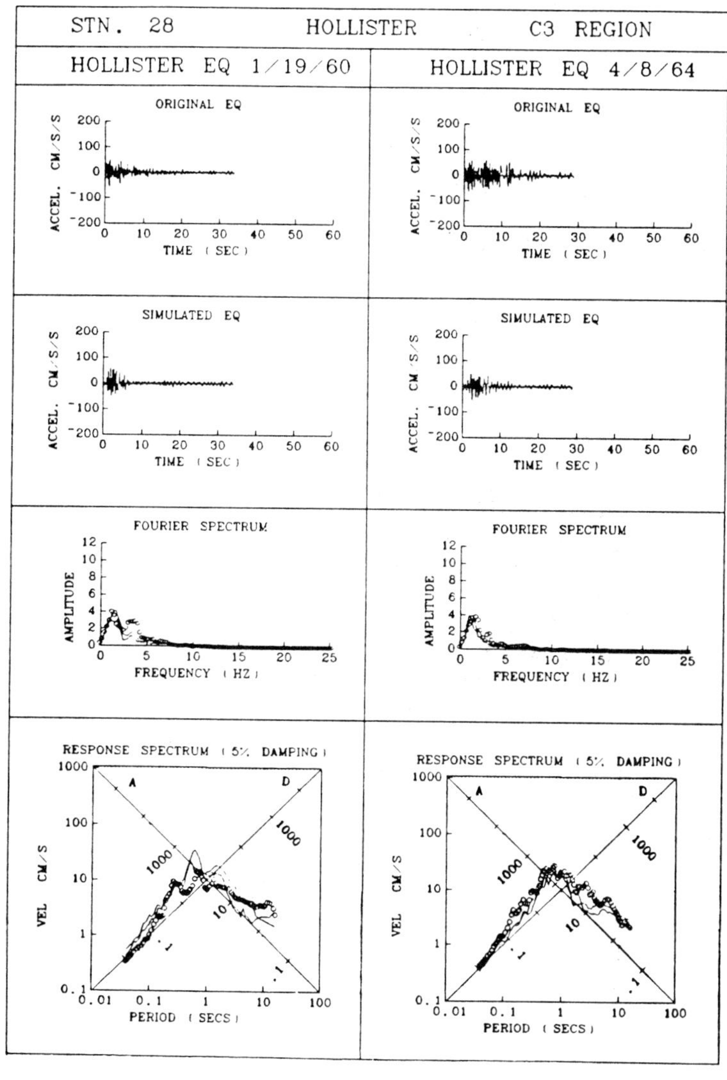

ooooo simulations ——— observed

Fig. 6 Simulations of two earthquakes recorded at one station.

ACKNOWLEDGMENT

This research was supported by a grant from the National Center for Earthquake Engineering Research, SUNY, Buffalo, NY; this support is gratefully acknowledged.

REFERENCES

Box, G. E. P. and G. M. Jenkins (1976). Time Series Analysis Forecasting and Control, Holden-Day, San Francisco.

Cakmak, A. S., R. I. Sherif, and G. W. Ellis (1985). "Modelling Earthquake Ground Motions in California Using Parametric Time Series Methods," Soil Dynamics and Earthquake Engineering, Vol. 4, No. 3. 124-131.

Chang, M. K., J. W. Kwiatkowski, R. F. Nau, R. M. Oliver and K. S. Pister (1982). "ARMA Models for earthquake ground motions," Earthquake Engineering and Structural Dynamics, Vol. 10, 651-662.

Gersch, W. and G. Kitagawa (1985). "A Time Varying AR Coefficient Model for Modelling and Simulating Earthquake Ground Motion," Earthquake Engineering and Structural Dynamics, Vol. 13, 243-254.

Hudson, D. E. (ed.) (1976). Strong Motion Earthquake Accelerograms - Index Volume, Report No. EERL 76-02, EERL, California Institute of Technology, Pasadena, CA.

Jennings, P. C., G. W. Housner, and N. C. Tsai (1968). "Simulated Earthquake Ground Motion," California Inst. of Technology Tech. Report, Pasadena, CA.

Jurkevics, A. and T. J. Ulrych (1979). "Autoregressive Parameters for a Suite of Strong-Motion Accelerograms," Bull. Seism. Soc. Am., Vol. 69, 2025-2036.

Kozin, F. (1977). "Estimation and Modelling of Non-stationary Time Series," Proc. Symp. Computer Meth. in Engineering, University of Southern California.

Polhemus, N. W. and A. S. Cakmak (1981). "Simulation of Earthquake Ground Motions Using Autoregressive Moving Average (ARMA) Models," Earthquake Eng. and Structural Dynamics, Vol. 9, 343-354.

Quandt, R. E. and S. M. Goldfeld (1985). GQOPT4/I Reference Manual, Princeton University, Princeton, NJ.

Richter, C. F. (1935). "An Instrumental Earthquake Magnitude Scale," Bull. Seism. Soc. Am., Vol. 5, 1-32.

Street, R. L. (1978). "A Note on the Horizontal to Vertical Lg Wave-Amplitude Ration in Eastern United States," Earthquake Notes, Vol. 49(2), 15-20.

Terashima, T. (1968). "Magnitude of Microearthquakes and Spectra of Microearthquake Waves," Bull. Int. Inst. Seismol. Earthq. Eng., Vol. 5, 31-108.

Trifunac, M. D. and A. G. Brady (1975). "On the Correlation of Seismic Intensity Scales with the Peaks of Recorded Strong Ground Motion," Bull. Seismol. Soc. Am., Vol. 65, 307-321.

Space-Time Variation of Strong Earthquake Ground Motion

Chin-Hsiung Loh
Department of Civil Engineering, National Central University, Chung-Li, Taiwan 320, Republic of China

INTRODUCTION

There are many engineering analyses of spatial characteristics of seismic ground motions that have been published in the literature (Harada[4,5]; Loh[6]; Shinozuka[7]; Vanmarcke[9]). These analyses of differential seismic ground motions over short distances are of significant interest to both seismologists and earthquake engineers. To study the spatially variable seismic motion, it requires strong motion data obtained with a dense network of strong motion accelerographs. The SMART-1 array, that is located in the northeast corner of Taiwan (Lotung city), provides such opportunity to study the space-time variation of seismic waves. From high-resolution wavenumber spectral analysis (Capon[2]), the dispersive characteristics of waves and the directional dependence of coherence were first studied based on the analysis of array data. A stochastic model incorporating the physical properties of source, travel path and the local site amplification effects was then developed to simulate the differential ground motion. An inverse problem was presented to estimate the site amplification effect in terms of frequency by analyzing the data between array response and on rock site (extended station). From this study, the site-dependent spectrum was estimated. The characteristics of spatial variation of seismic waves and the local site effect play an important role to this stochastic ground motion model. Some earthquake events from SMART-1 array that are used in this study is shown in Table 1.

SPACE-TIME CORRELATION STRUCTURES

The spatial variation of ground movement may be just as important as the temporal variation. By assuming a homogeneous and stationary random field, the cross-correlation function is defined by

$$R_{12}(\vec{r},\tau) = \lim_{\substack{T\to\infty \\ S\to\infty}} \frac{1}{2\pi S} \int_{-T}^{T} \int_{S} U_1(\vec{x},t)\; U_2(\vec{x}+\vec{\xi},t+\tau)\; dS(\vec{x})\; dt \qquad (1)$$

where $dS = dx\ dy\ dz$; $\vec{\xi}$ is the vector of separation distance between two points; and τ is the time lag. By taking Fourier transforms of these correlation functions with respect to the space separation $\vec{r}$ and time lag τ, the frequency-wavenumber spectrum $P(\omega,\underset{\sim}{k})$ is obtained. The general expression for covariance at space lag $\vec{r}$ and the frequency-wavenumber spectrum is

$$C(\omega,\vec{r}) = \frac{1}{(2\pi)^2} \int_{-\infty}^{\infty} P(\omega,\underset{\sim}{k})\; \exp[i\underset{\sim}{k}\cdot\vec{r}]\; d\underset{\sim}{k} \qquad (2)$$

where $P(\omega,\underset{\sim}{k})$ is the frequency-wavenumber spectrum of seismic signal and $C(\omega,\vec{r})$ is the cross-spectral density function. This spectrum provides an estimate of the speed and azimuth of a plane-wave fit to the recorded wave field.

In case of azimuthal symmetry in the intersensor spacing, it is advantageous to do the inverse wavenumber transform in cylindrical coordinate and $C(\omega,r)$ and $P(\omega,k)$ are related through a Fourier Bessel transform,

$$C(\omega,r) = \frac{1}{2\pi} \int_0^{\infty} P(\omega,k)\; J_o(kr)\; k\; dk \qquad (3)$$

where $J_o(kr)$ is the Bessel function of zero order. For propagating seismic signal at a fixed phase velocity V_o, the frequency wavenumber spectrum can be replaced by

$$\rho(\omega,k) = \frac{2\pi}{k_o} P(\omega)\; [\delta(k-k_o) + \delta(k+k_o)] \qquad (4)$$

The Equation (3) can also be represented as

$$C(\omega,r) = P(\omega)\; J_o(k_o r) \qquad (5)$$

The correlation coefficient along space coordinate is defined as

$$\rho(r) = \frac{C(r)}{C(o)} = J_o(k_o r) = J_o(\frac{\omega r}{V_o}) \qquad (6)$$

where $$C(r) = \frac{1}{2\pi} \int_{-\infty}^{\infty} C(\omega,r)\; d\omega \qquad (7)$$

Coherence estimated can be computed between all pairs of the seismographs at several frequencies corresponding to the spectral peaks.

Frequency-Wavenumber Spectrum --- The power spectrum, $P(\omega, \underset{\sim}{k})$, in frequency-wavenumber space is a function of wavenumber vector $\underset{\sim}{k}$ ($k = f/v$) and frequency f. Results are given for fixed frequency, thus displaying the velocity and the direction of propagation of seismic waves. This should be expected if the signal sources are concentrated around one point in wavenumber space which actually means plane waves propagating from that sources, and if no dominant peaks in wavenumber spectrum then the wave field is more complex than simple plane waves. The data of Event-39 (SMART-1) were used to study this frequency-wavenumber spectrum. Figure 1 shows the frequency wavenumber spectral estimates for three time windows using the east-west motion of Event-39. The time windows are specified in Figure 2. The estimated wave propagating direction and wave velocity is also shown in this figure. Since there is almost a strong "dominating wavenumber" peak in wavenumber spectrum at frequency of 0.977 Hz, the wave field can be described quite well by a single plane wave in these specified time windows.

It is obvious that the coherence estimates in the direction of propagation is consistently higher if the plane waves do exist at each frequency. As was shown in Figure 3, the black dot shows the estimated coherence versus sensor separation of Event 39 at a frequency of 0.977 Hz which plane wave assumption was made. One reason for explaining the loss of coherence in the direction of propagation is due mainly by energy at the same frequency exhibiting slightly different velocities within the measurement interval. Such a situation can be represented as waves arriving over a segment of a solid arc in wavenumber domain. The arc is specified between the range ($k_o-\Delta k$, $k_o+\Delta k$) in the direction of y-axis. If only small spread angles and velocity ranges are considered, coherence can be evaluated as follows:

$$\gamma^2(\vec{r},f) = \frac{1}{2k_o \sin\theta} \int_{-k_o \sin\theta}^{k_o \sin\theta} F(k_y,f)\, \exp[-2\pi i(k_x x)]\, dk_x$$

$$\cdot \frac{1}{2\Delta k} \int_{k_o-\Delta k}^{k_o+\Delta k} F(k_y,f)\, \exp[-2\pi i(k_y y)]\, dk_y \qquad (8)$$

where x and y are the components of the vector separation $\vec{r}$. If $F(k_x,f)$ and $F(k_y,f)$ are defined as unity, the expression reduced to

$$\gamma^2(\vec{r},f) = \left|\frac{\sin(2\pi k_o x \sin\theta)}{2\pi k_o x \sin\theta}\right|^2 \cdot \left|\frac{\sin(2\pi \Delta k y)}{2\pi \Delta k y}\right|^2 \qquad (9)$$

For $x = 0$ the coherence normal to the wavefront is a function of the velocity scatter. Consider this case only, as was shown in Figure 3, the velocity spreads between $\pm$ 0.2 kms^{-1} fit quite well to the real data that explains the loss of coherence in the direction of propagation. If the velocity spread is much larger, i.e. $\pm$ 0.5 kms^{-1}, the coherence will decay faster.

Space Correlation Function --- The seismic ground displacement is known to have a considerable variation from point to point in a half space. For the study of correlation of ground displacement versus spatial separation, different earthquake mechanism provides different forms of spatial correlation. It is assumed that a possible analytical expression for the space correlation function $R_{12}(D)$ of random earthquake displacement with zero mean and variance A^2 takes a form

$$R_{12}(D) = A^2 \exp[-\tilde{a}\ |D|]\cos(2\pi k_o D) \tag{10}$$

where $\tilde{a}$ and k_o are two parameters that control the shape of the spatial correlation. Figure 4 shows the estimated spatial correlation of ground displacement for motion along epicenter direction. Based on Equation (10), the regression line of space-correlation function is also shown in Figure 4. For sensor separation smaller than 0.4 km the correlation of ground displacement is quite high, that is useful information for the seismic analysis of buried pipeline.

SITE-DEPENDENT SPECTRUM

When the seismic wave travels through the ground surface layer, it includes the effects of source function, travel path, attenuation in amplitude, phase change, and the local site characteristics. Generally, the local site effects play an important rule to the response of surface ground movement. To estimate this local site effect, the array site response as well as the response on rock site (outcrop, 4 km south from center station) are used. By assuming the waves, either SH, SV or P waves, propagate upward from rock base to the surface ground, the soil layer transfer function can be estimated as

$$H(\omega) = \frac{S_{XY}(\omega)}{S_{XX}(\omega)} \tag{11}$$

where X and Y denote the rock site and surface ground movements, respectively. $S_{XY}(\omega)$ and $S_{XX}(\omega)$ are the cross-spectrum and auto-spectrum. $H(\omega)$ can be identified as the transfer function of soil amplification of local site conditions.

The effects of variations in properties of soil deposit were studied first. For this purpose, the motion on station E02 (on outcrop) was used as base rock excitation. This assumption is quite correct because the outcrop is only 4 km from center station. Studies were made to compare the effects of velocity response spectra by considering the motion of rock outcrop as well as array response (station C00). The result is shown in Figure 5. In general, these results show the characteristics of soil deposit to the influence of velocity response spectrum. It seems that the local site effect is quite important and has significant effect on the surface ground movements. In order to study the site-dependent soil condition, the transfer function $H(\omega)$ of intermediate soil was estimated. Figure 6 shows the estimated transfer function of soil deposit based on the data of two different earthquakes with the technique of Equation (11). Although the earthquake source mechanism as well as path effect are different, this figure shows the same predominant frequency of side-dependent soil deposit. It is concluded that the local soil condition plays an important rule to the response of ground surface motion.

STOCHASTIC MODEL OF DIFFERENTIAL GROUND MOVEMENTS

In examining the differential motions, it was thought that the differential ground displacements may be partially explained as being a consequence of phase delay in a long-period wave propagating across two sites. From random field theory, the power spectral density function of relative displacement $U_D(r,t)$ between two points x_1 and x_2 during an earthquake is given as

$$S_{U_D}(r,\omega) = S_{U_1U_1}(\omega) + S_{U_2U_2}(\omega) - 2\,\mathrm{Re}[S_{U_1U_2}(r,\omega)] \qquad (12)$$

where $S_{U_1U_1}(\omega)$ and $S_{U_2U_2}(\omega)$ are the displacement power spectra at points x_1 and x_2 respectively, and $\mathrm{Re}[S_{U_1U_2}(r,\omega)]$ is the real part of the cross power spectrum between U_1 and U_2. From Equation (12), $S_{U_1U_1}(\omega)$ was chosen the same form as the local spectrum by defining $D(\omega) = S_{U_1U_1}(\omega)$, then

$$S_{U_D}(r,\omega) = S_{U_1U_1}(\omega)\,\{1 + \frac{S_{U_2U_2}(\omega)}{S_{U_1U_1}(\omega)} - 2\,\mathrm{Re}[R(r,\omega)]\} \qquad (13)$$

The spectrum ratio, $S_{U_2U_2}(\omega)/S_{U_1U_1}(\omega)$, may change the form because of different geometric spreading function and local soil amplification.

The following three items may significantly influence the estimation of power spectrum of ground displacement. This stochastic model of auto-spectrum is formulated and discussed on the basis of source spectrum, path effect and soil amplification respectively.

Source Spectrum --- The form adopted here to model the source acceleration spectrum of an earthquake is given as (Boore[1])

$$S_o(\omega) = C \cdot \left[\frac{\omega^2}{1+(\omega/\omega_c)^2}\right]^2 \cdot \frac{1}{1+(\omega/\omega_m)^8}$$

$$= C \cdot S(\omega,\omega_c) \cdot P(\omega,\omega_m) \tag{14}$$

where C is a constant, $P(\omega,\omega_m)$ is a high cut filter that controls the decay rate at high frequencies, and ω_c is the corner frequency. The parameters of the model can be estimated from the outcrop response through regression analysis. Different earthquake may show different value of parameters.

Path Effect --- The attenuation of seismic waves in amplitude and phase change that characterizes the effect of ground surface layer bottom on propagating seismic wave is given by (Savy[8])

$$A_j(\omega) = \exp[-\frac{t_j\omega}{2\pi Q}]\ \exp[-i\,t_j\,\omega\,(1-\frac{1}{\pi Q}\,\ell n\,|\frac{\omega}{\omega_o}|)] \tag{15}$$

where t_j is equivalent to the time that wave has traveled from the source to the point of interest, Q is the attenuation factor of wave passage and ω_o is the frequency of the filter. Combine Equations (14) and (15), the auto-spectrum at the base of soil layer deposit can be expressed as

$$S_g(\omega) = S_o(\omega)\ |A_j(\omega)|^2 \tag{16}$$

Except Equation (16), the modified Kanai and Tajimi form of rock spectrum may also be adopted as the auto-spectrum of base soil layer deposit (Clough and Penzien[3]), i.e.,

$$S_g(\omega) = S \cdot \frac{1+4\xi_g^{\,2}(\omega/\omega_g)^2}{[1-(\omega/\omega_g)^2]^2+4\,\xi_g^{\,2}(\omega/\omega_g)^2}$$

$$\cdot \frac{(\omega/\omega_1)^4}{[1-(\omega/\omega_1)^2]^2+4\xi_1^{\,2}(\omega/\omega_1)^2} \tag{17}$$

Stochastic Soil-Amplification --- Under the hypothesis of vertical S-wave propagation through infinite horizontal layers, one represents a soil deposit by a simple multi degree-of-freedom dynamic system with n masses and springs. The base of the deposit, corresponding to bedrock or firm ground, assumed to be infinitely rigid. It is postulated that the primary response of surface ground due to bedrock motion will be mostly in shear, therefore the equation of motion of horizontal soil layer is represented as

$$\ddot{u}(x,t) + \frac{C}{\rho}\dot{u}(x,t) = \frac{KG}{\rho}[u''(x,t) + 2\alpha u'(x,t)] - \ddot{u}_g \qquad (18)$$

where ρ is the density of material of soil deposit, C is the damping coefficient, G is the shear modulus of elasticity and α is a constant that consider the variation of shear modulus with respect to the depth. $\ddot{u}_g(t)$ is the acceleration at the base of soil layer deposit. Considering the case of free response, this equation can be solved by normal mode approach. The mode shapes as obtained by Equation (18) are,

$$\phi_n(\xi) = \exp[-\alpha\xi]\sin(\lambda_n^{\,2} - \alpha^2)^{\frac{1}{2}}\xi \;, \quad n=1,2,... \qquad (19)$$

where $\xi = x/L$ and $\lambda_n^{\,2} = \rho L^2/KG \cdot \omega_n^{\,2}$

The Fourier spectrum at the soil surface can be expressed as

$$U_t(\omega,\xi) = U_g(\omega)\,[1 + \sum_n \phi_n(\xi)\,H_n(\omega)] \qquad (20)$$

and $H_n(\omega)$ is the frequency response function of the form

$$H_n(\omega) = \frac{-\omega^2}{(\omega_n^{\,2} - \omega^2) + 2\,\xi_n\,\omega_n\,\omega_i} \qquad (21)$$

The approximate model analysis is feasible to represent the stochastic equivalent local soil amplification. Based on this, the auto-spectrum of surface ground motion can be shown as

$$S_{U_t}(\omega) = S_g(\omega)\,\{1 + \sum_n A_n^{\,2}\phi_n^{\,2}(\xi)\,[-1+|H_i(\omega)|^2]\} \qquad (22)$$

where A_n is the partification factor of n-th mode.

Combined Equations (14), (15) and (22), a stochastic model incorporating the physical properties of source, travel path, and the local site amplification characteristics was developed to simulate the ground motion. By employing the displacement power spectrum of the form, the spectrum ratio as well as the displacement power spectrum of relative ground movement can be simulated. The root-mean-square response of relative ground displacement can be evaluated through the integration of Equation (13) in frequency domain.

Figure 7 shows the estimated spectra ratio from two earthquake data. The solid line was estimated directly from the ratio of auto-spectrum between two separated surface ground motion. The dash line was calculated by the ratio of the estimated soil amplification function at two different sites. Two methods were proposed to the estimation of spectra ratio which show good agreement between these methods. This also confirms that the site-dependent spectrum play an important rule to the response of surface ground movements. Figure 8 shows the estimated Fourier Amplitude of ground motion on outcrop (Station E02). Equation (14) was also used to fit the rock site motion. The estimated parameter is also shown in this figure.

CONCLUSION

The SMART-1 array in Taiwan provides good records for the study of spatial characteristics of seismic waves. From the analysis of space-time correlation function as well as the frequency wavenumber spectrum, they provide strong evidence that good correlation on strong intensity portion of seismic waves may be modelled as plane wave propagation. An inverse problem is presented for estimating the site amplification factor in terms of period by analyzing the data between outcrop and array response. It is found that the site-amplification effect has important relations to the surface ground movements. A stochastic model for determining the differential ground movements was developed here. This model considers the source characteristics, attenuation of wave passages and the local site amplification from which the auto-spectrum of surface ground motion can be developed.

ACKNOWLEDGEMENT

The author wishes to express his deepest thanks to Dr. Y. T. Yeh, Acting Director of Institute of Earth Sciences, Academia Sinica, for providing the SMART-1 Data and discussions for his study.

REFERENCES

1. Boore, D. M. (1983), Stochastic Simulation of High-Frequency Ground Motions Based on Seismological Models of the Radiated Spectra, BSSA, Vol. 73, No. 6, pp. 1865-1894.

2. Capon, J. (1969), High-Resolution Frequency-Wavenumber Spectrum Analysis, Proceedings of IEEE, 57, pp. 1408-1418.

3. Clough R. W. and J. Penzien (1980), Dynamic of Structures, Chapter 28.

4. Harada, T. (1984), Probabilistic Modelling of Spatial Variation of Strong Earthquake Ground Displacements, Proceedings of the 8th World Conference on Earthquake Engineering, pp. 605-612.

5. Harada, T. (1986), Stochastic Analysis of Ground Response Variability for Seismic Design of Buried Lifeline Structures, Proceedings of 7th Japan Earthquake Engineering Symposium, Tokyo.

6. Loh, C. H. (1985), Analysis of the Spatial Variation of Seismic Waves and Ground Movements from SMART-1 Array Data, Earthquake Engineering and Structural Dynamics, Vol. 13, pp. 561-581.

7. Shinozuka, M. and T. Harada (1985), Spatial Variabilities of Seismic Ground Motions and Their Design Implications for Buried Lifelines Structures, Proceedings of the Trilateral Seminar-Workshop on Lifeline Earthquake Engineering, National Taiwan University, pp. 249-264.

8. Savy, J. B. (1981), A Geophysical Model of Strong Motion for Large Ensemble Generation, M.I.T. Research Report R81-6.

9. Vanmarcke, E. H. and R. S. Harichandran (1984), Models of the Spatial Variation of Ground Motion for Seismic Analysis of Structures, Proceedings of the 8th World Conference on Earthquake Engineering, pp. 597-604.

TABLE 1: Some earthquakes Recorded by SMART-1 Array.

EVENT	ORIGIN TIME	DEPTH km	MAG.	AZIM. deg.	DELTA km
39	1986-01-16	10.2	6.5	63.6	22.2
40	1986-05-20	15.8	6.5	195.0	67.9
41	1986-07-30	1.5	6.2	141.9	5.8

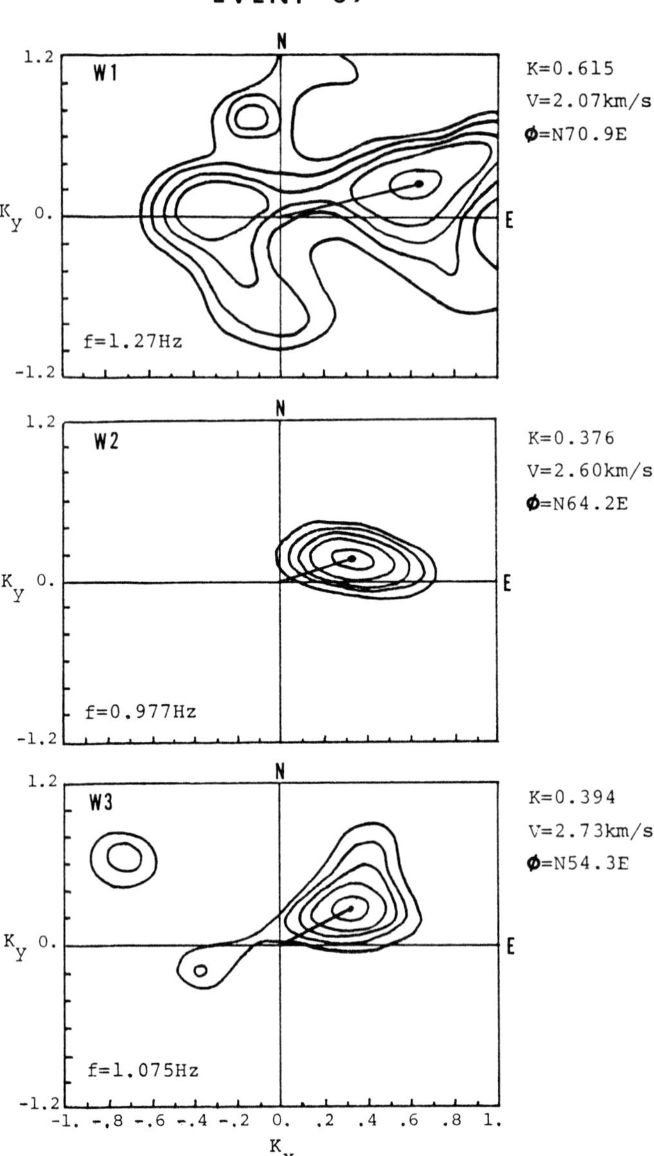

Fig.1: Frequency wave number spectral estimates for three time windows using the East-West motion of Event-39. Power levels are stippled to 5 dB below the maximum level. Contour intervals are 1 dB with respect to the maximum.

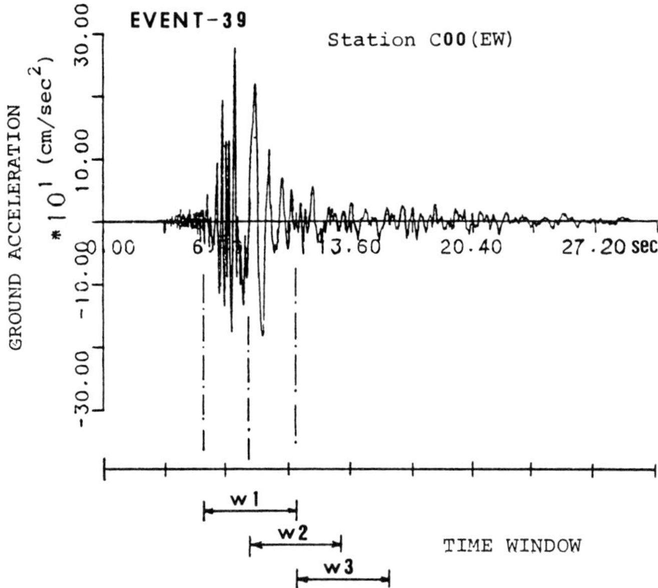

Fig.2: The specified three time wimdows for Frequency-Wavenumber spectral analysis.

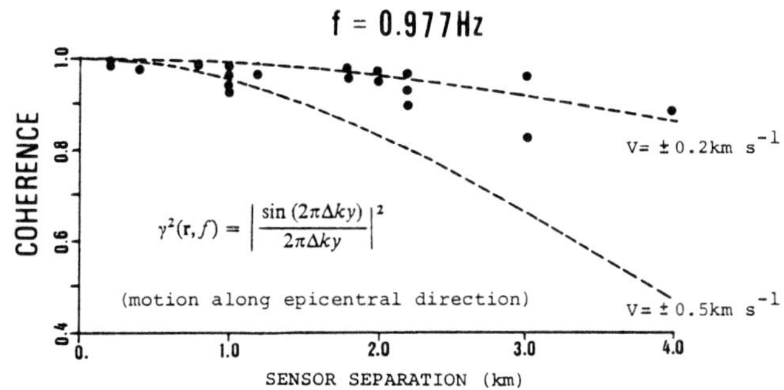

Fig.3: Estimation of spatial coherence at f=0.977Hz for motion along epicentral direction(Event-39).

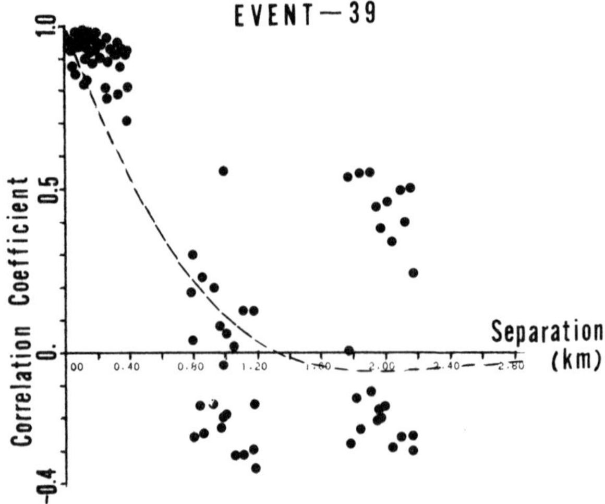

Fig.4: Theoretical and estimated correlation function of displacement records for motion along epicentral direction.

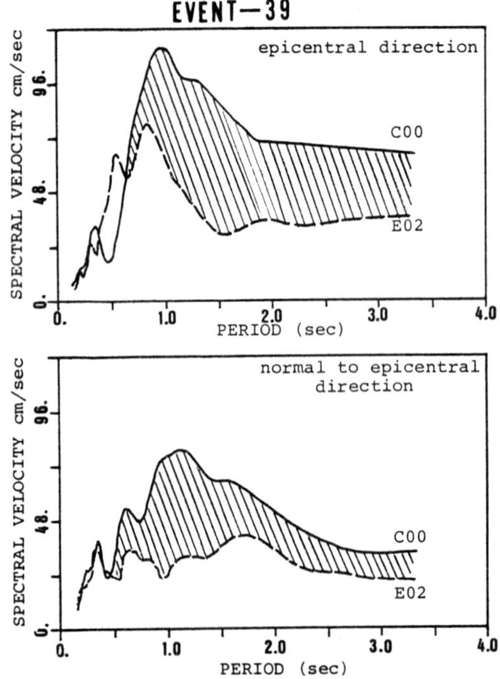

Fig.5: Computed range of response spectra for station C00 (alluvium site) and station E02 (rock site).

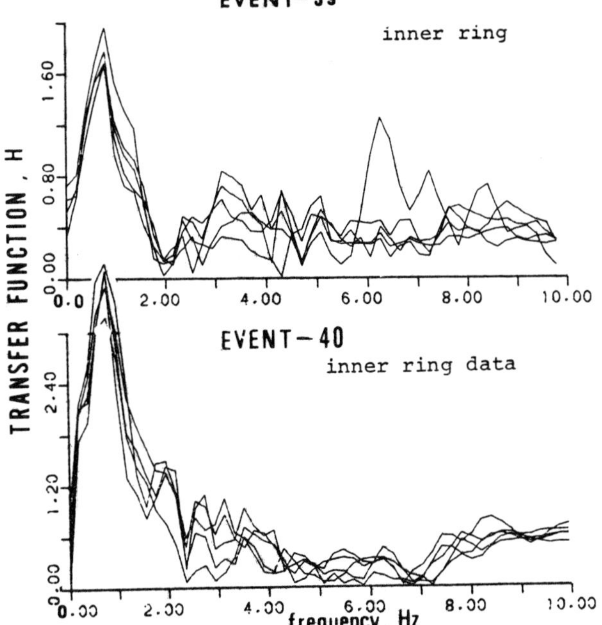

Fig. 6: Estimated transfer function of SMART-1 inner ring array site based on two different earthquake data (Event-39 and Event 40)

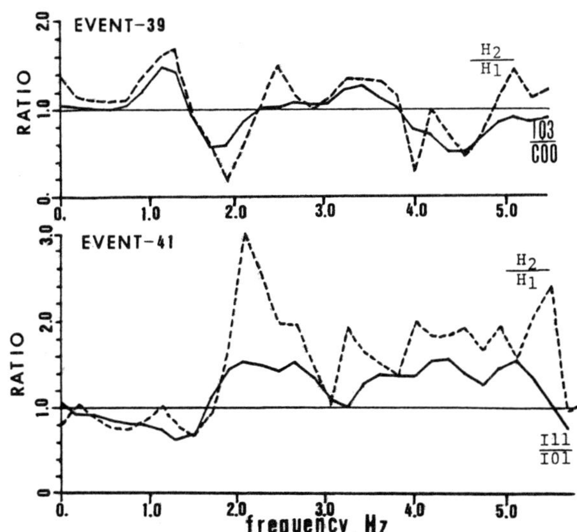

Fig. 7: Comparison of two distinct method of estimation on spectral ratio by using two different earthquake data.

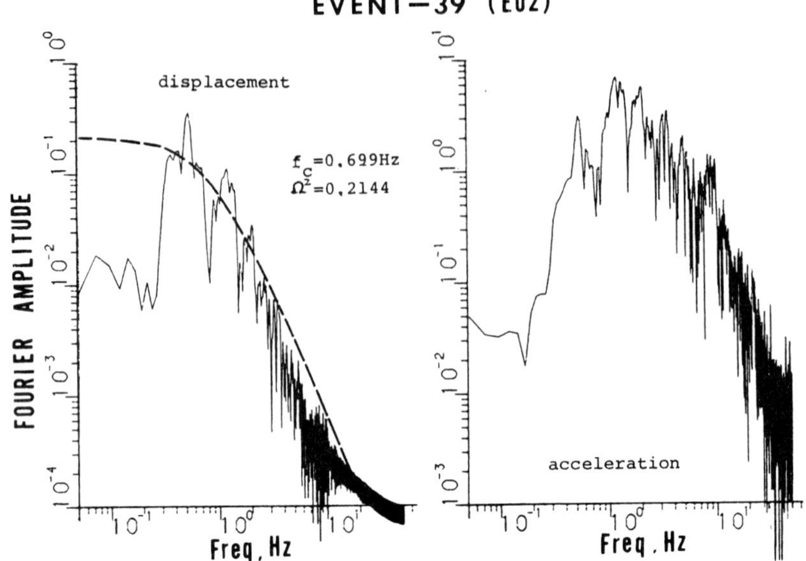

Fig.8: Fourier amplitute spectrum of ground displacement and acceleration of rock site(station E02). Smoothed curve: Model spectrum; Jagged curve: Estimated spectrum.

On the Attenuation of Macroseismic Intensity with Epicentral Distance

G. Grandori, F. Perotti, A. Tagliani
Dipartimento di Ingegneria Strutturale, Politecnico di Milano, Piazza Leonardo da Vinci, 32, Milano, Italy

INTRODUCTION

The relation between macroseismic intensity and epicentral distance is widely used in seismic hazard analysis. The attenuation of macroseismic intensity, I, actually depends both on epicentral distance D and on hypocentral depth h. Therefore, if h is not explicitly taken into account, the relation between I and D must refer to earthquakes having hypocentral depths which span a relatively small range.

Many formulas relating I and D have been suggested. Most of these are either equivalent to or special cases of the following two formulas [1]:

$$I_o - I = a_1 + b_1 \ln D + c_1 D, \qquad (1)$$

$$\ln I_o - \ln I = a_2 + b_2 \ln D + c_2 D, \qquad (2)$$

where I_o is the epicentral intensity.

As mentioned in [1], equations (1) and (2) derive from two different hypotheses about the relation between macroseismic intensity and seismic energy density.

As far as the use of the attenuation law in local hazard analysis is concerned, the following alternative procedures are generally applied.

a) Method of homogeneous zones. The method is based on the hypothesis that over the considered zone the mean annual number of earthquakes per unit area, λ'_o, is constant and that the intensity distribution

$$F_{I_o}(i) = \text{Prob}\ (I_o \leq i)$$

is also constant. $F_{Io}(i)$ and λ'_o are derived from historical data. It is then possible, using the attenuation law, to calculate λ and $F_I(i)$ at any given site.

b) Merely statistic method. The intensity I at the site for each historical earthquake is evaluated through the attenuation law. The series of local events so defined is then statistically interpreted in order to derive λ and $F_I(i)$ at the site.

In the application of formulas (1), (2) and of methods a), b), some difficulties may arise. In this paper such difficulties are discussed and some variants are proposed with particular reference to Italian conditions.

PRELIMINARY OBSERVATIONS ABOUT THE ATTENUATION LAW

As mentioned in [1], equations (1) and (2) "are based on the implicit assumption that the seismic energy is radiated from a point source. The equations can be expected to fail where distance is not large compared to the source dimensions".

This limitation is especially important in the evaluation of seismic hazard at a site which is close to, or inside a seismic region; and this is in general the most interesting case from the engineering point of view. Some changes in the structure of formulas (1), (2) seem therefore to be necessary.

Equation (1) implies that attenuation $I_o - I$ is independent of I_o. This is not always true. As mentioned in (1), "inspection of isoseismal maps will suggest that the rate of attenuation of intensity with distance is often more rapid for small than for large earthquakes".

As far as equation (2) is concerned, note that this is equivalent to

$$I_o - I = I_o \left[1 - \frac{1}{\exp\ (a_2 + b_2\ \ln D + c_2 D)}\ .\right] \qquad (3)$$

This means that, according to equation (2), the rate of attenuation of intensity with distance is more rapid for large than for small earthquakes. This may be true in some cases, but as previously mentioned real situations often show an opposite behaviour.

As an example, fig. 1 shows the correlations (1), (2) obtained using the coefficients a, b, c that have been defined in [1] for the "Cordilleran Province".

Mean values of equivalent radii D_i obtained from isoseismal maps of 19 italian earthquakes are plotted in fig. 2. The hypocentral depths of these earthquakes, evaluated according to Blake's formula [2], range between 10 and 15 km. Fig. 2 shows that, on the average, for the earthquakes here considered the rate of attenuation is more rapid for small than for large earthquake. This kind of behaviour cannot be interpreted by either formula (1) or (2).

PROPOSAL OF A VARIANT OF EQUATION (1).

Many variants of equations (1) and (2) have been proposed in order to adapt them to different conditions. We propose here a new variant of equation (1) which shows a very good fitness to the sample of fig. 2.

Assume as starting point the following simplified variant of equation (1):

$$i = I_o - I = a + b \ln D_i \qquad (4)$$

Equation (4) is equivalent to the relations used by Gutenberg and Richter [3] and Cornell [4].

In order to make the rate of attenuation depend on I_o it is sufficient to introduce D_i/D_o instead of D_i:

$$i = I_o - I = a + b \ln \frac{D_i}{D_o} \quad , \qquad (5)$$

where D_o is the equivalent radius of the highest mapped isoseismal line. From equation (5) we obtain

$$D_i = D_o \alpha e^{i/b} \quad , \qquad (6)$$

where $\alpha = e^{-a/b}$.

Obviously, if D_o is constant, equation (6) coincides with (5). However, if D_o depends on I_o all the radii D_i depend on I_o as well.

Assume the ratio $D_o(I_o=j)/D_o(I_o=j-1)$ to be constant (i.e. independent of j) and call Φ such constant. As a consequence

$$\frac{D_o\ (I_o = j)}{D_o\ (I_o=j-i)} = \frac{D_i\ (I_o = j)}{D_i\ (I_o=j-1)} = \Phi. \tag{7}$$

The constant Φ is a measure of the more (or less) rapid rate of attenuation of small earthquakes compared with large ones. If $\Phi > 1$ the trend is of the type of fig. 2, while $\Phi < 1$ would correspond to the behaviour represented by formula (2) in fig. 1.

The value of Φ for a given set of earthquakes can be calculated as the mean value of the ratios $D_i(I_o=j)/D_i(I_o=j-i)$ The data of the sample of fig. 2 lead to the mean value $\Phi = 1.36$.

Observe now that, for $i \geq 1$, equation (6) implies that the ratio

$$\Psi = \frac{D_{i+1} - D_i}{D_i - D_{i-1}} \tag{8}$$

does not depend either on i or on I_o. It is easy to prove that its value is

$$\Psi = e^{1/b} \tag{9}$$

The experimental value of Ψ can be calculated as the mean value of the ratios (8). For the sample of fig. 2 we obtain $\Psi\ (i \geq 1) = 1.58$.

As to the ratio

$$\Psi_o = \frac{D_1 - D_o}{D_o} \quad , \tag{10}$$

taking into account the condition

$$D_i = D_o \quad \text{for} \quad i = 0 \quad , \tag{11}$$

equation (6) gives

$$\Psi_o = \Psi - 1 \quad . \tag{12}$$

From a qualitative standpoint, equation (12) is acceptable. In fact the rate of attenuation is more rapid for small than for large distances. However equation (12) represents too rigid a constraint from a quantitative point of view. For instance, the sample of fig. 2 leads to a mean value $\Psi_o = 1.07$, while equation (12) would lead to $\Psi_o = 0.58$.

In order to overcome this inconsistency, it is convenient to modify equation (6) by adding a new coefficient:

$$D_i = D_o \beta e^{i/b} + \xi \quad , \tag{13}$$

The ratio Ψ is given again by equation (9), while the value of Ψ_o can now be adapted to experimental data.

Combining equations (13) and (10) with condition (11) we obtain:

$$D_i = D_o \left(1 + \Psi_o \frac{\Psi^i - 1}{\Psi - 1}\right) \quad , \tag{14}$$

i.e.

$$i = I_o - I = \frac{1}{\ln \Psi} \ln \left[1 + \frac{\Psi - 1}{\Psi_o} \left(\frac{D_i}{D_o} - 1\right) \right] . \tag{15}$$

If coefficients Ψ_o, Ψ, Φ are known, equations (7) and (15) completely define the attenuation law, provided that a reference value for D_o is established. This can be done as follows. Using equation (14) derive from the experimental values D_i a mean value $\bar{D}_o$ for each I_o. Impose then that the reference value of D_o through equation (7), minimizes the deviations (squared) from the mean values $\bar{D}_o$. For the sample of fig. 2 one obtains, in this way, $D_o\ (I_i = 10) = 9.3$ km.

In conclusion, the coefficients that define the attenuation law (15), with the help of equation (7), for the sample of fig. 2 are:

$$\Psi_o = 1.07 \quad , \quad \Psi = 1.58 \quad , \quad \Phi = 1.36 \quad ,$$

$$D_o\ (I_o = 10) = 9.3 \text{ km} \quad .$$

The attenuation law so defined is represented in fig. 3.

Using the data of the same sample, the coefficients of the formulas (1) and (2) have been calculated using the least square method. The results are shown in fig. 4.

INFLUENCE OF THE STRUCTURE OF THE ATTENUATION LAW ON THE EVALUATION OF LOCAL HAZARD

Consider a site located at the center of a hypothetical homogeneous seismic zone (fig. 5). The correlation between intensity and return period for this site has been calculated assuming alternatively the attenuation law (1), (2) and (15) with the coefficients derived from the sample of fig. 2. The results are in fig. 6.

The considerable differences clearly point out that the choice of the attenuation law is a crucial step in seismic hazard analysis.

ABOUT THE USE OF ATTENUATION LAW IN SEISMIC HAZARD ANALYSIS AT A SITE.

Consider a seismic region with area A. Suppose that the catalogue of historical earthquakes of the region is given and that for a certain number of the earthquakes the isoseismal map is known. In general for most of the earthquakes only I_o and epicentral coordinates are known. From the available isoseismal maps an attenuation law can be derived and assumed as suitable for the interpretation of all events.

At this point, as mentioned in the introduction, two alternative procedures are available in order to define the seismic hazard at a site.

As far as method a) is concerned, the hypothesis

$$F_{Io}(i) = \text{constant over } A \qquad (16)$$

cannot be avoided. In fact the size of a seismic region that can affect a given site is not very large and it is necessary to take all the events of the region into account in order to compute the distribution function $F_{Io}(i)$. We will assume that intensities I_o comply with hypothesis (16).

As to the spatial distribution of epicenters, if the hypothesis

$$\lambda'_o = \text{constant over } A \qquad (17)$$

is not verified (and this is often the case), the error in the evaluation of the intensity distribution $F_I(i)$ at the site is rather small. On the contrary, the error in the calculation of λ can be considerable.

As an example, a circular seismic region with radius 95 km has been considered. The distribution $F_{Io}(i)$ has been assumed

$$F_{Io}(i) = 1 - \exp(6.91 - 1.15\, I_o) \quad ; \quad I_o \geq 6$$

and the attenuation law (15) with $\Psi_o = 1$, $\Psi = 1.4$ has been used. For the sake of simplicity a constant value $D_o = 8$ km has been adopted.

The evaluation of $F_I(i)$ and λ at the center of the seismic region has been carried on under two different hypotheses:

1) λ'_o = constant over the region;
2) λ'_o = twice the average value used in 1) in the inner zone with radius 65 km; in this case the value of λ'_o in the remaining area has been obtained by imposing the total number λ_o of events in the region to be the same as in 1).

The maximum difference between the ordinates of $F_I(i)$ in the two cases is 6%, while the value of λ for the first hypothesis is about 50% of the value corresponding to the second one.

If the merely statistic method b) is used, the actual distribution of epicenters is automatically taken into account, so that the best estimate of λ is obtained. On the other hand, the evaluation of the distribution $F_I(i)$ becomes uncertain due to the fact that, at the site, the number of events becomes in general very small for the higher intensities.

In conclusion, method a) leads to a good estimate of $F_I(i)$ but may be unreliable for the calculation of λ. Method b) offers the best estimate of λ but is affected by large uncertainties in the evaluation of $F_I(i)$.

Thus, it seems reasonable to adopt a mixed procedure: derive $F_I(i)$ using method a) and calculate λ through method b).

REFERENCES

1. Howell B.F. and Schultz T.R. (1975), Attenuation of Modified Mercalli Intensity with Distance from the Epicenter, Bull. Seism. Soc. Am., Vol. 65. pp. 651-665.

2. Blake A. (1941), On the Estimation of Focal Depth from Macroseismic Data, Bull. Seism. Soc. Am., Vol. 31, pp. 225-231.

3. Gutenberg B, and Richter C.F. (1942), Earthquake Magnitude, Intensity, Energy and Acceleration, Bull. Seism. Soc. Am., Vol., 32, pp. 163-191.

4. Cornell C.A. (1968), Engineering Seismic Risk Analysis, Bull. Seism. Soc. Am., Vol. 58. pp. 1583-1606.

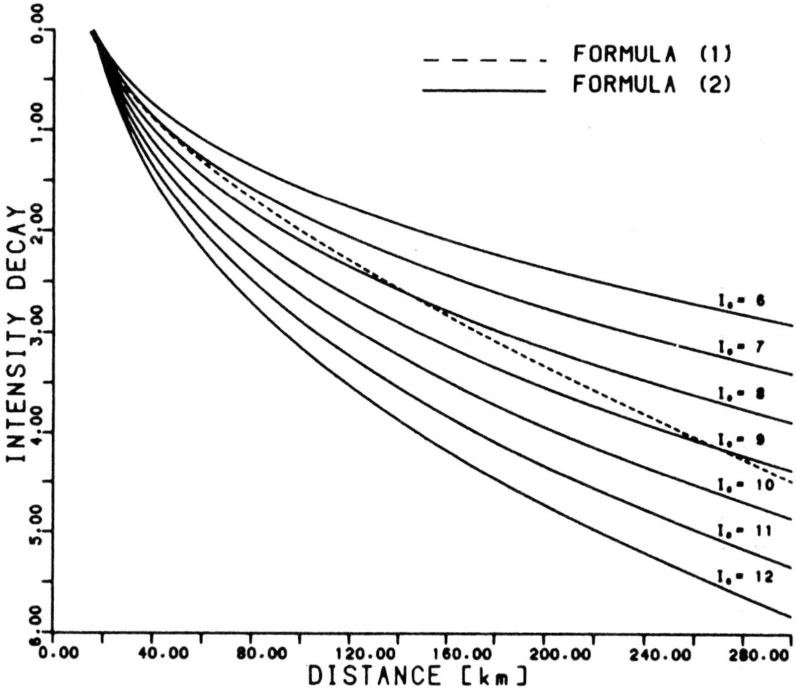

Figure 1. Intensity attenuation with epicentral distance for the "Cordilleran Province" following Ref. [1]

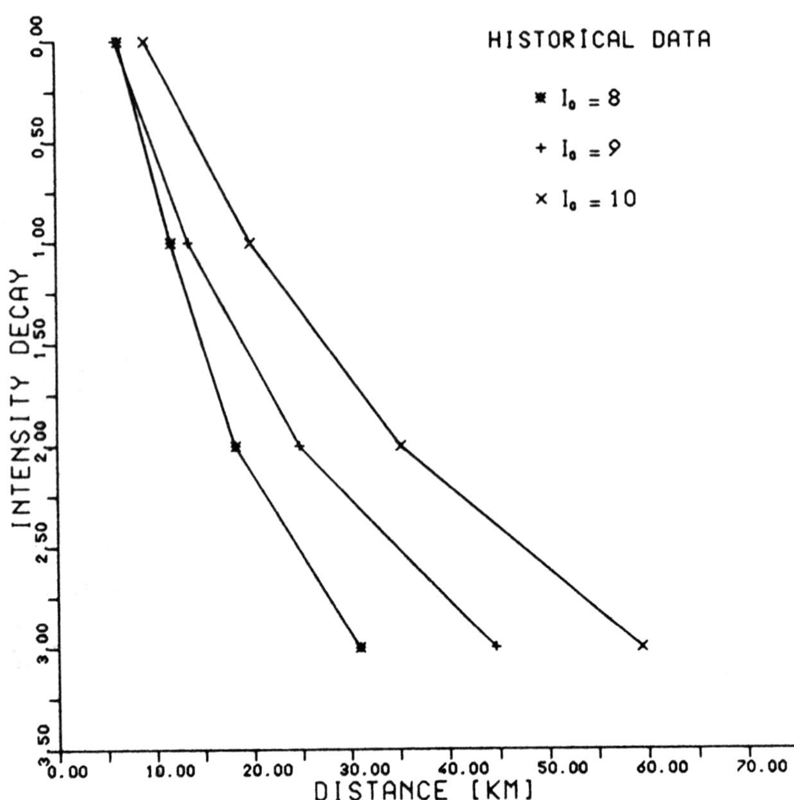

Figure 2. Average intensity attenuation with epicentral distance for 19 Italian earthquakes

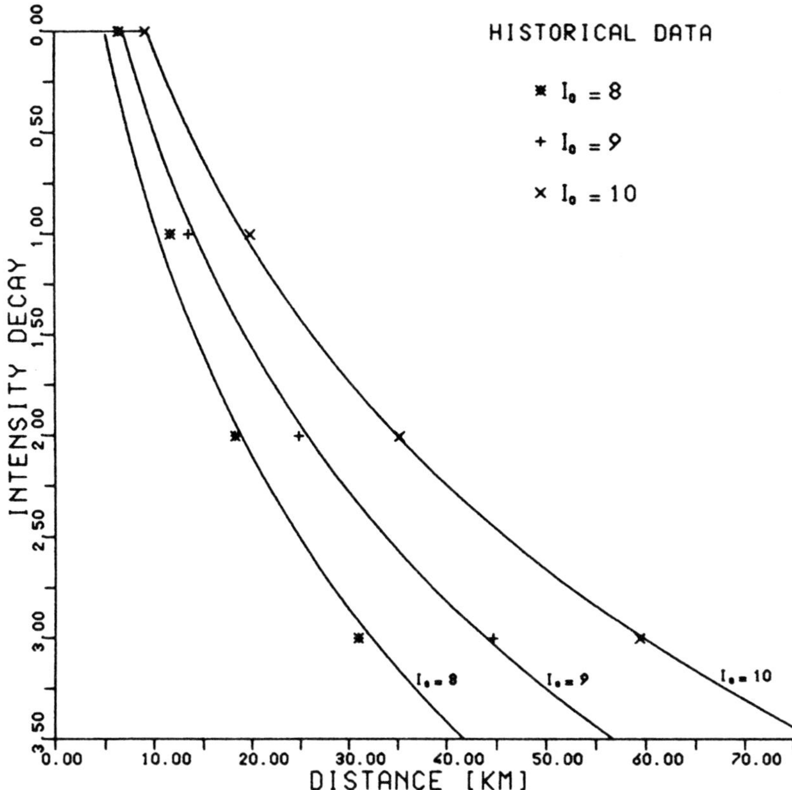

Figure 3. Interpretation of the sample of figure 2 using the proposed law

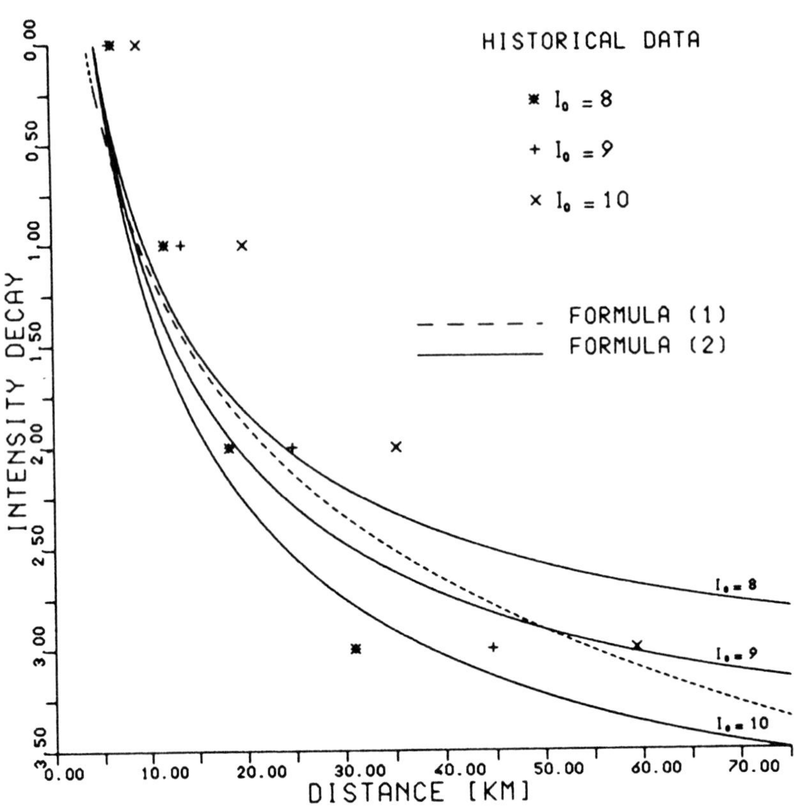

Figure 4. Interpretation of the sample of Figure 2 using formulas (1) and (2)

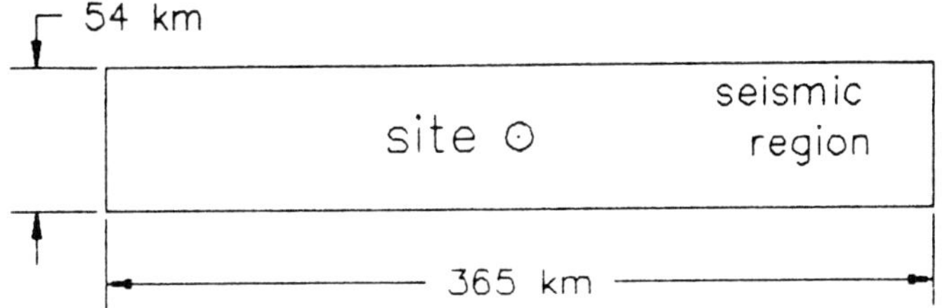

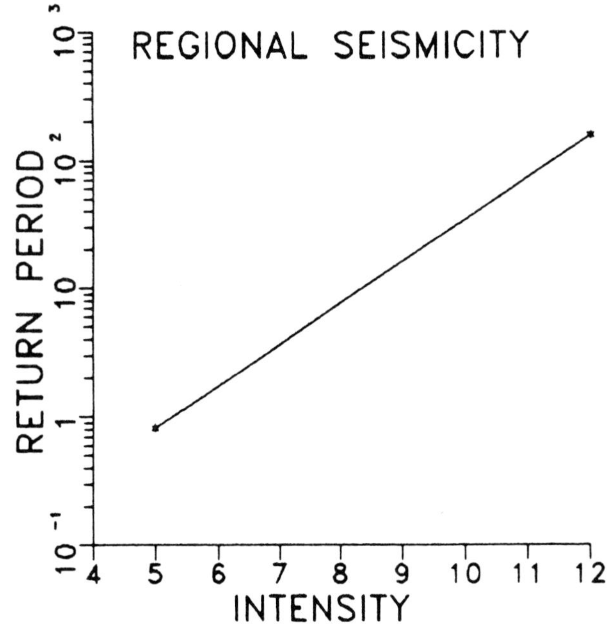

Figure 5. Hypothetical homogeneous seismic region

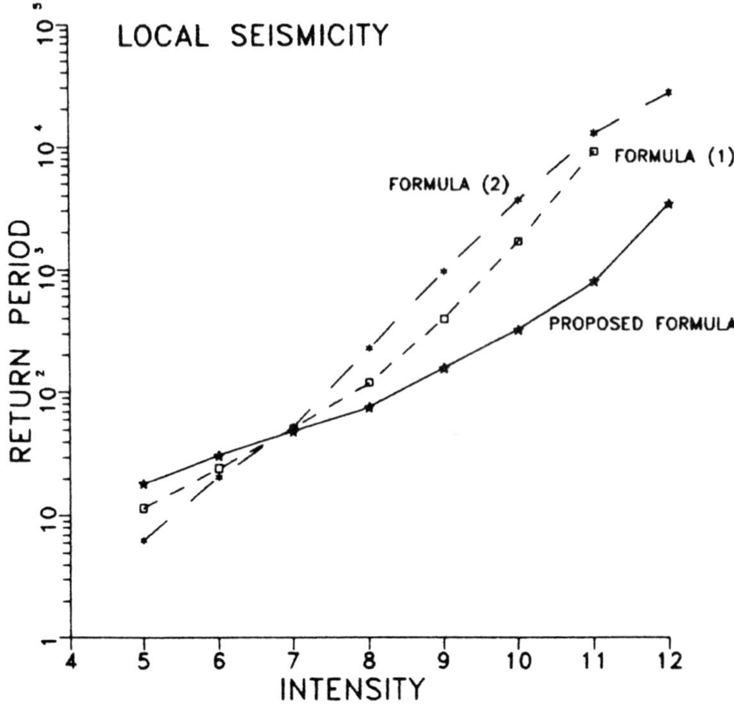

Figure 6. Local seismicity for the site of figure 5 using different attenuation laws

A Scaling Law of Source Parameters Associated with Strong-Motion Accelerograms

M. Kamiyama
Department of Civil Engineering, Tohoku Institute of Technology, Sendai Yagiyama Kasumicho, 982, Japan

INTRODUCTION

In earthquake engineering, faulting source models have recently become a significant tool for predicting strong ground motions. In an attempt to develop faulting model, Hartzell[1] presented a skilful technique in which earthquake ground motions are estimated by taking not only faulting source but also the other factors such as the earth response into consideration. Hartzell's technique is a semi-empirical method since a seismogram observed during a small earthquake is used as a kind of Green function for the faulting source of an objective large earthquake. Although he originally proposed the technique to estimate the ground displacement motions of periods over some seconds, it has also been recently applied in engineering field, where acceleration motions consisting mainly of periods less than some seconds are discussed about, because of its utility(Iida and Hakuno[2]). The estimation of acceleration motions, however, has not been yet conducted satisfactorily because the principle for superposing small earthquake records with regard to the faulting is not so valid as for the case of the displacement motions. Finding the inhomogeneous characteristics of faulting which stimulate acceleration motions, therefore, would be an important key to apply the technique to accelerograms. The purpose of this paper is to make clear general aspects of inhomogeneous faulting source and to obtain a simplified scaling law of source parameters by using a statistical analysis of acceleration spectra.

STATISTICAL ANALYSIS OF STRONG-MOTION SPECTRA

A statistical model for analyzing strong-motion spectra

As is well known, earthquake ground motions are mainly influenced by the three factors , namely, source characteristics, propagation path of waves, and local soil layers. The author proposed a statistical model, as follows, for quantitatively seperating the effects due to the three factors from the strong-motion spectra observed at ground surface(Kamiyama[3]).

$$\log_{10}V(T)=a(T)M+b(T)\log_{10}(\Delta+30)+c(T)D+d(T)+\sum_{i=1}^{N-1}A_i(T)S_i. \quad \text{------} \quad (1)$$

where V(T):motion spectra, M:earthquake magnitude, Δ :epicentral distance, D:focal depth, Si :dummy variable, a(T), b(T), c(T), d(T) ,and Ai(T):regression coefficients, T:period, and N:total number of observation sites.

Although Eq.(1) is such a model that above each factor is expressed simply in terms of earthquake magnitude, epicentral distance, focal depth, and dummy variables, it is very useful to obtain independently the effects due to each factor. The detailed derivation and physical meaning for the model were described fully in Ref.3. Here, to analyze the source characteristics in more detail, the following expression is given as a modified model by adding the power term of magnitude M to Eq.(1):

$$\log_{10}V(T)=a(T)M^2+b(T)M+c(T)\log_{10}(\Delta+30)+d(T)D+e(T)+\sum_{i=1}^{N-1}A_i(T)S_i. \quad \text{----}(2)$$

We use Eq.(2) as a statistical tool for deriving a scaling law of source spectra below. Even though various types of spectra can be applied to V(T) in Eq.(2), velocity response spectra with no damping were used in this paper because they are easily calculated and also nearly equivalent to Fourier acceleration spectra which are necessary to investigate the relation between spectra and source parameters of earthquake.

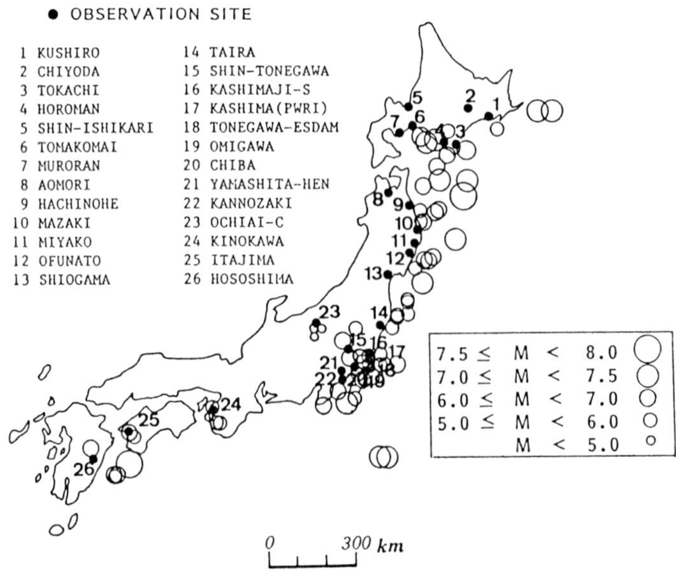

Fig.1 Observation sites and earthquake origins for the strong-motion accelerograms.

Strong motion accelerograms used for the statistical analysis

The statistical model of Eq.(2) was applied to the strong-motion accelerograms described in Ref.3. These are composed of 228 horizontal accelerograms having maximum acceleration more than 20 gal(Tsuchida et al.[4];Iwasaki et al.[5]). The observation sites and earthquake origins for these accelerograms are indicated in Fig.1. As shown in Fig.1, the observation sites are 26 in all, and these accelerograms were obtained on the conditions that earthquake magnitude M ranges from 4.1 to 7.9 (mean=6.4), epicentral distance from 3 to 350 km (mean=77.7 km), and focal depth from 0 to 130 km (mean=43 km).

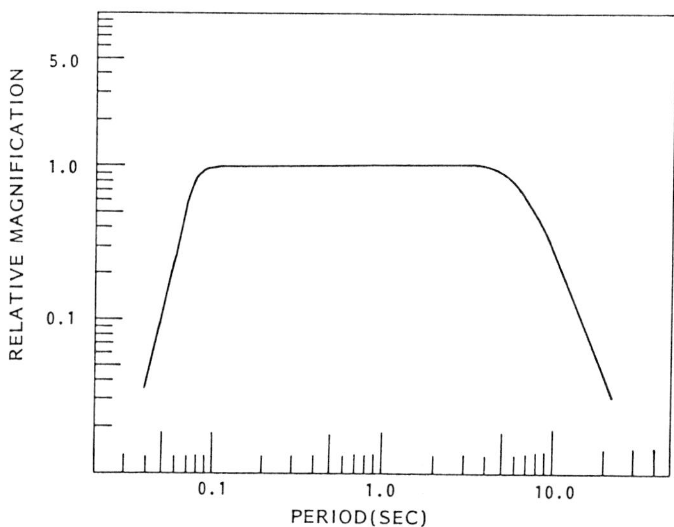

Fig.2 Overall frequency characteristics of the band pass filter.

In order to eradicate various numerical errors involved in the above accelerograms, a band pass filter proposed by Iai et al.[6] was employed. The overall frequency characteristic of the filter is shown in Fig.2. Fig.2 indicates that the spectra used here must be discussed within the flat part of periods ranging from about 0.09 sec to about 4.2 sec. The limited period band should be emphasized, in particular, when the so-called corner frequency of spectra is examined.

Statistical analysis results and averaged source spectra

Velocity response spectra with no damping (h=0.0,h:damping factor) were calculated from the accelerograms mentioned above, and were statistically analyzed in accordance with Eq.(2). In the statistical analysis, as explained in Ref.3, it is needed to pick up one site from all the observation sites as a base site for obtaining amplification factors due to individual local site condition with physical meaning. We selected OFUNATO site, which is labelled 12 in Fig.1, as the base site for the same reason described in Ref.3. We can find very rigid slate at OFUNATO site and its S-wave velocity is inferred from many geological materials to be about 1 ~ 2 km/sec. Accordingly, V(T) estimated by using the first five terms except the sixth of the right hand side in Eq.(2) corresponds to the response spectra at a rock site having such a rigidity. The regression coefficients a(T),b(T),c(T),d(T) and e(T) resulting from the statistical analysis of Eq.(2) are given in Fig.3. On the other hand, the coefficients Ai(T) (i=1 ~ N-1) which signify the amplification factors of spectra for each observation site were gotten as the values almost similar to the ones of Eq.(1). The present aim is to obtain the rock site spectra in connection with earthquake source rather than the amplification factors owing to individual site condition, so the results of Ai(T) are not discussed here.

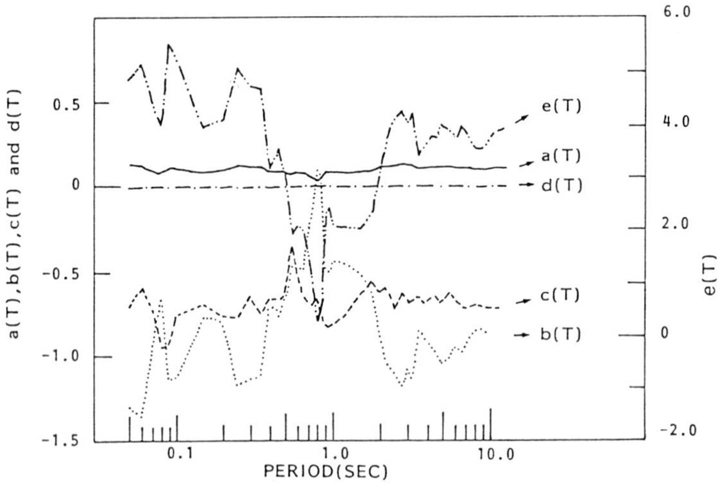

Fig.3 Regression coefficients resulting from the statistical analysis.

The regression coefficients shown in Fig.3 make it possible to estimate the velocity response spectra (h=0.0) at the rock site according to Eq.(2) provided that M, Δ and D are available. Fig.4 presents the rock site spectra which were estimated varying earthquake magnitude while keeping epicentral distance and focal depth constant(Δ =77.7 km, D=43 km). Since these constant distance and depth are equal to the mean values of each parameter used in the statistical analysis, Fig.4 can be understood as the source spectra which are normalized statistically by the mean values of epicentral distance and focal depth and moreover are scaled in terms of earthquake magnitude. In other words, Fig.4 would be regarded as a kind of scaling law of the source spectra deduced statistically from the past strong-motion accelerograms.

Fig.4 reveals the fact that there seems to be two corner periods in short and long period domains even though a little different trend is found according to magnitude M, and that the spectra have nearly flat configurations between the two corners. In addition to being almost the same values around 0.09 sec irrespective of magnitude M , the shorter corner periods are nearly equal to the shorter cutoff period in Fig.2. Hence the shorter corner periods here would be owing to the band pass filter applied to the accelerograms. As for corner period of spectra, Hanks[7] discussed about a higher corner frequency f_{max} in acceleration spectra. To discuss about f_{max} here, we need more wide-ranged spectra, especially, in shorter period domain, but such discussion is beyond our urgent aims. The longer corner periods, on the other hand, exist around 0.5 sec for the magnitudes of M=5 or M=6, and seem to become longer as magnitude gets larger. In addition, the existence of the longer corner periods becomes rather invalid together with increasing magnitude. In any case, the spectra shown in Fig.4 may reflect the characteristics peculiar to source mechanism.

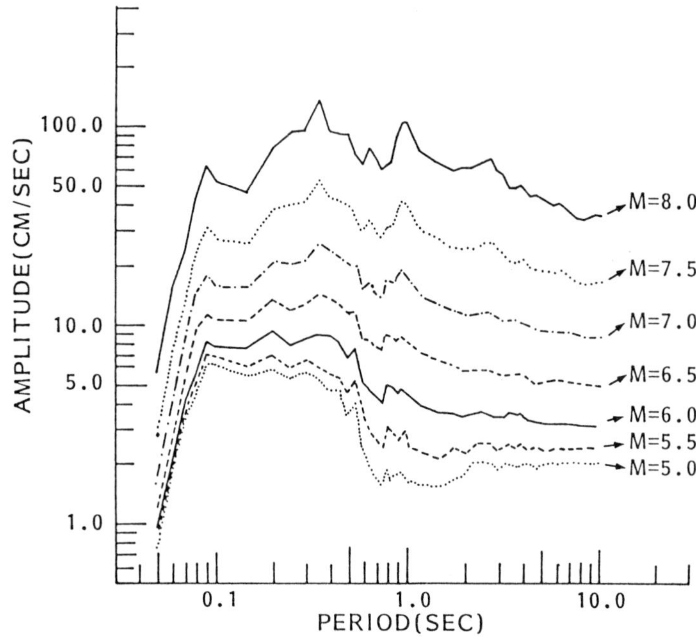

Fig.4 Rock site spectra statistically scaled with earthquake magnitude M.
(Epicentral distance Δ =77.7 km)
(Focal depth D=43 km)

It is almost impossible for the conventional faulting model like Haskell's one[8] to explain such high spectral amplitude in shorter period domain as shown in Fig.4. This drawback to the conventional models has lead to the recent inhomogeneous faulting models which have non-uniform parameters on their fault planes so that acceleration waves are more strongly originated in shorter period domain(Hirasawa[9]). Although several heterogeneous faulting models have been presented by many researchers, the so-called specific barrier model by Papageorgiou and Aki[10] is not only comparatively simple but also easy to understand physically. Furthermore it can effectively offer the expected values of source spectra . Thus this model can satisfy our requirements, and it is adopted as the theoretical model to explain our statistical spectra.

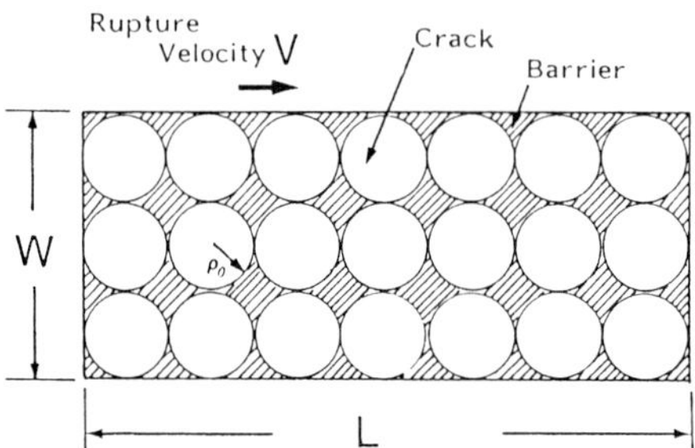

Fig.5 Papageorgiou and Akis' specific barrier model of faulting.

Papageorgiou and Aki considered such an idealized rectangular faulting model with the sizes of length L and width W as illustrated in Fig.5. Their model consists of the localized circular cracks with constant radius ρ_0 , which are assumed to slip according to the manner modeled by Sato and Hirasawa[11], and barriers distributed over the fault plane. Suppose now that above idealized inhomogeneous fault exists as shown in Fig.6. Then, following Papageorgiou and Aki, we can derive the expected value $<|F(\omega)|>$ for horizontal acceleration Fourier spectrum at a point, which has free surface and is r_0 away from the center of the fault, as follows(Kamiyama[12]):

$$<|F(\omega)|> = \frac{2}{\sqrt{2}} \frac{R^S}{4\pi\beta r_0} \sqrt{S} \left(\pi \frac{\Delta u'_{max}}{\rho_0}\right) v\beta C \qquad \text{---------} \quad (3)$$

where R^S : radiation pattern peculiar to shear dislocation, $S=L\times W$ (fault area), r_0 : distance from the center of the fault to the point, $\Delta u'_{max}$: local maximum dislocation of each crack, ρ_0 : radius of each crack, v:rupture velocity, β:shear wave velocity, and

$$C = \frac{1}{\sqrt{2}} \int_0^{\pi/2} \frac{\sqrt{1+k^2}}{1-k^2} \, d\theta ,$$

where $k = v \sin(\theta) / \beta$.

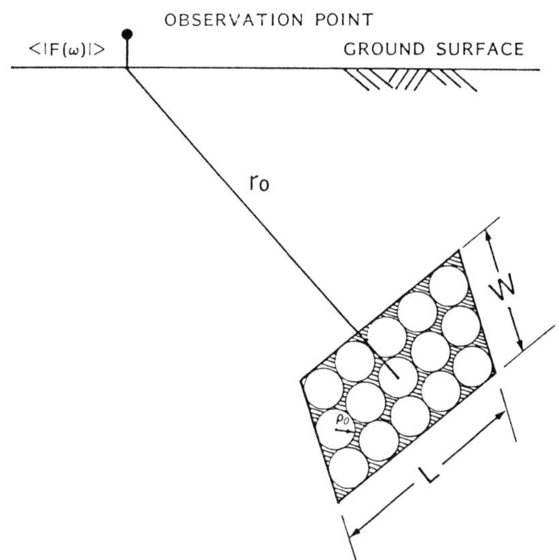

Fig.6 Schematic profile of faulting model and observation point.

SOURCE CHARACTERISTICS INFERRED FROM THE COMPARISON BETWEEN THE STATISTICAL SPECTRA AND THEORETICAL ONES

Eq.(3) makes it possible to predict the expected value of Fourier acceleration spectra $<|F(\omega)|>$ provided that some fault parameters like S, ρ_0, β and so on are available. Speaking conversely, however, it is also an expression to estimate the fault parameters from $<|F(\omega)|>$. As explained previously, the spectra in Fig.4 correspond to the $<|F(\omega)|>$ statistically scaled with earthquake magnitude M. Therefore, it follows that we can obtain how the unknown parameter ρ_0 is scaled with M by incorporating the spectra in Fig.4 with Eq.(3).

In the above discussion, it is an important problem whether or not the spectra in Fig.4 are composed of shear waves, since Eq.(3) was derived from only shear waves. By the way, Hanks [13] and Kamiyama [14] showed individually that shear waves dominate in shorter periods than about 1 ~ 2 sec, whereas surface waves are predominant in periods longer than that. As will be shown later, the spectral amplitudes in Fig.4 are dealt with almost within such period range of shear waves. So this suggests that there is little difference in waves between Eq.(3) and Fig.4 .

Among the parameters involved in Eq.(3), fault area and maximum dislocation of faulting were relatively well revealed from the past many earthquakes and they have been associated empirically with earthquake magnitude. Here, refering the empirical expression presented by Sato[15], the following expressions are obtained(Kamiyama[12]):

$$\sqrt{S} = 10^{0.5M-2.04} \quad , \text{----} \ (4) \qquad \Delta u'_{max} = \frac{6}{\pi} 10^{0.5M-1.4} \quad . \text{----} \ (5)$$

The substitution of Eqs.(4) and (5) into Eq.(3) gives

$$\rho_0 = \left(\frac{2}{\sqrt{2}} \frac{R^S}{4\pi r_0} 6\ v\ C\ 10^{M-3.44}\right) / <|F(\omega)|> \ . \text{----} \ (6)$$

We can estimate the values of ρ_0 varied according to M by using Eq.(6) as well as the spectra in Fig.4. In order to do so, we further need the constants R^S, β, v and r_0 appropriate for the spectra in Fig.4. These constants are determined as follows.

The Fourier spectra $<|F(\omega)|>$ in Fig.4 were obtained through the statistical dealing of the accelerograms from many observation sites distributed at random around various types of earthquakes as described previously. That is, the averaged radiation pattern between various sources and observation points may be proper to the present spectra. Thus R^S is assigned to be 0.63 which is the expected value of the radiation pattern on the focal sphere for shear dislocation. On the other hand, the statistical spectra in Fig.4 were also given as the spectra at the outcrop having shear wave velocity of 1 ~ 2 km/sec as well as due to the earthquake origin of focal depth D = 43 km. It would be quite difficult to allot an adequate shear wave velocity for such origin and observation site , but its precise value may be meaningless if we consider the degree of roughness for the other parameters. In the present case, accordingly, β = 3.5 km/sec is assumed in consideration of the averaged shear wave velocity of the crest around Japan. According to this assumption, v is set to be 2.52 km/sec using Geller's expression v/β =0.72[16]. Also r_0 =88.8 km was conditioned because the spectra in Fig.4 were obtained as Δ =77.7 km and D=43 km.

Table 1
Circular crack radius and total number of cracks obtained from the statistical spectra.

Magnitude	5.0	5.5	6.0	6.5	7.0	7.5	8.0
Period range for $<\|F(\omega)\|>$ (sec)	0.09 ~ 0.45	0.09 ~ 0.55	0.09 ~ 0.55	0.09 ~ 0.55	0.09 ~ 1.20	0.09 ~ 1.80	0.09 ~ 4.20
$<\|F(\omega)\|>$ (cm/sec)	5.54 ±0.57	6.12 ±0.64	8.11 ±0.64	12.10 ±1.24	18.24 ±3.23	35.26 ±7.20	71.55 ±19.56
Radius of cicular crack ρ_0 (km)	0.14 ±0.02	0.41 ±0.04	0.99 ±0.08	2.09 ±0.21	4.38 ±0.78	7.17 ±1.46	11.17 ±3.05
Number of crack n	103 ±11	39 ±4	22 ±2	15 ±2	13 ±2	13 ±3	17 ±5

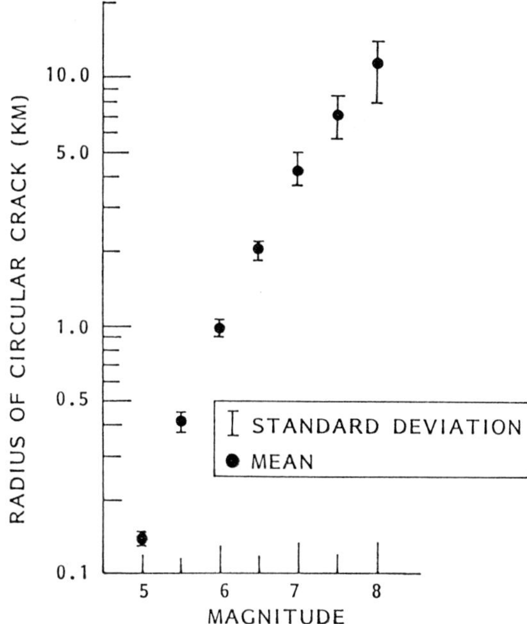

Fig.7 Relation between circular crack radius and earthquake magnitude.

By use of the above constants, we obtained Table 1 after calculating the average spectral amplitudes in the high level parts of each spectrum in Fig.4 and substituting them to Eq.(6). In Table 1, the mean and standard deviation for $<|F(\omega)|>$ and ρ_0 are tabulated together with the period range for averaging the spectral amplitudes. In addition, the mean value and standard deviation of total crack numbers, which are obtained by using ρ_0 - S relations, are also shown in Table 1. The period ranges for calculating the averaged spectral amplitude were determined in accordance to each magnitude M, considering the frequency characteristic of the filter shown in Fig.2 as well as the level and flatness of the spectral amplitudes.

The crack radius ρ_0 and total crack number n in Table 1 are plotted against the variations of M in Figs.7 and 8, respectively. We can see from Table 1 and Fig.7 that the radius of localized crack increases with earthquake size, having about 1.0 km for M=6 and about 11 km for M=8. From Fig.7 and Table 1, the regression expression between the radius ρ_0 and earthquake magnitude M is given by

$$\rho_0 = 10^{0.67M-3.87} \quad \text{-------------------- (7)}$$

From Table 1 and Fig.8, on the other hand, the total crack number n of faulting appears to be minimum around M=7, while it gets larger with decreasing magnitude and slightly larger with increasing magnitude. The increasing trend of total crack number is emphasized especially in smaller magnitude than 6. As mentioned above, the total crack number here was estimated by using the empirical relations of S-M in Eq.(4) as well as the values of ρ_0 in Table 1. However, fault area S of earthquake may be less reliable in such small magnitude because of the difficulty of estimation. For this reason, it should be further examined after due consideration whether the characteristics in the smaller magnitude are actually valid or not. In contrast to the above trend in the smaller magnitude, total crack number n seems to be rather stable in earthquake magnitude greater than 6. Earthquake damages, in general, are caused by such class of magnitude, so the total carck number in the domain of $M \geqq 6$ would be more important from the view point of engineering. In this domain, total crack numbers containing their statistical standard deviations are almost within the extent from 10 to 20, although there is an amount of scatter. When we consider the roughness of the present method, it would be relevant to conclude that the total crack number of faulting in such a size of earthquake as giving damages is relatively stable irrespective of magnitude and is from 10 to 20 at least or at most.

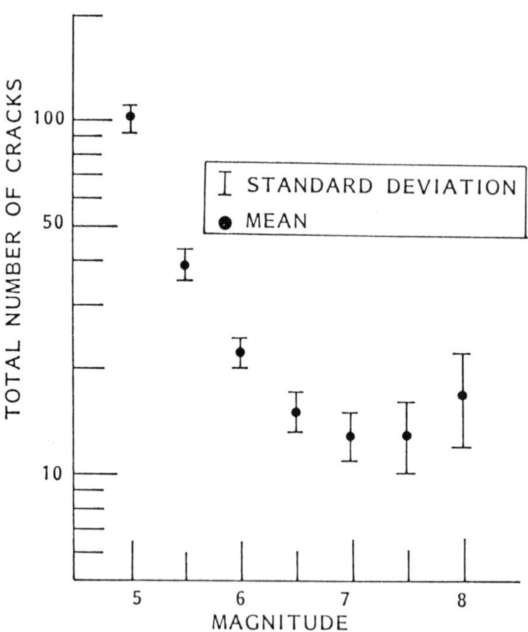

Fig.8 Relation between the total number of cracks and earthquake magnitude.

INTERPRETATION OF THE ESTIMATED SOURCE PARAMETER FROM THE PAST OBSERVED DATA

As explained previously, heterogeneous faulting model recently has become to arouse researcher's interest, so the observed data about the model are still little except Aki's data [17]. Aki obtained some values of the barrier interval of fault from the past representative earthquakes throughout the world by making use of geological materials or seismic ones. Here, Fig.7 is compared with Aki's results in order to investigate its validity. Aki's results are plotted in Fig.9 after arranging as ρ_0 -M relation. In Fig.9 the similar relations obtained by the present study are also shown. Fig.9 indicates that the present study is relatively compatible with Aki's results, even though there is a little difference between the both. Generally speaking, the parameters of heterogeneous fault such as ρ_0 are very difficult to exactly seize and besides Aki's results were deduced from some very old earthquakes. Hence, the compatibility between the both would not be bad if we consider such ambiguities belonging possibly to Aki's results.

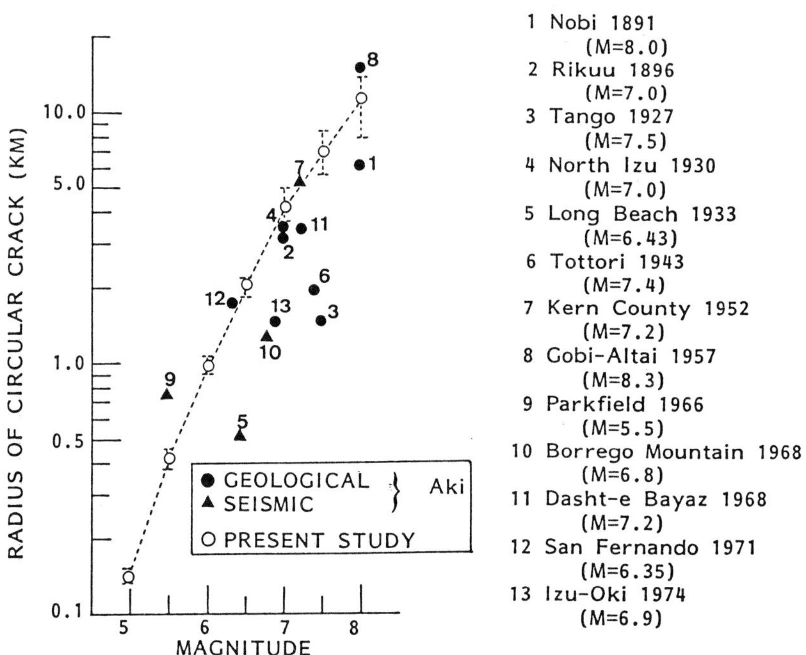

Fig.9 Comparison of the present study with Aki's results for the crack radius - earthquake magnitude relation.

CONCLUDING REMARKS

The principal conclusions drawn from this study are summarized as follows.

(1) The acceleration spectra statistically obtained from the strong-motion records have nearly flat spectral configuration between the two corner periods,and the configuration can be explained quantitatively well by the inhomogeneous faulting model of Papageorgiou and Aki.

(2) The conjunction of the statistical spectra and the inhomogeneous faulting theory lead to a scaling law about the localized crack radius of inhomogeneous fault as presented in Eq.(7). According to the law, the radius increases with earthquake magnitude M, and for instance, it has a size of about 1.0 km for M=6 and about 11.0 km for M=8.

(3) The total number of cracks for inhomogeneous fault is relatively stable regardless of earthquake magnitude, particularly, in earthquake size of causing damages to structures. That is, the total crack number is from 10 to 20 at most or at least between M=6 and M=8.

(4) The ρ_0 -M relation obtained by the present study was confirmed to be consistant to the data observed geologically or seismologically during the past representative earthquakes.

Papageorgiou and Akis' model is such a very idealized model that individual crack distributed on the fault plane is assumed to have an equal size, and therefore rather different results may be derived if the inequality of crack size is considered. In this point, the above results should be regarded as an averaged aspect of inhomogeneous fault. Even if so limited, however, the presented results would be useful to estimate acceleration ground motions with aid of faulting source model, for example, Hartzell's method.

REFERENCE

1. Hartzell,S.H.(1978), Earthquake Aftershocks as Green's Functions, Geophys. Letters , Vol.5,No.1,pp.1-4
2. Iida,M. and M. Hakuno(1983), The Synthesis of the Accelearion Wave in a Great Earthquake by Small Earthquake Records,Proc. of JSCE, No.329, pp.57-68(in Japanese)
3. Kamiyama,M. and E. Yanagisawa(1986), A Statistical Model for Estimating Response Spectra of Strong Earthquake Ground Motions with Emphasis on Local Soil Conditions, Soils and Foundations, Vol.26, No.2, pp. 16-32
4. Tsuchida,H. et al.. Annual Report on Strong-Motion Earthquake Records in Japanese Ports(1986-1983), Technical Note of the Port and Harbour Research Institute No.98 No.458, The Port and Harbour Research Institute, Ministry of Transport
5. Iwasaki, T. et al.. Strong-Motion Acceleration Records from Public Works in Japan No.1 No.8(1978-1983), Technical Note of the Public Works Research Institute, Vol.32- 38, The Public Works Research Institute, Ministry of Construction
6. Iai, S., E. Kurata and H. Tsuchida(1978), Digitization and

Correction of Strong-motion Accelerograms, Technical Note of the Port and Harbour Research Institute, No.286, pp.5- 56, (in Japanese)

7. Hanks,T.C.(1982), fmax , Bull.Seism.Soc.Am.Vol.72, pp.1867 -1879
8. Haskell,N.A.(1964), Total Energy Spectral Density of Elastic Wave Radiation from Propagating Faults, Bull.Seism.Soc.Am., Vol.54,no.6,pp.1811-1841
9. Hirasawa,T.(1978), Prediction of Maximum Acceleration due to a Stochastic Source Model, Report of Special Research on National Disasters, No.A-54-3, pp. 35 45,1979 (in Japanese) Vol.21, pp.415-431
10. Papageorgiou, A.S. and K. Aki(1983), A Specific Barrier Model for the Quantitative Description of Inhomogeneous Faulting and the Prediction of Strong Ground Motion, Part and Part , Bull.Seism.Soc.Am.Vol.73,pp.693-722, pp. 953-978
11. Sato,T. and T. Hirasawa(1973), Body Wave Spectra from Propagation Shear Cracks, J. Phys. Earth. Vol.21, pp.415-431
12. Kamiyama, M.(1987), Earthquake Source Characteristics Infered from the Statistically Analyzed Spectra of Strong Motions with Aid of Dynamic Model of Faulting, Proc. of JSCE (in print)
13. Hanks,T.C.(1975), Strong Ground Motion of the San Fernando, California, Earthquake: Ground Displacements, Bull. Seism. Soc. Am. Vol.65, pp.193-225
14. Kamiyama, M.(1985). A Study on Strong Earthquake Ground Motions with Application to Prediction, Doctoral thesis, Tohoku Univ.(in Japanese)
15. Sato,R.(1979), Theoretical Basis on Relationships between Focal Parameters and Earthquake Magnitude, J. Phys. Earth Vol.27, pp.353-372
16. Geller,R.J.(1976), Scaling Relations for Earthquake Source Parameters and Magnitudes, Bull. Seism. Soc. Am. Vol. 66,No.5, pp.1501-1523
17. Aki,K.(1984), Asperities , Barriers Characteristic Earthquake and Strong Motion Prediction, J. Geophys. Res. Vol.89, No.B7, pp.5867-5872

A Non-Linear Site Response Analysis for Earthquake Loading Using an Explicit Time-Stepping Finite Element Code

J. Evans, S. Hendry, J.C. Miles and J. Pappin
Ove Arup & Partners, 13 Fitzroy Street, London W1P 6BQ, England
R. Stubbs
H.M. Nuclear Installations Inspectorate, Health & Safety Executive, St Peter's House, Bootle, Merseyside L20 3LZ, England

INTRODUCTION

It has been recognised for a long time that the free surface ground motion resulting from an earthquake can be strongly influenced by the specific geological nature of the site at which it occurs. As a result, it is not usual for "site specific" seismic studies to be carried out where new plant are to be constructed, particularly if the plant is of a hazardous nature.

The nature of a site specific study is, however, complex. By definition, it is necessary to take account of the particular geological make-up of the site; the soil properties; the effects of soil/structure interaction and the effects of different types of seismic waves and their mutual interference.

This paper singles out the first two of the above parameters, (i.e. the particular geological make-up of the site and the soil properties) and sets out to demonstrate the influence which they can exert over the predicted ground surface motions. Special attention is paid to the non-linear nature of the soil properties and a number of sensitivity studies are referred to in order to demonstrate the effects which uncertainty in soil properties might have on the predicted free surface results.

It is emphasised that the purpose of the paper is to describe a numerical approach which has been found useful in assessing the response of different sites, rather than to report on the findings of any one particular site. The motivation for the work described is that, without some type of numerate approach to the problem, it is not possible to make confident assertions about the characteristics of a site. Too many parameters are involved for simple rules of thumb to be universally applied.

Throughout the paper, the zero period acceleration (ZPA) is the parameter which is used for the purposes of comparing one calculation with another.

THE SITE

This study was carried out for a site which is shown schematically in Figure 1. It consists of bedrock (considered to be rigid) at a depth of 88m below the ground surface, overlain by strata of clay and sand at the levels shown in the figure.

At each level, a stress-strain curve of the type shown in Figure 2 has been assigned to the soil. There is always some uncertainty in assigning materials properties to soil and so a range of properties is indicated in Figure 2. Between the upper and lower bounds a "best estimate" curve has been indicated. The sensitivity of the site response to uncertainties in the soil properties has been investigated by using a range of soil characteristic between the upper and lower bounds. The results are presented in the following sections.

THE MATHEMATICAL MODEL

The mathematical model adopted for the study is shown in Figure 3. It consists of a simple soil column built out of eight noded brick elements using the finite element program DYNA-3D (Ref. 1). This program enables non-linear soil properties to be modelled at the same time as following the dynamic transient nature of the seismic excitation.

Using DYNA-3D, the stress-strain characteristics of the soil were modelled using a multi-linear approximation to the true characteristics as shown in Figure 4. This approximation allows the soil behaviour to be represented during loading and unloading cycles and, as a consequence, it allows mechanisms which gives rise to hysteretic energy losses (or damping) in the soil to be modelled directly. The multi-linear material property is not a standard option within DYNA-3D; it was achieved in this case by overlaying a series of elastic/perfectly-plastic elements in the same location at each level in the model and connecting them together at their common nodes. It is worth noting that this approach allows "looping" to occur as shown in Figure 4. This is an important means of dissipating energy which is not always dealt with satisfactorily.

The input motion was applied as a velocity time-history at the base of the soil column. The input excitation was arbitrarily chosen, since the main purpose of the excercise was to study the relationship between the bedrock (input) motion and the free surface (output) motion. An unmodified medium site time-history, representing a 10^{-4} event in the U.K., was used. The time-history input motion is shown in Figure 5.

Although the mathematical model is comprehensive in terms of its ability to model dynamic and non-linear effects, it must be emphasised that it suffers from a number of limitations that should not be overlooked. In particular, the following points should be noted:

i) The model admits only vertically propagating P-waves and S-waves. Horizontal and inclined waves have not been considered.

ii) The only form of excitation applied to the model was a vertically propagating S-wave.

iii) There is no allowance in the model for soil/structure interaction effects. This paper therefore confines itself to a study of free surface effects.

iv) The bedrock has been considered to be rigid and to be the source of the input excitation. In reality, this is not the case but, since the study was comparative, this assumption has no great significance.

There are no fundamental theoretical reasons why any of the factors above should be excluded from the analysis, since DYNA-3D could allow all of them to be accounted for if need be. However, practical considerations meant that the model had to be limited in its generality at the time when the work was carried out.

PARAMETRIC STUDIES

Parametric studies were carried out in two phases.

i) Phase 1 To examine the degree of site amplification or attenuation which might be present for different intensities of earthquake. This exercise was carried out keeping all the soil properties at their "best estimate" values and varying the intensity of the bedrock input motion from the equivalent of 0.01'g' through to 0.35'g'.

ii) Phase 2 To examine the degree to which amplification or attenuation may be dependent on soil properties under strong motion inputs. This was carried out by applying the equivalent of a 0.35'g' earthquake at bedrock level and then varying the soil properties within the upper and lower bounds suggested in Figure 2. Variations were also carried out with high and low hysteretic damping ratios in the soil layers. (This was achieved by changing the curvature of the stress/strain characteristics for each soil layer).

Phase 1 Results

The results of this study are shown in Figure 6; the peak ground surface acceleration is plotted against the peak bedrock acceleration. It can be seen from this diagram that the site amplifies the bedrock motion at low intensities of excitation, but attenuates it at higher levels of excitation. This is in keeping with findings from practice (eg. as reported by Seed and Idriss - Ref.2), and it occurs because the degree of soil damping is related to the magnitude of the soil strain. Lower levels of bedrock excitation do not cause large soil strains and, under these conditions, there is very little energy loss in the soil. It behaves essentially as an undamped elastic column excited at its foundation. As the level of excitation grows, however, the strains in the soil become larger and the non-linear nature of the material becomes more apparent. A lot of energy can be absorbed by soil hysteresis and gradually the tendency of the site to amplify bedrock motions is reduced. These effects become more apparent as the input intensity increases and eventually the site begins to attenuate the bedrock motion.

Phase 2 Results

The results of this study are shown in Table 1. In all cases the input motion at bedrock level was scaled to the equivalent of 0.35'g'.

The time-history input signal at bedrock level is shown in Figure 6 and the time history output signal computed at the ground surface in each case is shown in Figure 7. The properties of the soil strata were varied systematically as follows:

i) Case A All sand layers and all clay layers set at their "best estimate" properties.

ii) Case B All sand layers and all clay layers given upper bound properties.

iii) Case C All sand layers and all clay layers given upper bound properties but the stress-strain characteristics in the lower strata modelled in greater detail (ie. with a better multi-linear approximation to the time curve).

iv) Case D All sand layers given properties between the best estimate and upper bound curves. Properties for the lower strata similar to Case C above but modified to give higher hysteretic damping than before.

v) Case E The same soil characteristics as Case D but with the properties of the lower strata modelled to give lower hysteretic damping than Case C.

Model	Peak Acceleration
A - Best estimate properties	3.08 m/s^2 (0.314g)
B - Upper bound properties	5.00 m/s^2 (0.512g)
C - As (B), but refined modelling of lower strata	4.64 m/s^2 (0.473g)
D - Between (A) & (B), but with high hysteretic damping	2.99 m/s^2 (0.305g)
E - Between (A) & (B), but with low hysteretic damping	4.14 m/s^2 (0.422g)

Table 1 Peak surface accelerations for a 0.35g earthquake at the bedrock

The difference between the soil curves for high damping and low damping as referred to in Cases D and E is shown by the stress-strain curves for clay shown in Figure 8. The shapes of the curves were adjusted to give damping ratios close to the upper bounds and lower bounds defined by Seed et al for sands and clays (Ref. 3).

It can be seen from Table 1 that the peak accelerations predicted at the ground surface vary considerably. It is particularly sensitive to soil strength. At one extreme, with all soil properties set to their upper bound values, the site exhibits considerable amplification, with a peak acceleration of 0.5'g' being predicted. With a 'best estimate' of the soil properties, the site attenuates the bedrock motion and a peak surface acceleration of only 0.3'g' is predicted.

It was found that the surface response is not only sensitive to soil strength properties. It is also sensitive to the shape of the stress-strain curves, particularly in the region 0-1% strain. This is demonstrated by comparing the results of Cases (B) and (C) where the only difference between the two calculations was in the shapes of the stress-strain curves as shown in Figure 9. This is a significant change in result for an apparently minor change in input data. However, the reason for the sensitivity is easy to see if reference is made to Figure 10. This figure records the stress- strain histories in the clay for Cases (B) and (C). Clearly, there has been a lot of cycling at strains up to 0.5%.

This effect is more pronounced when Cases (D) and (E) are compared. In Case (D) the curves lie close to an upper bound to the damping ratio Vs strain relationship, while in Case (E) the curves are close to a lower bound. The stress-strain histories generated in the sand and the clay for these two cases are shown in Figure 11. This figure shows that the precise degree of curvature of the stress-strain curve greatly influences the degree of hysteretic loss in the soil and so it is not surprising that relatively minor changes to the shape of the curve had a significant influence on the computed results. It does, however underline the fact that careful consideration of the curve shape is very important before making site specific predictions.

CONCLUSIONS

The results of the above studies give rise to the following conclusions.

First, the Phase 1 studies clearly show that the ability of a site to amplify or attenuate an input motion at bedrock level is dependent not only upon its geological nature, but also upon the intensity of earthquake to which it is subject. This reflects findings reported from real earthquakes and lends confidence to the results generated using the non-linear finite element approach which has been adopted here.

Second, the free surface accelerations predicted by the Phase 2 studies have shown a marked sensitivity to relatively minor variations in soil properties (ie. variations which lie within the band of confidence which might be applied to all soil property data). It has been shown that a site can change from an attenuating site into an amplifying site by introducing a change of curvature to the soil stress-strain curve in the region 0-1% strain. This highlights the need for care when using this type of site specific analysis to justify the safety of a plant, since in many cases, the data to define the precise shape of the soil stress-strain curve either does not exist or has a low degree of confidence associated with it.

From the foregoing, it is clear that any generalised statements about the characteristics of a deep, soft site must be treated with caution. Whether a site amplifies or attenuates the bedrock motion is dependent on a number of factors. For this reason, it is suggested that site-specific analyses should always be carried out if any special claims are to be made regarding the effect of the site on the free-surface motions.

It seems prudent to suggest that sensitivity studies should always be carried out over a range of soil strengths and damping ratios in order to sensibly bracket the spectrum of results which can be achieved theoretically.

ACKNOWLEDGEMENTS

The authors wish to thank Her Majesty's Nuclear Installation Inspectorate for permission to publish this work. The views expressed herein are not necessarily the views of the N.I.I.

REFERENCES

1. Hallquist, J. (1982). Non-linear Dynamic Analysis of Solids in Three Dimensions, UCID-19592. Lawrence Livermore Laboratories, California.

2. Seed, H.B. and Idriss, I.M. Ground Motions and Soil Liquefaction During Earthquakes. Engineering Monographs on Earthquake Criteria Structural Design and Strong Motion Records; Earthquake Engineering Research Institute.

3. Seed, H.B. and Idriss, I.M. (1970), Soil Moduli and Damping Factors for Dynamic Response Analysis. Earthquake Engineering Research Centre (Report No.EERC 70-10).

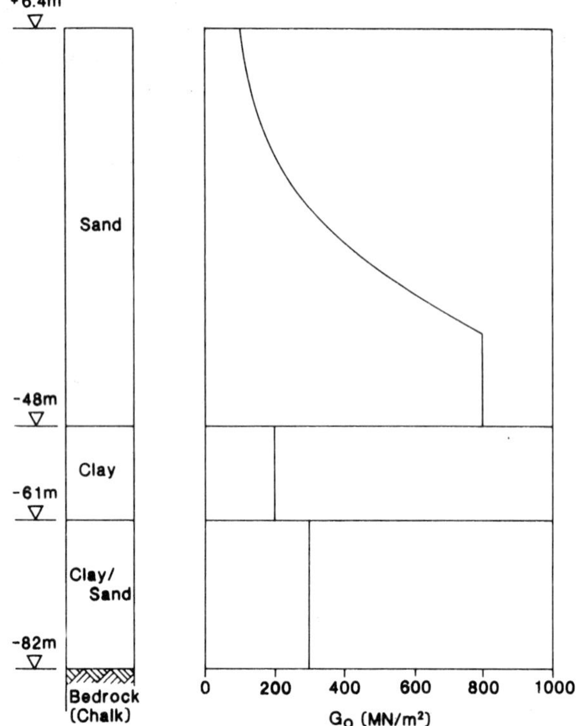

Figure 1. Soil Profile

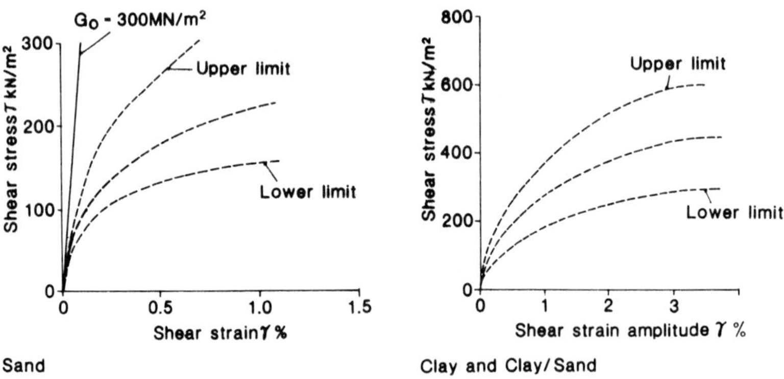

Figure 2. Typical stress-strain curves

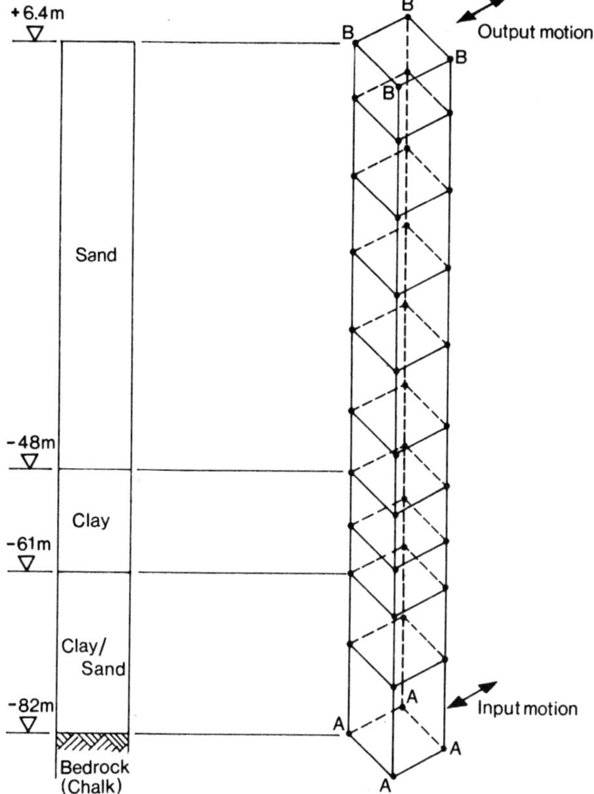

Figure 3. Finite element model of site

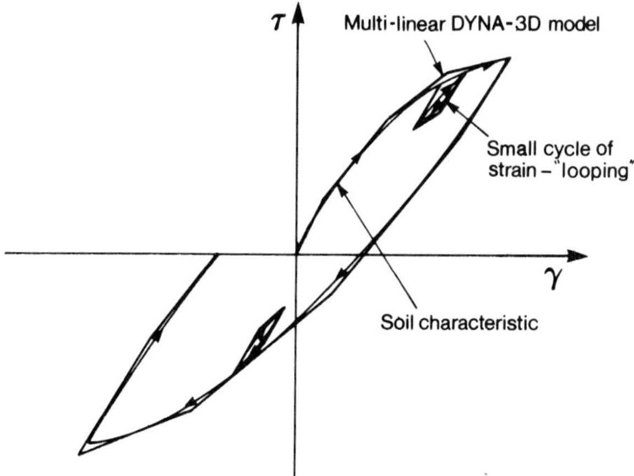

Figure 4. Modelling non-linear stress-strain properties for soil

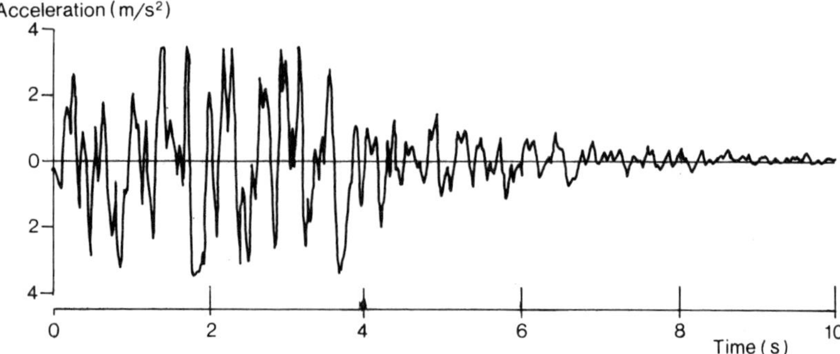

Figure 5. Bedrock (input) accelerations

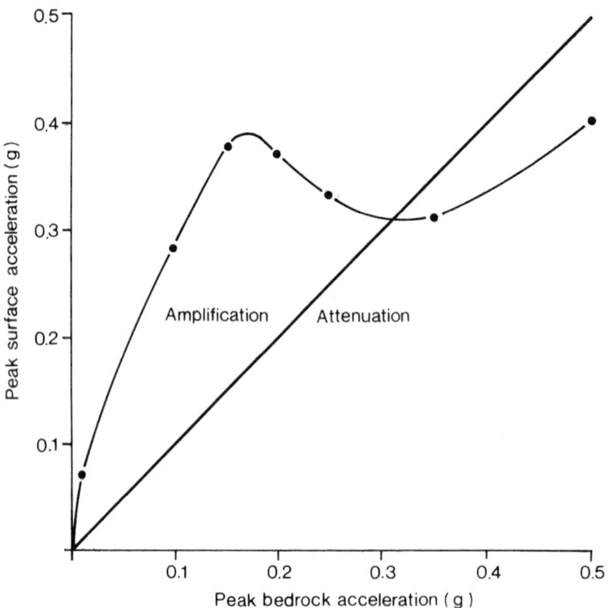

Figure 6. Peak surface acceleration as a function of bedrock acceleration

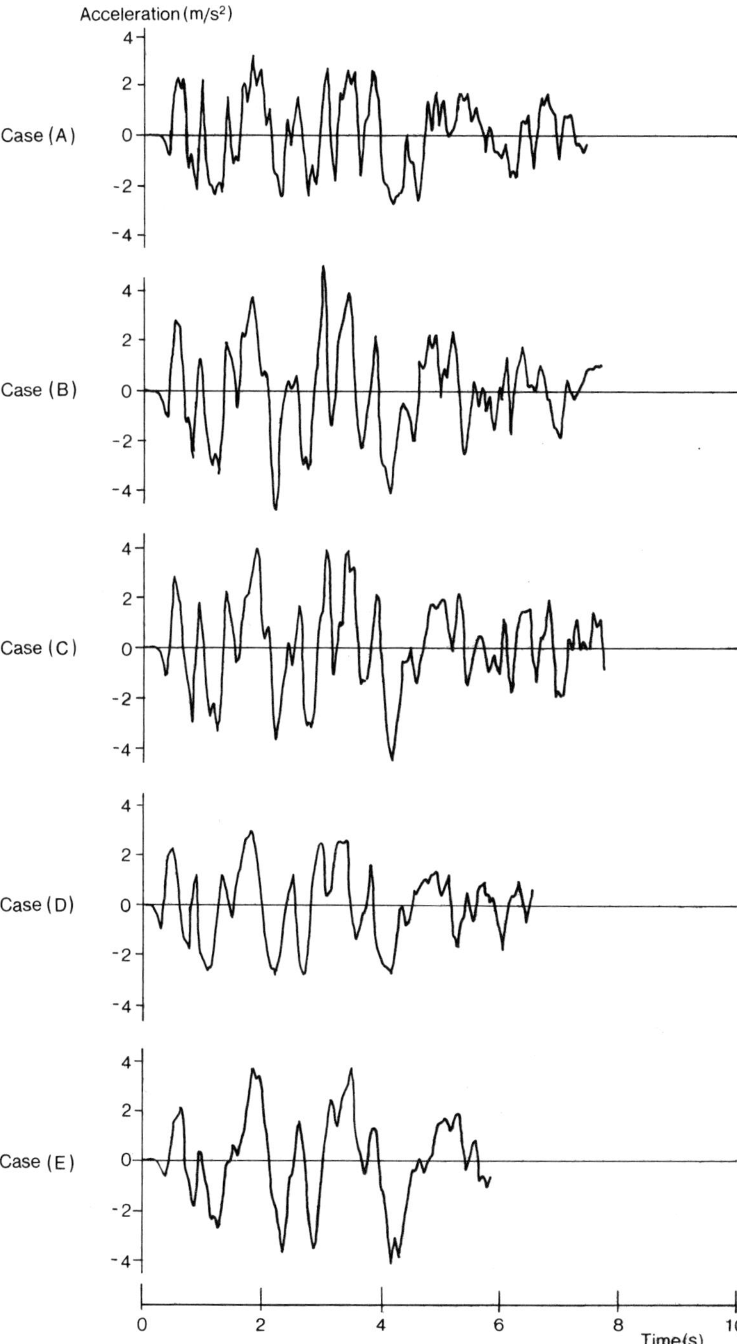

Figure 7. Computed surface accelerations

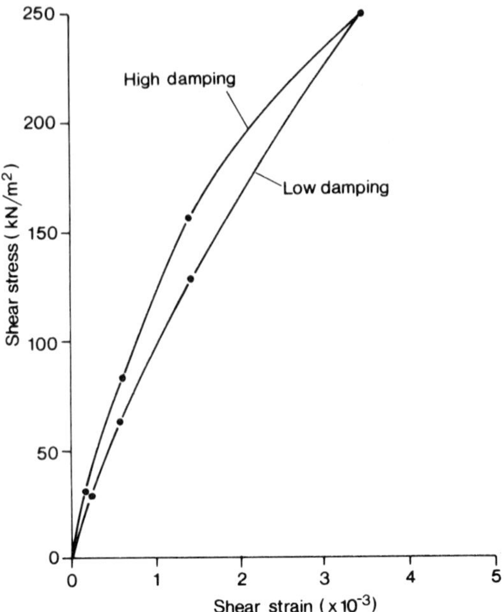

Figure 8. Stress - strain curves in clay

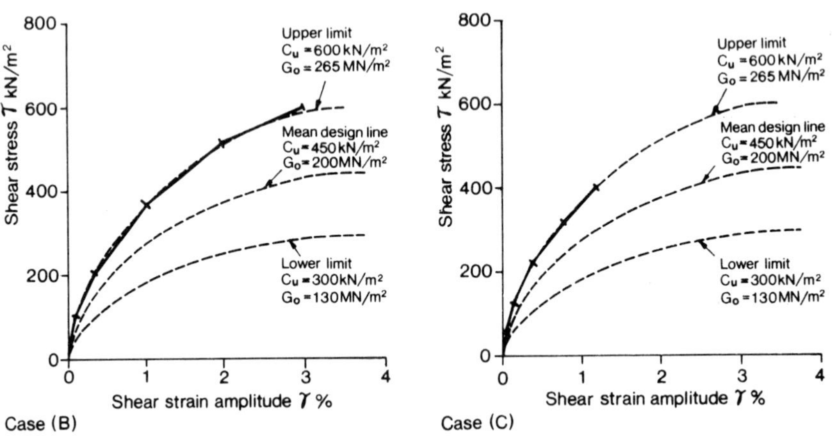

Figure 9. Stress strain approximation for cases (B) and (C)

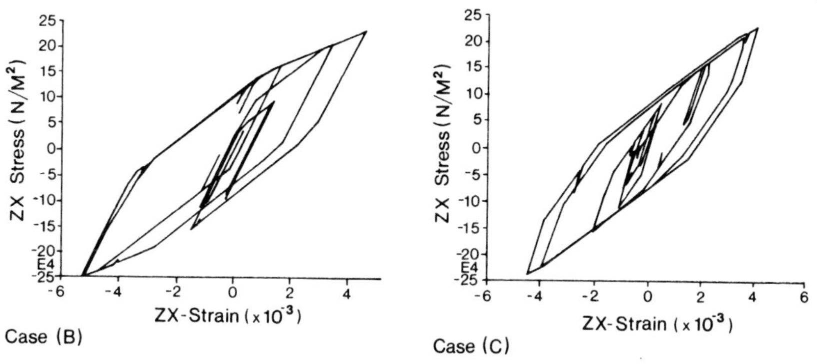

Figure 10. Stress-strain histories in Clay strata for cases (B) and (C)

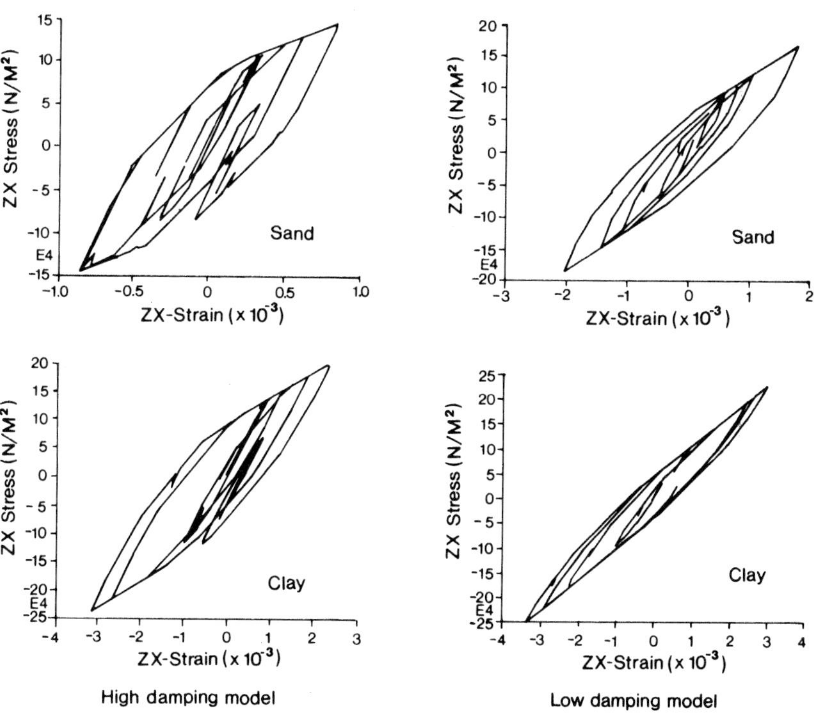

Figure 11. Stress-strain histories in cases (D) and (E)